OXFORD MATHEMATICAL MONOGRAPHS

OXFORD MATHEMATICAL MONOGRAPHS

Books in the series

J.W.P. Hirschfeld: *Finite projective spaces of three dimensions*
D.E. Edmunds and W.D. Evans: *Spectral theory and differential operators*
A. Pressley and G. Segal: *Loop groups, paperback*
L. Evens: *Cohomology of groups*
P.N. Hoffman and J.F. Humphreys: *Projective representations of the symmetric groups: Q-Functions and Shifted Tableaux*
B. Amberg, S. Franciosi, and F. de Giovanni: *Products of groups*
M.E. Gurtin: *Thermomechanics of evolving phase boundaries in the plane*
J. Faraut and A. Koranyi: *Analysis on symmetric cones*
B.L.R. Shawyer and B.B. Watson: *Borel's methods of summability*
P. Lancaster and L. Rodman: *Algebraic Riccati equations*
Jacques Thévenaz: *G-algebras and modular representation theory*
Hans-Joachim Baues: *Homotopy type and homology*
P.D. D'Eath: *Black holes: gravitational interactions*
R. Lowen: *Approach spaces: the missing link in the topology–uniformity–metric triad*
Nguyen Dinh Cong: *Topological dynamics of random dynamical systems*
S.K. Donaldson and P.B. Kronheimer: *The geometry of four-manifolds, paperback*
N.M.J. Woodhouse: *Geometric quantization, second edition, paperback*
J.W.P. Hirschfeld: *Projective geometries over finite fields, second edition*
David E. Evans and Yasuyuki Kawahigashi: *Quantum symmetries of operator algebras*
Norbert Klingen: *Arithmetical similarities: Prime decomposition and finite group theory*
K. Matsuzaki and M. Taniguchi: *Hyperbolic manifolds and Kleinian groups*
I.G. Macdonald: *Symmetric functions and Hall polynomials, second edition, paperback*
Isabelle Catto, Claude Le Bris, and Pierre-Louis Lions: *Mathematical Theory of Thermodynamic Limits: Thomas-Fermi type models*
D. McDuff and D. Salamon: *Introduction to symplectic topology, paperback*
M. Holschneider: *Wavelets: An analysis tool, paperback*
William M. Goldman: *Complex hyperbolic geometry*
Charles J. Colbourn and Alexander Rosa: *Triple systems*
V. A. Kozlov, V. G. Maz'ya and A. B. Movchan: *Asymptotic analysis of fields in multi-structures*
Gérard A. Maugin: *Nonlinear waves in elastic crystals*
George Dassios and Ralph Kleinman: *Low frequency scattering*
L. Ambrosio, N. Fusco and D. Pallara: *Functions of bounded variation and free discontinuity problems*
S.Y. Slavyanov and W. Lay: *Special Functions: A Unified Theory based on Singularities*
D. Joyce: *Compact Manifolds with Special Holonomy*
A. Carbone and S. Semmes: *A graphic apology for symmetry and implicitness*
Johann Boos: *Classical and modern methods in summability*
Nigel Higson and John Roe: *Analytic K-Homology*
S. Semmes: *Some novel types of fractal geometry*
Tadeusz Iwaniec and Gaven Martin: *Geometric Function Theory and Nonlinear Analysis*
Gerald W. Johnson and Michel L. Lapidus: *The Feynman Integral and Feynman 's Operational Calculus, paperback*
Terry Lyons and Zhongmin Qian: *System Control and Rough Paths*
Andrew Ranicki: *Algebraic and Geometric Surgery*
Leon Ehrenpreis: *The Radon Transform*
John C. Lennox and Derek J.S. Robinson: *The Theory of Infinite Soluble Groups*
A.A. Ivanov: *The Fourth Janko Group*
D. Huybrechts: *Fourier-Mukai transforms in algebraic geometry*

Fourier–Mukai transforms in algebraic geometry

D. HUYBRECHTS
Mathematisches Institut Universität Bonn

CLARENDON PRESS • OXFORD
2006

This book has been printed digitally and produced in a standard specification in order to ensure its continuing availability

OXFORD
UNIVERSITY PRESS

Great Clarendon Street, Oxford OX2 6DP

Oxford University Press is a department of the University of Oxford.
It furthers the University's objective of excellence in research, scholarship,
and education by publishing worldwide in

Oxford New York

Auckland Cape Town Dar es Salaam Hong Kong Karachi
Kuala Lumpur Madrid Melbourne Mexico City Nairobi
New Delhi Shanghai Taipei Toronto

With offices in

Argentina Austria Brazil Chile Czech Republic France Greece
Guatemala Hungary Italy Japan South Korea Poland Portugal
Singapore Switzerland Thailand Turkey Ukraine Vietnam

Oxford is a registered trade mark of Oxford University Press
in the UK and in certain other countries

Published in the United States
by Oxford University Press Inc., New York

Reprinted 2009

ISBN 978-0-19-929686-6

PREFACE

This book is based on a course given at the Institut de Mathématiques de Jussieu in 2004 and once more in 2005. It was conceived as a first specialized course in algebraic geometry. A student with a basic knowledge in algebraic geometry, e.g. a student having worked through the first three chapters of Hartshorne's book [45], should be able to follow the lectures without much trouble. Occasionally, notions from other areas, e.g. singular cohomology, Hodge theory, abelian varieties, K3 surfaces, were needed, which were then presented in a rather ad hoc manner, tailor-made for the purposes of the course. With a few exceptions full proofs are given. The exercises included in the text should help the reader to gain a working knowledge of the subject.

What is this book about? Its principal character is the derived category of coherent sheaves on a smooth projective variety. Derived categories of this type have been known for many years. Although widely accepted as the right framework for any kind of derived functors, e.g. cohomology groups, higher direct images, etc., they were usually considered as rather formal objects without much interesting internal structure. Contrary to the cohomology and the Chow ring of a projective variety X, the derived category of coherent sheaves as an invariant of X had not been investigated thoroughly. This has changed drastically over the last ten years.

The origin of the theory as treated here however goes back to celebrated papers by Mukai, more than twenty years ago. He constructed geometrically motivated equivalences between derived categories of non-isomorphic varieties. Also, over many years the Moscow school had constantly worked on the description of coherent sheaves on homogenous varieties, e.g. the projective space, Grassmannians, etc. On the other hand, Kontsevich's homological mirror symmetry has revived the interest in these questions outside the small circle of experts. Roughly, Kontsevich proposed to view mirror symmetry as an equivalence of the derived category of coherent sheaves of certain projective varieties with the Fukaya category associated to the symplectic geometry of the mirror variety. Although we deliberately do not enter into the details of this relation, it is this point of view that motivates and in some sense explains many of the central results as well as open problems in this area.

The derived category turns out to be a very reasonable invariant. Due to results of Bondal and Orlov one knows that it determines the variety whenever the canonical bundle is either ample or anti-ample. If this was true without any assumptions on the positivity of the canonical bundle, the theory would be without much interest. However, there is a region in the classification of projective varieties where the derived category turns out to be less rigid without getting completely out of hand. The most prominent example was observed by Mukai in the very first paper on the subject. He showed that the Poincaré bundle induces an equivalence between the derived category of an abelian variety A and the derived category of its dual $\widehat{A}$ (which in general is not isomorphic to A). These results, to be discussed in detail in various chapters, naturally lead to the question under which conditions two smooth projective varieties give rise to equivalent derived categories. This is the central theme of this book.

One word on the choice of the material. Everything that did not have a distinctive geometric touch has been left out. In particular, questions related to representation theory, e.g. of quivers, or to modules over (non-commutative) rings, have not been touched upon. This choice is due to personal taste, limitations by a one semester course and my own ignorance in some of these areas.

We refrain from giving a lengthy introduction to the contents of every chapter. A glance at the table of contents will give a first impression of which topics are treated, and the remarks at the beginning of each chapter provide more details. The reader familiar with the general yoga of derived categories and derived functors may go directly to Chapter 4 or 5 and come back to some of the background material collected in the first three chapters whenever needed.

Acknowledgements: I am intellectually indebted to A. Bondal, T. Bridgeland, Y. Kawamata, S. Mukai, and D. Orlov. The overwhelming part of the theory as presented here is due to them. The idea that this text could help to stimulate newcomers to pursue research originated by them was the driving force during the preparation of these notes.

I am particularly grateful to the Institut de Mathématiques de Jussieu for giving me (twice) the opportunity to teach the course this book is based on. The intellectual atmosphere at the institute has been very stimulating throughout the whole project and I have fond memories of all the discussions I had with my colleagues at the IJM during this time. In particular, I wish to thank J. Le Potier and R. Rouquier.

Many people have made comments on earlier versions of the book. I wish to thank U. Görtz, M. Nieper-Wißkirchen, K. Oguiso, D. Ploog, P. Stellari, R. Thomas, and D. Ben-Zvi. I am most grateful to Bahoua Fu and Justin Sawon who carefully worked through the book and sent me detailed lists of corrections, questions, and remarks.

CONTENTS

1

TRIANGULATED CATEGORIES

The reader familiar with the basic notions of abelian and derived categories may only need to browse through this section or skip it altogether. It will be much more interesting to come back to the specific results discussed here when, in the later chapters, they are actually applied to geometrically concrete problems. However, the reader not feeling completely at ease with the formal language of category theory should work through this chapter in order to be well prepared for everything that follows.

We hope that separating results from category theory from the other chapters rather than blending them in later when used, will help readers to understand which part of the theory is really geometrical and which is more formal.

On the other hand, this chapter is not meant as a thorough introduction to the subject. We only present those parts of the theory that are relevant in our context.

We will not worry about any kind of set theoretical issues and will always assume we remain in a given universe (or, as put in [39, p.58], 'that all the required hygiene regulations are obeyed').

1.1 Additive categories and functors

We suppose that the reader is familiar with the notion of a category and of a functor between two categories. For the reader's convenience we briefly recall a few central notions. If not otherwise stated all functors are covariant.

Definition 1.1 *Let $\mathcal{A}$ and $\mathcal{B}$ be two categories. A functor $F : \mathcal{A} \longrightarrow \mathcal{B}$ is* full *if for any two objects $A, B \in \mathcal{A}$ the induced map*

$$F : \mathrm{Hom}(A, B) \longrightarrow \mathrm{Hom}(F(A), F(B))$$

is surjective. The functor F is called faithful *if this map is injective for all $A, B \in \mathcal{A}$.*

A *morphism* $F \longrightarrow F'$ between two functors $F, F' : \mathcal{A} \longrightarrow \mathcal{B}$ is given by morphisms $\varphi_A \in \mathrm{Hom}(F(A), F'(A))$ for any object $A \in \mathcal{A}$ which are functorial in A, i.e. $F'(f) \circ \varphi_A = \varphi_B \circ F(f)$ for any $f : A \longrightarrow B$.

Definition 1.2 *Two functors $F, F' : \mathcal{A} \longrightarrow \mathcal{B}$ are* isomorphic *if there exists a morphism of functors $\varphi : F \longrightarrow F'$ such that for any object $A \in \mathcal{A}$ the induced morphism $\varphi_A : F(A) \longrightarrow F'(A)$ is an isomorphism (in $\mathcal{B}$).*

Equivalently, F and F' are isomorphic if there exist functor morphisms $\varphi : F \longrightarrow F'$ and $\psi : F' \longrightarrow F$ with $\varphi \circ \psi = \mathrm{id}$ and $\psi \circ \varphi = \mathrm{id}$.

Definition 1.3 *A functor $F : \mathcal{A} \longrightarrow \mathcal{B}$ is called an* equivalence *if there exists a functor $F^{-1} : \mathcal{B} \longrightarrow \mathcal{A}$ such that $F \circ F^{-1}$ is isomorphic to $\mathrm{id}_{\mathcal{B}}$ and $F^{-1} \circ F$ is isomorphic to $\mathrm{id}_{\mathcal{A}}$. One calls F^{-1} an* inverse *or, sometimes,* quasi-inverse *of F.*

Two categories $\mathcal{A}$ and $\mathcal{B}$ are called equivalent *if there exists an equivalence $F : \mathcal{A} \longrightarrow \mathcal{B}$.*

Clearly, any equivalence is fully faithful. A partial converse is provided by

Proposition 1.4 *Let $F : \mathcal{A} \longrightarrow \mathcal{B}$ be a fully faithful functor. Then F is an equivalence if and only if every object $B \in \mathcal{B}$ is isomorphic to an object of the form $F(A)$ for some $A \in \mathcal{A}$.*

Proof In order to define the inverse functor F^{-1}, one chooses for any $B \in \mathcal{B}$ an object $A_B \in \mathcal{A}$ together with an isomorphism $\varphi_B : F(A_B) \xrightarrow{\sim} B$. Then, let

$$F^{-1} : \mathcal{B} \longrightarrow \mathcal{A}$$

be the functor that associates to any object $B \in \mathcal{B}$ this distinguished object $A_B \in \mathcal{A}$ and for which $F^{-1} : \mathrm{Hom}(B_1, B_2) \longrightarrow \mathrm{Hom}(F^{-1}(B_1), F^{-1}(B_2))$ is given by the composition of

$$\mathrm{Hom}(B_1, B_2) \xrightarrow{\sim} \mathrm{Hom}(F(A_{B_1}), F(A_{B_2})), \quad f \longmapsto \varphi_{B_2}^{-1} \circ f \circ \varphi_{B_1}$$

and the inverse of the bijection

$$F : \mathrm{Hom}(A_{B_1}, A_{B_2}) \xrightarrow{\sim} \mathrm{Hom}(F(A_{B_1}), F(A_{B_2})).$$

The isomorphisms $F \circ F^{-1} \simeq \mathrm{id}_{\mathcal{B}}$ and $F^{-1} \circ F \simeq \mathrm{id}_{\mathcal{A}}$ are the ones that are naturally induced by the isomorphisms φ_B. □

The proposition immediately yields the

Corollary 1.5 *Any fully faithful functor $F : \mathcal{A} \longrightarrow \mathcal{B}$ defines an equivalence between $\mathcal{A}$ and the full subcategory of $\mathcal{B}$ of all objects $B \in \mathcal{B}$ isomorphic to $F(A)$ for some $A \in \mathcal{A}$.* □

In the following proposition we let $\mathrm{Fun}(\mathcal{A})$ be the category of all contravariant functors, i.e. the objects are functors $F : \mathcal{A}^{\mathrm{op}} \longrightarrow \mathbf{Set}$ and the morphisms are functor morphisms. Consider the natural functor

$$\mathcal{A} \longrightarrow \mathrm{Fun}(\mathcal{A}), \qquad A \longmapsto \mathrm{Hom}(\ \ , A)$$

Proposition 1.6 **(Yoneda lemma)** *This functor $\mathcal{A} \longrightarrow \mathrm{Fun}(\mathcal{A})$ defines an equivalence of $\mathcal{A}$ with the full subcategory of* representable functors *F, i.e. functors isomorphic to some $\mathrm{Hom}(\ \ , A)$. In particular, $A \longmapsto \mathrm{Hom}(\ \ , A)$ is fully faithful.*

Proof See [39, II.3]. □

We will rarely work with completely arbitrary categories. All our categories will at least be additive.

Definition 1.7 *A category* $\mathcal{A}$ *is an* additive category *if for every two objects* $A, B \in \mathcal{A}$ *the set* $\mathrm{Hom}(A, B)$ *is endowed with the structure of an abelian group such that the following three conditions are satisfied:*

i) *The compositions* $\mathrm{Hom}(A_1, A_2) \times \mathrm{Hom}(A_2, A_3) \longrightarrow \mathrm{Hom}(A_1, A_3)$ *written as* $(f, g) \longmapsto g \circ f$ *are bilinear.*

ii) *There exists a* zero object $0 \in \mathcal{A}$, *i.e. an object* 0 *such that* $\mathrm{Hom}(0, 0)$ *is the trivial group with one element.*

iii) *For any two objects* $A_1, A_2 \in \mathcal{A}$ *there exists an object* $B \in \mathcal{A}$ *with morphisms* $j_i : A_i \longrightarrow B$ *and* $p_i : B \longrightarrow A_i$, $i = 1, 2$, *which make* B *the* direct sum *and the* direct product *of* A_1 *and* A_2.

We tacitly assume the usual compatibilities $p_i \circ j_i = \mathrm{id}$, $p_2 \circ j_1 = p_1 \circ j_2 = 0$, and $j_1 \circ p_1 + j_2 \circ p_2 = \mathrm{id}$, which hold automatically up to automorphisms of B.

Exercise 1.8 Show that for any object $A \in \mathcal{A}$ in an additive category $\mathcal{A}$ there exist unique morphisms $0 \longrightarrow A$ and $A \longrightarrow 0$. The existence of such an object 0 in a category $\mathcal{A}$ is of course equivalent to ii).

A functor $F : \mathcal{A} \longrightarrow \mathcal{B}$ between additive categories $\mathcal{A}$ and $\mathcal{B}$ will usually be assumed to be *additive*, i.e. the induced maps $\mathrm{Hom}(A, B) \longrightarrow \mathrm{Hom}(F(A), F(B))$ are group homomorphisms.

Everything that has been said so far carries over to additive categories. In particular, an additive functor $F : \mathcal{A} \longrightarrow \mathcal{B}$ which is an equivalence is in fact an additive equivalence, i.e. the inverse functor F^{-1} is additive as well. The Yoneda lemma is modified as follows: For an additive category $\mathcal{A}$ we let $\mathrm{Fun}(\mathcal{A})$ be the category of contravariant additive(!) functors $F : \mathcal{A} \longrightarrow \mathbf{Ab}$, where $\mathbf{Ab}$ is the category of abelian groups. Then the Yoneda lemma in the form of Proposition 1.6 remains valid.

We will go one step further. As the categories we will eventually be interested in have geometric origin, i.e. are defined in terms of certain varieties over some base field, we usually deal with the following special type of additive categories. In the following we denote by k an arbitrary field.

Definition 1.9 *A* k-linear category *is an additive category* $\mathcal{A}$ *such that the groups* $\mathrm{Hom}(A, B)$ *are* k*-vector spaces and such that all compositions are* k*-bilinear.*

Additive functors between two k*-linear additive categories over a common base field* k *will be assumed to be* k-linear, *i.e. for any two objects* $A, B \in \mathcal{A}$ *the map* $F : \mathrm{Hom}(A, B) \longrightarrow \mathrm{Hom}(F(A), F(B))$ *is* k*-linear.*

Once again, everything that has been mentioned before carries over literally to additive categories over a field. Usually we will state all abstract results for additive categories, but in the applications everything will be over a base field. In principle, though, it could happen that two k-linear categories are equivalent as ordinary additive categories without being equivalent as k-linear categories.

The Yoneda lemma can again be adjusted to the situation: this time, one considers the category of contravariant k-linear functors from $\mathcal{A}$ into the category $\mathbf{Vec}(k)$ of k-vector spaces.

Definition 1.10 *An additive category $\mathcal{A}$ is called* abelian *if also the following condition holds true:*

iv) *Every morphism $f \in \mathrm{Hom}(A, B)$ admits a kernel and a cokernel and the natural map $\mathrm{Coim}(f) \longrightarrow \mathrm{Im}(f)$ is an isomorphism.*

Recall that the image $\mathrm{Im}(f)$ is a kernel for a cokernel $B \longrightarrow \mathrm{Coker}(f)$ and the coimage $\mathrm{Coim}(f)$ is a cokernel for a kernel $\mathrm{Ker}(f) \longrightarrow A$. So, condition iv) says that for any $f : A \longrightarrow B$ there exists the following diagram

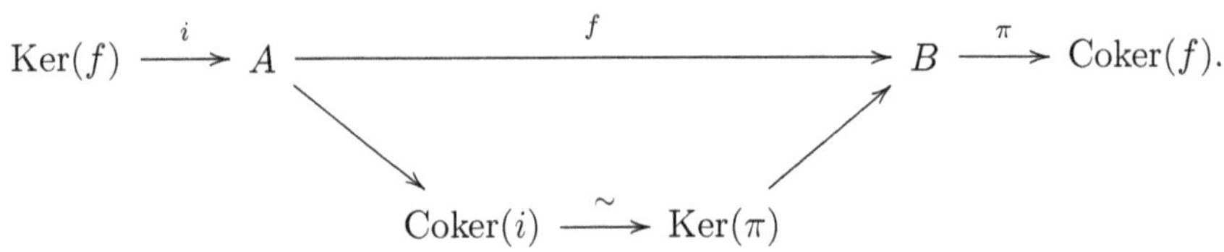

In particular, the notion of exact sequences is usually only considered for abelian categories. We recall that a sequence

$$A_1 \xrightarrow{f_1} A_2 \xrightarrow{f_2} A_3$$

is called *exact* if and only if $\mathrm{Ker}(f_2) = \mathrm{Im}(f_1)$.

Examples 1.11 i) Let R be a commutative ring. Then the category $\mathbf{Mod}(R)$ of R-modules is abelian. The full subcategory of finitely generated modules is abelian as well.

ii) Let X be a topological space. Then the category of sheaves of abelian groups $\mathbf{Sh}(X)$ is abelian. If a sheaf of commutative rings on X is fixed, then the subcategory of sheaves of modules over this sheaf of rings is again abelian.

iii) Let X be a scheme. Then the categories $\mathbf{Coh}(X)$ and $\mathbf{Qcoh}(X)$ of all coherent respectively quasi-coherent sheaves on X are both abelian.

Suppose $F : \mathcal{A} \longrightarrow \mathcal{B}$ is an additive functor between abelian categories. In particular, any sequence

$$A_1 \xrightarrow{f_1} A_2 \xrightarrow{f_2} A_3$$

with $f_2 \circ f_1 = 0$ (or, in other words, $\mathrm{Im}(f_1) \subset \mathrm{Ker}(f_2)$) is mapped to

$$F(A_1) \xrightarrow{F(f_1)} F(A_2) \xrightarrow{F(f_2)} F(A_3)$$

again with $F(f_2) \circ F(f_1) = F(f_2 \circ f_1) = 0$.

Definition 1.12 *The functor F is* left (right) exact *if any short exact sequence*

$$0 \longrightarrow A_1 \xrightarrow{f_1} A_2 \xrightarrow{f_2} A_3 \longrightarrow 0$$

is mapped to a sequence

$$0 \longrightarrow F(A_1) \xrightarrow{F(f_1)} F(A_2) \xrightarrow{F(f_2)} F(A_3) \longrightarrow 0$$

which is exact except possibly in $F(A_3)$ (respectively in $F(A_1)$). The functor is exact *if it is left and right exact.*

Exercise 1.13 Show that a functor F is left exact if and only if any exact sequence $0 \longrightarrow A_1 \longrightarrow A_2 \longrightarrow A_3$ (no surjectivity on the right!) induces an exact sequence $0 \longrightarrow F(A_1) \longrightarrow F(A_2) \longrightarrow F(A_3)$.

Examples 1.14 i) Let $\mathcal{A}$ be an abelian category and $A_0 \in \mathcal{A}$. Then

$$\mathrm{Hom}(A_0, \ \ ) : \mathcal{A} \longrightarrow \mathbf{Ab}$$

is a left exact functor. The contravariant functor

$$\mathrm{Hom}(\ \ , A_0) : \mathcal{A} \longrightarrow \mathbf{Ab}$$

is also left exact. (Left exactness of a contravariant functor $F : \mathcal{A} \longrightarrow \mathbf{Ab}$ means by definition left exactness of the covariant functor $F : \mathcal{A}^{\mathrm{op}} \longrightarrow \mathbf{Ab}$.)

ii) Recall that an object $P \in \mathcal{A}$ is called *projective* if $\mathrm{Hom}(P, \ \ )$ is right exact (and hence exact). An object $I \in \mathcal{A}$ is called *injective* if $\mathrm{Hom}(\ \ , I)$ is right exact (and hence exact).

iii) Free modules over a ring R are projective objects in $\mathbf{Mod}(R)$. But (locally) free sheaves in $\mathbf{Coh}(X)$ are almost never projective.

Definition 1.15 *Let $F : \mathcal{A} \longrightarrow \mathcal{B}$ be a functor between arbitrary categories.*

A functor $H : \mathcal{B} \longrightarrow \mathcal{A}$ is right adjoint *to F (one writes $F \dashv H$) if there exist isomorphisms*

$$\mathrm{Hom}(F(A), B) \simeq \mathrm{Hom}(A, H(B)) \tag{1.1}$$

for any two objects $A \in \mathcal{A}$ and $B \in \mathcal{B}$ which are functorial in A and B.

A functor $G : \mathcal{B} \longrightarrow \mathcal{A}$ is left adjoint *to F (one writes $G \dashv F$) if there exist isomorphisms* $\mathrm{Hom}(B, F(A)) = \mathrm{Hom}(G(B), A)$ *for any two objects $A \in \mathcal{A}$ and $B \in \mathcal{B}$ which are functorial in A and B.*

Clearly, H is right adjoint to F if and only if F is left adjoint to H.

Remarks 1.16 i) Suppose $F \dashv H$. Then $\mathrm{id}_{F(A)} \in \mathrm{Hom}(F(A), F(A))$ induces a morphism $A \longrightarrow H(F(A))$. The naturality of isomorphisms in the definition of the adjoint functor ensures that these morphisms define a functor morphism

$$h : \mathrm{id}_{\mathcal{A}} \longrightarrow H \circ F.$$

In the same vein, inserting $A = H(B)$ in (1.1) yields a canonical morphism $F(H(B)) \longrightarrow B$ and, therefore, a functor morphism

$$g : F \circ H \longrightarrow \mathrm{id}_{\mathcal{B}}.$$

ii) Using the Yoneda lemma 1.6, one verifies that a left (or right) adjoint functor, if it exists at all, is unique up to isomorphism. More explicitly, for two right adjoint functors H and H' of F one defines an isomorphism $H \simeq H'$ which for any $B \in \mathcal{B}$ is given as the image of the identity under the functorial isomorphism $\mathrm{Hom}(H(B), H(B)) \simeq \mathrm{Hom}(F(H(B)), B) \simeq \mathrm{Hom}(H(B), H'(B))$.

iii) If F is an additive functor (in particular, $\mathcal{A}$ and $\mathcal{B}$ are additive), then one requires the isomorphisms (1.1) to be isomorphisms of abelian groups. Similar, if everything is k-linear, then also these isomorphisms are required to be k-linear. A priori, one cannot exclude the pathological case of an adjoint functor that is not additive, although the functor itself is. This can only occur if the isomorphism in (1.1) is not a group homomorphism.

iv) If $\mathcal{A}$ and $\mathcal{B}$ are abelian categories and $F : \mathcal{A} \longrightarrow \mathcal{B}$ is left adjoint to $H : \mathcal{B} \longrightarrow \mathcal{A}$, then F is right exact and H is left exact. Note that even when F is left and right exact, its right adjoint is in general only left exact.

Exercise 1.17 Suppose $F \dashv H$. Show that

$$f \longmapsto \left(A \xrightarrow{h_A} H(F(A)) \xrightarrow{H(f)} H(B) \right)$$

describes the adjunction morphism $\mathrm{Hom}(F(A), B) = \mathrm{Hom}(A, H(B))$.

Exercise 1.18 Prove assertion iv) above.

Exercise 1.19 Suppose $F \dashv H$. Show that for the induced morphisms $g : F \circ H \longrightarrow \mathrm{id}$ and $h : \mathrm{id} \longrightarrow H \circ F$ the composition

$$H \xrightarrow{h_{H(\ )}} (H \circ F) \circ H = H \circ (F \circ H) \xrightarrow{H(g)} H$$

is the identity. See [72, IV.1] and [39, II.3] for a converse.

Examples 1.20 Let $f : X \longrightarrow Y$ be a morphism between two noetherian schemes X and Y. Then the pull-back functor

$$f^* : \mathbf{Qcoh}(Y) \longrightarrow \mathbf{Qcoh}(X)$$

is right exact and taking the direct image

$$f_* : \mathbf{Qcoh}(X) \longrightarrow \mathbf{Qcoh}(Y)$$

is left exact. Moreover, $f^* \dashv f_*$. If f is proper, the same holds for the categories of coherent sheaves on X and Y.

Lemma 1.21 *Let $F : \mathcal{A} \longrightarrow \mathcal{B}$ be a functor and $G \dashv F$ … the induced functor morphism $g : G \circ F \longrightarrow \mathrm{id}_{\mathcal{A}}$ induces for any A, B … the following commutative diagram*

$$\begin{array}{ccc} \mathrm{Hom}(A,B) & & \\ \downarrow{\scriptstyle \circ g_A} & \searrow{\scriptstyle F} & \\ \mathrm{Hom}(G(F(A)),B) & \xrightarrow{\sim} & \mathrm{Hom}(F(A),F(B)). \end{array}$$

Here, the isomorphism is given by adjunction.

Similarly, if $F \dashv H$ then the natural functor morphism $h : \mathrm{id}_{\mathcal{A}} \longrightarrow H \circ F$ induces for all $A, B \in \mathcal{A}$ the following commutative diagram:

$$\begin{array}{ccc} \mathrm{Hom}(A,B) & \xrightarrow{h_B \circ} & \mathrm{Hom}(A,H(F(B))) \\ & \searrow{\scriptstyle F} & \downarrow{\scriptstyle \wr} \\ & & \mathrm{Hom}(F(A),F(B)). \end{array}$$

Again, the isomorphism is given by adjunction.

Proof As $G \dashv F$, the following diagram commutes for all $f : A \longrightarrow B$ and all $C \in \mathcal{B}$:

$$\begin{array}{ccc} \mathrm{Hom}(G(C),A) & \xrightarrow{\sim} & \mathrm{Hom}(C,F(A)) \\ \downarrow{\scriptstyle f\circ} & \circlearrowleft & \downarrow{\scriptstyle F(f)\circ} \\ \mathrm{Hom}(G(C),B) & \xrightarrow{\sim} & \mathrm{Hom}(C,F(B)). \end{array}$$

Applied to $C = F(A)$ it yields

$$\begin{array}{ccc} \mathrm{Hom}(G(F(A)),A) & \xrightarrow{\sim} & \mathrm{Hom}(F(A),F(A)) \\ \downarrow & \circlearrowleft & \downarrow \\ \mathrm{Hom}(G(F(A)),B) & \xrightarrow{\sim} & \mathrm{Hom}(F(A),F(B)). \end{array}$$

Clearly, the vertical homomorphism on the right sends $\mathrm{id}_{F(A)}$ to $F(f)$. On the other hand, its image under

$$\mathrm{Hom}(F(A),F(A)) \simeq \mathrm{Hom}(G(F(A)),A) \longrightarrow \mathrm{Hom}(G(F(A)),B)$$

is just $f \circ g_A$.

This proves the commutativity of the lower triangle. The commutativity of the upper triangle is proved similarly. □

Corollary 1.22 *Suppose a fully faithful functor $F : \mathcal{A} \longrightarrow \mathcal{B}$ admits a left adjoint $G \dashv F$. Then the natural functor morphism*

$$g : G \circ F \xrightarrow{\sim} \mathrm{id}_{\mathcal{A}}$$

is an isomorphism.

Similarly, if a fully faithful functor $F : \mathcal{A} \longrightarrow \mathcal{B}$ admits a right adjoint $F \dashv H$, then the natural functor morphism

$$h : \mathrm{id}_{\mathcal{A}} \xrightarrow{\sim} H \circ F$$

is an isomorphism.

Proof Since $F : \mathrm{Hom}(A, B) \longrightarrow \mathrm{Hom}(F(A), F(B))$ is bijective, the commutativity of the diagram above proves that $G \circ F \longrightarrow \mathrm{id}_{\mathcal{A}}$ induces bijections

$$\mathrm{Hom}(A, B) \xrightarrow{\sim} \mathrm{Hom}((G \circ F)(A), B)$$

for all A and B. By the Yoneda lemma 1.6, this shows that $G \circ F \longrightarrow \mathrm{id}_{\mathcal{A}}$ is an isomorphism. The proof of the second statement is similar. □

The same arguments also show the converse:

Corollary 1.23 *Let $F : \mathcal{A} \longrightarrow \mathcal{B}$ and $G : \mathcal{B} \longrightarrow \mathcal{A}$ be two functors such that $G \dashv F$. If the induced functor morphism $G \circ F \longrightarrow \mathrm{id}_{\mathcal{A}}$ is an isomorphism, then F is fully faithful.*

Similarly, if $F \dashv H$ such that $\mathrm{id}_{\mathcal{A}} \longrightarrow H \circ F$ is an isomorphism, then F is fully faithful. □

Remark 1.24 In short, if $F \dashv H$, then:

$$F \text{ is fully faithful} \iff h : \mathrm{id}_{\mathcal{A}} \xrightarrow{\sim} H \circ F$$

and if $G \dashv F$, then:

$$F \text{ is fully faithful} \iff g : G \circ F \xrightarrow{\sim} \mathrm{id}_{\mathcal{A}}.$$

Exercise 1.25 Suppose $G \dashv F \dashv H$ and F fully faithful. Construct a canonical homomorphism $H \longrightarrow G$.

In many cases, adjoint functors exist. The case that interests us most is the case of equivalences. Here, the existence of left and right adjoints is granted by the following general result.

Proposition 1.26 *Let $F : \mathcal{A} \longrightarrow \mathcal{B}$ be an equivalence of categories. Then F admits a left adjoint and a right adjoint. More precisely, if $F' : \mathcal{B} \longrightarrow \mathcal{A}$ is an inverse functor of F then $F \dashv F' \dashv F$.*

Proof Very roughly, this is due to the following sequence of functorial isomorphisms

$$\mathrm{Hom}(F(A), B) \simeq \mathrm{Hom}(F'(F(A)), F'(B)) \simeq \mathrm{Hom}(A, F'(B)),$$

where we use $F'(F(A)) \simeq A$. Details are left to the diligent reader. □

Remark 1.27 These results justify the approach that is usually followed when proving the equivalence of certain categories: Suppose F is a functor that is hoped to be an equivalence and that admits a left adjoint $G \dashv F$ (or right adjoint $F \dashv H$). Then one checks whether the adjunction morphism $G \circ F \longrightarrow \mathrm{id}$ (respectively $\mathrm{id} \longrightarrow H \circ F$) is bijective. If so, the functor F is fully faithful. Eventually, one has to ensure that any object in the target category is isomorphic to an object in the image of F.

Definition 1.28 *Let $\mathcal{A}$ be a k-linear category. A* Serre functor *is a k-linear equivalence $S : \mathcal{A} \longrightarrow \mathcal{A}$ such that for any two objects $A, B \in \mathcal{A}$ there exists an isomorphism*

$$\eta_{A,B} : \mathrm{Hom}(A, B) \xrightarrow{\ \sim\ } \mathrm{Hom}(B, S(A))^*$$

(of k-vector spaces) which is functorial in A and B.

We write the induced pairing as

$$\mathrm{Hom}(B, S(A)) \times \mathrm{Hom}(A, B) \longrightarrow k \ , \ \ (f, g) \longmapsto \langle f | g \rangle.$$

Remark 1.29 In the original paper by Bondal and Kapranov [13] an additional condition was required, namely that for any two objects $A, B \in \mathcal{A}$ the following diagram commutes:

$$\begin{array}{ccc} \mathrm{Hom}(A,B) & \xrightarrow{\quad \eta_{A,B} \quad} & \mathrm{Hom}(B,S(A))^* \\ {\scriptstyle S}\downarrow & \circlearrowleft & \uparrow{\scriptstyle S^*} \\ \mathrm{Hom}(S(A),S(B)) & \xrightarrow{\ \eta_{S(A),S(B)}\ } & \mathrm{Hom}(S(B),S^2(A))^*. \end{array}$$

It turns out that this is automatically satisfied.[1] Indeed, inserting the additional diagonal arrow $\eta^*_{B,S(A)} : \mathrm{Hom}(S(A), S(B)) \longrightarrow \mathrm{Hom}(B, S(A))^*$ induced by the

[1] Thanks to Raphael Rouquier for explaining this to me.

defining property of a Serre functor, one reduces to the commutativity of the two triangles. More precisely, what we denote by $\eta^*_{B,S(A)}$ is in fact the composition of $\mathrm{Hom}(S(A),S(B)) \longrightarrow \mathrm{Hom}(S(A),S(B))^{**}$ with the actual $\eta^*_{B,S(A)}$. Thus one has to show that

$$\begin{array}{ccc}
\mathrm{Hom}(A,B) & & \\
 & \searrow^{\eta_{A,B}} & \\
S\Big\downarrow & & \mathrm{Hom}(B,S(A))^* \\
 & \nearrow_{\eta^*_{B,S(A)}} & \\
\mathrm{Hom}(S(A),S(B)) & &
\end{array}$$

is commutative or, equivalently, that for $f \in \mathrm{Hom}(B,S(A))$ and $g \in \mathrm{Hom}(A,B)$ one has $\langle f|g\rangle = \langle S(g)|f\rangle$. Since η is functorial in the second variable, we have the commutative diagram

$$\begin{array}{ccc}
\mathrm{Hom}(A,B) & \xrightarrow{\ \eta_{A,B}\ } & \mathrm{Hom}(B,S(A))^* \\
\circ g \Big\uparrow & \circlearrowleft & \Big\uparrow (S(g)\circ\ )^* \\
\mathrm{Hom}(B,B) & \xrightarrow{\ \eta_{B,B}\ } & \mathrm{Hom}(B,S(B))^*.
\end{array}$$

Applied to $\mathrm{id} \in \mathrm{Hom}(B,B)$ it yields $\langle f|g\rangle = \langle (S(g)\circ f)|\mathrm{id}\rangle$. We next claim that $\langle (S(g)\circ f)|\mathrm{id}\rangle = \langle S(g)|f\rangle$, which can be seen by commutativity of the analogous diagram (which uses functoriality of η in the first variable)

$$\begin{array}{ccc}
\mathrm{Hom}(B,B) & \xrightarrow{\ \eta_{B,B}\ } & \mathrm{Hom}(B,S(B))^* \\
f\circ \Big\downarrow & \circlearrowleft & \Big\downarrow (\ \circ f)^* \\
\mathrm{Hom}(B,S(A)) & \xrightarrow{\ \eta_{B,S(A)}\ } & \mathrm{Hom}(S(A),S(B))^*.
\end{array}$$

In order to avoid any trouble with the dual, one usually assumes that all Hom's in $\mathcal{A}$ are finite-dimensional. Under this hypothesis it is easy to see that a Serre functor, if it exists, is unique up to isomorphism. More generally one has the following

Lemma 1.30 *Let $\mathcal{A}$ and $\mathcal{B}$ be k-linear categories over a field k with finite-dimensional Hom's. If $\mathcal{A}$ and $\mathcal{B}$ are endowed with a Serre functor $S_{\mathcal{A}}$, respectively $S_{\mathcal{B}}$, then any k-linear equivalence*

$$F : \mathcal{A} \longrightarrow \mathcal{B}$$

commutes with Serre duality, i.e. there exists an isomorphism

$$F \circ S_{\mathcal{A}} \simeq S_{\mathcal{B}} \circ F.$$

Proof This is an application of the Yoneda lemma 1.6: since F is fully faithful, one has for any two objects $A, B \in \mathcal{A}$

$$\mathrm{Hom}(A, S(B)) \simeq \mathrm{Hom}(F(A), F(S(B))) \text{ and } \mathrm{Hom}(B, A) \simeq \mathrm{Hom}(F(B), F(A)).$$

Together with the two isomorphisms

$$\mathrm{Hom}(A, S(B)) \simeq \mathrm{Hom}(B, A)^* \text{ and } \mathrm{Hom}(F(B), F(A)) \simeq \mathrm{Hom}(F(A), S(F(B)))^*,$$

this yields a functorial (in A and B) isomorphism

$$\mathrm{Hom}(F(A), F(S(B))) \simeq \mathrm{Hom}(F(A), S(F(B))).$$

Using the hypothesis that F is an equivalence and, in particular, that any object in $\mathcal{B}$ is isomorphic to some $F(A)$, one concludes that there exists a functor isomorphism $F \circ S_{\mathcal{A}} \simeq S_{\mathcal{B}} \circ F$. □

Remark 1.31 Let $F : \mathcal{A} \longrightarrow \mathcal{B}$ be a functor between k-linear categories $\mathcal{A}$ and $\mathcal{B}$ endowed with Serre functors $S_{\mathcal{A}}$, respectively $S_{\mathcal{B}}$. Then

$$G \dashv F \quad \Rightarrow F \dashv S_{\mathcal{A}} \circ G \circ S_{\mathcal{B}}^{-1}.$$

(As before we assume that all Hom's are finite-dimensional.)

Indeed, under the given assumptions we have the following functorial isomorphisms:

$$\begin{aligned} \mathrm{Hom}(A_1, (S_{\mathcal{A}} \circ G \circ S_{\mathcal{B}}^{-1})(A_2)) &\simeq \mathrm{Hom}((G \circ S_{\mathcal{B}}^{-1})(A_2), A_1)^* \\ &\simeq \mathrm{Hom}(S_{\mathcal{B}}^{-1}(A_2), F(A_1))^* \\ &\simeq \mathrm{Hom}(F(A_1), S_{\mathcal{B}}(S_{\mathcal{B}}^{-1}(A_2))) \\ &\simeq \mathrm{Hom}(F(A_1), A_2). \end{aligned}$$

A similar argument allows the construction of a left adjoint if a right adjoint $F \dashv H$ is given. In particular, for functors between categories with Serre functors the existence of the left or the right adjoint implies the existence of the other one.

1.2 Triangulated categories and exact functors

Triangulated categories, the kind of categories we will be interested in throughout, were introduced independently and around the same time by Puppe [99] and in Verdier's thesis [118] under the supervision of Grothendieck. We recommend [39, 61, 88] for a more in-depth reading.

Let us start right away with the definition of a triangulated category.

Definition 1.32 *Let $\mathcal{D}$ be an additive category. The structure of a* triangulated category *on $\mathcal{D}$ is given by an additive equivalence*

$$T : \mathcal{D} \longrightarrow \mathcal{D},$$

the shift functor, *and a set of* distinguished triangles

$$A \longrightarrow B \longrightarrow C \longrightarrow T(A)$$

subject to the axioms TR1–TR4 *below.*

Before actually explaining the axioms TR, let us introduce the notation $A[1] := T(A)$ for any object $A \in \mathcal{D}$ and $f[1] := T(f) \in \mathrm{Hom}(A[1], B[1])$ for any morphism $f \in \mathrm{Hom}(A, B)$. Similarly, one writes $A[n] := T^n(A)$ and $f[n] := T^n(f)$ for $n \in \mathbb{Z}$. Thus, a triangle will also be denoted by $A \to B \to C \to A[1]$.

A morphism between two triangles is given by a commutative diagram

$$\begin{array}{ccccccc} A & \longrightarrow & B & \longrightarrow & C & \longrightarrow & A[1] \\ \downarrow{\scriptstyle f} & & \downarrow{\scriptstyle g} & & \downarrow{\scriptstyle h} & & \downarrow{\scriptstyle f[1]} \\ A' & \longrightarrow & B' & \longrightarrow & C' & \longrightarrow & A'[1]. \end{array}$$

It is an isomorphism if f, g, and h are isomorphisms.

Here are the axioms for a triangulated category:

TR1 i) Any triangle of the form

$$A \xrightarrow{\mathrm{id}} A \longrightarrow 0 \longrightarrow A[1]$$

is distinguished.

ii) Any triangle isomorphic to a distinguished triangle is distinguished.

iii) Any morphism $f : A \to B$ can be completed to a distinguished triangle

$$A \xrightarrow{f} B \longrightarrow C \longrightarrow A[1].$$

TR2 The triangle

$$A \xrightarrow{f} B \xrightarrow{g} C \xrightarrow{h} A[1]$$

is a distinguished triangle if and only if

$$B \xrightarrow{g} C \xrightarrow{h} A[1] \xrightarrow{-f[1]} B[1]$$

is a distinguished triangle.

TR3 Suppose there exists a commutative diagram of distinguished triangles with vertical arrows f and g:

$$\begin{array}{ccccccc} A & \longrightarrow & B & \longrightarrow & C & \longrightarrow & A[1] \\ {\scriptstyle f}\downarrow & & {\scriptstyle g}\downarrow & & {\scriptstyle h}\,\vdots & & \downarrow{\scriptstyle f[1]} \\ A' & \longrightarrow & B' & \longrightarrow & C' & \longrightarrow & A'[1]. \end{array}$$

Then the diagram can be completed to a commutative diagram, i.e. to a morphism of triangles, by a (not necessarily unique) morphism $h : C \longrightarrow C'$.

TR4 This is the axiom that is most complicated to state (and to print). It is called the octahedron axiom. As it is never used explicitly in this book (and implicitly only once, namely in the proof of Orlov's theorem 5.14), we refrain from including it here and refer to the literature for the precise formulation. (In fact, this axiom is missing in Puppe's definition, so that he deals rather with pre-triangulated categories.)

To give the reader nevertheless an impression of what this axiom is about, recall that for nested inclusions, of say abelian groups, $A \subset B \subset C$, there exists a canonical isomorphism $C/B \simeq (C/A)/(B/A)$. If one replaces the short exact sequences $A \longrightarrow B \longrightarrow B/A$, $A \longrightarrow C \longrightarrow C/A$, and $B \longrightarrow C \longrightarrow C/B$ by distinguished triangles in a triangulated category, then TR4 roughly requires $B/A \longrightarrow C/A \longrightarrow C/B$ to be distinguished as well (cf. [61, Ch.1.4]).[2]

The first two axioms TR1 and TR2 seem very natural. Essentially, they are saying that the set of distinguished triangles is preserved under shift and isomorphisms and that there are enough distinguished triangles available. The third one, TR3, seems a little less so, due to the non-uniqueness of the completing morphism.

Note, a priori we have not required that in a triangle $A \longrightarrow B \longrightarrow C \longrightarrow A[1]$ the composition $A \longrightarrow C$ is zero. But this can be easily deduced by combining TR1 and TR3.

Exercise 1.33 Prove the last statement.

Proposition 1.34 *Let $A \longrightarrow B \longrightarrow C \longrightarrow A[1]$ be a distinguished triangle in a triangulated category $\mathcal{D}$. Then for any object $A_0 \in \mathcal{D}$ the following induced sequences are exact:*

$$\mathrm{Hom}(A_0, A) \longrightarrow \mathrm{Hom}(A_0, B) \longrightarrow \mathrm{Hom}(A_0, C)$$

$$\mathrm{Hom}(C, A_0) \longrightarrow \mathrm{Hom}(B, A_0) \longrightarrow \mathrm{Hom}(A, A_0).$$

[2] Thanks for D. Ben-Zvi for this interpretation.

Proof Suppose $f : A_0 \longrightarrow B$ composed with $B \longrightarrow C$ is the trivial morphism $A_0 \longrightarrow B \longrightarrow C$. Then apply TR1 and TR3 to

$$\begin{array}{ccccc} A_0 & \xrightarrow{\text{id}} & A_0 & \longrightarrow & 0 \\ \vdots & & \downarrow f & & \downarrow \\ A & \longrightarrow & B & \longrightarrow & C, \end{array}$$

which allows us to lift f to a morphism $A_0 \longrightarrow A$.

The proof for the second assertion is similar. □

Remark 1.35 Due to TR2, $\text{Hom}(A_0, B) \longrightarrow \text{Hom}(A_0, C) \longrightarrow \text{Hom}(A_0, A[1])$ is exact as well and similarly for $\text{Hom}(\ \ , A_0)$. Thus, one obtains in fact long exact sequences.

Exercise 1.36 Suppose $A \longrightarrow B \longrightarrow C \longrightarrow A[1]$ is a distinguished triangle. Show that $A \longrightarrow B$ is an isomorphism if and only if $C \simeq 0$.

Exercise 1.37 Consider a morphism of distinguished triangles

$$\begin{array}{ccccccc} A & \longrightarrow & B & \longrightarrow & C & \longrightarrow & A[1] \\ f\downarrow & & g\downarrow & & h\downarrow & & \downarrow f[1] \\ A' & \longrightarrow & B' & \longrightarrow & C' & \longrightarrow & A'[1]. \end{array}$$

Show that if two of the vertical morphisms f, g, and h are isomorphisms then so is the third. Also note that f and g might be zero without h being so.

Exercise 1.38 Let $A \longrightarrow B \longrightarrow C \longrightarrow A[1]$ be a distinguished triangle in a triangulated category $\mathcal{D}$. Suppose that $C \longrightarrow A[1]$ is trivial. Show that then the triangle is split, i.e. is given by a direct sum decomposition $B \simeq A \oplus C$.

Definition 1.39 *An additive functor*

$$F : \mathcal{D} \longrightarrow \mathcal{D}'$$

between triangulated categories $\mathcal{D}$ and $\mathcal{D}'$ is called exact *if the following two conditions are satisfied:*

i) *There exists a functor isomorphism*

$$F \circ T_{\mathcal{D}} \xrightarrow{\sim} T_{\mathcal{D}'} \circ F.$$

ii) *Any distinguished triangle*

$$A \longrightarrow B \longrightarrow C \longrightarrow A[1]$$

in $\mathcal{D}$ is mapped to a distinguished triangle

$$F(A) \longrightarrow F(B) \longrightarrow F(C) \longrightarrow F(A)[1]$$

in $\mathcal{D}'$, where $F(A[1])$ is identified with $F(A)[1]$ via the functor isomorphism in i).

Remark 1.40 Once again, the notions of a triangulated category and of an exact functor have to be adjusted when one is interested in additive categories over a field k. In this case, the shift functor should be k-linear and one usually considers only k-linear exact functors.

Also note that in this case the two long exact cohomology sequences in Proposition 1.34 associated to a distinguished triangle are long exact sequences of k-vector spaces.

Compare the following proposition with Remark 1.16, iv).

Proposition 1.41 *Let $F : \mathcal{D} \longrightarrow \mathcal{D}'$ be an exact functor between triangulated categories. If $F \dashv H$, then $H : \mathcal{D}' \longrightarrow \mathcal{D}$ is exact.*

Similarly, if $G \dashv F$ then $G : \mathcal{D}' \longrightarrow \mathcal{D}$ is exact. See [13, 92].

Proof Let us first show that the adjoint functor H commutes with the shift functors T and T' on $\mathcal{D}$, respectively $\mathcal{D}'$. Since F is an exact functor, one has isomorphisms $F \circ T \simeq T' \circ F$ and $F \circ T^{-1} \simeq {T'}^{-1} \circ F$.

This yields the following functorial isomorphisms

$$\begin{aligned} \mathrm{Hom}(A, H(T'(B))) &\simeq \mathrm{Hom}(F(A), T'(B)) \simeq \mathrm{Hom}({T'}^{-1}(F(A)), B) \\ &\simeq \mathrm{Hom}(F(T^{-1}(A)), B) \simeq \mathrm{Hom}(T^{-1}(A), H(B)) \\ &\simeq \mathrm{Hom}(A, T(H(B))). \end{aligned}$$

As everything is functorial, the Yoneda lemma yields an isomorphism

$$H \circ T' \xrightarrow{\sim} T \circ H.$$

Next, we have to show that H maps a distinguished triangle in $\mathcal{D}'$ to a distinguished triangle in $\mathcal{D}$. Let $A \longrightarrow B \longrightarrow C \longrightarrow A[1]$ be a distinguished triangle in $\mathcal{D}'$. The induced morphism $H(A) \longrightarrow H(B)$ can be completed to a distinguished triangle

$$H(A) \longrightarrow H(B) \longrightarrow C_0 \longrightarrow H(A)[1].$$

Here we tacitly use $H(A[1]) \simeq H(A)[1]$ given by the above isomorphism $H \circ T' \simeq T \circ H$.

Using the adjunction morphisms $F(H(A)) \longrightarrow A$ and $F(H(B)) \longrightarrow B$ and the assumption that F is exact, one obtains a commutative diagram of distinguished triangles

$$\begin{array}{ccccccc}
F(H(A)) & \longrightarrow & F(H(B)) & \longrightarrow & F(C_0) & \longrightarrow & F(H(A))[1] \\
\downarrow & & \downarrow & & \vdots\ \xi & & \downarrow \\
A & \longrightarrow & B & \longrightarrow & C & \longrightarrow & A[1],
\end{array}$$

which can be completed by the dotted arrow according to axiom TR3.

Applying H to the whole diagram and using the adjunction $h : \mathrm{id} \longrightarrow H \circ F$, yields

$$\begin{array}{ccccccc}
H(A) & \longrightarrow & H(B) & \longrightarrow & C_0 & \longrightarrow & H(A)[1] \\
\downarrow & & \downarrow & & \downarrow h_{C_0} & & \downarrow \\
HFHA & \longrightarrow & HFHB & \longrightarrow & HFC_0 & \longrightarrow & HFHA[1] \\
\downarrow & & \downarrow & & \downarrow H(\xi) & & \downarrow \\
H(A) & \longrightarrow & H(B) & \longrightarrow & H(C) & \longrightarrow & H(A)[1].
\end{array}$$

Here, the curved vertical arrows are in both cases the identity morphisms (see Exercise 1.19). To conclude one would like to apply Exercise 1.37, but we are not allowed to use that $H(A) \longrightarrow H(B) \longrightarrow H(C) \longrightarrow H(A)[1]$ is distinguished. But using adjunction we know that for any A_0 the sequence

$$\begin{array}{ccccc}
\mathrm{Hom}(A_0, H(B)) & \longrightarrow & \mathrm{Hom}(A_0, H(C)) & \longrightarrow & \mathrm{Hom}(A_0, H(A)[1]) \\
\simeq \mathrm{Hom}(F(A_0), B) & & \simeq \mathrm{Hom}(F(A_0), C) & & \simeq \mathrm{Hom}(F(A_0), A[1])
\end{array}$$

is exact. Then we obtain $\mathrm{Hom}(A_0, C_0) \simeq \mathrm{Hom}(A_0, H(C))$ for all A_0 and hence $H(\xi) \circ h_{C_0} : C_0 \xrightarrow{\sim} H(C)$. Thus, $H(A) \longrightarrow H(B) \longrightarrow H(C) \longrightarrow H(A)[1]$ is isomorphic to the distinguished triangle $H(A) \longrightarrow H(B) \longrightarrow C_0 \longrightarrow H(A)[1]$, so it is itself distinguished by TR1. □

A subcategory $\mathcal{D}' \subset \mathcal{D}$ of a triangulated category is a *triangulated subcategory* if $\mathcal{D}'$ admits the structure of a triangulated category such that the inclusion is exact. If $\mathcal{D}' \subset \mathcal{D}$ is a full subcategory, then it is a triangulated subcategory if and only if $\mathcal{D}'$ is invariant under the shift functor and for any distinguished triangle $A \longrightarrow B \longrightarrow C \longrightarrow A[1]$ in $\mathcal{D}$ with $A, B \in \mathcal{D}'$ the object C is isomorphic to an object in $\mathcal{D}'$.

Definition 1.42 *A full triangulated subcategory $\mathcal{D}' \subset \mathcal{D}$ is called* admissible *if the inclusion has a right adjoint $\pi : \mathcal{D} \longrightarrow \mathcal{D}'$, i.e. there exist functorial isomorphisms $\mathrm{Hom}_{\mathcal{D}}(A, B) \simeq \mathrm{Hom}_{\mathcal{D}'}(A, \pi(B))$ for all $A \in \mathcal{D}'$ and $B \in \mathcal{D}$.*

The orthogonal complement *of a(n admissible) subcategory $\mathcal{D}' \subset \mathcal{D}$ is the full subcategory $\mathcal{D}'^{\perp}$ of all objects $C \in \mathcal{D}$ such that $\mathrm{Hom}(B, C) = 0$ for all $B \in \mathcal{D}'$.*

More accurately, the orthogonal complement as defined above should be called the right orthogonal complement. One similarly defines the left orthogonal complement, but this will never be used. So, orthogonal in this book will always mean right orthogonal.

Remarks 1.43 i) The right adjoint functor $\pi : \mathcal{D} \longrightarrow \mathcal{D}'$ for an admissible full triangulated subcategory $\mathcal{D}' \subset \mathcal{D}$ is exact by Proposition 1.41.

ii) The orthogonal complement of an admissible subcategory is a triangulated subcategory.

Indeed, the condition $\mathrm{Hom}(B, C) = 0$ for all $B \in \mathcal{D}'$ yields

$$\mathrm{Hom}(B, C[i]) \simeq \mathrm{Hom}(B[-i], C) = 0$$

for all $B \in \mathcal{D}'$, as $\mathcal{D}'$ is invariant under shift. Thus, if

$$C_1 \longrightarrow C_2 \longrightarrow C_3 \longrightarrow C_1[1]$$

is a distinguished triangle in $\mathcal{D}$ with $C_1, C_2 \in \mathcal{D}'^{\perp}$ then the long exact sequence obtained from applying $\mathrm{Hom}(B, \ )$ shows that also $C_3 \in \mathcal{D}'^{\perp}$.

iii) More explicitly, one shows that a full triangulated subcategory $\mathcal{D}' \subset \mathcal{D}$ is admissible if and only if for all $A \in \mathcal{D}$ there exists a distinguished triangle

$$B \longrightarrow A \longrightarrow C \longrightarrow B[1]$$

with $B \in \mathcal{D}'$ and $C \in \mathcal{D}'^{\perp}$. This goes as follows.

Suppose $\mathcal{D}'$ is admissible. The adjunction property of π allows us to associate to the identity in $\mathrm{Hom}_{\mathcal{D}'}(\pi(A), \pi(A))$ a morphism $B := \pi(A) \longrightarrow A$ which we may complete to a distinguished triangle

$$B \longrightarrow A \longrightarrow C \longrightarrow B[1].$$

In order to see that indeed $C \in \mathcal{D}'^{\perp}$, one applies $\mathrm{Hom}(B', \ )$ and uses that for all $B' \in \mathcal{D}'$

$$\mathrm{Hom}(B', B) \simeq \mathrm{Hom}(B', \pi(A)) \simeq \mathrm{Hom}(B', A).$$

Conversely, if such a distinguished triangle is given for any A, then one defines the functor $\pi : \mathcal{D} \longrightarrow \mathcal{D}'$ by $\pi(A) = B$. Now use $\mathrm{Hom}(B, C) = 0$ for $B \in \mathcal{D}'$ and $C \in \mathcal{D}'^{\perp}$ to show that B does not depend (up to isomorphism) on the choice of the triangle. Similarly, one shows that π is well-defined for morphisms.

iv) Admissible subcategories occur whenever there is a fully faithful exact functor $F : \mathcal{D}' \longrightarrow \mathcal{D}$ that admits a right adjoint. Indeed, in this case the functor F defines an equivalence between $\mathcal{D}'$ and an admissible subcategory of $\mathcal{D}$.

Exercise 1.44 Let $A \in \mathcal{D}$ be an object in a triangulated category $\mathcal{D}$. Show that

$$A^{\perp} := \{B \in \mathcal{D} \mid \mathrm{Hom}(A, B[i]) = 0 \text{ for all } i \in \mathbb{Z}\}$$

is a triangulated subcategory. If $\langle A \rangle$ denotes the smallest triangulated subcategory containing A, then $A^\perp \simeq \langle A \rangle^\perp$.

The notion of equivalence that will be important for us is the following.

Definition 1.45 *Two triangulated categories $\mathcal{D}$ and $\mathcal{D}'$ are* equivalent *if there exists an exact equivalence $F : \mathcal{D} \longrightarrow \mathcal{D}'$.*

If $\mathcal{D}$ is a triangulated category, the set $\mathrm{Aut}(\mathcal{D})$ *of isomorphism classes of equivalences $F : \mathcal{D} \longrightarrow \mathcal{D}$ forms the* group of autoequivalences of $\mathcal{D}$.

We conclude this section by a discussion of Serre functors in the context of triangulated categories. As it turns out, Serre functors and triangulated structures are always compatible. In the geometric situation considered later, this will be obvious, for the Serre functors there will by construction be exact. (So, the reader mainly interested in geometry may safely skip the not so easy proof of the following proposition.)

Proposition 1.46 **(Bondal, Kapranov)** *Any Serre functor on a triangulated category over a field k is exact. See [13].*

Proof For simplicity we shall assume that all Hom's are finite-dimensional.

By Lemma 1.30, a Serre functor S commutes with the shift functor T. It remains to show that under S a distinguished triangle $A \xrightarrow{f} B \xrightarrow{g} C \xrightarrow{h} A[1]$ is mapped to a distinguished triangle. In a first step, one completes $S(A) \longrightarrow S(B)$ to a distinguished triangle

$$S(A) \xrightarrow{S(f)} S(B) \xrightarrow{\varphi} C_0 \xrightarrow{\psi} S(A)[1].$$

Next, one tries to construct a commutative diagram

$$\begin{array}{ccccccc} S(A) & \longrightarrow & S(B) & \xrightarrow{\varphi} & C_0 & \xrightarrow{\psi} & S(A)[1] \\ \downarrow{=} & & \downarrow{=} & & \vdots\,\xi & & \downarrow{=} \\ S(A) & \longrightarrow & S(B) & \xrightarrow{S(g)} & S(C) & \xrightarrow{S(h)} & S(A[1]). \end{array}$$

The compatible long exact sequences, induced by applying $\mathrm{Hom}(D,\ )$ to the horizontal sequences, and the Yoneda lemma would then show that $\xi : C_0 \longrightarrow S(C)$ must be an isomorphism (see Exercise 1.37). (Note that the long sequence induced by the bottom sequence is dual to the long exact sequence induced by applying $\mathrm{Hom}(\ , D)$ to the distinguished triangle $A \longrightarrow B \longrightarrow C \longrightarrow A[1]$ and hence itself exact.)

It remains to prove the existence of ξ. Via Serre duality ξ can be considered as a linear form $\langle \xi| \quad \rangle$ on $\mathrm{Hom}(C, C_0)$. The two conditions ξ needs to satisfy are expressed as

$$\text{i) } \xi \circ \varphi = S(g) \quad \text{and} \quad \text{ii) } S(h) \circ \xi = \psi.$$

Clearly, i) holds if and only if $\langle \xi \circ \varphi|\alpha\rangle = \langle S(g)|\alpha\rangle$ for any $\alpha \in \mathrm{Hom}(C, S(B))$. By functoriality of the Serre functor $\langle \xi \circ \varphi|\alpha\rangle = \langle \xi|\varphi \circ \alpha\rangle$. Similarly, $\langle S(g)|\alpha\rangle = \langle \alpha|g\rangle = \langle \mathrm{id}_B|\alpha \circ g\rangle$, where the first equality is taken from Remark 1.29. Hence, condition i) is equivalent to

$$\text{i}') \ \langle \xi|\varphi \circ \quad \rangle = \langle \mathrm{id}_B| \quad \circ g\rangle \in \mathrm{Hom}(C, S(B))^*.$$

The condition ii) can be equivalently written as $\langle S(h) \circ \xi|\beta\rangle = \langle \psi|\beta\rangle$ for any $\beta \in \mathrm{Hom}(A[1], C_0)$. Again using functoriality and Remark 1.29, $\langle S(h) \circ \xi|\beta\rangle = \langle S(h)|\xi \circ \beta\rangle = \langle \xi \circ \beta|h\rangle = \langle \xi|\beta \circ h\rangle$ and $\langle \psi|\beta\rangle = \langle \mathrm{id}_A|\psi \circ \beta\rangle$. Hence, ii) is equivalent to

$$\text{ii}') \ \langle \xi| \quad \circ h\rangle = \langle \mathrm{id}_A|\psi \circ \quad \rangle \in \mathrm{Hom}(A[1], C_0)^*.$$

Thus, in order to ensure the existence of the desired ξ or, equivalently, of the linear form $\langle \xi| \quad \rangle : \mathrm{Hom}(C, C_0) \longrightarrow k$, it suffices to show that for any $\alpha \in \mathrm{Hom}(C, S(B))$ and any $\beta \in \mathrm{Hom}(A[1], C_0)$ one has

$$\text{If} \quad \varphi \circ \alpha = \beta \circ h, \quad \text{then} \quad \langle \mathrm{id}_B|\alpha \circ g\rangle = \langle \mathrm{id}_A|\psi \circ \beta\rangle.$$

Firstly, TR3 shows that there exists a commutative diagram

$$\begin{array}{ccccccccc}
B & \xrightarrow{g} & C & \xrightarrow{h} & A[1] & \xrightarrow{-f[1]} & B[1] & \xrightarrow{-g[1]} & C[1] \\
\downarrow{\scriptstyle\gamma} & & \downarrow{\scriptstyle\alpha} & & \downarrow{\scriptstyle\beta} & & \vdots{\scriptstyle\gamma[1]} & & \downarrow{\scriptstyle\alpha[1]} \\
S(A) & \xrightarrow{-S(f)} & S(B) & \xrightarrow{\varphi} & C_0 & \xrightarrow{\psi} & S(A)[1] & \xrightarrow{-S(f)[1]} & S(B)[1].
\end{array}$$

But then, using functoriality of the Serre pairing $\mathrm{Hom}(\quad, B) \simeq \mathrm{Hom}(B, S(\quad))^*$, one obtains

$$\langle \mathrm{id}_B|\alpha \circ g\rangle = -\langle \mathrm{id}_B|S(f) \circ \gamma\rangle = -\langle S(f)|\gamma\rangle = -\langle \gamma|f\rangle.$$

Similarly, one finds

$$\langle \mathrm{id}_A|\psi \circ \beta\rangle = -\langle \mathrm{id}_A, \gamma[1] \circ f[1]\rangle = -\langle \gamma[1], f[1]\rangle = -\langle \gamma|f\rangle$$

and hence $\langle \mathrm{id}_B|\alpha \circ g\rangle = \langle \mathrm{id}_A|\psi \circ \beta\rangle$. □

1.3 Equivalences of triangulated categories

In this section we discuss criteria that allow us to decide whether a given exact functor is fully faithful or even an equivalence. This continues the discussion of Remark 1.27 in the context of triangulated categories.

Let us begin with the definition of a spanning class. In many geometric situations, spanning classes (sometimes even several ones) are given naturally (cf. Proposition 3.17 or Corollary 3.19).

Definition 1.47 *A collection Ω of objects in a triangulated category $\mathcal{D}$ is a* spanning class *of $\mathcal{D}$ (or* spans *$\mathcal{D}$) if for all $B \in \mathcal{D}$ the following two conditions hold:*

i) *If* $\mathrm{Hom}(A, B[i]) = 0$ *for all* $A \in \Omega$ *and all* $i \in \mathbb{Z}$, *then* $B \simeq 0$.
ii) *If* $\mathrm{Hom}(B[i], A) = 0$ *for all* $A \in \Omega$ *and all* $i \in \mathbb{Z}$, *then* $B \simeq 0$.

Exercise 1.48 Suppose the triangulated category $\mathcal{D}$ is endowed with a Serre functor. Show that the two conditions i) and ii) in the definition are equivalent. So, in the presence of a Serre functor it suffices to require one of the two.

Proposition 1.49 *Let $F : \mathcal{D} \longrightarrow \mathcal{D}'$ be an exact functor between triangulated categories with left and right adjoints: $G \dashv F \dashv H$.*

Suppose Ω is a spanning class of $\mathcal{D}$ such that for all objects $A, B \in \Omega$ and all $i \in \mathbb{Z}$ the natural homomorphisms

$$F : \mathrm{Hom}(A, B[i]) \longrightarrow \mathrm{Hom}(F(A), F(B[i]))$$

are bijective. Then F is fully faithful. See [18, 92].

Proof First recall that H and G are both exact due to Proposition 1.41. This will be used throughout.

We shall use the following commutative diagram (see Lemma 1.21):

$$\begin{array}{ccc} \mathrm{Hom}(A,B) & \xrightarrow{h_B \circ} & \mathrm{Hom}(A, H(F(B))) \\ {\scriptstyle \circ g_A}\downarrow & \searrow^{F} & \downarrow{\scriptstyle \wr} \\ \mathrm{Hom}(G(F(A)), B) & \xrightarrow{\sim} & \mathrm{Hom}(F(A), F(B)) \end{array} \tag{1.2}$$

for arbitrary $A, B \in \mathcal{D}$.

We first show that for any $A \in \Omega$ the homomorphism $g_A : G(F(A)) \longrightarrow A$ is an isomorphism. In order to see this, choose a distinguished triangle

$$G(F(A)) \xrightarrow{g_A} A \longrightarrow C \longrightarrow G(F(A))[1].$$

Applying Hom(, B) for an arbitrary $B \in \mathcal{D}$ induces a long exact sequence which combined with the commutative lower triangle yields

$$\begin{array}{ccccccc} \mathrm{Hom}(C, B[i]) & \longrightarrow & \mathrm{Hom}(A, B[i]) & \xrightarrow{\circ g_A} & \mathrm{Hom}(G(F(A)), B[i]) & \longrightarrow \\ & & & \searrow^{F} & \downarrow \wr & \\ & & & & \mathrm{Hom}(F(A), F(B)[i]). & \end{array}$$

If $B \in \Omega$, then $F : \mathrm{Hom}(A, B[i]) \longrightarrow \mathrm{Hom}(F(A), F(B)[i])$ is bijective by assumption. Hence, $\mathrm{Hom}(C, B[i]) = 0$ for all $i \in \mathbb{Z}$ and all $B \in \Omega$. Since Ω spans $\mathcal{D}$, one finds $C \simeq 0$ and, therefore, $g_A : G(F(A)) \xrightarrow{\sim} A$.

Note that this immediately implies that for $A \in \Omega$ and any $B \in \mathcal{D}$ in fact all homomorphisms in (1.2) are bijections, in particular

$$h_B \circ : \mathrm{Hom}(A, B) \xrightarrow{\sim} \mathrm{Hom}(A, H(F(B))).$$

This applied to $B \in \mathcal{D}$ and using a distinguished triangle of the form (again use TR1 for its existence)

$$B \xrightarrow{h_B} H(F(B)) \longrightarrow C \longrightarrow B[1]$$

shows that $\mathrm{Hom}(A, C[i]) = 0$ for all $i \in \mathbb{Z}$ and all $A \in \Omega$. Hence, $C \simeq 0$ and, thus, $h_B : B \xrightarrow{\sim} H(F(B))$. In particular,

$$h_B \circ : \mathrm{Hom}(A, B) \xrightarrow{\sim} \mathrm{Hom}(A, H(F(B)))$$

for any $A \in \mathcal{D}$. Using the commutativity of the upper triangle in (1.2), this proves that $F : \mathrm{Hom}(A, B) \xrightarrow{\sim} \mathrm{Hom}(F(A), F(B))$ is bijective for all $A, B \in \mathcal{D}$, i.e. F is fully faithful. (This last step in the proof was also stated as Corollary 1.23.) □

Suppose we already know that the functor is fully faithful. What do we need to know in order to be able to decide whether it is in fact an equivalence? The following lemma provides a first criterion, whose assumption however is difficult to check. Building upon the arguments used in its proof we shall, however, deduce Proposition 1.54, which turns out to be extremely useful.

Lemma 1.50 *Let $F : \mathcal{D} \longrightarrow \mathcal{D}'$ be a fully faithful exact functor between triangulated categories and suppose that F has a right adjoint $F \dashv H$. Then F is an equivalence if and only if for any $C \in \mathcal{D}'$ the triviality of $H(C)$, i.e. $H(C) \simeq 0$, implies $C \simeq 0$.*

Proof By Corollary 1.22 one knows that for any A the adjunction morphism $h_A : A \longrightarrow HF(A)$ is an isomorphism.

In order to prove the assertion, one has to verify that also the adjunction morphism $g_B : FH(B) \to B$ is an isomorphism for any $B \in \mathcal{D}'$. Indeed, H would be an inverse of F in this case. Note that Corollary 1.22 does not apply, because we don't know whether H is fully faithful.

For any $B \in \mathcal{D}'$ the morphism $g_B : FH(B) \to B$ can be completed to a distinguished triangle

$$FH(B) \longrightarrow B \longrightarrow C \longrightarrow FH(B)[1].$$

Since H is exact by Proposition 1.41, we obtain a distinguished triangle in $\mathcal{D}$

$$HFH(B) \xrightarrow{H(g_B)} H(B) \longrightarrow H(C) \longrightarrow HFH(B)[1].$$

Since by Exercise 1.19 one knows that $H(g_B) \circ h_{H(B)} = \mathrm{id}_{H(B)}$ and, therefore, that $H(g_B)$ is an isomorphism, this shows $H(C) \simeq 0$. Hence, by assumption $C \simeq 0$ which in turn shows that g_B is an isomorphism. □

Exercise 1.51 State and prove the analogous statement for a left adjoint functor $G \dashv F$.

Definition 1.52 *A triangulated category $\mathcal{D}$ is* decomposed into triangulated subcategories $\mathcal{D}_1, \mathcal{D}_2 \subset \mathcal{D}$ *if the following three conditions are satisfied:*

i) *Both categories $\mathcal{D}_1$ and $\mathcal{D}_2$ contain objects non-isomorphic to 0.*

ii) *For all $A \in \mathcal{D}$ there exists a distinguished triangle*

$$B_1 \longrightarrow A \longrightarrow B_2 \longrightarrow B_1[1]$$

with $B_i \in \mathcal{D}_i$, $i = 1, 2$.

iii) $\mathrm{Hom}(B_1, B_2) = \mathrm{Hom}(B_2, B_1) = 0$ *for all $B_1 \in \mathcal{D}_1$ and $B_2 \in \mathcal{D}_2$.*

A triangulated category that cannot be decomposed is called indecomposable.

Later, we will see that the derived category of an integral scheme is indecomposable (see Proposition 3.10).

Exercise 1.53 Show that condition ii) in the presence of iii) just says that A is the direct sum of B_1 and B_2. In particular, the definition is symmetric in $\mathcal{D}_1$ and $\mathcal{D}_2$ despite the chosen order in ii).

Proposition 1.54 *Let $F : \mathcal{D} \to \mathcal{D}'$ be a fully faithful exact functor between triangulated categories. Suppose that $\mathcal{D}$ contains objects not isomorphic to 0 and that $\mathcal{D}'$ is indecomposable.*

Then F is an equivalence of categories if and only if F has a left adjoint $G \dashv F$ and a right adjoint $F \dashv H$ such that for any object $B \in \mathcal{D}'$ one has: $H(B) \simeq 0$ implies $G(B) \simeq 0$. See [18].

Proof In order to prove the proposition, one introduces two full triangulated subcategories $\mathcal{D}'_1, \mathcal{D}'_2 \subset \mathcal{D}'$. The first one, $\mathcal{D}'_1$, is the image of F, i.e. the full subcategory of all objects B isomorphic to some $F(A)$. Equivalently, $\mathcal{D}'_1$ is the full subcategory of objects $B \in \mathcal{D}'$ with $F(H(B)) \simeq B$ (induced by adjunction).

Indeed, if $B \simeq F(A)$, then

$$H(B) \simeq H(F(A)) \simeq A,$$

for F is fully faithful. Thus,

$$B \simeq F(A) \simeq F(H(B)).$$

The second category, $\mathcal{D}'_2$, consists of all $C \in \mathcal{D}'$ with $H(C) \simeq 0$. Clearly, both are triangulated subcategories of $\mathcal{D}'$.

The arguments in the proof of the previous lemma show that any $B \in \mathcal{D}'$ can be decomposed by a distinguished triangle

$$B_1 \longrightarrow B \longrightarrow B_2 \longrightarrow B[1]$$

with $B_i \in \mathcal{D}'_i$.

Furthermore, for all $B_1 \in \mathcal{D}'_1$ and $B_2 \in \mathcal{D}'_2$ we have

$$\mathrm{Hom}(B_1, B_2) \simeq \mathrm{Hom}(F(H(B_1)), B_2) \simeq \mathrm{Hom}(H(B_1), H(B_2)) = 0$$

and

$$\mathrm{Hom}(B_2, B_1) \simeq \mathrm{Hom}(B_2, F(H(B_1)) \simeq \mathrm{Hom}(G(B_2), H(B_1)) = 0,$$

for $H(B_2) \simeq 0$ by assumption implies $G(B_2) = 0$.

Since $\mathcal{D}'$ is indecomposable, either $\mathcal{D}'_1$ or $\mathcal{D}'_2$ is trivial, i.e. one of the two contains only objects isomorphic to 0. Suppose $\mathcal{D}'_1$ is trivial. Then for any $A \in \mathcal{D}$, the image $F(A)$ and hence $H(F(A))$ is trivial. As F is fully faithful, this proves $A \simeq H(F(A)) \simeq 0$ for all A, which contradicts the non-triviality of $\mathcal{D}$.

Hence, $\mathcal{D}'_2$ must be trivial. This proves that $\mathcal{D}'_1 \subset \mathcal{D}'$ is an equivalence, i.e. for every object $B \in \mathcal{D}'$ adjunction yields $F(H(B)) \simeq B$. Thus, H is a quasi-inverse of F. □

Remark 1.55 The proposition can be best applied when $G = H$. This particular case will in fact occur in the applications. So, if F is fully faithful and $H \dashv F \dashv H$ then F is an equivalence whenever $\mathcal{D}'$ is indecomposable.

The following is a combination of the two propositions in the presence of Serre functors.

Corollary 1.56 *Let $F : \mathcal{D} \longrightarrow \mathcal{D}'$ be an exact functor between triangulated categories $\mathcal{D}$ and $\mathcal{D}'$ with left adjoint $G \dashv F$ and right adjoint $F \dashv H$. Furthermore assume that Ω is a spanning class of $\mathcal{D}$ satisfying the following conditions:*

i) *For all $A, B \in \Omega$ the natural morphisms*

$$\mathrm{Hom}(A, B[i]) \longrightarrow \mathrm{Hom}(F(A), F(B)[i])$$

are bijective for all $i \in \mathbb{Z}$.

ii) *The categories $\mathcal{D}$ and $\mathcal{D}'$ admit Serre functors $S_{\mathcal{D}}$, respectively $S_{\mathcal{D}'}$ such that for all $A \in \Omega$ one has $F(S_{\mathcal{D}}(A)) \simeq S_{\mathcal{D}'}(F(A))$.*

iii) *The category $\mathcal{D}'$ is indecomposable and $\mathcal{D}$ is non-trivial.*

Then F is an equivalence. See [22].

Proof The first condition ensures by Proposition 1.49 that F is fully faithful. In order to apply Proposition 1.54, one has to verify the condition that $H(B) \simeq 0$ implies $G(B) \simeq 0$. This is done as follows.

Suppose $H(B) \simeq 0$. Using adjunction and the compatibility of the Serre functors with F, one finds for any $A \in \Omega$:

$$\begin{aligned} 0 = \mathrm{Hom}(A, H(B)) &\simeq \mathrm{Hom}(F(A), B) \simeq \mathrm{Hom}(B, S_{\mathcal{D}'}(F(A)))^* \\ &\simeq \mathrm{Hom}(B, F(S_{\mathcal{D}}(A)))^* \simeq \mathrm{Hom}(G(B), S_{\mathcal{D}}(A))^* \\ &\simeq \mathrm{Hom}(A, G(B)). \end{aligned}$$

Hence, $\mathrm{Hom}(A, G(B)) = 0$ for all $A \in \Omega$ and, therefore, $G(B) \simeq 0$. Note that the argument actually shows $G \simeq H$. □

Due to Remark 1.31, it suffices to assume the existence of only one of the adjoint functors in the above corollary.

1.4 Exceptional sequences and orthogonal decompositions

In the geometric context, the derived categories in question will usually be indecomposable (see Proposition 3.10). However, there are geometrically relevant situations where one can decompose the derived category in a weaker sense. This leads to the abstract notion of semi-orthogonal decompositions of a triangulated category, the topic of this section. Any full exceptional sequence yields such a semi-orthogonal decomposition, so we will discuss this notion first.

Definition 1.57 *An object $E \in \mathcal{D}$ in a k-linear triangulated category $\mathcal{D}$ is called* exceptional *if*

$$\mathrm{Hom}(E, E[\ell]) = \begin{cases} k & \text{if } \ell = 0 \\ 0 & \text{if } \ell \neq 0. \end{cases}$$

An exceptional sequence *is a sequence $E_1, \ldots, E_n$ of exceptional objects such that* $\mathrm{Hom}(E_i, E_j[\ell]) = 0$ *for all $i > j$ and all ℓ. In other words*

$$\mathrm{Hom}(E_i, E_j[\ell]) = \begin{cases} k & \text{if } \ell = 0, i = j \\ 0 & \text{if } i > j \text{ or if } l \neq 0, i = j. \end{cases}$$

An exceptional sequence is full *if $\mathcal{D}$ is generated by $\{E_i\}$, i.e. any full triangulated subcategory containing all objects E_i is equivalent to $\mathcal{D}$ (via the inclusion).*

Lemma 1.58 *Let $\mathcal{D}$ be a k-linear triangulated category such that for any $A, B \in \mathcal{D}$ the vector space $\bigoplus_i \mathrm{Hom}(A, B[i])$ is finite-dimensional.*

If $E \in \mathcal{D}$ is exceptional, then the objects $\bigoplus E[i]^{\oplus j_i}$ form an admissible triangulated subcategory $\langle E \rangle$ of $\mathcal{D}$.

Proof We leave it to the reader to check that $\langle E \rangle$ is indeed triangulated. In order to see that it is admissible one considers for any object $A \in \mathcal{D}$ the canonical morphism

$$\bigoplus \mathrm{Hom}(E, A[i]) \otimes E[-i] \longrightarrow A,$$

which can be completed to a distinguished triangle

$$\bigoplus \mathrm{Hom}(E, A[i]) \otimes E[-i] \longrightarrow A \longrightarrow B.$$

Using that E is exceptional, one finds $\mathrm{Hom}(E, B[i]) = 0$. Hence, $B \in \langle E \rangle^{\perp}$ (cf. iii) Remark 1.43). □

The concept of a (full) exceptional sequence is generalized by the following

Definition 1.59 *A sequence of full admissible triangulated subcategories*

$$\mathcal{D}_1, \ldots, \mathcal{D}_n \subset \mathcal{D}$$

is semi-orthogonal *if for all $i > j$*

$$\mathcal{D}_j \subset \mathcal{D}_i^{\perp}.$$

Such a sequence defines a semi-orthogonal decomposition *of $\mathcal{D}$ if $\mathcal{D}$ is generated by the $\mathcal{D}_i$, i.e. via inclusion $\mathcal{D}$ is equivalent to the smallest full triangulated subcategory of $\mathcal{D}$ containing all of them.*

Examples 1.60 i) Let $\mathcal{D}' \subset \mathcal{D}$ be an admissible full triangulated subcategory (cf. Definition 1.42). Then $\mathcal{D}_1 := \mathcal{D}'^{\perp}, \mathcal{D}_2 := \mathcal{D}' \subset \mathcal{D}$ is a semi-orthogonal decomposition of $\mathcal{D}$.

ii) Let $E_1, \ldots, E_n$ be an exceptional sequence in $\mathcal{D}$. Then the admissible triangulated subcategories (see Lemma 1.58)

$$\mathcal{D}_1 := \langle E_1 \rangle, \ldots, \mathcal{D}_n := \langle E_n \rangle$$

form a semi-orthogonal sequence.

If the exceptional sequence is full, then $\mathcal{D}_1, \ldots, \mathcal{D}_n \subset \mathcal{D}$ is a semi-orthogonal decomposition.

Lemma 1.61 *Any semi-orthogonal sequence of full admissible triangulated subcategories $\mathcal{D}_1, \ldots, \mathcal{D}_n \subset \mathcal{D}$ generates $\mathcal{D}$, i.e. defines a semi-orthogonal decomposition of $\mathcal{D}$, if and only if any object $A \in \mathcal{D}$ with $A \in \mathcal{D}_i^{\perp}$ for all $i = 1, \ldots, n$ is trivial.*

Proof Suppose $\mathcal{D}_1, \ldots, \mathcal{D}_n \subset \mathcal{D}$ is a semi-orthogonal decomposition. For any $A_0 \in \mathcal{D}$ one defines the full triangulated subcategory ${}^\perp A_0$ of all objects $A \in \mathcal{D}$ with $\mathrm{Hom}(A, A_0[i]) = 0$ for all $i \in \mathbb{Z}$ (cf. Exercise 1.44).

If $A_0 \in \bigcap \mathcal{D}_i^\perp$, then $\mathcal{D}_1, \ldots, \mathcal{D}_n \subset {}^\perp A_0$. Hence, ${}^\perp A_0 = \mathcal{D}$ and, in particular, $A_0 \in {}^\perp A_0$. The latter yields $\mathrm{Hom}(A_0, A_0) = 0$ and thus $A_0 \simeq 0$.

Let us now assume that $\bigcap \mathcal{D}_i^\perp = \{0\}$. For simplicity we assume $n = 2$ and leave the general case to the reader. We have to show that any $A_0 \in \mathcal{D}$ is contained in the triangulated subcategory generated by $\mathcal{D}_1$ and $\mathcal{D}_2$. Since $\mathcal{D}_2$ is admissible, one finds a distinguished triangle

$$A \longrightarrow A_0 \longrightarrow B \longrightarrow A[1]$$

with $A \in \mathcal{D}_2$ and $B \in \mathcal{D}_2^\perp$. The latter can be decomposed further by a distinguished triangle

$$C \longrightarrow B \longrightarrow C' \longrightarrow C[1]$$

with $C \in \mathcal{D}_1$ and $C' \in \mathcal{D}_1^\perp$ (use that $\mathcal{D}_1$ is admissible).

As $C \in \mathcal{D}_1 \subset \mathcal{D}_2^\perp$ and $B \in \mathcal{D}_2^\perp$, one finds $C' \in \mathcal{D}_2^\perp$. Hence, $C' \in \mathcal{D}_1^\perp \cap \mathcal{D}_2^\perp$, which implies $C' \simeq 0$ by assumption. Thus, $B \simeq C \in \mathcal{D}_1$. But then A_0 sits in a distinguished triangle with the other two objects being in $\mathcal{D}_1$, respectively $\mathcal{D}_2$. □

Exercise 1.62 Let $\mathcal{D}_1, \mathcal{D}_2 \subset \mathcal{D}$ be a semi-orthogonal decomposition of length two. Show that the inclusion $\mathcal{D}_1 \subset \mathcal{D}_2^\perp$ is an equivalence. More generally, if $\mathcal{D}_1, \ldots, \mathcal{D}_n \subset \mathcal{D}$ is a semi-orthogonal decomposition, then $\mathcal{D}_1 \subset \langle \mathcal{D}_2, \ldots, \mathcal{D}_n \rangle^\perp$ is an equivalence.

Exercise 1.63 Suppose $\mathcal{D}_1, \mathcal{D}_2 \subset \mathcal{D}$ is a semi-orthogonal decomposition of a triangulated category $\mathcal{D}$. Show that any object $A \in \mathcal{D}$ with $\mathrm{Hom}(A, B) = 0$ for all $B \in \mathcal{D}_1$ is isomorphic to an object in $\mathcal{D}_2$.

2

DERIVED CATEGORIES: A QUICK TOUR

This chapter is meant as a reminder. Many arguments are only sketched, if at all, and the reader not familiar with the material or feeling uncomfortable with certain aspects of it should go back to the literature.

At the same time we encourage the reader familiar with the basic notions of the theory to go as early as possible to Chapter 3 or even Chapter 4. Passing quickly to the results by Bondal, Bridgeland, Orlov, et al. on derived categories of coherent sheaves, the topic of this course, one gets to know derived categories from a more geometric point of view and this might help to digest the formal aspects of the general machinery.

2.1 Derived category of an abelian category

In this section we shall recall the fundamental aspects of derived categories. We begin by stating the existence of the derived category as a theorem, and explain the technical features, necessary for any calculation, later on. Derived functors will only be discussed in Section 2.2.

In the sequel, we will mostly be interested in the derived category of the abelian category of (coherent) sheaves or of modules over a ring. We recommend the textbooks [39, 61, 70] for more details and other examples.

Remark 2.1 Often, an object in a given abelian category $\mathcal{A}$ is studied in terms of its resolutions. To be more specific, recall that by definition a coherent sheaf $\mathcal{F}$ on a scheme X can locally be given by finitely many generators and finitely many relations. In other words, at least locally there exists an exact sequence $\mathcal{O}_X^{\oplus m_1} \longrightarrow \mathcal{O}_X^{\oplus m_2} \longrightarrow \mathcal{F} \longrightarrow 0$. On a smooth projective variety X any coherent sheaf $\mathcal{F}$ admits a locally free resolution of length $n = \dim(X)$, i.e. there exists an exact sequence of the form

$$0 \longrightarrow \mathcal{E}_n \longrightarrow \cdots \longrightarrow \mathcal{E}_1 \longrightarrow \mathcal{E}_0 \longrightarrow \mathcal{F} \longrightarrow 0$$

where all $\mathcal{E}_i$ are locally free coherent sheaves. Thus, in order to study arbitrary coherent sheaves on X one may switch to locally free sheaves and complexes of those.

More generally, when working with an abelian category, it is often necessary and natural to allow also complexes of objects in $\mathcal{A}$. This leads to the notion of the category of complexes.

Let us briefly recall the definition of the *category of complexes* Kom($\mathcal{A}$) of an abelian category $\mathcal{A}$. A complex in $\mathcal{A}$ consists of a diagram of objects and morphisms in $\mathcal{A}$ of the form

$$\cdots \longrightarrow A^{i-1} \xrightarrow{d^{i-1}} A^i \xrightarrow{d^i} A^{i+1} \xrightarrow{d^{i+1}} \cdots$$

satisfying $d^i \circ d^{i-1} = 0$ or, equivalently, $\mathrm{Im}(d^{i-1}) \subset \mathrm{Ker}(d^i)$, for all $i \in \mathbb{Z}$.

A morphism $f : A^\bullet \longrightarrow B^\bullet$ between two complexes $A^\bullet$ and $B^\bullet$ is given by a commutative diagram

$$\begin{array}{ccccccccc}
\cdots \longrightarrow & A^{i-1} & \xrightarrow{d_A^{i-1}} & A^i & \xrightarrow{d_A^i} & A^{i+1} & \xrightarrow{d_A^{i+1}} & \cdots \\
& \downarrow{\scriptstyle f^{i-1}} & & \downarrow{\scriptstyle f^i} & & \downarrow{\scriptstyle f^{i+1}} & & \\
\cdots \longrightarrow & B^{i-1} & \xrightarrow{d_B^{i-1}} & B^i & \xrightarrow{d_B^i} & B^{i+1} & \xrightarrow{d_B^{i+1}} & \cdots
\end{array}$$

Definition 2.2 *The* category of complexes Kom($\mathcal{A}$) *of an abelian category* $\mathcal{A}$ *is the category whose objects are complexes* $A^\bullet$ *in* $\mathcal{A}$ *and whose morphisms are morphisms of complexes.*

Proposition 2.3 *The category of complexes* Kom($\mathcal{A}$) *of an abelian category is again abelian.*

Proof The proof is straightforward. E.g. the zero object in Kom($\mathcal{A}$) is the complex $\ldots \longrightarrow 0 \longrightarrow 0 \longrightarrow 0 \longrightarrow \ldots$ and the kernel of a morphism $f : A^\bullet \longrightarrow B^\bullet$ is the complex of the kernels $\mathrm{Ker}(f^i)$, $i \in \mathbb{Z}$. □

Also note that mapping an object $A \in \mathcal{A}$ to the complex $A^\bullet$ with $A^0 = A$ and $A^i = 0$ for $i \neq 0$ identifies $\mathcal{A}$ with a full subcategory of Kom($\mathcal{A}$).

The complex category Kom($\mathcal{A}$) has two more features: cohomology and shift. Let us start out with the *shift functor.*

Definition 2.4 *Let* $A^\bullet \in$ Kom($\mathcal{A}$) *be a given complex. Then* $A^\bullet[1]$ *is the complex with* $(A^\bullet[1])^i := A^{i+1}$ *and differential* $d^i_{A[1]} := -d_A^{i+1}$.

The shift $f[1]$ *of a morphism of complexes* $f : A^\bullet \longrightarrow B^\bullet$ *is the complex morphism* $A^\bullet[1] \longrightarrow B^\bullet[1]$ *given by* $f[1]^i := f^{i+1}$.

The following fact is easily verified.

Corollary 2.5 *The shift functor* T : Kom($\mathcal{A}$) $\longrightarrow$ Kom($\mathcal{A}$), $A^\bullet \longmapsto A^\bullet[1]$ *defines an equivalence of abelian categories.* □

More precisely, the inverse functor T^{-1} is given by $A^\bullet \longmapsto A^\bullet[-1]$, where, more generally, $A^\bullet[k]^i = A^{k+i}$ and $d^i_{A[k]} = (-1)^k d_A^{i+k}$ for any $k \in \mathbb{Z}$.

Note, however, that Kom($\mathcal{A}$) endowed with the shift functor T does not define a triangulated category. Indeed, we would also have to give the class of distinguished triangles and the canonical choices, like short exact sequences or mapping cones, do not work.

Exercise 2.6 Prove that short exact sequences $0 \longrightarrow A^\bullet \longrightarrow B^\bullet \longrightarrow C^\bullet \longrightarrow 0$, which can be viewed as triangles with trivial $C^\bullet \longrightarrow A^\bullet[1]$, do not, in general, satisfy the conditions imposed on distinguished triangles in a triangulated category.

Recall that the *cohomology* $H^i(A^\bullet)$ of a complex $A^\bullet$ is the quotient

$$H^i(A^\bullet) := \frac{\mathrm{Ker}(d^i)}{\mathrm{Im}(d^{i-1})} \in \mathcal{A},$$

i.e. $H^i(A^\bullet) = \mathrm{Coker}(\mathrm{Im}(d^{i-1}) \longrightarrow \mathrm{Ker}(d^i))$. A complex $A^\bullet$ is *acyclic* if $H^i(A^\bullet) = 0$ for all $i \in \mathbb{Z}$. Any complex morphism $f : A^\bullet \longrightarrow B^\bullet$ induces natural homomorphisms

$$H^i(f) : H^i(A^\bullet) \longrightarrow H^i(B^\bullet).$$

Exercise 2.7 Suppose $F : \mathcal{A} \longrightarrow \mathcal{B}$ is an additive functor between abelian categories. Show that F is exact if and only of the image $F(A^\bullet)$ of any acyclic complex $A^\bullet$ in $\mathcal{A}$ is an acyclic complex in $\mathcal{B}$.

Remark 2.8 Proposition 2.3 allows us to speak of short exact sequences in Kom($\mathcal{A}$). One of the fundamental facts in homological algebra says that any short exact sequence

$$0 \longrightarrow A^\bullet \longrightarrow B^\bullet \longrightarrow C^\bullet \longrightarrow 0$$

induces a long exact sequence

$$\cdots \longrightarrow H^i(A^\bullet) \longrightarrow H^i(B^\bullet) \longrightarrow H^i(C^\bullet) \longrightarrow H^{i+1}(A^\bullet) \longrightarrow \cdots.$$

See [39] or any standard textbook on homological algebra. The construction of the boundary morphism $H^i(C^\bullet) \longrightarrow H^{i+1}(A^\bullet)$ is easier if one allows oneself to work with modules over a ring. The case of length two complexes runs under the name 'snake lemma'.

The induced map for the cohomology objects is used to define quasi-isomorphisms, which play a central rôle in the passage to the derived category.

Definition 2.9 *A morphism of complexes* $f : A^\bullet \longrightarrow B^\bullet$ *is a* quasi-isomorphism *(or* qis, *for short) if for all* $i \in \mathbb{Z}$ *the induced map* $H^i(f) : H^i(A^\bullet) \longrightarrow H^i(B^\bullet)$ *is an isomorphism.*

Note that a resolution as considered in Remark 2.1 gives rise to a quasi-isomorphism $(\mathcal{E}_n \to \ldots \to \mathcal{E}_1 \to \mathcal{E}_0) \to \mathcal{F}$. This explains why it is desirable not to distinguish between quasi-isomorphic complexes.

The central idea for the definition of the derived category is this: quasi-isomorphic complexes should become isomorphic objects in the derived category. We shall begin our discussion with the following existence theorem. Details of the construction are provided by the subsequent discussion.

Theorem 2.10 *Let $\mathcal{A}$ be an abelian category and let* $\mathrm{Kom}(\mathcal{A})$ *be its category of complexes. Then there exists a category* $\mathrm{D}(\mathcal{A})$*, the* derived category of $\mathcal{A}$*, and a functor*

$$Q : \mathrm{Kom}(\mathcal{A}) \longrightarrow \mathrm{D}(\mathcal{A})$$

such that:

i) *If $f : A^\bullet \to B^\bullet$ is a quasi-isomorphism, then $Q(f)$ is an isomorphism in* $\mathrm{D}(\mathcal{A})$.
ii) *Any functor $F : \mathrm{Kom}(\mathcal{A}) \to \mathcal{D}$ satisfying property* i) *factorizes uniquely over* $Q : \mathrm{Kom}(\mathcal{A}) \to \mathrm{D}(\mathcal{A})$*, i.e. there exists a unique functor (up to isomorphism)* $G : \mathrm{D}(\mathcal{A}) \to \mathcal{D}$ *with $F \simeq G \circ \mathrm{Q}$:*

$$\begin{array}{ccc} \mathrm{Kom}(\mathcal{A}) & \xrightarrow{\quad Q \quad} & \mathrm{D}(\mathcal{A}) \\ & {\scriptstyle F}\searrow \quad \swarrow{\scriptstyle G} & \\ & \mathcal{D}. & \end{array}$$

As stated, the theorem is a pure existence result. In order to be able to work with the derived category, we have to understand which objects become isomorphic under $Q : \mathrm{Kom}(\mathcal{A}) \to \mathrm{D}(\mathcal{A})$ and, more complicated, how to represent morphisms in the derived category. Explaining this, will at the same time provide a proof for the above theorem. Moreover, we shall observe the following facts.

Corollary 2.11 i) *Under the functor $Q : \mathrm{Kom}(\mathcal{A}) \to \mathrm{D}(\mathcal{A})$ the objects of the two categories* $\mathrm{Kom}(\mathcal{A})$ *and* $\mathrm{D}(\mathcal{A})$ *are identified* .

ii) *The cohomology objects $H^i(A^\bullet)$ of an object $A^\bullet \in \mathrm{D}(\mathcal{A})$ are well-defined objects of the abelian category $\mathcal{A}$.*

iii) *Viewing any object in $\mathcal{A}$ as a complex concentrated in degree zero yields an equivalence between $\mathcal{A}$ and the full subcategory of* $\mathrm{D}(\mathcal{A})$ *that consists of all complexes $A^\bullet$ with $H^i(A^\bullet) = 0$ for $i \neq 0$.*

Contrary to the category of complexes $\mathrm{Kom}(\mathcal{A})$, the derived category $\mathrm{D}(\mathcal{A})$ is in general not abelian, but it is always triangulated. The shift functor indeed descends to $\mathrm{D}(\mathcal{A})$ and a natural class of distinguished triangles can be found, as will be explained shortly.

Suppose $C^\bullet \longrightarrow A^\bullet$ is a quasi-isomorphism. As the derived category is to be constructed in a way that any quasi-isomorphism becomes an isomorphism, any morphism of complexes $C^\bullet \longrightarrow B^\bullet$ will have to count as a morphism $A^\bullet \longrightarrow B^\bullet$ in the derived category. This leads to the definition of morphisms in the derived category as diagrams of the form

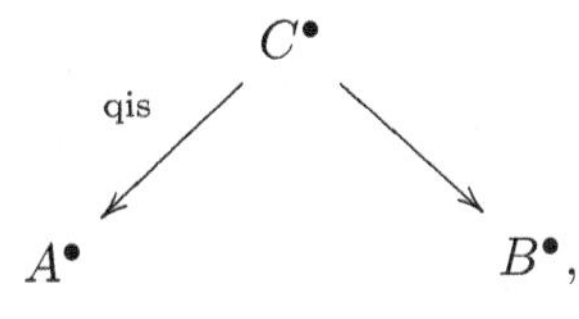

where $C^\bullet \longrightarrow A^\bullet$ is a quasi-isomorphism.

In order to make this a sensible definition of morphisms, one has to explain when two such roofs are considered equal and how to define the composition in the derived category. The natural context for both problems is the homotopy category of complexes. This will be an intermediate step in passing from $\mathrm{Kom}(\mathcal{A})$ to $\mathrm{D}(\mathcal{A})$:

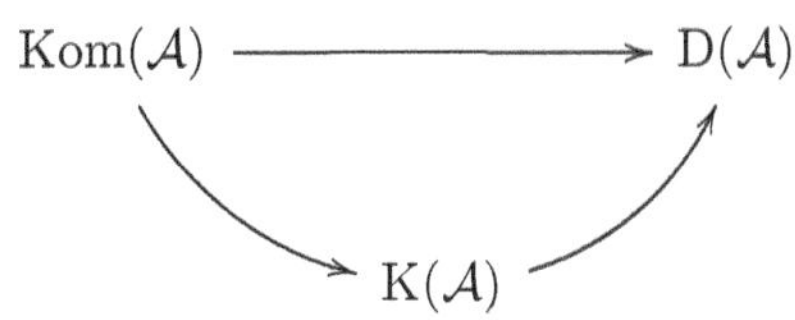

By abuse of notation, we shall again write $Q : \mathrm{K}(\mathcal{A}) \longrightarrow \mathrm{D}(\mathcal{A})$ for the natural functor.

Definition 2.12 *Two morphisms of complexes*

$$f, g : A^\bullet \longrightarrow B^\bullet$$

are called homotopically equivalent, $f \sim g$, *if there exists a collection of homomorphisms* $h^i : A^i \longrightarrow B^{i-1}$, $i \in \mathbb{Z}$, *such that*

$$f^i - g^i = h^{i+1} \circ d_A^i + d_B^{i-1} \circ h^i.$$

The homotopy category of complexes $\mathrm{K}(\mathcal{A})$ *is the category whose objects are the objects of* $\mathrm{Kom}(\mathcal{A})$, *i.e.* $\mathrm{Ob}(\mathrm{K}(\mathcal{A})) = \mathrm{Ob}(\mathrm{Kom}(\mathcal{A}))$, *and morphisms* $\mathrm{Hom}_{\mathrm{K}(\mathcal{A})}(A^\bullet, B^\bullet) := \mathrm{Hom}_{\mathrm{Kom}(\mathcal{A})}(A^\bullet, B^\bullet)/\sim$.

That the definition makes sense, e.g. that the composition is well-defined in $\mathrm{K}(\mathcal{A})$, follows from the following assertions which are all easily verified.

Proposition 2.13 i) *Homotopy equivalence between morphisms* $A^\bullet \longrightarrow B^\bullet$ *of complexes is an equivalence relation.*

ii) *Homotopically trivial morphisms form an 'ideal' in the morphisms of* $\mathrm{Kom}(\mathcal{A})$.

iii) *If* $f \sim g : A^\bullet \longrightarrow B^\bullet$, *then* $H^i(f) = H^i(g)$ *for all* i.

iv) *If* $f : A^\bullet \longrightarrow B^\bullet$ *and* $g : B^\bullet \longrightarrow A^\bullet$ *are given such that* $f \circ g \sim \mathrm{id}_B$ *and* $g \circ f \sim \mathrm{id}_A$, *then* f *and* g *are quasi-isomorphisms and, more precisely,* $H^i(f)^{-1} = H^i(g)$. □

Remark 2.14 Note that the definition of $\mathrm{K}(\mathcal{A})$ makes sense for any additive category. This will be needed later when we consider the full subcategory of all injective objects in a given abelian category (cf. Proposition 2.40).

Now comes the precise definition of the derived category. The first step is to describe the objects of $\mathrm{D}(\mathcal{A})$. This is easy, we simply set

$$\mathrm{Ob}(\mathrm{D}(\mathcal{A})) := \mathrm{Ob}(\mathrm{K}(\mathcal{A})) = \mathrm{Ob}(\mathrm{Kom}(\mathcal{A})).$$

The set of morphism $\mathrm{Hom}_{\mathrm{D}(\mathcal{A})}$ between two complexes $A^\bullet$ and $B^\bullet$ viewed as objects in $\mathrm{D}(\mathcal{A})$ is the set of all equivalence classes of diagrams of the form

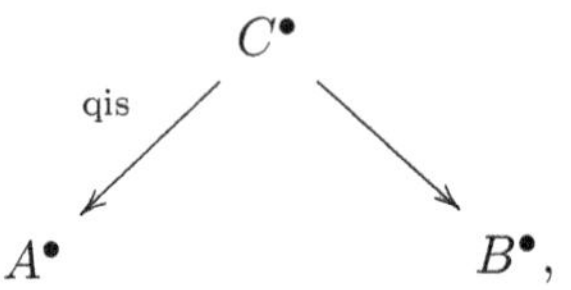

where $C^\bullet \longrightarrow A^\bullet$ is a quasi-isomorphism. Two such diagrams are equivalent if they are dominated in the homotopy category $\mathrm{K}(\mathcal{A})$(!) by a third one of the same sort, i.e. there exists a commutative diagram in $\mathrm{K}(\mathcal{A})$ of the form

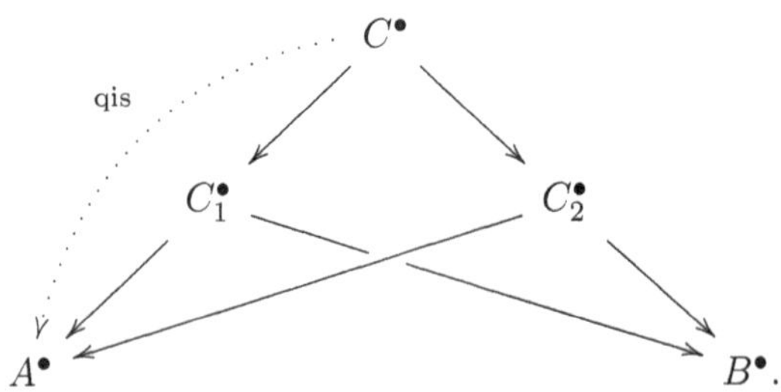

(In particular, the compositions $C^\bullet \longrightarrow C_1^\bullet \longrightarrow A^\bullet$ and $C^\bullet \longrightarrow C_2^\bullet \longrightarrow A^\bullet$ are homotopy equivalent. Thus, since the first one is a qis, also the latter one is. Why commutativity is required only in $\mathrm{K}(\mathcal{A})$, i.e. up to homotopy, and not in $\mathrm{Kom}(\mathcal{A})$ will become clear later (cf. proof of Proposition 2.16). Roughly, if the stronger condition is imposed, the composition of two such roofs could no longer be defined.)

In this way, we have defined objects and morphisms of our category $\mathrm{D}(\mathcal{A})$, but we still have to check a number of properties. In particular, we have to define the composition of two morphisms. If two morphisms

$$\begin{array}{ccc} & C_1^\bullet & \\ {\scriptstyle \text{qis}}\swarrow & & \searrow \\ A^\bullet & & B^\bullet \end{array} \qquad \text{and} \qquad \begin{array}{ccc} & C_2^\bullet & \\ {\scriptstyle \text{qis}}\swarrow & & \searrow \\ B^\bullet & & C^\bullet \end{array}$$

are given, we want the composition of both be given by a commutative (in the homotopy category $\mathrm{K}(\mathcal{A})$!) diagram of the form

$$\begin{array}{ccccc} & & C_0^\bullet & & \\ & {\scriptstyle \text{qis}}\swarrow & & \searrow & \\ & C_1^\bullet & & C_2^\bullet & \\ {\scriptstyle \text{qis}}\swarrow & & \searrow \quad {\scriptstyle \text{qis}}\swarrow & & \searrow \\ A^\bullet & & B^\bullet & & C^\bullet. \end{array} \tag{2.1}$$

There are two obvious problems: one has to ensure that such a diagram exists and that it is unique up to equivalence.

Both things hold true, but we need to introduce the mapping cone in order to explain why. The mapping cone will as well play a central rôle in the definition of the triangulated structure on $\mathrm{K}(\mathcal{A})$ and $\mathrm{D}(\mathcal{A})$ (see Proposition 2.24).

Definition 2.15 *Let $f : A^\bullet \longrightarrow B^\bullet$ be a complex morphism. Its* mapping cone *is the complex $\mathrm{C}(f)$ with*

$$\mathrm{C}(f)^i = A^{i+1} \oplus B^i \quad \text{and} \quad d^i_{\mathrm{C}(f)} := \begin{pmatrix} -d_A^{i+1} & 0 \\ f^{i+1} & d_B^i \end{pmatrix}.$$

(Note that in the literature one finds different conventions for the definition of the differential $d_{\mathrm{C}(f)}$, e.g. f^{i+1} with an extra sign.)

The reader easily checks that the mapping cone is a complex. Moreover, there exist two natural complex morphisms

$$\tau : B^\bullet \longrightarrow \mathrm{C}(f) \qquad \text{and} \qquad \pi : \mathrm{C}(f) \longrightarrow A^\bullet[1]$$

given by the natural injection $B^i \longrightarrow A^{i+1} \oplus B^i$ and the natural projection $A^{i+1} \oplus B^i \longrightarrow A^\bullet[1]^i = A^{i+1}$, respectively. The composition $B^\bullet \longrightarrow \mathrm{C}(f) \longrightarrow A^\bullet[1]$ is trivial and the composition $A^\bullet \longrightarrow B^\bullet \longrightarrow \mathrm{C}(f)$ is homotopic to the trivial map. In fact, $B^\bullet \longrightarrow \mathrm{C}(f) \longrightarrow A^\bullet[1]$ is a short exact sequence of complexes. In particular, we obtain the long exact cohomology sequence (cf. Remark 2.8)

$$H^i(A^\bullet) \longrightarrow H^i(B^\bullet) \longrightarrow H^i(\mathrm{C}(f)) \longrightarrow H^{i+1}(A^\bullet) \longrightarrow \quad \cdots$$

Also, by construction any commutative diagram can be completed as follows

$$\begin{array}{ccccccc} A_1^\bullet & \xrightarrow{f_1} & B_1^\bullet & \longrightarrow & \mathrm{C}(f_1) & \longrightarrow & A_1^\bullet[1] \\ \downarrow & \circlearrowleft & \downarrow & & \vdots & & \downarrow \\ A_2^\bullet & \xrightarrow{f_2} & B_2^\bullet & \longrightarrow & \mathrm{C}(f_2) & \longrightarrow & A_2^\bullet[1]. \end{array}$$

This probably reminds the reader of axiom TR3. The following proposition should be viewed in light of axiom TR2. (In fact, the triangles defined by the mapping cone will form the distinguished triangles in the homotopy and in the derived category, cf. Proposition 2.24). It will also be crucial for defining the composition of morphisms in the derived category.

Proposition 2.16 *Let $f : A^\bullet \longrightarrow B^\bullet$ be a morphism of complexes and let $\mathrm{C}(f)$ be its mapping cone that comes with the two natural morphisms $\tau : B^\bullet \longrightarrow \mathrm{C}(f)$ and $\pi : \mathrm{C}(f) \longrightarrow A^\bullet[1]$. Then there exists a complex morphism $g : A^\bullet[1] \longrightarrow \mathrm{C}(\tau)$ which is an isomorphism in $\mathrm{K}(\mathcal{A})$ and such that the following diagram is commutative in $\mathrm{K}(\mathcal{A})$:*

$$\begin{array}{ccccccc} B^\bullet & \xrightarrow{\tau} & \mathrm{C}(f) & \xrightarrow{\pi} & A^\bullet[1] & \xrightarrow{-f} & B^\bullet[1] \\ \downarrow = & & \downarrow = & & \vdots\, g & & \downarrow = \\ B^\bullet & \xrightarrow{\tau} & \mathrm{C}(f) & \xrightarrow{\tau_\tau} & \mathrm{C}(\tau) & \xrightarrow{\pi_\tau} & B^\bullet[1]. \end{array}$$

Proof The morphism $g : A^\bullet[1] \longrightarrow \mathrm{C}(\tau)$ is easy to define: We let

$$A^\bullet[1]^i = A^{i+1} \longrightarrow \mathrm{C}(\tau)^i = B^{i+1} \oplus \mathrm{C}(f)^i = B^{i+1} \oplus A^{i+1} \oplus B^i$$

be the map $(-f^{i+1}, \mathrm{id}, 0)$. We leave it to the reader to verify that this is indeed a complex morphism.

The inverse g^{-1} in $\mathrm{K}(\mathcal{A})$ can be given as the projection onto the middle factor.

The commutativity (in $\mathrm{Kom}(\mathcal{A})$!) of the diagram

$$\begin{array}{ccc} A^\bullet[1] & \xrightarrow{-f} & B^\bullet[1] \\ g \downarrow & & \downarrow \mathrm{id} \\ \mathrm{C}(\tau) & \xrightarrow{\pi_\tau} & B^\bullet[1] \end{array}$$

is straightforward. (But note the annoying sign, which is in fact responsible for the sign in TR2.)

The diagram

$$\begin{array}{ccc} \mathrm{C}(f) & \xrightarrow{\pi} & A^\bullet[1] \\ {\scriptstyle \mathrm{id}}\downarrow & & \downarrow{\scriptstyle g} \\ \mathrm{C}(f) & \xrightarrow{\tau_\tau} & \mathrm{C}(\tau) \end{array}$$

does not commute in $\mathrm{Kom}(\mathcal{A})$, but it does commute up to homotopy. To prove this, one first checks that $g \circ g^{-1}$ is indeed homotopic to the identity and then uses $g^{-1} \circ \tau_\tau = \pi$. For the details see [61, 1.4]. □

Let us first show how to use the construction of the mapping cone in order to compose two morphisms in the derived category. In order to do this, we consider a quasi-isomorphism $f : A^\bullet \longrightarrow B^\bullet$ and an arbitrary morphism $g : C^\bullet \longrightarrow B^\bullet$.

Proposition 2.17 *There exists a commutative diagram in* $\mathrm{K}(\mathcal{A})$

$$\begin{array}{ccc} C_0^\bullet & \xrightarrow{\text{qis}} & C^\bullet \\ \downarrow & & \downarrow{\scriptstyle g} \\ A^\bullet & \xrightarrow[\text{qis}]{f} & B^\bullet. \end{array}$$

Proof Note that the existence of a commutative diagram (even in the complex category and even without $A^\bullet \longrightarrow B^\bullet$ being a qis) is trivial. The difficulty consists in constructing it such that $C_0^\bullet \longrightarrow C^\bullet$ is a qis.

The idea is to make use of a commutative diagram of the form

$$\begin{array}{ccccccc} \mathrm{C}(\tau \circ g)[-1] & \longrightarrow & C^\bullet & \longrightarrow & \mathrm{C}(f) & \longrightarrow & \mathrm{C}(\tau \circ g) \\ \vdots & & {\scriptstyle g}\downarrow & & \| & & \vdots \\ A^\bullet & \xrightarrow{f} & B^\bullet & \xrightarrow{\tau} & \mathrm{C}(f) & \longrightarrow & A^\bullet[1]. \end{array}$$

Due to the previous proposition we know that $B^\bullet \xrightarrow{\tau} \mathrm{C}(f) \longrightarrow A^\bullet[1]$ in $\mathrm{K}(\mathcal{A})$ is isomorphic to the triangle $B^\bullet \xrightarrow{\tau} \mathrm{C}(f) \longrightarrow \mathrm{C}(\tau)$. Then use the natural morphism $\mathrm{C}(\tau \circ g) \longrightarrow \mathrm{C}(\tau)$.

Using the long exact cohomology sequences, one proves that the morphism $C_0^\bullet := \mathrm{C}(\tau \circ g)[-1] \longrightarrow C^\bullet$ is a quasi-isomorphism. □

The proposition is central as its immediate consequence is

Corollary 2.18 *The composition of roofs as proposed by (2.1) exists and is well-defined.*

Proof Apply Proposition 2.17 to

$$\begin{array}{ccc} & & C_1^\bullet \\ & & \downarrow \\ C_2^\bullet & \xrightarrow[\text{qis}]{} & B^\bullet \end{array}$$

in (2.1). We leave it to the reader to show that the equivalence class of the appearing roof is unique. □

Exercise 2.19 One might be tempted to define $C_0^\bullet$ more directly as the fibred sum of $A^\bullet$ and $C^\bullet$ over $B^\bullet$. Find an example that shows that this in general does not work. (E.g. try a surjection for $B^\bullet = B^0 \twoheadrightarrow B^1$ with kernel A and $C^\bullet = B^1[-1]$.)

Exercise 2.20 Show that a complex $A^\bullet$ is isomorphic to 0 in $\mathcal{D}(\mathcal{A})$ if and only if $H^i(A^\bullet) \simeq 0$ for all i. On the other hand, find an example of a complex morphism $f : A^\bullet \longrightarrow B^\bullet$ such that $H^i(f) = 0$ for all i, but without f being trivial in $\mathcal{D}(\mathcal{A})$. See [39, 44].

In fact, f is zero in $\mathrm{D}(\mathcal{A})$ if and only if there exists a qis $g : C^\bullet \longrightarrow A^\bullet$ such that $f \circ g$ is homotopically zero.

Exercise 2.21 Check that $\mathrm{D}(\mathcal{A})$ is an additive category.

Remark 2.22 Behind the construction of the derived category there is a general procedure, called *localization*. Roughly, one constructs the localization of a category with respect to a localizing class of morphisms. In our case, these are the quasi-isomorphisms. It turns out that quasi-isomorphisms indeed form a localizing class in $\mathrm{K}(\mathcal{A})$ (but not in $\mathrm{Kom}(\mathcal{A})$!). For details see [39, 61].

Definition 2.23 *A triangle*

$$A_1^\bullet \longrightarrow A_2^\bullet \longrightarrow A_3^\bullet \longrightarrow A_1^\bullet[1]$$

in $\mathrm{K}(\mathcal{A})$ *(respectively in* $\mathrm{D}(\mathcal{A})$*) is called* distinguished *if it is isomorphic in* $\mathrm{K}(\mathcal{A})$ *(respectively in* $\mathrm{D}(\mathcal{A})$*) to a triangle of the form*

$$A^\bullet \xrightarrow{f} B^\bullet \xrightarrow{\tau} \mathrm{C}(f) \xrightarrow{\pi} A^\bullet[1]$$

with f a complex morphism.

Proposition 2.24 *Distinguished triangles given as in Definition 2.23 and the natural shift functor for complexes $A^\bullet \longmapsto A^\bullet[1]$ make the homotopy category of complexes $K(\mathcal{A})$ and the derived category $\mathrm{D}(\mathcal{A})$ of an abelian category into a triangulated category.*

Moreover, the natural functor $Q_{\mathcal{A}} : \mathrm{K}(\mathcal{A}) \longrightarrow \mathrm{D}(\mathcal{A})$ *is an exact functor of triangulated categories.*

Proof Again we refer to the literature, e.g. [39, IV.2]. As before, the mapping cone plays a central rôle in the verification of the axioms TR. Note that there is the additional difficulty in the derived category that an isomorphism of two triangles is not given by honest morphisms. □

Exercise 2.25 Let $\mathcal{A} := \mathrm{Vec}_f(k)$ be the abelian category of finite-dimensional vector spaces over a field k. Show that $\mathrm{D}(\mathcal{A})$ is equivalent to $\prod_{i\in\mathbb{Z}} \mathcal{A}$. More precisely, any complex $A^\bullet \in \mathrm{D}(\mathcal{A})$ is isomorphic to its cohomology complex $\bigoplus H^i(A^\bullet)[-i]$ (with trivial differentials).

Exercise 2.26 Show more generally that the assertion in the last exercise holds true, whenever the abelian category $\mathcal{A}$ is semi-simple, i.e. such that any short exact sequence in $\mathcal{A}$ splits. See [39, III.2.3].

Exercise 2.27 Suppose $0 \longrightarrow A \xrightarrow{f} B \longrightarrow C \longrightarrow 0$ is a short exact sequence in an abelian category $\mathcal{A}$. Show that under the full embedding $\mathcal{A} \hookrightarrow \mathrm{K}(\mathcal{A})$ (or $\mathcal{A} \hookrightarrow \mathrm{D}(\mathcal{A})$) this becomes a distinguished triangle $A \longrightarrow B \longrightarrow C \xrightarrow{\delta} A[1]$ in $\mathrm{K}(\mathcal{A})$ (respectively $\mathrm{D}(\mathcal{A})$) with δ given as the composition of the inverse (in $\mathrm{K}(\mathcal{A})$ respectively $\mathrm{D}(\mathcal{A})$) of the quasi-isomorphism $\mathrm{C}(f) \longrightarrow C$ and the natural morphism $\mathrm{C}(f) \longrightarrow A[1]$.

Conversely, if $A \longrightarrow B \longrightarrow C \longrightarrow A[1]$ is a distinguished triangle with objects $A, B, C \in \mathcal{A}$, then $0 \longrightarrow A \longrightarrow B \longrightarrow C \longrightarrow 0$ is a short exact sequence in $\mathcal{A}$.

Exercise 2.28 Suppose $A^\bullet \longrightarrow B^\bullet \longrightarrow C^\bullet \longrightarrow A^\bullet[1]$ is a distinguished triangle in the derived category $\mathrm{D}(\mathcal{A})$. Show that it naturally induces a long exact sequence $\ldots \longrightarrow H^i(A^\bullet) \longrightarrow H^i(B^\bullet) \longrightarrow H^i(C^\bullet) \longrightarrow H^{i+1}(A^\bullet) \longrightarrow \ldots$ (cf. Remark 2.8).

By definition, complexes in the categories $\mathrm{K}(\mathcal{A})$ and $\mathrm{D}(\mathcal{A})$ are unbounded, but often it is more convenient to work with bounded ones.

Definition 2.29 *Let* $\mathrm{Kom}^*(\mathcal{A})$, *with* $* = +, -$, *or* b, *be the category of complexes* $A^\bullet$ *with* $A^i = 0$ *for* $i \ll 0$, $i \gg 0$, *respectively* $|i| \gg 0$.

By dividing out first by homotopy equivalence and then by quasi-isomorphisms one obtains the categories $\mathrm{K}^*(\mathcal{A})$ and $\mathrm{D}^*(\mathcal{A})$ with $* = +, -$, or b. Let us consider the natural functors $\mathrm{D}^*(\mathcal{A}) \longrightarrow \mathrm{D}(\mathcal{A})$ given by just forgetting the boundedness condition.

Proposition 2.30 *The natural functors* $\mathrm{D}^*(\mathcal{A}) \longrightarrow \mathrm{D}(\mathcal{A})$, *where* $* = +, -$, *or* b, *define equivalences of* $\mathrm{D}^*(\mathcal{A})$ *with the full triangulated subcategories of all complexes* $A^\bullet \in \mathrm{D}(\mathcal{A})$ *with* $H^i(A^\bullet) = 0$ *for* $i \ll 0$, $i \gg 0$, *respectively* $|i| \gg 0$.

Proof The idea is the following (see Exercise 2.31). Suppose $H^i(A^\bullet) = 0$ for $i > i_0$. Then the commutative diagram

$$\begin{array}{ccccccccccc}
\cdots & \longrightarrow & A^{i_0-2} & \longrightarrow & A^{i_0-1} & \longrightarrow & \mathrm{Ker}(d_A^{i_0}) & \longrightarrow & 0 & \longrightarrow & \cdots \\
 & & \downarrow{\scriptstyle =} & & \downarrow{\scriptstyle =} & & \big\downarrow & & \downarrow & & \\
\cdots & \longrightarrow & A^{i_0-2} & \longrightarrow & A^{i_0-1} & \longrightarrow & A^{i_0} & \longrightarrow & A^{i_0+1} & \longrightarrow & \cdots
\end{array}$$

defines a quasi-isomorphism between a complex in $\mathrm{K}^-(\mathcal{A})$ and $A^\bullet$.

Similarly, if $H^i(A^\bullet) = 0$ for $i < i_0$, one considers

$$\begin{array}{ccccccccc}
\cdots & \longrightarrow & A^{i_0-1} & \longrightarrow & A^{i_0} & \longrightarrow & A^{i_0+1} & \longrightarrow & \cdots \\
 & & \downarrow & & \downarrow & & \downarrow{\scriptstyle =} & & \\
\cdots & \longrightarrow & 0 & \longrightarrow & \mathrm{Coker}(d^{i_0-1}) & \longrightarrow & A^{i_0+1} & \longrightarrow & \cdots
\end{array}$$

For details see [61]. Note that the statement is about the derived and not about the homotopy category. Indeed, it should be clear from the two diagrams that in order to replace a cohomologically bounded complex by a bounded complex one needs to pass via a roof. □

The same arguments prove iii) in Corollary 2.11 saying that $\mathcal{A}$ is canonically equivalent to the full subcategory of all objects $A^\bullet \in \mathrm{D}(\mathcal{A})$ with $H^i(A^\bullet) \simeq 0$ for $i \neq 0$ (cf. [39, III.5]).

Exercise 2.31 Let $A^\bullet$ be a complex with $H^i(A^\bullet) = 0$ for $i > m$. Show that $A^\bullet$ is quasi-isomorphic (and hence isomorphic as an object in $\mathrm{D}(\mathcal{A})$) to a complex $B^\bullet$ with $B^i = 0$ for $i > m$.

State and prove an analogous statement for a complex $A^\bullet$ with $H^i(A^\bullet) = 0$ for $i < m$.

Exercise 2.32 Let $A^\bullet$ be a complex with $m := \max\{i \mid H^i(A^\bullet) \neq 0\} < \infty$. Show that there exists a morphism

$$\varphi : A^\bullet \longrightarrow H^m(A^\bullet)[-m]$$

in the derived category such that $H^m(\varphi) : H^m(A^\bullet) \longrightarrow H^m(A^\bullet)$ equals the identity.

Similarly, if $m := \min\{i \mid H^i(A^\bullet) \neq 0\} > -\infty$, then there exists a morphism $\varphi : H^m(A^\bullet)[-m] \longrightarrow A^\bullet$ with $H^m(\varphi) = \mathrm{id}$.

Exercise 2.33 Suppose $H^i(A^\bullet) = 0$ for $i < i_0$. Show that there exists a distinguished triangle

$$H^{i_0}(A^\bullet)[-i_0] \longrightarrow A^\bullet \xrightarrow{\ \varphi\ } B^\bullet \longrightarrow H^{i_0}(A^\bullet)[1 - i_0]$$

in $\mathrm{D}(\mathcal{A})$ with $H^i(B^\bullet) = 0$ for $i \leq i_0$ and φ inducing isomorphisms $H^i(A^\bullet) \simeq H^i(B^\bullet)$ for $i > i_0$. State and prove the analogous result for a complex $A^\bullet$ with $H^i(A^\bullet) = 0$ for $i > i_0$.

Due to the very construction of the derived category, it is sometimes quite cumbersome to do explicit calculations there. Often, however, it is possible to work with a very special class of complexes for which morphisms in the derived category and in the homotopy category are the same thing. Depending on the kind of functors one is interested in, the notion of injective, respectively, projective objects will be crucial. Both concepts were recalled in Examples 1.14. Note that they are dual in the sense that an object $I \in \mathcal{A}$ is injective if and only if the same object considered as an object of the opposite category $\mathcal{A}^{\mathrm{op}}$ is projective.

Definition 2.34 *An abelian category $\mathcal{A}$ contains* enough injective *(respectively* enough projective*) objects if for any object $A \in \mathcal{A}$ there exists an injective morphism $A \longrightarrow I$ with $I \in \mathcal{A}$ injective (respectively a surjective morphism $P \longrightarrow A$ with $P \in \mathcal{A}$ projective).*

An *injective resolution* of an object $A \in \mathcal{A}$ is an exact sequence

$$0 \longrightarrow A \longrightarrow I^0 \longrightarrow I^1 \longrightarrow I^2 \longrightarrow \cdots$$

with all $I^i \in \mathcal{A}$ injective. Similarly, a *projective resolution* of A consists of an exact sequence

$$\cdots \longrightarrow P^{-2} \longrightarrow P^{-1} \longrightarrow P^0 \longrightarrow A \longrightarrow 0$$

with projective objects $P^i \in \mathcal{A}$. In other words, an injective resolution is given by a complex $I^\bullet$ and a quasi-isomorphism $A \longrightarrow I^\bullet$ where $I^i = 0$ for $i < 0$ and all I^i injective. A projective resolution can be explained similarly.

Clearly, if $\mathcal{A}$ contains enough injectives, any object $A \in \mathcal{A}$ admits an injective resolution. More generally one has

Proposition 2.35 *Suppose $\mathcal{A}$ is an abelian category with enough injectives. For any $A^\bullet \in \mathrm{K}^+(\mathcal{A})$ there exist a complex $I^\bullet \in \mathrm{K}^+(\mathcal{A})$ with $I^i \in \mathcal{A}$ injective objects and a quasi-isomorphism $A^\bullet \longrightarrow I^\bullet$.*

Proof As $A^\bullet$ is a bounded below complex, we can proceed by induction as follows. Suppose we have constructed a morphism

$$f_i : A^\bullet \longrightarrow (\ldots I^{i-1} \longrightarrow I^i \longrightarrow 0 \longrightarrow \ldots)$$

such that $H^j(f_i)$ is bijective for $j < i$ and injective for $j = i$. As the notation suggests, the objects I^j are injective. Then one constructs a complex morphism

$$f_{i+1} : A^\bullet \longrightarrow (\ldots I^{i-1} \longrightarrow I^i \longrightarrow I^{i+1} \longrightarrow 0 \longrightarrow \ldots)$$

which now induces bijective maps $H^j(f_{i+1})$ for $j \leq i$ and an injective map $H^{i+1}(f_{i+1})$.

We only indicate the first step. For the complete proof see [39, III.5] or [61, I.1.1.7]. Suppose $A^\bullet$ is of the form $0 \longrightarrow A^0 \longrightarrow A^1 \longrightarrow A^2 \longrightarrow \ldots$. By assumption, there exists an injective object I^0 and a monomorphism $A^0 \longrightarrow I^0$. The induced morphism $f_0 : A^\bullet \longrightarrow (I^0 \longrightarrow 0 \longrightarrow \ldots)$ has the property that $H^i(f_0)$ is an isomorphism for $i < 0$ and injective for $i = 0$.

The definition of I^1 and the morphism $I^0 \longrightarrow I^1$ is easy: Consider the object $(I^0 \oplus A^1)/A^0$ and choose an injective object I^1 containing it. The morphisms $I^0 \longrightarrow I^1$ and $A^1 \longrightarrow I^1$ are the obvious ones. The cohomological properties are readily verified.

(The same idea works in principle also for the definition of I^{i+1} as an injective object containing $(I^i \oplus A^{i+1})/A^i$, but one has in addition to ensure that $I^{i-1} \longrightarrow I^i \longrightarrow I^{i+1}$ be zero, which makes the general case more technical.) □

Corollary 2.36 *Suppose $\mathcal{A}$ is an abelian category with enough injectives. Any $A^\bullet \in \mathrm{D}(\mathcal{A})$ with $H^i(A^\bullet) = 0$ for $i \ll 0$ is isomorphic (as an object of the derived category) to a complex $I^\bullet$ of injective objects I^i with $I^i = 0$ for $i \ll 0$.*

Proof By Proposition 2.30 we may assume that $A^i = 0$ for $i \ll 0$. Then use the proposition. □

Exercise 2.37 Spell out the dual statements for a category with enough projectives.

Lemma 2.38 *Suppose $A^\bullet \longrightarrow B^\bullet$ is a quasi-isomorphism between two complexes $A^\bullet, B^\bullet \in \mathrm{K}^+(\mathcal{A})$. Then for any complex $I^\bullet$ of injective objects I^i with $I^i = 0$ for $i \ll 0$ the induced map*

$$\mathrm{Hom}_{\mathrm{K}(\mathcal{A})}(B^\bullet, I^\bullet) \xrightarrow{\sim} \mathrm{Hom}_{\mathrm{K}(\mathcal{A})}(A^\bullet, I^\bullet)$$

is bijective.

Proof Completing $B^\bullet \longrightarrow A^\bullet$ to a distinguished triangle in the homotopy category $\mathrm{K}^+(\mathcal{A})$ and using the long exact $\mathrm{Hom}(\ \ , I^\bullet)$-sequence reduces the claim to the assertion that $\mathrm{Hom}_{\mathrm{K}(\mathcal{A})}(C^\bullet, I^\bullet) = 0$ for any acyclic complex $C^\bullet$, i.e. a complex quasi-isomorphic to 0 (cf. Exercise 2.20).

A homotopy between any complex morphism $g : C^\bullet \longrightarrow I^\bullet$ and the zero map can be explicitly constructed by standard algebraic homology methods, see [57, I.6].

The principal idea is the following: The desired homotopy $(h^i : C^i \longrightarrow I^{i-1})$ is constructed by induction. If the h^j have already been given for $j \leq i$, then consider $g^i - d_I^{i-1} \circ h^i : C^i \longrightarrow I^i$, which factorizes over $C^i/C^{i-1} \longrightarrow I^i$. Due to the injectivity of I^i, this lifts to a morphism $h^{i+1} : C^{i+1} \longrightarrow I^i$, in other words $g^i - d_I^{i-1} \circ h^i = h^{i+1} \circ d_C^i$. □

Lemma 2.39 *Let $A^\bullet, I^\bullet \in \mathrm{Kom}^+(\mathcal{A})$ such that all I^i are injective. Then*

$$\mathrm{Hom}_{\mathrm{K}(\mathcal{A})}(A^\bullet, I^\bullet) = \mathrm{Hom}_{\mathrm{D}(\mathcal{A})}(A^\bullet, I^\bullet).$$

Proof Clearly, there is a natural map $\mathrm{Hom}_{\mathrm{K}(\mathcal{A})}(A^\bullet, I^\bullet) \longrightarrow \mathrm{Hom}_{\mathrm{D}(\mathcal{A})}(A^\bullet, I^\bullet)$ and we have to show that for any morphism

$$\begin{array}{ccc} & B^\bullet & \\ {}^{\mathrm{qis}}\swarrow & & \searrow \\ A^\bullet & & I^\bullet \end{array}$$

in $\mathrm{D}(\mathcal{A})$ there exists a unique morphism of complexes $A^\bullet \longrightarrow I^\bullet$ making the diagram commutative up to homotopy.

In other words, one has to show that for any quasi-isomorphism $B^\bullet \longrightarrow A^\bullet$ in $\mathrm{Kom}^+(\mathcal{A})$ the induced map $\mathrm{Hom}_{\mathrm{K}(\mathcal{A})}(A^\bullet, I^\bullet) \longrightarrow \mathrm{Hom}_{\mathrm{K}(\mathcal{A})}(B^\bullet, I^\bullet)$ is bijective. This is Lemma 2.38. □

In the following proposition we consider the full additive subcategory $\mathcal{I} \subset \mathcal{A}$ of all injectives of an abelian category $\mathcal{A}$. As for an abelian category, $\mathrm{K}^*(\mathcal{I})$ can be defined (cf. Remark 2.14) and is again triangulated.

The composition of the inclusion $\mathcal{I} \subset \mathcal{A}$ with the natural exact functor $Q_\mathcal{A}$: $\mathrm{K}^*(\mathcal{A}) \longrightarrow \mathrm{D}^*(\mathcal{A})$ yields the natural exact functor $\iota : \mathrm{K}^*(\mathcal{I}) \longrightarrow \mathrm{D}^*(\mathcal{A})$.

Proposition 2.40 *Suppose that $\mathcal{A}$ contains enough injectives, i.e. any object in $\mathcal{A}$ can be embedded into an injective one. Then the natural functor*

$$\iota : \mathrm{K}^+(\mathcal{I}) \longrightarrow \mathrm{D}^+(\mathcal{A})$$

is an equivalence.

Proof Let us first check that the functor $\mathrm{K}^+(\mathcal{I}) \longrightarrow \mathrm{D}^+(\mathcal{A})$, even without the hypothesis, is fully faithful. We have to verify two things. Firstly, if a complex morphism $f : I^\bullet \longrightarrow J^\bullet$ of complexes of injectives I^i, J^j is zero in $\mathrm{D}(\mathcal{A})$ then f is homotopic to zero. Secondly, any morphism in $\mathrm{D}(\mathcal{A})$ can be completed to a commutative diagram in $\mathrm{K}(\mathcal{A})$

$$\begin{array}{ccc} & C^\bullet & \\ {}^{\mathrm{qis}}\swarrow & & \searrow \\ I^\bullet & \cdots\cdots\cdots\to & J^\bullet. \end{array}$$

Both assertions follow from the two lemmas preceding the proposition. (So, it is the injectivity of $J^\bullet$ that matters for both parts.) Note that there is a subtlety here, as the complex $C^\bullet$ might a priori only be cohomologically bounded below, but of course for $A^\bullet$ bounded below the qis $C^\bullet \longrightarrow A^\bullet$ factorizes over

$(0 \longrightarrow C^i/\mathrm{Im}(d^{i-1}) \longrightarrow C^{i+1} \longrightarrow \dots) \longrightarrow A^\bullet$ for $i \ll 0$. (Implicitly, the same problem appeared in the proof of Proposition 2.30.)

In order to see that the given functor is not only fully faithful but indeed essentially surjective one applies Proposition 2.35. □

In the course of the proof we have indeed used the assumption that all complexes are bounded below. For the analogous statement replacing injective by projective, one would have to work in $\mathrm{D}^-(\mathcal{A})$ for the same reason.

Remark 2.41 Eventually, we will be interested in the bounded derived category of coherent sheaves $\mathrm{D^b}(\mathbf{Coh}(X))$. However, there are reasons that will oblige us to work with bigger abelian categories and/or with unbounded derived categories. We add a few more comments about how this comes about.

As in the process of deriving functors one has to work with injective resolutions and those are almost never bounded, derived functors are often defined in the bigger categories of unbounded (or only partially bounded) complexes. Only a posteriori are they then restricted to the smaller categories.

It is not only that we would like to stay in the bounded derived category, but also have to work with unbounded complexes, we also have to leave the abelian category $\mathbf{Coh}(X)$ we are primarily interested in and work in the bigger ones $\mathbf{Qcoh}(X)$ or even $\mathbf{Sh}(X)$. The reason is essentially the same: we want to replace coherent sheaves by their injective resolutions, but there are almost no coherent injective sheaves. We will come back to this question, but the reader might keep in mind for the following discussion the inclusions of abelian categories $\mathbf{Coh}(X) \subset \mathbf{Qcoh}(X) \subset \mathbf{Sh}(X)$.

The general context can be set as follows. Consider a full abelian subcategory $\mathcal{A} \subset \mathcal{B}$ of an abelian category $\mathcal{B}$. Then there are two derived categories $\mathrm{D}(\mathcal{A})$ and $\mathrm{D}(\mathcal{B})$ with an obvious exact functor $\mathrm{D}(\mathcal{A}) \longrightarrow \mathrm{D}(\mathcal{B})$ between them.

One might wonder whether this functor defines an equivalence between $\mathrm{D}(\mathcal{A})$ and the full subcategory of $\mathrm{D}(\mathcal{B})$ containing those complexes whose cohomology is in $\mathcal{A}$. This does not hold, as in general $\mathrm{D}(\mathcal{A}) \longrightarrow \mathrm{D}(\mathcal{B})$ is neither full nor faithful. Fortunately, in the geometric situation, e.g. passing from $\mathrm{D^b}(\mathbf{Sh}(X))$ to $\mathrm{D^b}(\mathbf{Qcoh}(X))$, things are better behaved, as shown by the next proposition. Recall that a *thick subcategory* $\mathcal{A}$ of an abelian category $\mathcal{B}$ is a full abelian subcategory such that any extension in $\mathcal{B}$ of objects in $\mathcal{A}$ is again in $\mathcal{A}$.

Proposition 2.42 *Let $\mathcal{A} \subset \mathcal{B}$ be a thick subcategory and suppose that any $A \in \mathcal{A}$ can be embedded in an object $A' \in \mathcal{A}$ which is injective as an object of $\mathcal{B}$.*

Then the natural functor $\mathrm{D}(\mathcal{A}) \longrightarrow \mathrm{D}(\mathcal{B})$ induces an equivalence

$$\mathrm{D}^+(\mathcal{A}) \xrightarrow{\sim} \mathrm{D}^+_{\mathcal{A}}(\mathcal{B})$$

of $\mathrm{D}^+(\mathcal{A})$ and the full triangulated subcategory $\mathrm{D}^+_{\mathcal{A}}(\mathcal{B}) \subset \mathrm{D}^+(\mathcal{B})$ of complexes with cohomology in $\mathcal{A}$.

Analogously, one has $\mathrm{D^b}(\mathcal{A}) \simeq \mathrm{D^b_{\mathcal{A}}}(\mathcal{B})$.

Proof The assumption that $\mathcal{A}$ is a thick subcategory is clearly necessary in order to ensure that $\mathrm{D}^+_{\mathcal{A}}(\mathcal{B})$ is a triangulated subcategory.

The idea of the proof is to replace any complex bounded below in $\mathcal{B}$ with cohomology in $\mathcal{A}$ by an injective resolution contained in $\mathcal{A}$. For the proof we refer to the literature [39, 61]. This is very similar to Proposition 2.35. □

Exercise 2.43 Prove that under the assumption of the proposition $\mathrm{D}^+_{\mathcal{A}}(\mathcal{B})$ is indeed a full triangulated subcategory of $\mathrm{D}^+(\mathcal{B})$.

2.2 Derived functors

Let $F : \mathcal{A} \longrightarrow \mathcal{B}$ be an additive functor between abelian categories. If F is not exact, the image of an acyclic complex in $\mathcal{A}$, i.e. one that becomes trivial in $\mathrm{D}(\mathcal{A})$, is not, in general, acyclic (see Exercise 2.7). Thus, the naive extension of F to a functor between the derived categories $\mathrm{D}(\mathcal{A}) \longrightarrow \mathrm{D}(\mathcal{B})$ does not make sense, except when F is an exact functor. It is straightforward to verify the following slightly more general lemma. The equivalence of the two conditions uses the cone construction.

Lemma 2.44 *Let $F : \mathrm{K}^*(\mathcal{A}) \longrightarrow \mathrm{K}^*(\mathcal{B})$ be an exact functor of triangulated categories. Then F naturally induces a commutative diagram*

$$\begin{array}{ccc} \mathrm{K}^*(\mathcal{A}) & \longrightarrow & \mathrm{K}^*(\mathcal{B}) \\ \downarrow & & \downarrow \\ \mathrm{D}^*(\mathcal{A}) & \longrightarrow & \mathrm{D}^*(\mathcal{B}) \end{array}$$

if one of the following two conditions holds true.

i) *Under F a quasi-isomorphism is mapped to a quasi-isomorphism.*
ii) *The image of an acyclic complex is again acyclic.* □

(Note that in the lemma the functor need not come from a functor between the abelian categories.) If the functor $F : \mathcal{A} \longrightarrow \mathcal{B}$ is not exact (or if $F : \mathrm{K}^*(\mathcal{A}) \longrightarrow \mathrm{K}^*(\mathcal{B})$ does not satisfy i) or ii)), a more complicated construction is needed in order to induce a natural functor between the derived categories. The new functor, called the derived functor, will not produce a commutative diagram as in Lemma 2.44, but it has the advantage to encode more information even when applied to an object in the abelian category. Roughly, it explains why the original functor fails to be exact.

In order to ensure existence of the derived functor, we will always have to assume some kind of exactness. For a left exact functor $F : \mathcal{A} \longrightarrow \mathcal{B}$ (see Definition 1.12) one constructs the right derived functor

$$RF : \mathrm{D}^+(\mathcal{A}) \longrightarrow \mathrm{D}^+(\mathcal{B})$$

and for a right exact functor $F : \mathcal{A} \longrightarrow \mathcal{B}$ one constructs the left derived functor

$$LF : \mathrm{D}^-(\mathcal{A}) \longrightarrow \mathrm{D}^-(\mathcal{B}).$$

Both constructions are completely analogous and we shall discuss only RF. In the applications, however, there is a difference in that $\mathbf{Qcoh}(X)$ for a smooth projective variety (in fact, for any noetherian scheme) always contains enough injectives, but never enough projectives (well, except if X is a collection of points).

Before explaining the general construction, we provide a list of most of the left, respectively right exact functors that will be used in the geometric context. We will come back to this list in Section 3.3. In this section only the abstract machinery is described.

Examples 2.45 i) Let X be a topological space. Then the global section functor

$$\Gamma(X,\ \ ) : \mathbf{Sh}(X) \longrightarrow \mathbf{Ab}$$

is a left exact functor. Similarly, for a scheme X it defines left exact functors $\Gamma(X,\ \ ) : \mathbf{Qcoh}(X) \longrightarrow \mathbf{Ab}$ and $\Gamma(X,\ \ ) : \mathbf{Coh}(X) \longrightarrow \mathbf{Ab}$. If X is a projective variety over a field k, then this becomes the left exact functor

$$\Gamma(X,\ ) : \mathbf{Coh}(X) \longrightarrow \mathbf{Vec}_f(k)$$

into the category of finite-dimensional vector spaces (see [45, II, 5.19]).

ii) Let $f : X \longrightarrow Y$ be a continuous map. Then the direct image functor $f_* : \mathbf{Sh}(X) \longrightarrow \mathbf{Sh}(Y)$ is left exact. Similarly, if $f : X \longrightarrow Y$ is a morphism of schemes, then

$$f_* : \mathbf{Qcoh}(X) \longrightarrow \mathbf{Qcoh}(Y)$$

is left exact. If $f : X \longrightarrow Y$ is a proper morphism of noetherian schemes, then the direct image defines a left exact functor (see [45, III, 8.8])

$$f_* : \mathbf{Coh}(X) \longrightarrow \mathbf{Coh}(Y).$$

This in particular applies to any morphism of projective varieties.

Note that in general, $\Gamma(Y,\ \ ) \circ f_* = \Gamma(X,\ \ )$ and $f_* = \Gamma(X,\ \ )$ if f is the projection onto a point.

iii) Suppose X is a scheme and $\mathcal{F} \in \mathbf{Qcoh}(X)$. Then

$$\mathrm{Hom}(\mathcal{F},\ \ ) : \mathbf{Qcoh}(X) \longrightarrow \mathbf{Ab}$$

is left exact. For a coherent sheaf $\mathcal{F}$ on a projective variety X over a field k one has the left exact functor

$$\mathrm{Hom}(\mathcal{F},\ \ ) : \mathbf{Coh}(X) \longrightarrow \mathbf{Vec}_f(k).$$

iv) Consider, as before, a quasi-coherent sheaf $\mathcal{F}$ on a scheme X. Then the sheaf of homomorphisms from $\mathcal{F}$ defines a left exact functor

$$\mathcal{H}om(\mathcal{F},\ \ ) : \mathbf{Qcoh}(X) \longrightarrow \mathbf{Qcoh}(X).$$

If $\mathcal{F}$ is coherent, then $\mathcal{H}om(\mathcal{F},\ \ )$ takes coherent sheaves to coherent ones. Also note that $\Gamma \circ \mathcal{H}om(\mathcal{F},\ \ ) = \mathrm{Hom}(\mathcal{F},\ \ )$.

We leave it to the reader to write down the analogous statements for the contravariant functors $\mathrm{Hom}(\ \ ,\mathcal{F})$ and $\mathcal{H}om(\ \ ,\mathcal{F})$.

v) Let X be a topological space endowed with a sheaf of commutative rings $\mathcal{R}$. Consider the abelian category of sheaves of $\mathcal{R}$-modules $\mathbf{Sh}_{\mathcal{R}}(X)$. If $\mathcal{F} \in \mathbf{Sh}_{\mathcal{R}}(X)$, then

$$\mathcal{F} \otimes_{\mathcal{R}} (\ \ ) : \mathbf{Sh}_{\mathcal{R}}(X) \longrightarrow \mathbf{Sh}_{\mathcal{R}}(X)$$

is a right exact functor.

vi) Let $f : X \longrightarrow Y$ be a continuous map. Then the inverse image defines an exact functor

$$f^{-1} : \mathbf{Sh}(Y) \longrightarrow \mathbf{Sh}(X) \quad \text{and} \quad f^{-1} : \mathbf{Sh}_{\mathcal{R}}(Y) \longrightarrow \mathbf{Sh}_{f^{-1}\mathcal{R}}(X),$$

where $\mathcal{R}$ is any sheaf of rings on Y.

If $f : X \longrightarrow Y$ is a morphism of schemes, one defines

$$f^* := (\mathcal{O}_X \otimes_{f^{-1}\mathcal{O}_Y}\ \ ) \circ f^{-1} : \mathbf{Qcoh}(Y) \longrightarrow \mathbf{Qcoh}(X).$$

This is a right exact functor, as it is the composition of the exact functor $f^{-1} : \mathbf{Sh}_{\mathcal{O}_Y}(Y) \longrightarrow \mathbf{Sh}_{f^{-1}\mathcal{O}_Y}(X)$ and the right exact functor $\mathcal{O}_X \otimes_{f^{-1}\mathcal{O}_Y} (\ \ ) : \mathbf{Sh}_{f^{-1}\mathcal{O}_Y}(X) \longrightarrow \mathbf{Sh}_{\mathcal{O}_X}(X)$. (To be very precise, the latter takes a priori values in $\mathbf{Sh}_{f^{-1}\mathcal{O}_Y}(X)$, but the tensor product $\mathcal{O}_X \otimes_{f^{-1}\mathcal{O}_Y} \mathcal{F}$ comes with a natural $\mathcal{O}_X$-module structure.)

Clearly, the inverse image in this sense maps a coherent sheaf to a coherent sheaf, i.e.

$$f^* : \mathbf{Coh}(Y) \longrightarrow \mathbf{Coh}(X).$$

To conclude, we recall that $f^* \dashv f_*$. Due to the general fact (see Remark 1.16), this shows once more that f^* and f_* are right, respectively left exact.

Now, back to the abstract setting. We let $F : \mathcal{A} \longrightarrow \mathcal{B}$ be a left exact functor of abelian categories. Furthermore, we assume that $\mathcal{A}$ contains enough injectives. In particular, we will use the equivalence $\iota : \mathrm{K}^+(\mathcal{I}_{\mathcal{A}}) \longrightarrow \mathrm{D}^+(\mathcal{A})$ naturally induced by the functor $Q_{\mathcal{A}} : \mathrm{K}^+(\mathcal{A}) \longrightarrow \mathrm{D}^+(\mathcal{A})$ (see Proposition 2.40). By ι^{-1} we denote a quasi-inverse of ι given by choosing a complex of injective objects quasi-isomorphic to any given complex that is bounded below. Thus, we have the diagram

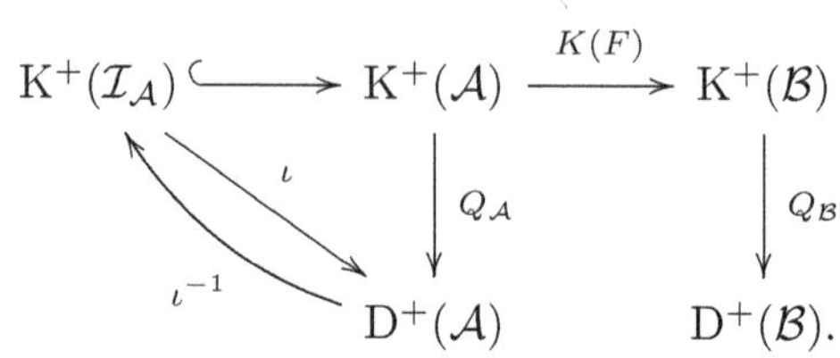

Here, $K(F)$ is the functor that maps $(\ldots \to A^{i-1} \to A^i \to A^{i+1} \to \ldots)$ to $(\ldots \to F(A^{i-1}) \to F(A^i) \to F(A^{i+1}) \to \ldots)$ which is well-defined for the homotopy categories.

Definition 2.46 *The* right derived functor *of F is the functor*

$$RF := Q_{\mathcal{B}} \circ K(F) \circ \iota^{-1} : D^+(\mathcal{A}) \longrightarrow D^+(\mathcal{B}).$$

Let us list some of the main properties of the right derived functor:

Proposition 2.47 i) *There exists a natural morphism of functors*

$$Q_{\mathcal{B}} \circ K(F) \longrightarrow RF \circ Q_{\mathcal{A}}.$$

ii) *The right derived functor $RF : D^+(\mathcal{A}) \to D^+(\mathcal{B})$ is an exact functor of triangulated categories.*

iii) *Suppose $G : D^+(\mathcal{A}) \to D^+(\mathcal{B})$ is an exact functor. Then any functor morphism $Q_{\mathcal{B}} \circ K(F) \to G \circ Q_{\mathcal{A}}$ factorizes through a unique functor morphism $RF \to G$.*

Proof i) Let $A^\bullet \in D^+(\mathcal{A})$ and $I^\bullet := \iota^{-1}(A^\bullet)$. The natural transformation $\mathrm{id} \to \iota \circ \iota^{-1}$ yields a functorial morphism $A^\bullet \to I^\bullet$ in $D^+(\mathcal{A})$, which itself is given by some roof $A^\bullet \xleftarrow{\text{qis}} C^\bullet \to I^\bullet$. Using the injectivity of I^i yields a unique morphism $A^\bullet \to I^\bullet$ in $K(\mathcal{A})$ (see Lemma 2.38), which, moreover, is independent of the choice of $C^\bullet$. Altogether, one obtains a functorial morphism $K(F)(A^\bullet) \to K(F)(I^\bullet) = RF(A^\bullet)$.

ii) The category $K^+(\mathcal{I}_{\mathcal{A}})$ is triangulated and $\iota : K^+(\mathcal{I}_{\mathcal{A}}) \to D^+(\mathcal{A})$ is clearly exact. Thus, also the inverse functor ι is exact (cf. Proposition 1.41). Hence, RF is the composition of three exact functors and, therefore, itself exact.

iii) See [39, III.6.11]. □

These properties determine the right derived functor RF of a left exact functor F up to unique isomorphism.

Definition 2.48 *Let* $RF : \mathrm{D}^+(\mathcal{A}) \longrightarrow \mathrm{D}^+(\mathcal{B})$ *be the right derived functor of a left exact functor* $F : \mathcal{A} \longrightarrow \mathcal{B}$. *Then for any complex* $A^\bullet \in \mathrm{D}^+(\mathcal{A})$ *one defines:*

$$R^iF(A^\bullet) := H^i(RF(A^\bullet)) \in \mathcal{B}.$$

Remark 2.49 The induced additive functors

$$R^iF : \mathcal{A} \longrightarrow \mathcal{B}$$

are the *higher derived functors* of F.

Note that $R^iF(A) = 0$ for $i < 0$ and $R^0F(A) \simeq F(A)$ for any $A \in \mathcal{A}$. Indeed, if

$$A \longrightarrow I^0 \longrightarrow I^1 \longrightarrow \cdots$$

is an injective resolution, then $R^iF(A) = H^i(\ldots \longrightarrow F(I^0) \longrightarrow F(I^1) \longrightarrow \ldots)$ and, in particular,

$$R^0F(A) = \mathrm{Ker}(F(I^0) \longrightarrow F(I^1)) = F(A),$$

as F is left exact.

An object $A \in \mathcal{A}$ is called F*-acyclic* if $R^iF(A) \simeq 0$ for $i \neq 0$.

Corollary 2.50 *Under the above assumptions any short exact sequence*

$$0 \longrightarrow A \longrightarrow B \longrightarrow C \longrightarrow 0$$

in the abelian category $\mathcal{A}$ *gives rise to a long exact sequence*

$$0 \longrightarrow F(A) \longrightarrow F(B) \longrightarrow F(C) \longrightarrow R^1F(A) \longrightarrow \cdots$$

$$\cdots \longrightarrow R^iF(B) \longrightarrow R^iF(C) \longrightarrow R^{i+1}F(A) \longrightarrow \cdots$$

Proof According to Exercise 2.27, any short exact sequence in $\mathcal{A}$ gives rise to a distinguished triangle $A \longrightarrow B \longrightarrow C \longrightarrow A[1]$ and hence to a distinguished triangle $RF(A) \longrightarrow RF(B) \longrightarrow RF(C) \longrightarrow RF(A)[1]$. One concludes by using Exercise 2.28. □

Remark 2.51 Going through the above arguments, one finds that our hypothesis can be weakened. This remark explains two possible and useful ways to do so. Firstly, the functor might only be given between the homotopy categories (and not between the abelian categories) and, secondly, we might have to work with abelian categories which do not contain enough injectives.

• Let us give the most general statement right away (see [44, II, 5.1]): Suppose $\mathcal{A}$ and $\mathcal{B}$ are abelian categories and

$$F : \mathrm{K}^+(\mathcal{A}) \longrightarrow \mathrm{K}(\mathcal{B})$$

is an exact functor (recall that both categories are triangulated). Then the right derived functor

$$RF : \mathrm{D}^+(\mathcal{A}) \longrightarrow \mathrm{D}(\mathcal{B})$$

satisfying i)–iii) of Proposition 2.47 exists whenever there exists a triangulated subcategory $\mathcal{K}_F \subset \mathrm{K}^+(\mathcal{A})$ which is *adapted* to F, i.e. which satisfies the following two conditions:

i) If $A^\bullet \in \mathcal{K}_F$ is acyclic, i.e. $H^i(A^\bullet) = 0$ for any i, then $F(A^\bullet)$ is acyclic.
ii) Any $A^\bullet \in \mathrm{K}^+(\mathcal{A})$ is quasi-isomorphic to a complex in $\mathcal{K}_F$.

We will need this more general statement, e.g. in order to define the derived functor $R\mathrm{Hom}^\bullet(A^\bullet, B^\bullet)$ or, for a more geometrical example, the left derived functor of the tensor product of two sheaves. In the first case, we start out with a functor that only lives on the level of complexes and in the latter we have to deal with the fact that the category of sheaves over a ring in general does not contain enough projectives.

• If the functor F is given on the level of the abelian categories, but finding enough injectives is problematic (or simply impossible), then the following approach towards the derived functor, as a special case of the general one above, often works. See [39, III.6].

Suppose $F : \mathcal{A} \longrightarrow \mathcal{B}$ is a left exact functor. In this situation one defines 'adapted' already on the level of the abelian categories. A class of objects $\mathcal{I}_F \subset \mathcal{A}$ stable under finite sums is *F-adapted* if the following conditions hold true:

i) If $A^\bullet \in \mathrm{K}^+(\mathcal{A})$ is acyclic with $A^i \in \mathcal{I}_F$ for all i, then $F(A^\bullet)$ is acyclic.
ii) Any object in $\mathcal{A}$ can be embedded into an object of $\mathcal{I}_F$.

Under these conditions, the localization of $\mathrm{K}^+(\mathcal{I}_F)$ by quasi-isomorphisms between complexes with objects in $\mathcal{I}_F$ is equivalent to $\mathrm{D}^+(\mathcal{A})$. This is due to ii). Condition i) ensures that the image under F of a quasi-isomorphism of complexes in $\mathcal{I}_F$ is again a quasi-isomorphism. Hence, we may define $K(F)$ on the localization of $\mathrm{K}^+(\mathcal{I}_F)$ (cf. Lemma 2.44). The right derived functor RF of F is then defined in the same way by using these two facts.

Note that if $\mathcal{A}$ contains enough injectives, then the class of injective objects $\mathcal{I}_\mathcal{A}$ is F-adapted for any left exact functor F. In this case, we may enlarge $\mathcal{I}_\mathcal{A}$ by all F-acyclic objects, i.e. by those objects $A \in \mathcal{A}$ with $R^iF(A) = 0$ for $i \neq 0$. This yields a larger adapted class for F.

Exercise 2.52 Let $\mathcal{I}_F$ be an F-adapted class, e.g. the class $\mathcal{I}_\mathcal{A}$ of all injective objects in an abelian category with enough injectives. Show that enlarging $\mathcal{I}_F$ by all F-acyclic ones yields again an F-adapted class. Use Corollary 2.50.

Exercise 2.53 Let $\mathcal{I}_F$ be an adapted class for a left exact functor $F : \mathcal{A} \longrightarrow \mathcal{B}$. Show that $R^iF(A) = 0$ for all $i \neq 0$ and all $A \in \mathcal{I}_F$.

Exercise 2.54 Suppose we know that the right derived functor RF exists. Let $A^\bullet$ be a complex of F-acyclic objects. Show that $RF(A^\bullet) \simeq K(F)(A^\bullet)$.

In other words, in order to compute $RF(A^\bullet)$ of an arbitrary complex $A^\bullet$ it suffices to find a qis $A^\bullet \simeq I^\bullet \in \mathrm{K}^+(\mathcal{A})$ such that all I^i are F-acyclic. Then $RF(A^\bullet) \simeq K(F)(I^\bullet)$.

Exercise 2.55 Write down the conditions on a class of objects to be adapted to a right exact functor. (This actually is the situation one needs to consider for the tensor product.)

In Section 3.1 we will study in detail a number of derived functors in the geometric setting, e.g. higher direct images. Here, we shall stay in the general situation and only consider the covariant functor $\mathrm{Hom}(A, \ ) : \mathcal{A} \longrightarrow \mathbf{Ab}$ for an arbitrary object $A \in \mathcal{A}$ and its contravariant relative $\mathrm{Hom}(\ , A)$. Clearly, $\mathrm{Hom}(A, \ )$ is left exact and if $\mathcal{A}$ contains enough injectives, one defines

$$\mathrm{Ext}^i(A, \ ) := H^i \circ R\mathrm{Hom}(A, \ ).$$

It turns out that these Ext-groups can be interpreted purely in terms of certain homomorphism groups within the derived category:

Proposition 2.56 *Suppose $A, B \in \mathcal{A}$ are objects of an abelian category containing enough injectives. Then there are natural isomorphisms*

$$\mathrm{Ext}^i_{\mathcal{A}}(A, B) \simeq \mathrm{Hom}_{\mathrm{D}(\mathcal{A})}(A, B[i]),$$

where A and B are considered as complexes concentrated in degree zero.

Proof Suppose

$$B \longrightarrow I^0 \longrightarrow I^1 \longrightarrow \cdots$$

is an injective resolution of B. By construction, $R\mathrm{Hom}(A, B)$ as an object of the derived category $\mathrm{D}^+(\mathbf{Ab})$ is isomorphic to the complex $(\mathrm{Hom}(A, I^i))_{i\in\mathbb{N}}$. Therefore, $\mathrm{Ext}^i(A, B)$ is the cohomology of this complex.

A morphism $f \in \mathrm{Hom}(A, I^i)$ is a *cycle*, i.e. it is contained in the kernel of $\mathrm{Hom}(A, I^i) \longrightarrow \mathrm{Hom}(A, I^{i+1})$, if and only if f defines a morphism of complexes $f : A \longrightarrow I^\bullet[i]$. This morphism of complexes is homotopically trivial if and only if f is a boundary, i.e. in the image of $\mathrm{Hom}(A, I^{i-1}) \longrightarrow \mathrm{Hom}(A, I^i)$.

Hence, $\mathrm{Ext}^i(A, B) \simeq \mathrm{Hom}_{\mathrm{K}(\mathcal{A})}(A, I^\bullet[i])$. Since $I^\bullet$ is a complex of injectives, we have $\mathrm{Hom}_{\mathrm{K}(\mathcal{A})}(A, I^\bullet[i]) = \mathrm{Hom}_{\mathrm{D}(\mathcal{A})}(A, I^\bullet[i])$ (see Lemma 2.39). Using $B \simeq I^\bullet$ as objects of $\mathrm{D}^+(\mathcal{A})$, this proves the assertion. □

Remarks 2.57 i) The above arguments can easily be generalized to a description of $\mathrm{Ext}^i_{\mathcal{A}}(A^\bullet, B^\bullet)$. Suppose $A^\bullet \in \mathrm{Kom}(\mathcal{A})$. Then, the exact functor

$$\mathrm{Hom}^\bullet(A^\bullet, \ ) : \mathrm{K}^+(\mathcal{A}) \longrightarrow \mathrm{K}(\mathbf{Ab})$$

associates to a complex $B^\bullet$ the *inner hom* $\mathrm{Hom}^\bullet(A^\bullet, B^\bullet)$.

By definition $\mathrm{Hom}^\bullet(A^\bullet, B^\bullet)$ is the complex with

$$\mathrm{Hom}^i(A^\bullet, B^\bullet) := \bigoplus \mathrm{Hom}(A^k, B^{k+i}) \quad \text{and} \quad d(f) := d_B \circ f - (-1)^i f \circ d_A.$$

The full triangulated subcategory of complexes of injectives is adapted to this functor (always under the assumption that $\mathcal{A}$ has enough injectives) and we may define

$$R\mathrm{Hom}^\bullet(A^\bullet, \ \ ) : \mathrm{D}^+(\mathcal{A}) \longrightarrow \mathrm{D}(\mathbf{Ab})$$

(see Remark 2.51). Then set

$$\mathrm{Ext}^i(A^\bullet, B^\bullet) := H^i(R\mathrm{Hom}^\bullet(A^\bullet, B^\bullet)).$$

The arguments to prove Proposition 2.56 can be adapted to this more general situation. One obtains natural isomorphisms

$$\mathrm{Ext}^i(A^\bullet, B^\bullet) \simeq \mathrm{Hom}_{\mathrm{D}(\mathcal{A})}(A^\bullet, B^\bullet[i]). \tag{2.2}$$

ii) It is noteworthy, that by (2.2) these Ext-groups only depend on $A^\bullet$ as an element of the derived category. Indeed, if $A_1^\bullet \longrightarrow A_2^\bullet$ is a qis, then the induced morphism $R\mathrm{Hom}^\bullet(A_2^\bullet, B^\bullet) \longrightarrow R\mathrm{Hom}^\bullet(A_1^\bullet, B^\bullet)$ yields isomorphisms on the cohomology and is therefore an isomorphism in $\mathrm{D}(\mathbf{Ab})$. Hence, by the principle applied already in Lemma 2.44 the functor $R\mathrm{Hom}^\bullet(\ \ , B^\bullet)$ descends to the derived category and we have thus defined a bifunctor

$$\mathrm{D}(\mathcal{A})^{\mathrm{op}} \times \mathrm{D}^+(\mathcal{A}) \longrightarrow \mathrm{D}(\mathbf{Ab}),$$

which is exact in each variable. (Alternatively, it suffices to check that for an acyclic complex $A^\bullet$ the complex $R\mathrm{Hom}^\bullet(A^\bullet, B^\bullet)$ is acyclic which follows directly from (2.2).)

iii) Suppose that $\mathcal{A}$ is a category with enough projective objects. Then one defines for any complex $B^\bullet \in \mathrm{Kom}(\mathcal{A})$ the right derived functor of the left exact functor $\mathrm{Hom}(\ \ , B^\bullet) : \mathrm{K}^-(\mathcal{A})^{\mathrm{op}} \longrightarrow \mathrm{K}(\mathbf{Ab})$. This yields the exact functor

$$R\mathrm{Hom}^\bullet(\ \ , B^\bullet) : \mathrm{D}^-(\mathcal{A})^{\mathrm{op}} \longrightarrow \mathrm{D}(\mathbf{Ab}).$$

Using similar arguments as in ii), one finds that it only depends on $B^\bullet$ as an object in the derived category and thus defines again a bifunctor which is exact in the two variables

$$\mathrm{D}^-(\mathcal{A})^{\mathrm{op}} \times \mathrm{D}(\mathcal{A}) \longrightarrow \mathrm{D}(\mathbf{Ab}).$$

If $\mathcal{A}$ has enough projectives and enough injectives, then the two bifunctors in ii) and iii) give rise to the same bifunctor (cf. [44, I, 6.3])

$$R\mathrm{Hom}^\bullet(\ \ ,\ \ ) : \mathrm{D}^-(\mathcal{A})^{\mathrm{op}} \times \mathrm{D}^+(\mathcal{A}) \longrightarrow \mathrm{D}(\mathbf{Ab}).$$

That the cohomology of both yields the same, follows from the above remarks.

Also note that in a category $\mathcal{A}$ with enough injectives, but possibly not enough projectives, due to ii) the derived functor $R\mathrm{Hom}^\bullet(\ \ ,B^\bullet) : \mathrm{D}^-(\mathcal{A})^{\mathrm{op}} \longrightarrow \mathrm{D}(\mathbf{Ab})$ nevertheless exists if $B^\bullet$ is bounded below.

iv) Composition in the derived category naturally leads to composition for Ext-groups:

$$\mathrm{Ext}^i(A^\bullet, B^\bullet) \times \mathrm{Ext}^j(B^\bullet, C^\bullet) \longrightarrow \mathrm{Ext}^{i+j}(A^\bullet, C^\bullet),$$

where we assume for simplicity $A^\bullet, B^\bullet, C^\bullet \in \mathrm{D}^+(\mathcal{A})$.

Indeed, elements in

$$\mathrm{Ext}^i(A^\bullet, B^\bullet) \simeq \mathrm{Hom}_{\mathrm{D}(\mathcal{A})}(A^\bullet, B^\bullet[i])$$

and

$$\mathrm{Ext}^j(B^\bullet, C^\bullet) \simeq \mathrm{Hom}_{\mathrm{D}(\mathcal{A})}(B^\bullet, C^\bullet[j]) = \mathrm{Hom}_{\mathrm{D}(\mathcal{A})}(B^\bullet[i], C^\bullet[i+j])$$

can be composed to elements in $\mathrm{Ext}^{i+j}(A^\bullet, C^\bullet) \simeq \mathrm{Hom}_{\mathrm{D}(\mathcal{A})}(A^\bullet, C^\bullet[i+j])$.

Proposition 2.58 *Let $F_1 : \mathcal{A} \longrightarrow \mathcal{B}$ and $F_2 : \mathcal{B} \longrightarrow \mathcal{C}$ be two left exact functors of abelian categories. Assume that there exist adapted classes $\mathcal{I}_{F_1} \subset \mathcal{A}$ and $\mathcal{I}_{F_2} \subset \mathcal{B}$ for F_1, respectively F_2 such that $F_1(\mathcal{I}_{F_1}) \subset \mathcal{I}_{F_2}$.*

Then the derived functors $RF_1 : \mathrm{D}^+(\mathcal{A}) \longrightarrow \mathrm{D}^+(\mathcal{B})$, $RF_2 : \mathrm{D}^+(\mathcal{B}) \longrightarrow \mathrm{D}^+(\mathcal{C})$, and $R(F_2 \circ F_1) : \mathrm{D}^+(\mathcal{A}) \longrightarrow \mathrm{D}^+(\mathcal{C})$ exist and there is a natural isomorphism

$$R(F_2 \circ F_1) \simeq RF_2 \circ RF_1.$$

Proof The existence of RF_1 and RF_2 follows from the assumptions. Moreover, since $F_1(\mathcal{I}_{F_1}) \subset \mathcal{I}_{F_2}$, the class $\mathcal{I}_{F_1}$ is also adapted to the composition $F_2 \circ F_1$ and, hence, $R(F_2 \circ F_1)$ exists as well.

A natural morphism $R(F_2 \circ F_1) \longrightarrow RF_2 \circ RF_1$ is given by the universality property of the derived functor $R(F_2 \circ F_1)$.

If $A^\bullet \in \mathrm{D}^+(\mathcal{A})$ is isomorphic to a complex $I^\bullet \in \mathrm{K}^+(\mathcal{I}_{F_1})$, then this morphism

$$\begin{array}{ccc} R(F_2 \circ F_1)(A^\bullet) & \longrightarrow & R(F_2)((RF_1)(A^\bullet)) \\ \simeq (K(F_2) \circ K(F_1))(I^\bullet) & & \simeq R(F_2)(K(F_1)(I^\bullet)) \\ & & \simeq K(F_2)(K(F_1)(I^\bullet)) \end{array}$$

is an isomorphism. □

Remarks 2.59 i) Suppose that $\mathcal{A}$ and $\mathcal{B}$ both contain enough injectives. Then the assumption of the proposition is satisfied if $F_1(\mathcal{I}_\mathcal{A}) \subset \mathcal{I}_\mathcal{B}$, but this might be difficult to verify. We might, however, enlarge $\mathcal{I}_\mathcal{B}$ by all F_2-acyclic objects in $\mathcal{B}$, i.e. by objects B with $R^iF_2(B) = 0$ for $i \neq 0$. This yields a new adapted class $\mathcal{I}_{F_2}$ (see Exercise 2.52) and we then only have to show that for any injective object $I \in \mathcal{I}_\mathcal{A} \subset \mathcal{A}$ the image $F_1(I) \in \mathcal{B}$ is F_2-acyclic. The proposition is often applied in this form.

ii) The result also holds true for the derived functors of exact functors between the homotopy categories. In this case, one starts with exact functors $F_1 : \mathrm{K}^+(\mathcal{A}) \longrightarrow \mathrm{K}^+(\mathcal{B})$ and $F_2 : \mathrm{K}^+(\mathcal{B}) \longrightarrow \mathrm{K}(\mathcal{C})$, for which one assumes the existence of adapted triangulated subcategories $\mathcal{K}_{F_1} \subset \mathrm{K}^+(\mathcal{A})$ and $\mathcal{K}_{F_2} \subset \mathrm{K}^+(\mathcal{B})$ such that $F_1(\mathcal{K}_{F_1}) \subset \mathcal{K}_{F_2}$.

2.3 Spectral sequences

In this short section we will explain how spectral sequences occur whenever two derived functors are composed. We will not enter the very technical details of this machinery, but hope to provide at least the amount necessary to follow the applications in the later sections.

The main data of a *spectral sequence* in an abelian category $\mathcal{A}$ is a collection of objects

$$(E_r^{p,q}, E^n),\ n,p,q,r \in \mathbb{Z}, r \geq 1,$$

and morphisms

$$d_r^{p,q} : E_r^{p,q} \longrightarrow E_r^{p+r,q-r+1}$$

subject to the following conditions.

i) $d_r^{p+r,q-r+1} \circ d_r^{p,q} = 0$ for all p,q,r, which thus yields a complex $E_r^{p+\bullet r,q-\bullet r+\bullet}$.

ii) There are isomorphisms

$$E_{r+1}^{p,q} \simeq H^0(E_r^{p+\bullet r,q-\bullet r+\bullet})$$

which are part of the data.

iii) For any (p,q) there exists an r_0 such that $d_r^{p,q} = d_r^{p-r,q+r-1} = 0$ for $r \geq r_0$. In particular, $E_r^{p,q} \simeq E_{r_0}^{p,q}$ for all $r \geq r_0$. This object is called $E_\infty^{p,q}$.

iv) There is a decreasing filtration

$$\ldots F^{p+1}E^n \subset F^pE^n \subset \ldots \subset E^n,$$

such that

$$\bigcap F^pE^n = 0 \text{ and } \bigcup F^pE^n = E^n,$$

and isomorphisms

$$E_\infty^{p,q} \simeq F^pE^{p+q}/F^{p+1}E^{p+q}.$$

So, in some sense the objects $E_r^{p,q}$ converge towards subquotients of a certain filtration of E^n. Usually, all the objects of one layer, say $E_r^{p,q}$ with r fixed, are explicitly given. Then one writes

$$E_r^{p,q} \Rightarrow E^{p+q}.$$

(In the definition of the spectral sequence, we may as well just require the terms $E_r^{p,q}$ be given only for $r \geq m$ for some m. The information is just the same. In fact, in the applications the spectral sequences are often given for $r \geq 2$.)

It may help to visualize these data as follows. For $r = 1$ one has the following system of horizontal complexes:

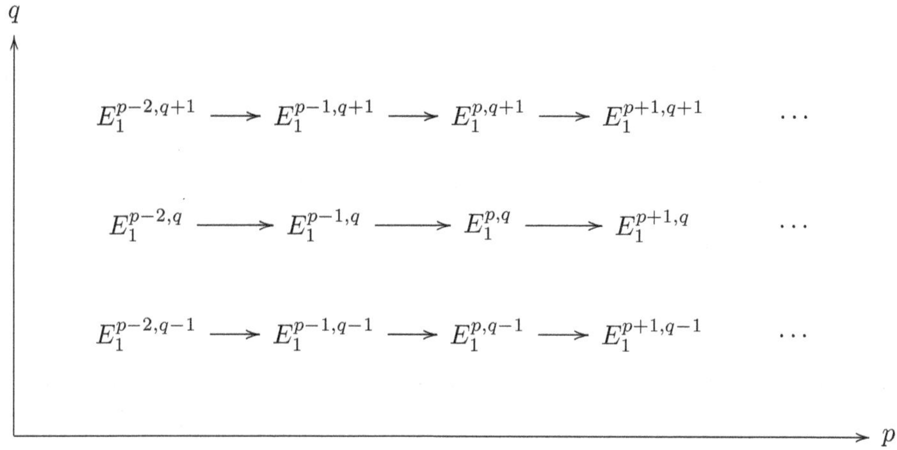

For $r = 2$ it looks like this:

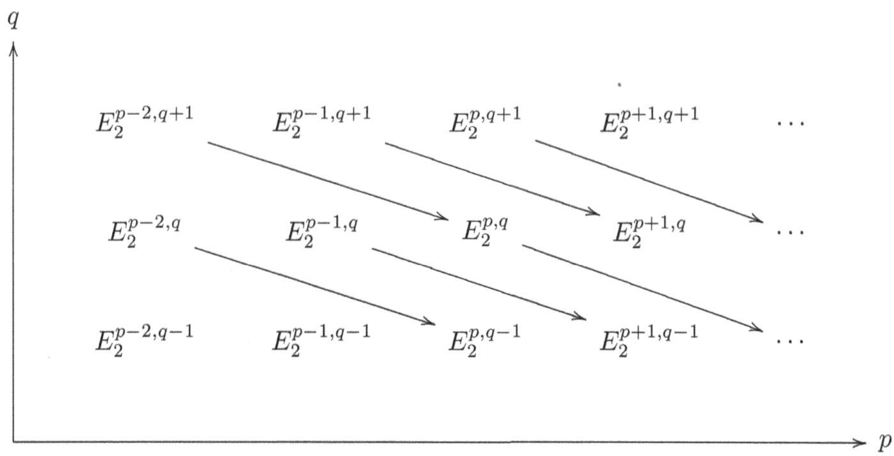

Remark 2.60 In most of the applications one does not go beyond E_2 or E_3. In the easiest situation the argument will go like this: For some reason one knows that all differentials on the E_2-level are trivial. Hence, E^n admits a filtration the subquotients of which are isomorphic to $E_2^{p,n-p}$. E.g. if the objects are all vector spaces, then this yields a non-canonical isomorphism $E^n = \bigoplus E_2^{p,n-p}$.

Sometimes, one just knows the vanishing of $d_r^{p,q}$ and $d_r^{p-r,q+r-1}$ for some (p,q) (e.g. for the simple reason that all $E_r^{p+r,q-r+1}$ are trivial). In this case, the non-vanishing $E_2^{p,q} \neq 0$ implies $E^{p+q} \neq 0$.

The standard source for spectral sequences are 'nice' filtrations on complexes. Such 'nice' filtrations occur naturally on the total complex of a double complex concentrated in, e.g. the first quadrant. We briefly sketch this part of the theory.

Definition 2.61 *A* double complex $K^{\bullet,\bullet}$ *consists of objects* $K^{i,j}$ *for* $i,j \in \mathbb{Z}$ *and morphisms*

$$d_I^{i,j} : K^{i,j} \longrightarrow K^{i+1,j} \quad \text{and} \quad d_{II}^{i,j} : K^{i,j} \longrightarrow K^{i,j+1}$$

satisfying

$$d_I^2 = d_{II}^2 = d_I d_{II} + d_{II} d_I = 0.$$

The total complex $K^\bullet := \mathrm{tot}(K^{\bullet,\bullet})$ *of a double complex* $K^{\bullet,\bullet}$ *is the complex* $K^n = \bigoplus_{i+j=n} K^{i,j}$ *with* $d = d_I + d_{II}$.

In particular, $K^{i,\bullet}$ and $K^{\bullet,j}$ form complexes for all i, j

Examples 2.62 The complex $\mathrm{Hom}^\bullet(A^\bullet, B^\bullet)$ is the total complex of the natural double complex $K^{i,j} := \mathrm{Hom}(A^{-i}, B^j)$ endowed with the two differentials $d_I = (-1)^{j-i+1} d_A$ and $d_{II} = d_B$. There is absolutely no guarantee for the signs here. One finds all kinds of sign conventions in the literature, in the definition of a double complex, as well as in the construction of its total complex, and in the definition of $\mathrm{Hom}^\bullet(\ ,\ )$. Mostly, the differences in the signs are of no importance, but getting it coherent is troublesome.

On the total complex $K^\bullet$ of a double complex $K^{\bullet,\bullet}$ there exists a natural decreasing filtration (in fact, there are two natural ones due to the symmetry of the situation):

$$F^\ell K^n := \bigoplus_{j \geq \ell} K^{n-j,j}, \tag{2.3}$$

which satisfies $d_I(F^\ell K^n) \subset F^\ell(K^{n+1})$.

This lends itself to the following generalization.

Definition 2.63 *A* filtered complex *is a complex* $K^\bullet$ *together with a decreasing filtration* $\ldots F^\ell K^n \subset F^{\ell-1} K^n \subset \ldots \subset K^n$ *for all the objects* K^n *such that* $d^n(F^\ell K^n) \subset F^\ell K^{n+1}$.

$$\begin{array}{ccccccccc}
\cdots & \longrightarrow & K^{n-1} & \longrightarrow & K^{n} & \longrightarrow & K^{n+1} & \longrightarrow & \cdots \\
 & & \uparrow & & \uparrow & & \uparrow & & \\
\cdots & \longrightarrow & F^{\ell}(K^{n-1}) & \longrightarrow & F^{\ell}(K^{n}) & \longrightarrow & F^{\ell}(K^{n+1}) & \longrightarrow & \cdots \\
 & & \uparrow & & \uparrow & & \uparrow & & \\
\cdots & \longrightarrow & F^{\ell+1}(K^{n-1}) & \longrightarrow & F^{\ell+1}(K^{n}) & \longrightarrow & F^{\ell+1}(K^{n+1}) & \longrightarrow & \cdots
\end{array}$$

Back to the case of the total complex of a double complex. Clearly, the graded objects $\mathrm{gr}^{\ell} K^n = F^{\ell}K^n/(F^{\ell+1}K^n) = K^{n-\ell,\ell}$ form the complex $K^{\bullet,\ell}[-\ell]$ (up to the global sign $(-1)^{\ell}$). Hence, $H^k(\mathrm{gr}^{\ell}(K^{\bullet})) = H^{k-\ell}(K^{\bullet,\ell})$ and the cohomology of this complex (with respect to the surviving d_{II}) yields $H^{\ell}_{II}(H^{k-\ell}_{I}(K^{\bullet,\bullet}))$.

Assuming an additional finiteness condition, any filtered complex gives rise to a spectral sequence. More precisely, one has to assume that for each n there exist $\ell_+(n)$ and $\ell_-(n)$ with $F^{\ell}K^n = 0$ for $\ell \geq \ell_+(n)$ and $F^{\ell}K^n = K^n$ for $\ell \leq \ell_-(n)$.

In the situation of the double complex this translates to:

Proposition 2.64 *Suppose $K^{\bullet,\bullet}$ is a double complex such that for any n one has $K^{n-\ell,\ell} = 0$ for $|\ell| \gg 0$. Then the filtration (2.3) naturally induces a spectral sequence*

$$E_2^{p,q} = H^p_{II}H^q_I(K^{\bullet,\bullet}) \Rightarrow H^{p+q}(K^{\bullet}). \tag{2.4}$$

Remark 2.65 As mentioned, the proposition works more generally for filtered complexes. Moreover, it actually yields an E_1-spectral sequence

$$E_1^{p,q} = H^{p+q}(F^pK^{\bullet}/F^{p+1}K^{\bullet}).$$

In case of a double complex as above, this reads

$$E_1^{p,q} = H^{p+q}(K^{\bullet,p}[-p]) = H^q(K^{\bullet,p}) \Rightarrow H^{p+q}(K^{\bullet}).$$

The construction of the spectral sequence is explicit, but the verifications, although in principle elementary, are cumbersome. Let us just describe the objects $E_r^{p,q}$ and the filtration of the limit E^n. First, one introduces

$$Z_r^{p,q} := d^{-1}(F^{p+r}K^{p+q+1}) \cap (F^pK^{p+q})$$

and then sets

$$E_r^{p,q} := Z_r^{p,q}/\left(Z_{r-1}^{p+1,q-1} + dZ_{r-1}^{p-r+1,q+r-2}\right).$$

Eventually, $F^pH^n(K^{\bullet}) := \mathrm{Im}\left(H^n(F^pK^{\bullet}) \longrightarrow H^n(K^{\bullet})\right)$.

The general fact is crucial for the proof of the next proposition, which is often applied to the derived functors of functors between abelian categories and an object $A \in \mathcal{A}$. The following more general form turns out to be useful as well.

Proposition 2.66 *Let $F_1 : \mathrm{K}^+(\mathcal{A}) \longrightarrow \mathrm{K}^+(\mathcal{B})$ and $F_2 : \mathrm{K}^+(\mathcal{B}) \longrightarrow \mathrm{K}(\mathcal{C})$ be two exact functors. Suppose that $\mathcal{A}$ and $\mathcal{B}$ contain enough injectives and that the image under F_1 of a complex $I^\bullet$ with $I^i \in \mathcal{I}_\mathcal{A}$ is contained in an F_2-adapted triangulated subcategory $\mathcal{K}_{F_2}$.*

Then for any complex $A^\bullet \in \mathrm{D}^+(\mathcal{A})$ there exists a spectral sequence

$$E_2^{p,q} = R^pF_2(R^qF_1(A^\bullet)) \Rightarrow E^n = R^n(F_2 \circ F_1)(A^\bullet). \tag{2.5}$$

Surprisingly, this is even interesting in the case that F_1 is the identity. For any left exact functor F the spectral sequence (2.5) then reads

$$E_2^{p,q} = R^pF(H^q(A^\bullet)) \Rightarrow R^{p+q}F(A^\bullet). \tag{2.6}$$

The special case (2.6) is the key to the general result: Suppose $A^\bullet \in \mathrm{D}^+(\mathcal{A})$ is isomorphic to a complex $I^\bullet \in \mathrm{K}^+(\mathcal{I}_{F_1})$. Then $RF_1(A^\bullet) \simeq F_1(I^\bullet)$ and

$$R^pF_2(R^qF_1(A^\bullet)) \simeq R^pF_2(H^q(F_1(I^\bullet))).$$

On the other hand,

$$\begin{aligned} R^n(F_2 \circ F_1)(A^\bullet) &\simeq H^n(R(F_2 \circ F_1)(A^\bullet)) \simeq H^n(RF_2(RF_1(A^\bullet))) \\ &\simeq H^n(RF_2(F_1(I^\bullet))) \simeq R^nF_2(F_1(I^\bullet)). \end{aligned}$$

Thus, it suffices to prove (2.6) for which we will have to write down the appropriate double complex. This double complex is provided by the *Cartan–Eilenberg resolution* of $A^\bullet$.

The Cartan–Eilenberg resolution is a double complex $C^{\bullet,\bullet}$ together with a morphism $A^\bullet \longrightarrow C^{\bullet,0}$ such that:

i) $C^{i,j} = 0$ for $j < 0$.
ii) The sequences

$$A^i \longrightarrow C^{i,0} \longrightarrow C^{i,1} \longrightarrow \cdots$$

are injective resolutions of A^i inducing injective resolutions of $\mathrm{Ker}(d_A^i)$, $\mathrm{Im}(d_A^i)$, and $H^i(A^\bullet)$.
iii) The sequences $C^{\bullet,j}$ are split for all j, i.e. all short exact sequences

$$0 \longrightarrow \mathrm{Ker} d_I^{i,j} \longrightarrow C^{i,j} \longrightarrow \mathrm{Im}(d_I^{i,j}) \longrightarrow 0$$

split.

A Cartan–Eilenberg resolution exists whenever the abelian category contains enough injectives (see [39, III.7]).

In our situation, we use the Cartan–Eilenberg resolution $C^{\bullet,\bullet}$ of $A^\bullet$ to define the double complex $K^{\bullet,\bullet}$ by $K^{i,j} := F(C^{i,j})$. Then $H_I^q(K^{\bullet,\ell}) = FH_I^q(C^{\bullet,\ell})$, because of iii). As $H_I^q(C^{\bullet,\ell})$ for fixed q and running ℓ defines an injective resolution of $H^q(A^\bullet)$ (see ii)), we obtain

$$H_{II}^p H_I^q(K^{\bullet,\bullet}) = R^pF(H^q(A^\bullet)).$$

The limit in the spectral sequence (2.4) is

$$\begin{aligned} H^{p+q}(\mathrm{tot}(K^{\bullet,\bullet})) &= H^{p+q}(F(\mathrm{tot}(C^{\bullet,\bullet}))) \stackrel{*}{=} H^{p+q}(RF(A^\bullet)) \\ &= R^{p+q}F(A^\bullet), \end{aligned}$$

where in $(*)$ we use the general fact that if $C^{\bullet,\bullet}$ is a double complex satisfying i) and such that there exists a complex morphism $A^\bullet \longrightarrow C^{\bullet,0}$ inducing resolutions $A^i \longrightarrow C^{i,0} \longrightarrow C^{i,1} \longrightarrow \ldots$, then the induced morphism $A^\bullet \longrightarrow \mathrm{tot}(C^{\bullet,\bullet})$ is a quasi-isomorphism.

Note that the finiteness assumption needed to ensure the convergence is satisfied due to our assumption that $A^\bullet \in \mathrm{K}^+(\mathcal{A})$.

Remark 2.67 One also has a spectral sequence

$$E_1^{p,q} = R^qF(A^p) \Rightarrow R^{p+q}F(A^\bullet),$$

which sometimes is very useful. This is a consequence of Remark 2.65, but with respect to the other filtration of the image under F of the Cartan–Eilenberg resolution $C^{\bullet,\bullet}$ of $A^\bullet$. Indeed, if $K^{p,q} = F(C^{p,q})$, then

$$H^q(K^{p,\bullet}) = H^q(F(C^{p,\bullet})) = R^qF(A^p).$$

In the geometric context we will often make use of the facts that shall be explained next. The reason behind this is that the derived category of complexes bounded from below very convenient for the definition of various derived functors, but slightly too big for other purposes.

Corollary 2.68 *Suppose* $F : \mathrm{K}^+(\mathcal{A}) \longrightarrow \mathrm{K}^+(\mathcal{B})$ *is an exact functor which admits a right derived functor* $RF : \mathrm{D}^+(\mathcal{A}) \longrightarrow \mathrm{D}^+(\mathcal{B})$ *and assume that* $\mathcal{A}$ *has enough injectives.*

i) *Suppose* $\mathcal{C} \subset \mathcal{B}$ *is a thick subcategory with* $R^iF(A) \in \mathcal{C}$ *for all* $A \in \mathcal{A}$ *and that there exists* $n \in \mathbb{Z}$ *with* $R^iF(A) = 0$ *for* $i < n$ *and all* $A \in \mathcal{A}$. *Then* RF *takes values in* $\mathrm{D}^+_{\mathcal{C}}(\mathcal{B})$, *i.e.*

$$RF : \mathrm{D}^+(\mathcal{A}) \longrightarrow \mathrm{D}^+_{\mathcal{C}}(\mathcal{B}).$$

ii) *If* $RF(A) \in \mathrm{D}^{\mathrm{b}}(\mathcal{B})$ *for any object* $A \in \mathcal{A}$, *then* $RF(A^\bullet) \in \mathrm{D}^{\mathrm{b}}(\mathcal{B})$ *for any complex* $A^\bullet \in \mathrm{D}^{\mathrm{b}}(\mathcal{A})$, *i.e.* RF *induces an exact functor*

$$RF : \mathrm{D}^{\mathrm{b}}(\mathcal{A}) \longrightarrow \mathrm{D}^{\mathrm{b}}(\mathcal{B}).$$

Proof Both assertions follow immediately from the spectral sequence $E_2^{p,q} = R^pF(H^q(A^\bullet)) \Rightarrow R^{p+q}F(A^\bullet)$. □

Remark 2.69 i) In many cases, the category $\mathrm{D}^+_{\mathcal{C}}(\mathcal{B})$ is equivalent to the derived category $\mathrm{D}^+(\mathcal{C})$. See Proposition 2.42.

ii) The assumption that there exist enough injectives can be weakened. In fact, assertion ii) holds true whenever the derived functor exists, but the spectral sequence cannot be applied directly, as it had been constructed in terms of a Cartan–Eilenberg resolution.

For example, let $A^\bullet \in \mathrm{D}^{\mathrm{b}}(\mathcal{A})$ be a bounded complex with $A^i = 0$ for $i > n$. Choose a quasi-isomorphism $A^\bullet \longrightarrow I^\bullet$ with F-acyclic objects I^i. Then,

$$A := \mathrm{Ker}(d_I^n) \longrightarrow I^n \longrightarrow I^{n+1} \longrightarrow I^{n+2} \longrightarrow \cdots$$

is an F-acyclic resolution of $A \in \mathcal{A}$. Thus, $F(I^{i-1}) \longrightarrow F(I^i) \longrightarrow F(I^{i+1})$ is exact for $i \gg 0$, for $R^iF(A) = 0$ for $i \gg 0$ by assumption. Hence, $R^iF(A^\bullet) = 0$ for $i \gg 0$ and, therefore, $RF(A^\bullet) \in \mathrm{D}^{\mathrm{b}}(\mathcal{B})$.

Examples 2.70 i) Let $A^\bullet, B^\bullet \in \mathrm{D}(\mathcal{A})$ with $B^\bullet$ bounded below and suppose that $\mathcal{A}$ has enough injectives. Then there exists a spectral sequence

$$E_2^{p,q} = \mathrm{Hom}_{\mathrm{D}(\mathcal{A})}(A^\bullet, H^q(B^\bullet)[p]) \Rightarrow \mathrm{Hom}_{\mathrm{D}(\mathcal{A})}(A^\bullet, B^\bullet[p+q]). \qquad (2.7)$$

Here we use (2.2) for the identification

$$R^p\mathrm{Hom}^\bullet(A^\bullet, H^q(B^\bullet)) \simeq \mathrm{Ext}^p(A^\bullet, H^q(B^\bullet)) \simeq \mathrm{Hom}_{\mathrm{D}(\mathcal{A})}(A^\bullet, H^q(B^\bullet)[p]).$$

ii) Similarly, if $\mathcal{A}$ contains enough projectives, such that we can compute $R^p\mathrm{Hom}(A^\bullet, B^\bullet)$ for $A^\bullet$ bounded above as the right derived functor of the contravariant functor $\mathrm{Hom}^\bullet(\ \ , B^\bullet) : \mathrm{K}^-(\mathcal{A})^{\mathrm{op}} \longrightarrow \mathrm{K}(\mathbf{Ab})$, then we can use the spectral sequence

$$E_2^{p,q} = \mathrm{Hom}_{\mathrm{D}(\mathcal{A})}(H^{-q}(A^\bullet), B^\bullet[p]) \Rightarrow \mathrm{Hom}_{\mathrm{D}(\mathcal{A})}(A^\bullet, B^\bullet[p+q]). \qquad (2.8)$$

In fact, we will use this spectral sequence also in the case when $\mathcal{A}$ only has enough injectives. Then we have to assume that $B^\bullet$ is bounded below. In this case we cannot apply Proposition 2.66 directly, but a similar argument via a double complex yields the result.

More precisely, one argues as follows. Suppose $A^\bullet \in \mathrm{K}^-(\mathcal{A})$ and $B^\bullet \in \mathrm{K}^+(\mathcal{A})$. Since we assume that $\mathcal{A}$ has enough injectives, there exists a qis $B^\bullet \longrightarrow I^\bullet \in \mathrm{K}^+(\mathcal{A})$ with all I^i injective. Then, following Example 2.62 we form the double complex $K^{i,j} := \mathrm{Hom}(A^{-i}, I^j)$ with differentials $d_I = (-1)^{j-i+1}d_A$ and $d_{II} = d_B$. With the notation introduced above, we find that the complex $(\mathrm{gr}^\ell K^\bullet, d_I)$ is just $\mathrm{Hom}(A^\bullet, I^\ell)[-\ell]$.

Since I^ℓ is injective, the functor $\mathrm{Hom}(\ \ , I^\ell)$ is exact and thus commutes with cohomology. Hence, $H^k(\mathrm{Hom}(A^\bullet, I^\ell)[-\ell]) = \mathrm{Hom}(H^{\ell-k}(A^\bullet), I^\ell)$. Therefore,

$$H^p_{II}H^q_I(K^{\bullet,\bullet}) = H^p(\mathrm{Hom}(H^{-q}(A^\bullet), I^\bullet) = \mathrm{Ext}^p(H^{-q}(A^\bullet), B^\bullet).$$

The finiteness condition on the filtration needed for the convergence of the spectral sequence is provided by the boundedness assumption on $A^\bullet$ and $B^\bullet$.

Eventually, the limit of the spectral sequence is identified as $H^n(\mathrm{tot}(K^{\bullet,\bullet})) = H^n(\mathrm{Hom}^\bullet(A^\bullet, I^\bullet)) \simeq \mathrm{Ext}^n(A^\bullet, B^\bullet)$.

Exercise 2.71 Assume the spectral sequences (2.7) and (2.8) exist. Suppose $A^\bullet, B^\bullet \in \mathrm{D^b}(\mathcal{A})$ with $H^i(A^\bullet) = 0$ for $i > 0$ and $H^i(B^\bullet) = 0$ for $i < 0$. Prove $\mathrm{Hom}(A^\bullet, B^\bullet) = 0$.

We conclude with a discussion of 'ample sequences' in abelian categories. Geometrically, as the name suggests, this concept is realized by powers of an ample line bundle on a projective variety. This will be explained in Section 3.2. Thus, the abelian category we have in mind is $\mathbf{Coh}(X)$.

Any ample sequence in an abelian category turns out to be a spanning class in the associated derived category. So, in spirit the main result, Proposition 2.73, belongs in Section 1.3, but as ampleness of sequences makes sense only for abelian categories and their derived categories, it is included here. You may skip this part and come back to it when proving Proposition 3.18.

Definition 2.72 *A sequence of objects $L_i \in \mathcal{A}$, $i \in \mathbb{Z}$, in a k-linear abelian category $\mathcal{A}$ is called* ample *if for any object $A \in \mathcal{A}$ there exists an integer $i_0(A)$ such that for $i < i_0(A)$ the following conditions are satisfied:*

i) *The natural morphism* $\mathrm{Hom}(L_i, A) \otimes_k L_i \longrightarrow A$ *is surjective.*
ii) *If $j \neq 0$, then* $\mathrm{Hom}(L_i, A[j]) = 0$.
iii) $\mathrm{Hom}(A, L_i) = 0$.

In order to define the tensor product $V \otimes_k L$ (where V is a vector space and $L \in \mathcal{A}$) as an object in $\mathcal{A}$ properly, we either have to assume that V is finite-dimensional or that infinite direct sums exist in $\mathcal{A}$. Later, our category $\mathcal{A}$ will be $\mathbf{Coh}(X)$ of a projective variety over a field k and hence all $\mathrm{Hom}(L_i, A)$ are indeed finite-dimensional.

Proposition 2.73 *Let L_i, $i \in \mathbb{Z}$ be an ample sequence in a k-linear abelian category $\mathcal{A}$ of finite homological dimension. Then, considered as objects in the derived category $\mathrm{D^b}(\mathcal{A})$, the L_i span $\mathrm{D^b}(\mathcal{A})$.*

Before giving the proof let us first recall what it means to be of *finite homological dimension.* If $\mathcal{A}$ has enough injectives, then it means that there exists an integer ℓ such that $\mathrm{Ext}^i(A, B) = 0$ for all $A, B \in \mathcal{A}$ and all $i > \ell$. If we don't want to assume or don't want to use the existence of enough injectives, we can simply require that $\mathrm{Hom}_{\mathrm{D}(\mathcal{A})}(A, B[i]) = 0$ for all $A, B \in \mathcal{A}$ viewed as objects in the derived category $\mathrm{D}(\mathcal{A})$ and all $i > \ell$. That the two definitions are equivalent follows from Proposition 2.56.

If $\mathcal{A}$ has finite homological dimension, then for a fixed bounded complex $A^\bullet$ there exists $i_0(A^\bullet)$ such that $\mathrm{Hom}(A^\bullet, B[i]) = 0$ for all $i > i_0(A^\bullet)$ and all

$B \in \mathcal{A}$. A quick way to see this is by applying the spectral sequence (under the assumptions of enough injectives or any other ensuring its existence):

$$E_2^{p,q} = \mathrm{Hom}(H^{-q}(A^\bullet), B[p]) \Rightarrow \mathrm{Hom}(A^\bullet, B[p+q]).$$

The $E_2^{p,q}$-term is trivial for $p > \ell$. Hence, $\mathrm{Hom}(A^\bullet, B[i]) = 0$ for $i > i_0(A^\bullet) := \ell + \max\{m \mid H^{-m}(A^\bullet) \neq 0\}$.

(An elementary proof, i.e. one avoiding the spectral sequence, is also available. It eventually is enough to show that there are no complex morphisms $B^\bullet \longrightarrow B[i]$ for $i > i_0(A^\bullet)$ for any complex $B^\bullet$ with the same cohomology as $A^\bullet$. Splitting the complex $B^\bullet$ in short exact sequences yields the result. We leave the details of this argument to the reader.)

Proof The proof of the proposition consists of two steps proving the two conditions i) and ii) in Definition 1.47. (In fact, one proves a slightly stronger version of both. See Remark 2.75.) Only in the proof of i) do we need the extra assumption on the homological dimension.

i) Let $A^\bullet \in \mathrm{D^b}(\mathcal{A})$ such that $\mathrm{Hom}(L_i, A^\bullet[j]) = 0$ for all i and all j. Suppose $A^\bullet$ is non-trivial. Thus, we may assume that $A^\bullet$ is of the form

$$\cdots \longrightarrow 0 \longrightarrow A^n \longrightarrow A^{n+1} \longrightarrow \cdots$$

with $H^n(A^\bullet) \neq 0$ (cf. Exercise 2.31).

Hence, $\mathrm{Hom}(L_i, H^n(A^\bullet)) \hookrightarrow \mathrm{Hom}(L_i, A^\bullet[n]) = 0$ for all i. On the other hand, by condition i) of Definition 2.72, the evaluation map

$$\mathrm{Hom}(L_i, H^n(A^\bullet)) \otimes L_i \twoheadrightarrow H^n(A^\bullet)$$

is surjective for $i < i_0(H^n(A^\bullet))$. This yields a contradiction and, hence, $A^\bullet \simeq 0$.

ii) Let $A^\bullet \in \mathrm{D^b}(\mathcal{A})$ such that $\mathrm{Hom}(A^\bullet, L_i[j]) = 0$ for all i and all j. If Serre duality is available (e.g. if $\mathcal{A} = \mathbf{Coh}(X)$ with X smooth projective, see Theorem 3.12), then this case can be reduced to the preceding discussion and we obtain immediately $A^\bullet \simeq 0$.

If not, the argument is slightly more involved and runs as follows. We may assume that $A^\bullet$ is a bounded complex of the form $\longrightarrow A^{n-1} \longrightarrow A^n \longrightarrow 0 \longrightarrow \ldots$ with $H^n(A^\bullet) \neq 0$. The ampleness of $\{L_i\}$ allows us to construct a surjection

$$\mathrm{Hom}(L_i, H^n(A^\bullet)) \otimes L_i \twoheadrightarrow H^n(A^\bullet)$$

for any $i < i_0(H^n(A^\bullet))$. Its kernel will be called B_1. Since $\mathrm{Hom}(A^\bullet, L_i) = 0$, the long exact sequence induced by

$$0 \longrightarrow B_1 \longrightarrow \mathrm{Hom}(L_i, H^n(A^\bullet)) \otimes L_i \longrightarrow H^n(A^\bullet) \longrightarrow 0$$

yields an injection $\mathrm{Hom}(A^\bullet, H^n(A^\bullet)) \hookrightarrow \mathrm{Hom}(A^\bullet, B_1[1])$, for its kernel is a quotient of

$$\mathrm{Hom}(A^\bullet, \mathrm{Hom}(L_i, H^n(A^\bullet)) \otimes L_i) = \mathrm{Hom}(L_i, H^n(A^\bullet)) \otimes \mathrm{Hom}(A^\bullet, L_i) = 0.$$

Then one continues to proceed with B_1 in the same way, i.e. one finds a surjection $\mathrm{Hom}(L_i, B_1) \otimes L_i \twoheadrightarrow B_1$ and denotes its kernel by B_2. (One might have to pass to an even smaller i.) As before, the induced long exact cohomology sequence yields an injection

$$\mathrm{Hom}(A^\bullet, B_1[1]) \hookrightarrow \mathrm{Hom}(A^\bullet, B_2[2]),$$

because $\mathrm{Hom}(A^\bullet, L_i[j]) = 0$. Thus, recursively we obtain nested inclusions

$$\mathrm{Hom}(A^\bullet, H^n(A^\bullet)) \hookrightarrow \mathrm{Hom}(A^\bullet, B_1[1]) \hookrightarrow \mathrm{Hom}(A^\bullet, B_2[2]) \hookrightarrow \cdots$$

Since there exists a non-trivial morphism $A^\bullet \twoheadrightarrow H^n(A^\bullet)$, we obtain in this way for all $j > 0$ an object $B_j \in \mathcal{A}$ with $\mathrm{Hom}(A^\bullet, B_j[j]) \neq 0$. This contradicts the assumption on the homological dimension of $\mathcal{A}$ and its consequences explained above. □

Exercise 2.74 Go through the above proof again and show that we have actually only used condition i) in Definition 2.72. (The other ones will play their rôle in Chapter 4.)

Remark 2.75 The above proof shows slightly more. Namely, if for a given complex $A^\bullet$ and all $i \ll 0$ (depending on $A^\bullet$) one has $\mathrm{Hom}(A^\bullet, L_i[j]) = 0$ for all j, then $A^\bullet \simeq 0$. Similarly, for the vanishing of $\mathrm{Hom}(L_i[j], A^\bullet)$.

3

DERIVED CATEGORIES OF COHERENT SHEAVES

This chapter applies the general machinery of the last one to derived categories of sheaves on a scheme or a smooth projective variety. Most of the material is standard (Serre duality, higher direct images, etc.) but a few more recent results are blended in to prepare the stage for the sequel, e.g. we will prove a few results ensuring that the categories in question are accessible by the methods discussed in Chapter 1.

Section 3.1 introduces the category we are primarily interested in, the derived category of the abelian category of coherent sheaves. As an injective sheaf is almost never coherent, quasi-coherent sheaves cannot be avoided. The section contains first structure results, in particular for curves. Serre duality, a topic that will be taken up in greater generality in Section 3.4, will be stated in its derived version.

In Section 3.2 we prove that the structure sheaves of closed points and the powers of an ample line bundle provide examples of spanning classes.

All kinds of derived functors, important in the geometric context, are introduced and discussed in Section 3.3. Useful technical results, mostly concerning the various compatibilities between them, are stated and partially proven.

3.1 Basic structure

The derived category of coherent sheaves enters the stage. We show that it is indecomposable if and only if the scheme is connected. A derived version of Serre duality (see Theorem 3.12) is stated and used to show that on a curve objects in the derived category can always be written as direct sums of shifted sheaves (see Corollary 3.15).

The category we are primarily interested in is the category of coherent sheaves on a projective variety or, more generally, a (noetherian) scheme:

Definition 3.1 *Let X be a scheme. Its* derived category $\mathrm{D}^{\mathrm{b}}(X)$ *is by definition the bounded derived category of the abelian category* $\mathbf{Coh}(X)$, *i.e.*

$$\mathrm{D}^{\mathrm{b}}(X) := \mathrm{D}^{\mathrm{b}}(\mathbf{Coh}(X)).$$

Definition 3.2 *Two schemes X and Y defined over a field k are called* derived equivalent *(or, simply,* D-equivalent*) if there exists a k-linear exact equivalence $\mathrm{D}^{\mathrm{b}}(X) \simeq \mathrm{D}^{\mathrm{b}}(Y)$.*

Similarly, one introduces $\mathrm{D}(X)$, $\mathrm{D}^{+}(X)$, and $\mathrm{D}^{-}(X)$, but the hero of this course is the bounded derived category $\mathrm{D}^{\mathrm{b}}(X)$. Unfortunately, the underlying abelian category $\mathbf{Coh}(X)$ usually contains no non-trivial injective objects, so that in order to compute derived functors we have to pass to bigger abelian categories. Most often, we will work with the abelian category of quasi-coherent sheaves $\mathbf{Qcoh}(X)$, with its derived categories $\mathrm{D}^*(\mathbf{Qcoh}(X))$ with $* = \mathrm{b}, +, -$, and sometimes with the abelian category of $\mathcal{O}_X$-modules $\mathbf{Sh}_{\mathcal{O}_X}(X)$.

Whenever the scheme X is defined over a field k, the derived categories will tacitly be considered as k-linear categories.

Notation In order to avoid any possible confusion between sheaf cohomology $H^i(X, \mathcal{F})$ and the cohomology $H^i(\mathcal{F}^\bullet)$ of a complex of sheaves, we will from now on write $\mathcal{H}^i(\mathcal{F}^\bullet)$ for the latter.

Proposition 3.3 *On a noetherian scheme X any quasi-coherent sheaf $\mathcal{F}$ admits a resolution*

$$0 \longrightarrow \mathcal{F} \longrightarrow \mathcal{I}^0 \longrightarrow \mathcal{I}^1 \longrightarrow \cdots$$

by quasi-coherent sheaves $\mathcal{I}^i$ which are injective as $\mathcal{O}_X$-modules.

Proof For a proof see [44, II, 7.18]. □

Thus, in the case we are interested in, i.e. X a (smooth) projective variety over a field, the result applies. So due to Proposition 2.42, we can either think of $\mathrm{D}^*(\mathbf{Qcoh}(X))$ (with $* = \mathrm{b}, +$) as the bounded (resp. bounded below) derived category of $\mathbf{Qcoh}(X)$ or as the full triangulated subcategory of $\mathrm{D}^*(\mathbf{Sh}_{\mathcal{O}_X})$ of bounded (resp. bounded below) complexes with quasi-coherent cohomology:

Corollary 3.4 *For any noetherian scheme there are natural equivalences:*

$$\mathrm{D}^*(\mathbf{Qcoh}(X)) \simeq \mathrm{D}^*_{\mathrm{qcoh}}(\mathbf{Sh}_{\mathcal{O}_X}(X))$$

with $* = \mathrm{b}, +$. □

The passage from the quasi-coherent to the coherent world is trickier. Obviously, Proposition 2.42 does not apply for the simple reason that a finitely generated module is usually too small to be injective. In other words, we cannot hope to find an injective (in $\mathbf{Coh}(X)$ or $\mathbf{Qcoh}(X)$) resolution of a coherent sheaf by coherent sheaves. We nevertheless have the following result.

Proposition 3.5 *Let X be a noetherian scheme. Then the natural functor*

$$\mathrm{D}^{\mathrm{b}}(X) \longrightarrow \mathrm{D}^{\mathrm{b}}(\mathbf{Qcoh}(X))$$

defines an equivalence between the derived category $\mathrm{D}^{\mathrm{b}}(X)$ of X and the full triangulated subcategory $\mathrm{D}^{\mathrm{b}}_{\mathrm{coh}}(\mathbf{Qcoh}(X))$ of bounded complexes of quasi-coherent sheaves with coherent cohomology.

Proof Let $\mathcal{G}^\bullet$ be a bounded complex of quasi-coherent sheaves

$$\cdots \longrightarrow 0 \longrightarrow \mathcal{G}^n \longrightarrow \cdots \longrightarrow \mathcal{G}^m \longrightarrow 0 \longrightarrow \cdots$$

with coherent cohomology $\mathcal{H}^i$. Suppose $\mathcal{G}^i$ is coherent for $i > j$. Then apply Lemma 3.6 below to the surjections

$$d^j : \mathcal{G}^j \twoheadrightarrow \mathrm{Im}(d^j) \subset \mathcal{G}^{j+1} \quad \text{and} \quad \mathrm{Ker}(d^j) \twoheadrightarrow \mathcal{H}^j$$

which yield subsheaves $\mathcal{G}_1^j \subset \mathcal{G}^j$ and $\mathcal{G}_2^j \subset \mathrm{Ker}(d^j) \subset \mathcal{G}^j$, respectively. We may now replace $\mathcal{G}^j$ by the coherent sheaf generated by $\mathcal{G}_i^j$, $i = 1, 2$, and $\mathcal{G}^{j-1}$ by the pre-image of the new $\mathcal{G}^j$ under $\mathcal{G}^{j-1} \longrightarrow \mathcal{G}^j$. Clearly, the inclusion defines a quasi-isomorphism of the new complex to the old one and now $\mathcal{G}^i$ is coherent for $i \geq j$. □

Lemma 3.6 *If $\mathcal{G} \twoheadrightarrow \mathcal{F}$ is an $\mathcal{O}_X$-module homomorphism from a quasi-coherent sheaf $\mathcal{G}$ onto a coherent sheaf $\mathcal{F}$ on a noetherian scheme X, then there exists a coherent subsheaf $\mathcal{G}' \subset \mathcal{G}$ such that the composition $\mathcal{G}' \subset \mathcal{G} \twoheadrightarrow \mathcal{F}$ is still surjective.*

Proof The statement is clear for modules: For any surjection $M \twoheadrightarrow N$ with N finitely generated, there exists a finitely generated module $M' \subset M$ that projects onto N. Thus, locally the statement holds true. Covering X by finitely many open affines, the assertion reduces to the following well-known statement (cf. [45, II, Exc.5.15]). Let $U \subset X$ be an open subscheme of a noetherian scheme X and let $\mathcal{F} \subset \mathcal{G}|_U$ be a coherent subsheaf of the restriction of a quasi-coherent sheaf $\mathcal{G}$ on X. Then there exists a coherent subsheaf $\mathcal{F}' \subset \mathcal{G}$ such that $\mathcal{F}'|_U = \mathcal{F}$. □

Remark 3.7 i) Since **Qcoh**(X) has enough injectives (at least if X is noetherian), we can define $R\mathrm{Hom}^\bullet(\mathcal{E}^\bullet, \mathcal{F}^\bullet)$ for any $\mathcal{E}^\bullet \in \mathrm{D}(\mathbf{Qcoh}(X))$ and any $\mathcal{F}^\bullet \in \mathrm{D}^+(\mathbf{Qcoh}(X))$. In particular, this works for $\mathcal{E}^\bullet, \mathcal{F}^\bullet \in \mathrm{D}^{\mathrm{b}}(X)$.

Moreover, Remark 2.57 and Proposition 3.5 yield

$$\mathrm{Ext}^i(\mathcal{E}^\bullet, \mathcal{F}^\bullet) \simeq \mathrm{Hom}_{\mathrm{D}^{\mathrm{b}}(X)}(\mathcal{E}^\bullet, \mathcal{F}^\bullet[i])$$

for all $\mathcal{E}^\bullet, \mathcal{F}^\bullet \in \mathrm{D}^{\mathrm{b}}(X)$.

We will frequently use Examples 2.70, i.e. the spectral sequences (2.7) and (2.8). Thus, for any $\mathcal{E}^\bullet, \mathcal{F}^\bullet \in \mathrm{D}^{\mathrm{b}}(X)$ one has

$$E_2^{p,q} = \mathrm{Ext}^p(\mathcal{E}^\bullet, \mathcal{H}^q(\mathcal{F}^\bullet)) \Rightarrow \mathrm{Ext}^{p+q}(\mathcal{E}^\bullet, \mathcal{F}^\bullet) \tag{3.1}$$

and

$$E_2^{p,q} = \mathrm{Ext}^p(\mathcal{H}^{-q}(\mathcal{E}^\bullet), \mathcal{F}^\bullet) \Rightarrow \mathrm{Ext}^{p+q}(\mathcal{E}^\bullet, \mathcal{F}^\bullet). \tag{3.2}$$

ii) It is a deep fact that for a projective variety X over a field k the cohomology $H^i(X, \mathcal{F})$ of any coherent sheaf $\mathcal{F}$ is finite-dimensional (cf. [45, III, 5.2] and Theorem 3.21). This can be used to show that $\mathrm{Ext}^i(\mathcal{E}, \mathcal{F})$ is also of finite dimension for any two coherent sheaves $\mathcal{E}, \mathcal{F}$. By applying the spectral sequences (3.1) and (3.2), one easily sees that $\mathrm{Ext}^i(\mathcal{E}^\bullet, \mathcal{F}^\bullet)$ are actually finite-dimensional for any $\mathcal{E}^\bullet, \mathcal{F}^\bullet \in \mathrm{D}^{\mathrm{b}}(X)$.

Thus, in the sequel we will assume that $\mathrm{Hom}(\mathcal{E}^\bullet, \mathcal{F}^\bullet)$ is finite-dimensional for all $\mathcal{E}^\bullet, \mathcal{F}^\bullet \in \mathrm{D^b}(X)$ as soon as X is projective over a field. This in particular avoids all possible trouble with Serre functors in this category.

Before discussing any of the many geometric derived functors in Section 3.3, let us prove a few results (Propositions 3.10, 3.17, 3.18) on the structure of the derived category of coherent sheaves.

Definition 3.8 *The* support *of a complex $\mathcal{F}^\bullet \in \mathrm{D^b}(X)$ is the union of the supports of all its cohomology sheaves, i.e. it is the closed subset*

$$\mathrm{supp}(\mathcal{F}^\bullet) := \bigcup \mathrm{supp}(\mathcal{H}^i(\mathcal{F}^\bullet)).$$

Lemma 3.9 *Suppose $\mathcal{F}^\bullet \in \mathrm{D^b}(X)$ and $\mathrm{supp}(\mathcal{F}^\bullet) = Z_1 \sqcup Z_2$, where $Z_1, Z_2 \subset X$ are disjoint closed subsets. Then $\mathcal{F}^\bullet \simeq \mathcal{F}_1^\bullet \oplus \mathcal{F}_2^\bullet$ with $\mathrm{supp}(\mathcal{F}_j^\bullet) \subset Z_j$ for $j = 1, 2$.*

Proof One way to see this is by induction on the length of the complex. The assertion is clear for complexes of length zero, i.e. for shifts of arbitrary coherent sheaves, and will be proved in general by induction.

Let $\mathcal{F}^\bullet$ be a complex of length at least two. Suppose m is minimal with $0 \neq \mathcal{H}^m(\mathcal{F}^\bullet) =: \mathcal{H}$. The sheaf $\mathcal{H}$ may be decomposed as $\mathcal{H} \simeq \mathcal{H}_1 \oplus \mathcal{H}_2$ with $\mathrm{supp}(\mathcal{H}_j) \subset Z_j$. Consider the natural morphism $\mathcal{H}[-m] \longrightarrow \mathcal{F}^\bullet$ inducing the identity on the m-th cohomology and choose a distinguished triangle (see Exercise 2.33)

$$\mathcal{H}[-m] \longrightarrow \mathcal{F}^\bullet \longrightarrow \mathcal{G}^\bullet \longrightarrow \mathcal{H}[1-m].$$

The long exact cohomology sequence shows that $\mathcal{H}^q(\mathcal{G}^\bullet) = \mathcal{H}^q(\mathcal{F}^\bullet)$ for $q > m$ and $\mathcal{H}^q(\mathcal{G}^\bullet) = 0$ for $q \leq m$. Thus, the induction hypothesis applies to $\mathcal{G}^\bullet$ and we may write $\mathcal{G}^\bullet = \mathcal{G}_1^\bullet \oplus \mathcal{G}_2^\bullet$ with $\mathrm{supp}(\mathcal{H}^q(\mathcal{G}_j^\bullet)) \subset Z_j$ for all q. Next, use the spectral sequence

$$E_2^{p,q} = \mathrm{Hom}(\mathcal{H}^{-q}(\mathcal{G}_1^\bullet), \mathcal{H}_2[p]) \Rightarrow \mathrm{Hom}(\mathcal{G}_1^\bullet, \mathcal{H}_2[p+q]).$$

to prove $\mathrm{Hom}(\mathcal{G}_1^\bullet, \mathcal{H}_2[1-m]) = 0$. Indeed, $\mathcal{H}^{-q}(\mathcal{G}_1^\bullet)$ and $\mathcal{H}_2$ are coherent sheaves with disjoint support and, hence, all Ext-groups between them are trivial. (This is quite clear for Ext^0 and Ext^1. The general case can be verified by either using an injective or projective resolution of $\mathcal{H}_2$, respectively $\mathcal{H}^{-q}(\mathcal{G}_1^\bullet)$ or by using the local-to-global spectral sequence

$$E_2^{p,q} = H^p(X, \mathcal{E}xt^q(\ ,\ )) \Rightarrow \mathrm{Ext}^{p+q}(\ ,\ ),$$

that reduces the problem to a local statement, which then is obvious.)

Similarly, one finds $\mathrm{Hom}(\mathcal{G}_2^\bullet, \mathcal{H}_1[1-m]) = 0$. This proves that $\mathcal{F}^\bullet \simeq \mathcal{F}_1^\bullet \oplus \mathcal{F}_2^\bullet$, where the $\mathcal{F}_i^\bullet$ are chosen to complete $\mathcal{G}_j^\bullet \longrightarrow \mathcal{H}_j[1-m]$ to a distinguished triangle

$$\mathcal{F}_j^\bullet \longrightarrow \mathcal{G}_j^\bullet \longrightarrow \mathcal{H}_j[1-m] \longrightarrow \mathcal{F}_j^\bullet[1].$$

In particular, $\operatorname{supp}(\mathcal{F}_j^\bullet) \subset Z_j$. Once a direct sum decomposition is established, one necessarily has $\operatorname{supp}(\mathcal{F}_j^\bullet) = Z_j$. □

Proposition 3.10 *Let X be a noetherian scheme and let $\mathrm{D^b}(X)$ be its bounded derived category of coherent sheaves. Then $\mathrm{D^b}(X)$ is an indecomposable triangulated category if and only if X is connected. See [18].*

Proof Definition 1.52 explains what is meant by decomposing a triangulated category.

If X is not connected, e.g. $X = X_1 \sqcup X_2$, then we let $\mathcal{D}_1 := \mathrm{D^b}(X_1)$ and $\mathcal{D}_2 := \mathrm{D^b}(X_2)$. Applying the lemma allows us to write any $\mathcal{F}^\bullet$ as

$$\mathcal{F}^\bullet \simeq \mathcal{F}_1^\bullet \oplus \mathcal{F}_2^\bullet$$

with $\mathcal{F}_j^\bullet \in \mathrm{D^b}(X_j)$. Thus, ii) in Definition 1.52 is clear and i) and iii) are rather obvious.

Suppose now that X is connected and that $\mathrm{D^b}(X)$ is decomposed by $\mathcal{D}_1, \mathcal{D}_2 \subset \mathrm{D^b}(X)$. We shall derive a contradiction.

Consider $\mathcal{O}_X$ as an object in $\mathrm{D^b}(X)$ and its decomposition $\mathcal{O}_X = \mathcal{F}_1^\bullet \oplus \mathcal{F}_2^\bullet$ with $\mathcal{F}_j^\bullet \in \mathcal{D}_j$. We may assume that $\mathcal{F}_1^\bullet$ and $\mathcal{F}_2^\bullet$ are actually coherent sheaves, for their cohomology is concentrated in degree zero. Since the direct sum is an $\mathcal{O}_X$-module decomposition, they are ideal sheaves $\mathcal{F}_j^\bullet \simeq \mathcal{I}_{X_j}$ of certain closed subschemes $X_j \subset X$. Moreover, $\mathcal{O}_X = \mathcal{I}_{X_1} + \mathcal{I}_{X_2} \subset \mathcal{I}_{X_1 \cap X_2}$ and $\mathcal{I}_{X_1 \cup X_2} \subset \mathcal{I}_{X_1} \cap \mathcal{I}_{X_2} = 0$. Therefore, $X_1 \cap X_2 = \varnothing$ and $X_1 \cup X_2 = X$. Since X is connected, one of the two subschemes must be empty and hence either $\mathcal{O}_X \in \mathcal{D}_1$ or $\mathcal{O}_X \in \mathcal{D}_2$. Suppose $\mathcal{O}_X \in \mathcal{D}_1$.

If $x \in X$ is a closed point, then the decomposition of $k(x)$ with respect to $\mathcal{D}_1, \mathcal{D}_2 \subset \mathrm{D^b}(X)$ must be trivial. Hence, either $k(x) \in \mathcal{D}_1$ or $k(x) \in \mathcal{D}_2$. The existence of the non-trivial homomorphism $\mathcal{O}_X \longrightarrow k(x)$ excludes the latter. Thus, for any closed point $x \in X$ its structure sheaf $k(x)$ is contained in $\mathcal{D}_1$.

Suppose there exists a non-trivial $\mathcal{F}^\bullet \in \mathcal{D}_2$. Choose m maximal with $\mathcal{H}^m := \mathcal{H}^m(\mathcal{F}^\bullet) \neq 0$ and pick a closed point x in the support of $\mathcal{H}^m$. In particular, there exists a surjection $\mathcal{H}^m \twoheadrightarrow k(x)$, which one uses to construct a non-trivial morphism $\mathcal{F}^\bullet \longrightarrow k(x)[-m]$ in $\mathrm{D^b}(X)$ as follows: Consider the natural quasi-isomorphism

$$(\ldots \longrightarrow \mathcal{F}^{m-1} \longrightarrow \mathrm{Ker}(d^m) \longrightarrow 0 \longrightarrow \ldots) \longrightarrow \mathcal{F}^\bullet$$

(see Exercise 2.31) and compose its inverse (in $\mathrm{D^b}(X)$) with the non-trivial

$$(\ldots \longrightarrow \mathcal{F}^{m-1} \longrightarrow \mathrm{Ker}(d^m) \longrightarrow 0 \ldots) \longrightarrow \mathcal{H}^m[-m] \\ \longrightarrow k(x)[-m]$$

(cf. Exercise 2.32). As there are no non-trivial homomorphisms between objects in $\mathcal{D}_1$ and $\mathcal{D}_2$, this yields the contradiction. □

Let now X be a smooth projective variety over a field k and ω_X its canonical bundle. Often, ω_X is also called the *dualizing sheaf* of X, the reason for which will be explained now.

First note that for any locally free sheaf $\mathcal{M}$ the functor $\mathbf{Coh}(X) \longrightarrow \mathbf{Coh}(X)$ given by $\mathcal{F} \longmapsto \mathcal{F} \otimes \mathcal{M}$ is exact. In particular, it immediately descends to an exact functor of the derived categories $\mathrm{D}^*(X) \longrightarrow \mathrm{D}^*(X)$, for $* = \mathrm{b}, +, -$, which will be denoted $\mathcal{M} \otimes (\ \ )$. Other exact functors, available on any triangulated category, are given by the shift functors $[i] : \mathrm{D}^*(X) \longrightarrow \mathrm{D}^*(X)$ with $i \in \mathbb{Z}$.

Definition 3.11 *Let X be a smooth projective variety of dimension n. Then one defines the exact functor S_X as the composition*

$$\mathrm{D}^*(X) \xrightarrow{\ \omega_X \otimes (\ \ )\ } \mathrm{D}^*(X) \xrightarrow{\ [n]\ } \mathrm{D}^*(X),$$

where $* = \mathrm{b}, +, -$.

In view of the following result, S_X is called the *Serre functor* of X. (More accurately, Serre functors are called Serre functors because they formalize Serre duality.)

Theorem 3.12 **(Serre duality)** *Let X be a smooth projective variety over a field k. Then*

$$S_X : \mathrm{D}^{\mathrm{b}}(X) \longrightarrow \mathrm{D}^{\mathrm{b}}(X)$$

is a Serre functor in the sense of Definition 1.28.

More explicitly, Serre duality says that for any two complexes $\mathcal{E}^\bullet, \mathcal{F}^\bullet \in \mathrm{D}^{\mathrm{b}}(X)$ there exists a functorial isomorphism

$$\eta_{\mathcal{E}^\bullet, \mathcal{F}^\bullet} : \mathrm{Hom}_{\mathrm{D}(\mathcal{A})}(\mathcal{E}^\bullet, \mathcal{F}^\bullet) \xrightarrow{\ \sim\ } \mathrm{Hom}_{\mathrm{D}(\mathcal{A})}(\mathcal{F}^\bullet, \mathcal{E}^\bullet \otimes \omega_X[n])^*.$$

It is more commonly stated as isomorphisms for any $i \in \mathbb{Z}$

$$\mathrm{Ext}^i(\mathcal{E}^\bullet, \mathcal{F}^\bullet) \xrightarrow{\ \sim\ } \mathrm{Ext}^{n-i}(\mathcal{F}^\bullet, \mathcal{E}^\bullet \otimes \omega_X)^*,$$

which is functorial in $\mathcal{E}^\bullet$ and $\mathcal{F}^\bullet$. Use $\mathrm{Ext}^i(\mathcal{E}^\bullet, \mathcal{F}^\bullet) = \mathrm{Hom}_{\mathrm{D}(\mathcal{A})}(\mathcal{E}^\bullet, \mathcal{F}^\bullet[i])$ to confirm the equivalence of the two versions.

Proof One way to prove this is by using the standard Serre duality in the form $\mathrm{Ext}^i(\mathcal{F}, \omega_X) \simeq H^{n-i}(X, \mathcal{F})^*$ for a coherent sheaf $\mathcal{F}$ (cf. [45, II.7]). This is indeed a special case of the above assertion as $H^{n-i}(X, \mathcal{F}) = \mathrm{Ext}^{n-i}(\mathcal{O}_X, \mathcal{F})$.

Serre duality can also be seen as a particular case of the Grothendieck–Verdier duality (see Section 3.4), the proof of which is given in [44]. □

Serre duality for coherent sheaves as stated yields a quick proof for

Proposition 3.13 *Suppose $\mathcal{F}$ and $\mathcal{G}$ are coherent sheaves on a smooth projective variety X of dimension n. Then*

$$\mathrm{Ext}^i(\mathcal{F},\mathcal{G}) = 0 \quad \text{for} \quad i > n.$$

In other words, the homological dimension of $\mathbf{Coh}(X)$ *is n.*

Proof Simply note that $\mathrm{Ext}^i(\mathcal{F},\mathcal{G}) \simeq \mathrm{Ext}^{i-n}(\mathcal{G},\mathcal{F}\otimes\omega_X)^* = 0$ for negative $i-n$.

The homological dimension of $\mathbf{Coh}(X)$ can indeed not be smaller than n, for $\mathrm{Ext}^n(\mathcal{O}_X,\omega_X) \simeq \mathrm{Hom}(\mathcal{O}_X,\mathcal{O}_X)^* \neq 0$. □

Corollary 3.14 *Suppose X is a smooth projective variety. Then for any two complexes $\mathcal{E}^\bullet, \mathcal{F}^\bullet \in \mathrm{D}^{\mathrm{b}}(X)$ one has $R\mathrm{Hom}^\bullet(\mathcal{E}^\bullet,\mathcal{F}^\bullet) \in \mathrm{D}^{\mathrm{b}}(\mathbf{Ab})$.*

Proof Use Corollary 2.68, ii) or rather the spectral sequences (3.1) and (3.2). □

The following folklore result describes objects in the derived category of a smooth curve. It is another particularly nice consequence of the proposition.

Corollary 3.15 *Let C be a smooth projective curve. Then any object in $\mathrm{D}^{\mathrm{b}}(C)$ is isomorphic to a direct sum $\bigoplus \mathcal{E}_i[i]$, where the $\mathcal{E}_i$ are coherent sheaves on C.*

Proof We proceed by induction over the length of the complex. Suppose $\mathcal{E}^\bullet$ is a complex of length k with $\mathcal{H}^i(\mathcal{E}^\bullet) = 0$ for $i < i_0$. Then use Exercise 2.32 to find a distinguished triangle of the form

$$\mathcal{H}^{i_0}(\mathcal{E}^\bullet)[-i_0] \longrightarrow \mathcal{E}^\bullet \longrightarrow \mathcal{E}_1^\bullet \longrightarrow \mathcal{H}^{i_0}(\mathcal{E}^\bullet)[1-i_0]$$

with $\mathcal{E}_1^\bullet$ of length $k-1$ and with $\mathcal{H}^i(\mathcal{E}_1^\bullet) = 0$ for $i \le i_0$.

If this distinguished triangle splits, then $\mathcal{E}^\bullet \simeq \mathcal{E}_1^\bullet \oplus \mathcal{H}^{i_0}(\mathcal{E}^\bullet)[-i_0]$ and the induction hypothesis for $\mathcal{E}_1^\bullet$ allows us to conclude.

Thus it suffices to prove the vanishing $\mathrm{Hom}(\mathcal{E}_1^\bullet, \mathcal{H}^{i_0}(\mathcal{E}^\bullet)[1-i_0]) = 0$ (see Exercise 1.38). Write $\mathcal{E}_1^\bullet \simeq \bigoplus_{i>i_0} \mathcal{H}^i(\mathcal{E}_1^\bullet)[-i]$. Then

$$\mathrm{Hom}(\mathcal{E}_1^\bullet, \mathcal{H}^{i_0}(\mathcal{E}^\bullet)[1-i_0]) \simeq \bigoplus_{i>i_0} \mathrm{Ext}^{1+i-i_0}(\mathcal{H}^i(\mathcal{E}_1^\bullet), \mathcal{H}^{i_0}(\mathcal{E}^\bullet)) = 0,$$

as the homological dimension of a curve is one. □

Exercise 3.16 Convince yourself that the proof applies more generally to any abelian category of homological dimension ≤ 1.

Also in the the next section Serre duality will serve as an important technical tool. E.g. it will allow us to pass from cohomology groups like $\mathrm{Hom}(k(x),\ )$ to those of the form $\mathrm{Hom}(\ , k(x))$, which are easier to handle.

3.2 Spanning classes in the derived category

The two propositions in this section describe the two most common spanning classes in $\mathrm{D}^{\mathrm{b}}(X)$. Other more specific ones will be encountered in later chapters.

Proposition 3.17 *Let X be a smooth projective variety. Then the objects of the form $k(x)$ with $x \in X$ a closed point span the derived category $\mathrm{D}^{\mathrm{b}}(X)$.*

Proof See Definition 1.47 for the notion of a spanning class.

In order to prove the assertion, it suffices to show that for any non-trivial $\mathcal{F}^\bullet \in \mathrm{D}^{\mathrm{b}}(X)$ there exist closed points $x_1, x_2 \in X$ and integers i_1, i_2 such that

$$\mathrm{Hom}(\mathcal{F}^\bullet, k(x_1)[i_1]) \neq 0 \quad \text{and} \quad \mathrm{Hom}(k(x_2), \mathcal{F}^\bullet[i_2]) \neq 0.$$

Applying Serre duality $\mathrm{Hom}(k(x), \mathcal{F}^\bullet[i_2]) \simeq \mathrm{Hom}(\mathcal{F}^\bullet, k(x)[\dim(X) - i_2])^*$, we find that it is enough to ensure the existence of i_1 and x_1.

This is now proven in complete analogy to the arguments in the proof of Proposition 3.10, but this time, for a change, we use the spectral sequence (3.2)

$$E_2^{p,q} := \mathrm{Hom}(\mathcal{H}^{-q}, k(x)[p]) \Rightarrow \mathrm{Hom}(\mathcal{F}^\bullet, k(x)[p+q]),$$

where $\mathcal{H}^q := \mathcal{H}^q(\mathcal{F}^\bullet)$.

Since $\mathcal{F}^\bullet$ is non-trivial, there exists a maximal m such that $\mathcal{H}^m \neq 0$. With this choice of m all differentials with source $E_r^{0,-m}$ are trivial. As negative Ext-groups between coherent sheaves are always trivial, one has $E_2^{p,q} = 0$ for $p < 0$ and hence all differentials with target $E_r^{0,-m}$ are also trivial. Thus, $E_\infty^{0,-m} = E_2^{0,-m}$.

If we now choose a point x in the support of $\mathcal{H}^m$, then

$$E_\infty^{0,-m} = E_2^{0,-m} = \mathrm{Hom}(\mathcal{H}^m, k(x)) \neq 0$$

and hence $\mathrm{Hom}(\mathcal{F}^\bullet, k(x)[-m]) \neq 0$. □

There is another choice for a spanning class of the derived category of coherent sheaves on a projective variety provided by the powers of an ample line bundle. For the definition of an ample sequence see Definition 2.72. The geometric realization of this concept is provided by the following

Proposition 3.18 *Let X be a projective variety over a field. If L is an ample line bundle on X, then the powers L^i, $i \in \mathbb{Z}$, form an ample sequence in the abelian category* **Coh**(X). *See [15].*

Proof By definition (cf. [45, II.7]) an ample line bundle L has the property that for any coherent sheaf $\mathcal{F}$ there exists an n_0 such that for any $n \geq n_0$ the sheaf $\mathcal{F} \otimes L^n$ is globally generated. This means that

$$H^0(X, \mathcal{F} \otimes L^n) \otimes \mathcal{O}_X \twoheadrightarrow \mathcal{F} \otimes L^n$$

is surjective. Tensoring with L^{-n} and writing $H^0(X, \mathcal{F} \otimes L^n) = \mathrm{Hom}(L^{-n}, \mathcal{F})$ shows that the canonical map

$$\mathrm{Hom}(L^i, \mathcal{F}) \otimes L^i \twoheadrightarrow \mathcal{F}$$

is surjective for $i < i_0 := -n_0$. Therefore, condition i) in Definition 2.72 is satisfied.

To see ii), one invokes one of the fundamental theorems of Serre (cf. [45, III, 5.2]) saying that $H^i(X, \mathcal{F} \otimes L^n) = 0$ for any $i > 0$ and $n > n_0$ (the latter depending on $\mathcal{F}$), where L is an ample line bundle on a projective scheme over a noetherian ring. This proves ii) right away.

Let $\mathcal{F}$ be a coherent sheaf on X. Then $\mathrm{Hom}(\mathcal{F}, L^i)$ is finite-dimensional. We may suppose that L is very ample. Indeed, if not then pass to a very ample power L^k and prove the statement for the finite number of sheaves $\mathcal{F} \otimes L^i$, $i = 0, \ldots, k-1$.

So, we might assume that for any closed point $x \in X$ there exists a section $0 \neq s_x \in H^0(X, L)$ with $0 = s_x(x) \in L(x)$. If $0 \neq \varphi \in \mathrm{Hom}(\mathcal{F}, L^i)$, then there exists a closed point $x \in X$ with $\varphi(x) : \mathcal{F}(x) \twoheadrightarrow L^i(x)$ non-trivial. Hence, φ is not in the image of the inclusion $\mathrm{Hom}(\mathcal{F}, L^{i-1}) \xrightarrow{s_x} \mathrm{Hom}(\mathcal{F}, L^i)$ induced by applying $\mathrm{Hom}(\mathcal{F}, \ )$ to the short exact sequence

$$0 \longrightarrow L^{i-1} \xrightarrow{\cdot s_x} L^i \longrightarrow L^i|_{Z(s_x)} \longrightarrow 0.$$

As the spaces are all finite-dimensional, this only happens a finite number of times before they become trivial, i.e. $\mathrm{Hom}(\mathcal{F}, L^i) = 0$ for $i \ll 0$. This proves iii).

(The proof of the last part simplifies, if one restricts to smooth projective varieties over a field. Then we may use Serre duality and Serre vanishing to see that

$$\mathrm{Hom}(\mathcal{F}, L^i) \simeq H^{\dim(X)}(X, \mathcal{F} \otimes L^{-i} \otimes \omega_X)^* = 0$$

for $i \ll 0$.) □

Corollary 3.19 *If X is a smooth projective variety and L is an ample line bundle on X, then the powers L^i, $i \in \mathbb{Z}$, form a spanning class in the derived category* $\mathrm{D^b}(X)$.

Proof This is an immediate consequence of Propositions 2.73 and 3.18. We have only to recall that $\mathbf{Coh}(X)$ has finite homological dimension, whenever X is smooth (see Proposition 3.13).

Proposition 3.17 covers the case of zero-dimensional X, i.e. when X is a point. □

At this point, the reader could directly pass to the next chapter, where the important results of Bondal and Orlov on the classification of derived categories of varieties with ample (anti-)canonical bundle are presented. The material of Sections 3.3 and 3.4 on derived functors on $\mathrm{D^b}(X)$ will only be needed from Chapter 5 on (with one minor exception in the proof of Proposition 4.9).

3.3 Derived functors in algebraic geometry

In the following we shall discuss all derived functors that will be needed in the sequel. All assumptions will be carefully stated, but for certain details, we have to refer to the literature.

Cohomology Let X be a noetherian scheme over a field k. The global section functor

$$\Gamma : \mathbf{Qcoh}(X) \longrightarrow \mathbf{Vec}(k), \quad \mathcal{F} \longmapsto \Gamma(X, \mathcal{F})$$

is a left exact functor. Since $\mathbf{Qcoh}(X)$ has enough injectives (cf. Proposition 3.3), its right derived functor

$$R\Gamma : \mathrm{D}^+(\mathbf{Qcoh}(X)) \longrightarrow \mathrm{D}^+(\mathbf{Vec}(k))$$

exists. The higher derived functors are denoted

$$H^i(X, \mathcal{F}^\bullet) := R^i\Gamma(\mathcal{F}^\bullet).$$

For a sheaf $\mathcal{F}$ these are just the cohomology groups $H^i(X, \mathcal{F})$, $i = 0, 1, \ldots$ and for an honest complex $\mathcal{F}^\bullet$ they are sometimes called the *hypercohomology groups* $\mathbb{H}^i(X, \mathcal{F}^\bullet)$.

Since every complex of vector spaces splits, one has in fact

$$R\Gamma(\mathcal{F}^\bullet) \simeq \bigoplus H^i(X, \mathcal{F}^\bullet)[-i]$$

in $\mathrm{D}^+(\mathbf{Vec}(k))$.

The following is a special case of a general result for sheaves of abelian groups on noetherian topological spaces. Another generalization is provided by Theorem 3.22.

Theorem 3.20 **(Grothendieck)** *For any quasi-coherent sheaf $\mathcal{F}$ on a noetherian scheme X one has $H^i(X, \mathcal{F}) = 0$ for $i > \dim(X)$. See [45, III, 2.7].*

Together with Corollary 2.68 it shows that the functor $R\Gamma$ induces an exact functor

$$R\Gamma : \mathrm{D}^{\mathrm{b}}(\mathbf{Qcoh}(X)) \longrightarrow \mathrm{D}^{\mathrm{b}}(\mathbf{Vec}(k)).$$

Let us next restrict the functor of global sections to the full subcategory of coherent sheaves $\Gamma : \mathbf{Coh}(X) \longrightarrow \mathbf{Vec}(k)$.

Theorem 3.21 **(Serre)** *If $\mathcal{F}$ is a coherent sheaf on a projective scheme X over a field k, then all cohomology groups $H^i(X, \mathcal{F})$ are of finite dimension. See [45, III, 5.2].*

(In fact, it suffices that X is proper (cf. [43, III, 3.2.1]).) In particular, the functor of global sections yields the left exact functor $\Gamma : \mathbf{Coh}(X) \longrightarrow \mathbf{Vec}_f(k)$.

However, its derived functor cannot be constructed directly, as $\mathbf{Coh}(X)$ does not, in general, contain enough injectives. However, due to the theorem the right derived functor

$$\mathrm{D}^{\mathrm{b}}(X) \longrightarrow \mathrm{D}^{\mathrm{b}}(\mathbf{Vec}_f(k))$$

can be obtained as the composition of exact functors

$$\mathrm{D}^{\mathrm{b}}(X) \longrightarrow \mathrm{D}^{\mathrm{b}}(\mathbf{Qcoh}(X)) \longrightarrow \mathrm{D}^{\mathrm{b}}(\mathbf{Vec}(k)).$$

Here one uses Corollary 2.68 which works despite the fact that $\mathbf{Coh}(X)$ does not contain enough injective objects (cf. Remark 2.69, ii)). The existence of the spectral sequence needed in its proof can be ensured by viewing any coherent sheaf as a quasi-coherent one.

Note that under our assumption that X is noetherian, $\mathrm{D}^{\mathrm{b}}(X)$ is equivalent to the full subcategory of bounded complexes of quasi-coherent sheaves with coherent cohomology (cf. Proposition 3.5). We summarize the above discussion by the following diagram

$$\begin{array}{ccc}
\mathrm{D}^+(\mathbf{Qcoh}(X)) & \xrightarrow{\;R\Gamma\;} & \mathrm{D}^+(\mathbf{Vec}(k)) \\
\uparrow & & \uparrow \\
\mathrm{D}^{\mathrm{b}}(\mathbf{Qcoh}(X)) & \xrightarrow{\;\text{Thm. 3.20}\;} & \mathrm{D}^{\mathrm{b}}(\mathbf{Vec}(k)) \\
\uparrow & & \uparrow \\
\mathrm{D}^{\mathrm{b}}(X) & \xrightarrow[X \text{ proper}]{\;\text{Thm. 3.21}\;} & \mathrm{D}^{\mathrm{b}}(\mathbf{Vec}_f(k)).
\end{array}$$

Direct image Let $f : X \longrightarrow Y$ be a morphism of noetherian schemes. The direct image is a left exact functor

$$f_* : \mathbf{Qcoh}(X) \longrightarrow \mathbf{Qcoh}(Y).$$

Again, we use that $\mathbf{Qcoh}(X)$ has enough injectives in order to define the right derived functor

$$Rf_* : \mathrm{D}^+(\mathbf{Qcoh}(X)) \longrightarrow \mathrm{D}^+(\mathbf{Qcoh}(X)).$$

The *higher direct images* $R^i f_*(\mathcal{F}^\bullet)$ of a complex of sheaves $\mathcal{F}^\bullet$ are, by definition, the cohomology sheaves $\mathcal{H}^i(Rf_*(\mathcal{F}^\bullet))$ of $Rf_*(\mathcal{F}^\bullet)$. In particular, to any quasi-coherent sheaf $\mathcal{F}$ on X one associates the quasi-coherent sheaves $R^i f_* \mathcal{F}$ on Y.

If X is a noetherian scheme over a field k, then the global section functor $\Gamma : \mathbf{Qcoh}(X) \longrightarrow \mathbf{Vec}(k)$ is a special case of the direct image. Indeed, the direct image $f_*\mathcal{F}$ of a sheaf $\mathcal{F}$ under the structure morphism $f : X \longrightarrow \mathrm{Spec}(k)$ is nothing but Γ. So, in this case $R^i f_*(\mathcal{F}^\bullet) = H^i(X, \mathcal{F}^\bullet)$.

From this point of view the following result naturally generalizes Theorem 3.20. In fact, using [45, III, 8.1] it can be easily deduced from it.

Theorem 3.22 *For a quasi-coherent sheaf $\mathcal{F}$ on X and a morphism $f : X \longrightarrow Y$ of noetherian schemes the higher direct images $R^i f_*\mathcal{F}$ are trivial for $i > \dim(X)$.*

Thus, the higher direct image functor induces an exact functor (cf. Corollary 2.68)

$$Rf_* : \mathrm{D^b}(\mathbf{Qcoh}(X)) \longrightarrow \mathrm{D^b}(\mathbf{Qcoh}(Y)).$$

To stay in the coherent world, one has to use the following coherence criterion which generalizes Theorem 3.21.

Theorem 3.23 *If $f : X \longrightarrow Y$ is a projective (or proper) morphism of noetherian schemes, then the higher direct images $R^i f_*(\mathcal{F})$ of a coherent sheaf $\mathcal{F}$ on X are again coherent. See [45, III, 8.8] or [43, III, 3.2.1].*

Thus, using Proposition 3.5 once more we obtain for any proper morphism $f : X \longrightarrow Y$ of noetherian schemes the right derived functor

$$Rf_* : \mathrm{D^b}(X) \longrightarrow \mathrm{D^b}(Y)$$

as the composition of $\mathrm{D^b}(X) \longrightarrow \mathrm{D^b}(\mathbf{Qcoh}(X))$ and the derived direct image for quasi-coherent sheaves $Rf_* : \mathrm{D^b}(\mathbf{Qcoh}(X)) \longrightarrow \mathrm{D^b}(\mathbf{Qcoh}(Y))$.

Thus, summarizing the discussion by a diagram similar to the one given above for the global section functor we have

$$\begin{array}{ccc}
\mathrm{D^+}(\mathbf{Qcoh}(X)) & \xrightarrow{\quad Rf_* \quad} & \mathrm{D^+}(\mathbf{Qcoh}(Y)) \\
\uparrow & & \uparrow \\
\mathrm{D^b}(\mathbf{Qcoh}(X)) & \xrightarrow{\text{Thm. 3.22}} & \mathrm{D^b}(\mathbf{Qcoh}(Y)) \\
\uparrow & & \uparrow \\
\mathrm{D^b}(X) & \xrightarrow[f\ \text{proper}]{\text{Thm. 3.23}} & \mathrm{D^b}(Y).
\end{array}$$

Sometimes, another f_*-adapted class is useful. Recall that a sheaf $\mathcal{F}$ is *flabby* if for any open subset $U \subset X$ the restriction map $\mathcal{F}(X) \longrightarrow \mathcal{F}(U)$ is surjective.

The following lemma collects the standard facts about flabby sheaves. Together with Proposition 3.3, it immediately shows that the class of all flabby sheaves is f_*-adapted.

Lemma 3.24 *Any injective $\mathcal{O}_X$-sheaf is flabby. Any flabby sheaf $\mathcal{F}$ on X is f_*-acyclic for any morphism $f : X \longrightarrow Y$, i.e. $R^i f_*(\mathcal{F}) = 0$ for $i > 0$, and, moreover, $f_*\mathcal{F}$ is again flabby.*

Proof For the first assertion see [45, III.2].

The second one uses that $R^i f_*(\mathcal{F})$ is the sheaf associated to the presheaf $U \longmapsto H^i(f^{-1}(U), \mathcal{F}|_{f^{-1}(U)})$ (see [45, III.8]) and that a flabby sheaf is acyclic in the sense that the higher cohomology is trivial (see [45, III.2]). □

For a composition

$$X \xrightarrow{\ f\ } Y \xrightarrow{\ g\ } Z$$

of two morphisms one knows that $(g \circ f)_* = g_* \circ f_*$, from which we want to conclude

$$R(g \circ f)_* \simeq Rg_* \circ Rf_* : \mathrm{D^b}(\mathbf{Qcoh}(X)) \longrightarrow \mathrm{D^b}(\mathbf{Qcoh}(Z)).$$

In order to apply Proposition 2.58, we have to ensure the existence of an f_*-adapted class $\mathcal{I} \subset \mathbf{Qcoh}(X)$ such that $f_*(\mathcal{I})$ is contained in a g_*-adapted class.

We let $\mathcal{I}$ be the class of injective sheaves. Then $\mathcal{I}$ is f_* adapted, for $\mathbf{Qcoh}(X)$ has enough injectives (see Proposition 3.3). As the direct image of a flabby sheaf is again flabby, $f_*(\mathcal{I})$ is contained in the g_*-adapted class of all flabby sheaves.

Applying Proposition 2.66 leads to the *Leray spectral sequence*:

$$E_2^{p,q} = R^p g_*(R^q f_*(\mathcal{F}^\bullet)) \Rightarrow R^{p+q}(g \circ f)_*(\mathcal{F}^\bullet). \tag{3.3}$$

Any morphism $f : X \longrightarrow Y$ of noetherian schemes over a field k may be composed with the structure morphism $Y \longrightarrow \mathrm{Spec}(k)$. This yields

$$R\Gamma(Y,\ \ ) \circ Rf_* \simeq R\Gamma(X,\ \ )$$

and (3.3) becomes

$$E_2^{p,q} = H^p(Y, R^q f_*(\mathcal{F}^\bullet)) \Rightarrow H^{p+q}(X, \mathcal{F}^\bullet).$$

Another interesting special case of the Leray spectral sequence is deduced by considering the case that f is the identity. Then

$$E_2^{p,q} = R^p g_* \mathcal{H}^q(\mathcal{F}^\bullet) \Rightarrow R^{p+q} g_* \mathcal{F}^\bullet \tag{3.4}$$

and, even more special, for $g : X = Y \longrightarrow \mathrm{Spec}(k)$ one obtains

$$E_2^{p,q} = H^p(X, \mathcal{H}^q(\mathcal{F}^\bullet)) \Rightarrow H^{p+q}(X, \mathcal{F}^\bullet). \tag{3.5}$$

Local Homs Let $\mathcal{F} \in \mathbf{Qcoh}(X)$. Then

$$\mathcal{H}om(\mathcal{F}, \ \ ) : \mathbf{Qcoh}(X) \longrightarrow \mathbf{Qcoh}(X)$$

is a left exact functor.

Recall that $\mathcal{H}om(\mathcal{F}, \mathcal{E})$ is the sheaf

$$U \longmapsto \mathrm{Hom}(\mathcal{F}|_U, \mathcal{E}|_U)$$

of $\mathcal{O}_X$-modules (no sheafification needed here!) and that for $\mathcal{F}$ and $\mathcal{E}$ (quasi-)coherent the sheaf $\mathcal{H}om(\mathcal{F}, \mathcal{E})$ is also (quasi-)coherent (see [45, II.5]).

Due to the existence of enough injectives in $\mathbf{Qcoh}(X)$ (we always assume that X is noetherian), the derived functor

$$R\mathcal{H}om(\mathcal{F}, \ \ ) : \mathrm{D}^+(\mathbf{Qcoh}(X)) \longrightarrow \mathrm{D}^+(\mathbf{Qcoh}(X)) \tag{3.6}$$

exists. By definition

$$\mathcal{E}xt^i(\mathcal{F}, \mathcal{E}) := R^i\mathcal{H}om(\mathcal{F}, \mathcal{E})$$

for any quasi-coherent sheaves $\mathcal{E}$ and $\mathcal{F}$. At least for coherent sheaves $\mathcal{F}$, the definition is local in the sense that the stalk can be described by (cf. [45, III.6])

$$\mathcal{E}xt^i(\mathcal{F}, \mathcal{E})_x \simeq \mathrm{Ext}^i_{\mathcal{O}_{X,x}}(\mathcal{F}_x, \mathcal{E}_x).$$

This is essentially due to the commutativity of the diagram

$$\begin{array}{ccc} \mathbf{Qcoh}(X) & \xrightarrow{\mathcal{H}om(\mathcal{F}, \ \ )} & \mathbf{Qcoh}(X) \\ \downarrow & & \downarrow \\ \mathbf{Mod}(\mathcal{O}_{X,x}) & \xrightarrow[\mathrm{Hom}(\mathcal{F}_x, \ \ )]{} & \mathbf{Mod}(\mathcal{O}_{X,x}) \end{array}$$

for any locally free sheaf $\mathcal{F}$. In particular, $\mathcal{E}xt^i(\mathcal{E}, \mathcal{F})$ are coherent if $\mathcal{E}$ and $\mathcal{F}$ are so.

Restricting (3.6) to coherent sheaves yields the functor

$$R\mathcal{H}om(\mathcal{F}, \ \ ) : \mathrm{D}^+(X) \longrightarrow \mathrm{D}^+(X).$$

Recall that for a non-regular local ring A the $\mathrm{Ext}^i_A(M, \ \ )$ might be non-trivial even for $i \gg 0$. Thus, only for a regular scheme X do we obtain a functor between the bounded derived categories (cf. Proposition 3.26 below). Thus,

$$
\begin{array}{ccc}
\mathrm{D}^+(\mathbf{Qcoh}(X)) & \xrightarrow{R\mathcal{H}om(\mathcal{F},\ )} & \mathrm{D}^+(\mathbf{Qcoh}(X)) \\
\uparrow & & \uparrow \\
\mathrm{D}^+(X) & \xrightarrow{\mathcal{F}\ \text{coherent}} & \mathrm{D}^+(Y) \\
\uparrow & & \uparrow \\
\mathrm{D}^{\mathrm{b}}(X) & \xrightarrow{X\ \text{regular}} & \mathrm{D}^{\mathrm{b}}(X).
\end{array}
$$

The construction generalizes to complexes $\mathcal{F}^\bullet \in \mathrm{D}^-(X)$. One first defines for a complex $\mathcal{F}^\bullet$, that is bounded above, the exact functor

$$\mathcal{H}om^\bullet(\mathcal{F}^\bullet,\ ) : \mathrm{K}^+(\mathbf{Qcoh}(X)) \longrightarrow \mathrm{K}^+(\mathbf{Qcoh}(X)),$$

$$\mathcal{H}om^i(\mathcal{F}^\bullet, \mathcal{E}^\bullet) := \prod \mathcal{H}om(\mathcal{F}^p, \mathcal{E}^{i+p}) \ \text{ with } \ d = d_{\mathcal{E}} - (-1)^i d_{\mathcal{F}}.$$

In the sequel, we shall need the following fact which is readily deduced from the corresponding local statements for modules over a ring (cf. Remark 2.57, i)).

Lemma 3.25 *Let $\mathcal{E}^\bullet$ be a complex of injective sheaves. If $\mathcal{F}^\bullet$ or $\mathcal{E}^\bullet$ is acyclic, then $\mathcal{H}om^\bullet(\mathcal{F}^\bullet, \mathcal{E}^\bullet)$ is acyclic. See [44, II.3].* □

The assertion for acyclic $\mathcal{E}^\bullet$ shows that the class of complexes of injective sheaves is adapted for the functor $\mathcal{H}om^\bullet(\mathcal{F}^\bullet,\ )$. Hence, the right derived functor exists (see Remark 2.51):

$$R\mathcal{H}om(\mathcal{F}^\bullet,\ ) : \mathrm{D}^+(\mathbf{Qcoh}(X)) \longrightarrow \mathrm{D}^+(\mathbf{Qcoh}(X)).$$

In order to see that the functor descends to the derived category in the left argument, one applies the lemma for $\mathcal{F}^\bullet$ acyclic. (We may always assume that $\mathcal{E}^\bullet$ is a complex of injectives.) Then Lemma 2.44 shows that we have a bifunctor:

$$R\mathcal{H}om(\ ,\ ) : \mathrm{D}^-(\mathbf{Qcoh}(X))^{\mathrm{op}} \times \mathrm{D}^+(\mathbf{Qcoh}(X)) \longrightarrow \mathrm{D}^+(\mathbf{Qcoh}(X)),$$

which we use to introduce

$$\mathcal{E}xt^i(\mathcal{F}^\bullet, \mathcal{E}^\bullet) := R^i\mathcal{H}om(\mathcal{E}^\bullet, \mathcal{F}^\bullet).$$

As has been mentioned earlier, there are not many projective objects in the category of coherent sheaves. However, for the purpose of computing local Exts locally free sheaves are enough. More precisely, if $\mathcal{F}^\bullet$ is a complex of locally free sheaves $\mathcal{F}^i$, then $R\mathcal{H}om(\mathcal{F}^\bullet,\ )$ can be computed as $\mathcal{H}om(\mathcal{F}^\bullet,\ )$. This can be deduced from the local statement that says that for any complex $M^\bullet$ of free modules over a local ring A the cohomology $R\mathrm{Hom}(M^\bullet, \ )$ can be computed as $\mathrm{Hom}(M^\bullet, \ )$ (free modules are projective!).

Again, for regular schemes one can work with the bounded categories due to the following

Proposition 3.26 *If X is regular, then any $\mathcal{F}^\bullet \in \mathrm{D^b}(X)$ is isomorphic to a bounded complex $\mathcal{G}^\bullet \in \mathrm{D^b}(X)$ of locally free sheaves $\mathcal{G}^i$.*

The key ingredient for the proof is the fundamental fact that on regular schemes any coherent sheaf $\mathcal{G}$ admits a locally free resolution of finite length n. In fact, one can always assume $n \leq \dim(X)$.

Exercise 3.27 Try to imitate the sketch of the proof of Proposition 2.35 and prove the assertion of the above proposition.

Thus, the situation may be summarized by the diagram

$$\begin{array}{ccc}
\mathrm{D}^-(\mathbf{Qcoh}(X))^{\mathrm{op}} \times \mathrm{D}^+(\mathbf{Qcoh}(X)) & \xrightarrow{\ R\mathcal{H}om\ } & \mathrm{D}^+(\mathbf{Qcoh}(X)) \\
\uparrow & & \uparrow \\
\mathrm{D}^-(X)^{\mathrm{op}} \times \mathrm{D}^+(X) & \longrightarrow & \mathrm{D}^+(X) \\
\uparrow & & \uparrow \\
\mathrm{D^b}(X)^{\mathrm{op}} \times \mathrm{D^b}(X) & \xrightarrow{\ X \text{ regular}\ } & \mathrm{D^b}(X).
\end{array}$$

Similar to Example 2.70 one shows the existence of the following spectral sequences

$$E_2^{p,q} = \mathcal{E}xt^p(\mathcal{F}^\bullet, \mathcal{H}^q(\mathcal{E}^\bullet)) \Rightarrow \mathcal{E}xt^{p+q}(\mathcal{F}^\bullet, \mathcal{E}^\bullet) \tag{3.7}$$

$$E_2^{p,q} = \mathcal{E}xt^p(\mathcal{H}^{-q}(\mathcal{F}^\bullet), \mathcal{E}^\bullet) \Rightarrow \mathcal{E}xt^{p+q}(\mathcal{F}^\bullet, \mathcal{E}^\bullet). \tag{3.8}$$

Trace map The last observations allow us to define the *trace map*

$$\mathrm{tr}_{\mathcal{E}^\bullet} : R\mathcal{H}om(\mathcal{E}^\bullet, \mathcal{E}^\bullet) \longrightarrow \mathcal{O}_X$$

for any $\mathcal{E}^\bullet \in \mathrm{D^b}(X)$.

For simplicity we will assume that X is regular and replace $\mathcal{E}^\bullet$ by a bounded complex of locally free sheaves. Then $R\mathcal{H}om(\mathcal{E}^\bullet, \mathcal{E}^\bullet) \simeq \mathcal{H}om^\bullet(\mathcal{E}^\bullet, \mathcal{E}^\bullet)$.

By definition $\mathcal{H}om^0(\mathcal{E}^\bullet, \mathcal{E}^\bullet) = \bigoplus_i \mathcal{H}om(\mathcal{E}^i, \mathcal{E}^i)$ and the usual trace maps $\mathrm{tr}_{\mathcal{E}^i} : \mathcal{H}om(\mathcal{E}^i, \mathcal{E}^i) \longrightarrow \mathcal{O}_X$ for the locally free sheaves $\mathcal{E}^i$ give rise to the trace map

$$\mathrm{tr}_{\mathcal{E}^\bullet} := \bigoplus (-1)^i \mathrm{tr}_{\mathcal{E}^i}.$$

Exercise 3.28 Prove that $\mathrm{tr}_{\mathcal{E}^\bullet}$ does define a complex morphism.

Dual We define the dual $\mathcal{F}^{\bullet\vee}$ of a complex $\mathcal{F}^\bullet \in \mathrm{D}^-(\mathbf{Qcoh}(X))$ of quasi-coherent sheaves as

$$\mathcal{F}^{\bullet\vee} := R\mathcal{H}om(\mathcal{F}^\bullet, \mathcal{O}_X) \in \mathrm{D}^+(\mathbf{Qcoh}(X)).$$

$$\mathcal{F}^{\bullet\vee} := R\mathcal{H}om(\mathcal{F}^\bullet, \mathcal{O}_X) \in \mathrm{D}^+(\mathbf{Qcoh}(X)).$$

So, in general even for a sheaf $\mathcal{F}$ it is not simply the dual sheaf $\mathcal{H}om(\mathcal{F}, \mathcal{O}_X)$. E.g. if $\mathcal{F}$ is a coherent sheaf on a smooth variety with $\mathrm{codim}(\mathrm{supp}(\mathcal{F})) \geq d$, then $\mathcal{F}^\vee$ is a complex concentrated in degree $\geq d$, i.e. $\mathcal{H}^q(\mathcal{F}^\vee) = 0$ for $q < d$. This follows either from the local statement for $\mathrm{Ext}^q(M, A)$ (cf. [77, 15.E]) or from an argument using Serre duality (cf. [53, 1.1.6]).

The dual of the structure sheaf of a subvariety can be computed explicitly (see Corollary 3.40).

If we let $\mathcal{F}^\bullet$ be a complex

$$\cdots \longrightarrow \mathcal{F}^{i-1} \longrightarrow \mathcal{F}^i \longrightarrow \mathcal{F}^{i+1} \longrightarrow \cdots$$

with locally free sheaves $\mathcal{F}^i$, then $\mathcal{F}^{\bullet\vee}$ is obtained as

$$\ldots \mathcal{H}om(\mathcal{F}^{i+1}, \mathcal{O}_X) \longrightarrow \mathcal{H}om(\mathcal{F}^i, \mathcal{O}_X) \longrightarrow \mathcal{H}om(\mathcal{F}^{i-1}, \mathcal{O}_X) \ldots.$$

If X is regular, then $\mathcal{F}^{\bullet\vee} = R\mathcal{H}om(\mathcal{F}^\bullet, \mathcal{O}_X) \in \mathrm{D}^\mathrm{b}(X)$ for any $\mathcal{F}^\bullet \in \mathrm{D}^\mathrm{b}(X)$.

Tensor product Let us again start out with the sheaf version of the tensor product before explaining the more general notion of the derived tensor product of two complexes. So, let $\mathcal{F} \in \mathbf{Coh}(X)$. The tensor product defines the right exact functor

$$\mathcal{F} \otimes (\ \ ) : \mathbf{Coh}(X) \longrightarrow \mathbf{Coh}(X)$$

and we are interested in its left derived functor. To simplify the argument, we shall right away assume that X is a projective scheme over a field k.

Any coherent sheaf $\mathcal{E}$ admits a resolution by locally free sheaves or, in other words, for any $\mathcal{E}$ there exists a surjection

$$\mathcal{E}^0 \longrightarrow\!\!\!\!\!\rightarrow \mathcal{E}$$

with $\mathcal{E}^0$ locally free (see, e.g. Proposition 3.18). Moreover, if $\mathcal{E}^\bullet$ is an acyclic complex bounded above with all $\mathcal{E}^i$ locally free, then $\mathcal{F} \otimes \mathcal{E}^\bullet$ is still acyclic. (Reduce it to the local situation of an exact complex of free modules which stays exact if tensored by any module.)

These two facts show that the class of locally free sheaves in $\mathbf{Coh}(X)$ is adapted for the right exact functor $\mathcal{F}\otimes(\ )$ (cf. Remark 2.57). Thus, the left derived functor

$$\mathcal{F}\otimes^L(\ ):\mathrm{D}^-(X)\longrightarrow \mathrm{D}^-(X)$$

exists.

By definition,

$$\mathcal{T}or_i(\mathcal{F},\mathcal{E}):=\mathcal{H}^{-i}(\mathcal{F}\otimes^L\mathcal{E}).$$

If X is smooth of dimension n, then any coherent sheaf $\mathcal{E}$ admits a locally free resolution of length n (cf. Proposition 3.26). Hence, in this case $\mathcal{T}or_i(\mathcal{F},\mathcal{E})=0$ for $i>n$. Thus, by Corollary 2.68

$$\begin{array}{ccc} \mathrm{D}^-(X) & \xrightarrow{\mathcal{F}\otimes^L(\ )} & \mathrm{D}^-(X) \\ \uparrow & & \uparrow \\ \mathrm{D}^{\mathrm{b}}(X) & \xrightarrow{X \text{ smooth}} & \mathrm{D}^{\mathrm{b}}(X). \end{array}$$

Let us now pass to the more general situation. Consider a complex $\mathcal{F}^\bullet\in \mathrm{K}^-(\mathbf{Coh}(X))$ that is bounded above and define the exact functor

$$\mathcal{F}^\bullet\otimes(\ ):\mathrm{K}^-(\mathbf{Coh}(X))\longrightarrow \mathrm{K}^-(\mathbf{Coh}(X))$$

$$(\mathcal{F}^\bullet\otimes\mathcal{E}^\bullet)^i:=\bigoplus_{p+q=i}\mathcal{F}^p\otimes\mathcal{E}^q \quad\text{with}\quad d=d_{\mathcal{F}}\otimes 1+(-1)^i 1\otimes d_{\mathcal{E}}.$$

Thus, by definition $\mathcal{F}^\bullet\otimes\mathcal{E}^\bullet$ is the total complex of the double complex $K^{\bullet,\bullet}$ with $K^{p,q}=\mathcal{F}^p\otimes\mathcal{E}^q$ and the two differentials $d_I=d_{\mathcal{F}}\otimes 1$ and $d_{II}=(-1)^{p+q}1\otimes d_{\mathcal{E}}$.

In order to define the derived functor, one verifies that the subcategory of complexes of locally free sheaves is adapted to $\mathcal{F}^\bullet\otimes(\ )$. Since $\mathbf{Coh}(X)$ contains enough locally free sheaves for X projective, it remains to check that the image of an acyclic complex $\mathcal{E}^\bullet$ with all $\mathcal{E}^i$ locally free is again acyclic. To see this, one uses the spectral sequence

$$E_2^{p,q}=\mathcal{H}_I^p\mathcal{H}_{II}^q(K^{\bullet,\bullet})\Rightarrow\mathcal{H}^{p+q}(\mathcal{F}^\bullet\otimes\mathcal{E}^\bullet),$$

which is a consequence of Proposition 2.64 with the two filtrations interchanged. For $\mathcal{E}^\bullet$ acyclic and all $\mathcal{E}^i$ locally free the complex $\mathcal{F}^p\otimes\mathcal{E}^\bullet$ is acyclic for any p (this has been used earlier). Hence, $\mathcal{H}_{II}^q(\mathcal{F}^p\otimes\mathcal{E}^\bullet)=0$ and, therefore, $E_2^{p,q}=0$ for all q and all p. Therefore, also $\mathcal{F}^\bullet\otimes\mathcal{E}^\bullet$ is exact. Thus, the left derived functor

$$\mathcal{F}^\bullet\otimes^L(\ ):\mathrm{D}^-(X)\longrightarrow\mathrm{D}^-(X)$$

exists.

The analogous spectral sequence interchanging d_I and d_{II} shows that for a complex of locally free sheaves $\mathcal{E}^\bullet$ and an acyclic complex $\mathcal{F}^\bullet$ the tensor product $\mathcal{F}^\bullet \otimes \mathcal{E}^\bullet$ is again acyclic. In other words, the induced bifunctor

$$\mathrm{K}^-(\mathbf{Coh}(X)) \times \mathrm{D}^-(X) \longrightarrow \mathrm{D}^-(X)$$

need not be derived in the first factor and descends to the bifunctor $\otimes^L$ for the derived categories (cf. Lemma 2.44).

If X is smooth, then the functor is defined for the bounded derived categories. Indeed, any bounded complex of coherent sheaves is quasi-isomorphic to a bounded complex of locally free sheaves (see Proposition 3.26) and the tensor product of two of those is again bounded. Hence,

$$\begin{array}{ccc} \mathrm{D}^-(X) \times \mathrm{D}^-(X) & \xrightarrow{\otimes^L} & \mathrm{D}^-(X) \\ \uparrow & & \uparrow \\ \mathrm{D}^\mathrm{b}(X) \times \mathrm{D}^\mathrm{b}(X) & \xrightarrow{X \text{ smooth}} & \mathrm{D}^\mathrm{b}(X). \end{array}$$

Computing the derived tensor product $\mathcal{F}^\bullet \otimes^L \mathcal{E}^\bullet$ as the ordinary tensor product of complexes of locally free sheaves quasi-isomorphic to $\mathcal{F}^\bullet$ (respectively $\mathcal{E}^\bullet$) yields the following functorial isomorphisms

$$\mathcal{F}^\bullet \otimes^L \mathcal{E}^\bullet \simeq \mathcal{E}^\bullet \otimes^L \mathcal{F}^\bullet$$

$$\mathcal{F}^\bullet \otimes^L (\mathcal{E}^\bullet \otimes^L \mathcal{G}^\bullet) \simeq (\mathcal{F}^\bullet \otimes^L \mathcal{E}^\bullet) \otimes^L \mathcal{G}^\bullet.$$

Generalizing the above definition of the sheaf Tor one sets

$$\mathcal{T}\!or_i(\mathcal{F}^\bullet, \mathcal{E}^\bullet) := \mathcal{H}^{-i}(\mathcal{F}^\bullet \otimes^L \mathcal{E}^\bullet)$$

which can often be computed via the spectral sequence

$$E_2^{p,q} = \mathcal{T}\!or_{-p}(\mathcal{H}^q(\mathcal{F}^\bullet), \mathcal{E}^\bullet) \Rightarrow \mathcal{T}\!or_{-(p+q)}(\mathcal{F}^\bullet, \mathcal{E}^\bullet). \tag{3.9}$$

The argument that worked for right derived functors does not apply literally, as we have not said anything about Cartan–Eilenberg resolutions in this context (cf. Proposition 2.66). But an easy ad hoc argument goes as follows: We may assume that $\mathcal{E}^\bullet$ is a complex of locally free sheaves. Then $\mathcal{T}\!or_{-p}(\mathcal{H}^q(\mathcal{F}^\bullet), \mathcal{E}^\bullet)$ can be computed as the p-th cohomology of the complex $\mathcal{H}^q(\mathcal{F}^\bullet) \otimes \mathcal{E}^\bullet$. Similarly, $\mathcal{T}\!or_{-(p+q)}(\mathcal{F}^\bullet, \mathcal{E}^\bullet)$ can be computed as the $(p+q)$-th cohomology of the complex $\mathcal{F}^\bullet \otimes \mathcal{E}^\bullet$. The latter is the total complex of the natural double complex and the claimed spectral sequence corresponds to the standard spectral sequence for a double complex (see Proposition 2.64).

Inverse image Let $f : (X, \mathcal{O}_X) \longrightarrow (Y, \mathcal{O}_Y)$ be a morphism of ringed spaces. Then

$$f^* : \mathbf{Sh}_{\mathcal{O}_Y}(Y) \longrightarrow \mathbf{Sh}_{\mathcal{O}_X}(X)$$

is by definition the composition of the exact functor

$$f^{-1} : \mathbf{Sh}_{\mathcal{O}_Y}(Y) \longrightarrow \mathbf{Sh}_{f^{-1}(\mathcal{O}_Y)}(X)$$

and the right exact functor

$$\mathcal{O}_X \otimes_{f^{-1}(\mathcal{O}_Y)} (\ \ ) : \mathbf{Sh}_{f^{-1}(\mathcal{O}_Y)}(X) \longrightarrow \mathbf{Sh}_{\mathcal{O}_X}(X).$$

Thus, f^* is right exact and if $\mathcal{O}_X \otimes^L_{f^{-1}(\mathcal{O}_Y)} (\ \ )$ is the left derived functor of $\mathcal{O}_X \otimes_{f^{-1}(\mathcal{O}_Y)} (\ \ )$, then

$$Lf^* := \left(\mathcal{O}_X \otimes^L_{f^{-1}(\mathcal{O}_Y)} (\ \ )\right) \circ f^{-1} : \mathrm{D}^-(Y) \longrightarrow \mathrm{D}^-(X).$$

To be precise, the arguments of the previous paragraph are not quite sufficient to derive the tensor product $\mathcal{O}_X \otimes^L_{f^{-1}(\mathcal{O}_Y)} (\ \ )$, as we only explained how to derive the tensor product of $\mathcal{O}_X$-modules (on a projective scheme). But the more general situation is handled in the same way. Moreover, in most of our applications we don't even have to derive f^*, as f is often flat and, therefore, f^* exact.

Similarly to the spectral sequence (3.9) one obtains

$$E_2^{p,q} = L^p f^*(\mathcal{H}^q(\mathcal{E}^\bullet)) \Rightarrow L^{p+q} f^*(\mathcal{E}^\bullet), \tag{3.10}$$

where by definition $L^p f^*(\mathcal{F}^\bullet) = \mathcal{H}^p(Lf^*(\mathcal{F}^\bullet))$.

Let us mention two useful results that will serve as technical tools in the sequel.

Lemma 3.29 *Let $i : T \hookrightarrow X$ be a closed subscheme. Then for any $\mathcal{F}^\bullet \in \mathrm{D}^{\mathrm{b}}(X)$ one has*

$$\mathrm{supp}(\mathcal{F}^\bullet) \cap T = \mathrm{supp}(Li^*(\mathcal{F}^\bullet)).$$

Proof For the definition of the support of a complex see Definition 3.8.

One direction is easy. Indeed, if $x \notin \mathrm{supp}(\mathcal{F}^\bullet)$, then the restriction $\mathcal{F}^\bullet|_U$ of $\mathcal{F}^\bullet$ to an open neighbourhood $x \in U \subset X$ is trivial. Hence, also $Li_U^*(\mathcal{F}^\bullet|_U) = (Li^*(\mathcal{F}^\bullet))|_{U\cap T}$ is trivial, where $i_U : U \cap T \hookrightarrow U$. Thus, $x \notin \mathrm{supp}(Li^*(\mathcal{F}^\bullet))$. This proves $\mathrm{supp}(Li^*(\mathcal{F}^\bullet)) \subset \mathrm{supp}(\mathcal{F}^\bullet)$.

Conversely, let $x \in \mathrm{supp}(\mathcal{F}^\bullet)$. If i_0 is maximal with $x \in \mathrm{supp}(\mathcal{H}^{i_0}(\mathcal{F}^\bullet))$, then $\mathcal{T}or_0(\mathcal{H}^{i_0}(\mathcal{F}^\bullet), k(x)) \not\simeq 0$. As $\mathcal{T}or_{-p}(\mathcal{H}^q(\mathcal{F}^\bullet), k(x)) = 0$ for $p > 0$, the spectral sequence

$$E_2^{p,q} = \mathcal{T}or_{-p}(\mathcal{H}^q(\mathcal{F}^\bullet), k(x)) \Rightarrow \mathcal{T}or_{-(p+q)}(\mathcal{F}^\bullet, k(x)) = \mathcal{H}^{p+q}(\mathcal{F}^\bullet(x))$$

shows $\mathcal{H}^{i_0}(\mathcal{F}^\bullet(x)) \not\simeq 0$. In particular, $\mathcal{F}^\bullet(x) \not\simeq 0$, where $\mathcal{F}^\bullet(x)$ denotes the derived pull-back of $\mathcal{F}^\bullet$ under the embedding of the closed point $\{x\} \hookrightarrow X$.

Since deriving the composition of the pull-backs is isomorphic to the composition of the derived pull-backs, one has $\mathcal{F}^\bullet(x) \simeq (Li^*(\mathcal{F}^\bullet))(x)$. Hence, $x \in \mathrm{supp}(Li^*(\mathcal{F}^\bullet))$. □

Exercise 3.30 Prove the following fact, which was tacitly used in the above proof: Let $i_x : \{x\} \hookrightarrow X$ be the embedding of a closed point. Then for any complex $\mathcal{F}^\bullet$ one has $\mathcal{F}^\bullet(x) := Li_x^*\mathcal{F}^\bullet \neq 0$ if and only if $x \in \mathrm{supp}(\mathcal{F}^\bullet)$.

For the following lemma consider a morphism $S \longrightarrow X$. If $x \in X$ is a closed point, we denote by $i_x : S_x \hookrightarrow S$ the closed embedding of the fibre over x.

Lemma 3.31 *Suppose $\mathcal{Q} \in \mathrm{D^b}(S)$ and assume that for all closed points $x \in X$ the derived pull-back $Li_x^*\mathcal{Q} \in \mathrm{D^b}(S_x)$ is a complex concentrated in degree zero, i.e. a sheaf.*

Then $\mathcal{Q}$ is isomorphic to a sheaf which is flat over X.

Proof In order to verify the claim we will apply the spectral sequence (3.10) to the inclusion i_x and obtain

$$E_2^{p,q} = \mathcal{H}^p(Li_x^*\mathcal{H}^q(\mathcal{Q})) \Rightarrow \mathcal{H}^{p+q}(Li_x^*\mathcal{Q}).$$

By assumption the right hand side is trivial except possibly for $p + q = 0$. Choose m maximal with $\mathcal{H}^m(\mathcal{Q}) \neq 0$. Then there exists a closed point $x \in X$ with $E_2^{0,0} = \mathcal{H}^0(Li_x^*\mathcal{H}^m(\mathcal{Q})) \neq 0$ (this is just the ordinary pull-back). But this non-triviality survives the passing to the limit in the spectral sequence and hence $m = 0$. For the same reason, $E_2^{0,-1} = \mathcal{H}^{-1}(Li_x^*\mathcal{H}^0(\mathcal{Q}))$ with $x \in X$ arbitrary also survives and must, therefore, be trivial. This shows that the sheaf $\mathcal{H}^0(\mathcal{Q})$ is actually flat over X.

$$\begin{array}{ccccc}
0 & 0 & 0 & 0 & 0 \\
 & & & & \vdots \\
* & * & * & E_2^{0,m} & 0 \\
 & & & & \\
* & * & * & * & 0
\end{array}$$

(For the reader's convenience we recall the result from commutative algebra behind this: Let $A \longrightarrow B$ be a local ring homomorphism and M a B-module. In order to show that M is A-flat, one has to verify that for any finitely generated ideal $\mathfrak{a} \subset A$ the map $\mathfrak{a} \otimes M \longrightarrow M$ is injective. Of course, it suffices to show this for $\mathfrak{a} = \mathfrak{m}$, the maximal ideal of A. Suppose $0 \longrightarrow N_1 \longrightarrow N_2 \longrightarrow M \longrightarrow 0$ is a short exact sequence of A-modules.

The analogue of $\mathcal{H}^{-1}(Li_x^*\mathcal{H}^0(\mathcal{Q})) = 0$ for $\mathcal{H}^0(\mathcal{Q})$ replaced by M yields the injectivity of $N_1/\mathfrak{m}N_1 \longrightarrow N_2/\mathfrak{m}N_2$. If N_2 is A-flat, then $N_2 \otimes \mathfrak{m} \longrightarrow N_2$ is injective. Both statements together and the snake lemma readily yield the injectivity of $M \otimes \mathfrak{m} \longrightarrow M$.)

Also note that the flatness of $\mathcal{H}^0(\mathcal{Q})$ over X implies that the higher derived pull-backs $E_2^{p,0} = \mathcal{H}^p(Li_x^*\mathcal{H}^0(\mathcal{Q}))$ are trivial for $p < 0$.

The last thing one has to check is that there is no non-trivial cohomology below, i.e. that $\mathcal{H}^q(\mathcal{Q}) = 0$ for $q < 0$. Suppose not; then we choose m maximal among all $q < 0$ with $\mathcal{H}^q(\mathcal{Q}) \neq 0$ and x a closed point $x \in X$ in the support of $\mathcal{H}^m(\mathcal{Q})$.

$$
\begin{array}{ccccc}
0 & \cdots & 0 & E_2^{0,0} & 0 \\
\vdots & & & & \\
0 & \cdots & 0 & 0 & 0 \\
 & & & & \\
* & \cdots & * & E_2^{0,m} & 0 \\
 & & & & \\
* & & * & * & 0
\end{array}
$$

Since all $E_2^{-p,q} = \mathcal{H}^{-p}(Li_x^*\mathcal{H}^q(\mathcal{Q}))$ with $q > m$ and $p < 0$ are trivial, this would again yield the contradiction $E^m = \mathcal{H}^m(Li_x^*\mathcal{Q}) = \mathcal{H}^0(Li_x^*\mathcal{H}^m(\mathcal{Q})) \neq 0$ in the limit. □

Compatibilities i) Let $f : X \longrightarrow Y$ be a proper morphism of projective schemes over a field k. For any $\mathcal{F}^\bullet \in \mathrm{D^b}(X)$, $\mathcal{E}^\bullet \in \mathrm{D^b}(Y)$ there exists a natural isomorphism (*projection formula*):

$$
Rf_*(\mathcal{F}^\bullet) \otimes^L \mathcal{E}^\bullet \xrightarrow{\sim} Rf_*(\mathcal{F}^\bullet \otimes^L Lf^*(\mathcal{E}^\bullet)). \tag{3.11}
$$

This is a consequence of the classical projection formula for a locally free sheaf $\mathcal{E}$ and an arbitrary sheaf $\mathcal{F}$, which states $f_*(\mathcal{F} \otimes f^*\mathcal{E}) \simeq f_*(\mathcal{F}) \otimes \mathcal{E}$ (cf. [45, II.5]).

ii) Let $f : X \longrightarrow Y$ be a morphism of projective schemes and let $\mathcal{F}^\bullet, \mathcal{E}^\bullet \in \mathrm{D^b}(Y)$. Then there exists a natural isomorphism

$$
Lf^*(\mathcal{F}^\bullet) \otimes^L Lf^*(\mathcal{E}^\bullet) \xrightarrow{\sim} Lf^*(\mathcal{F}^\bullet \otimes^L \mathcal{E}^\bullet). \tag{3.12}
$$

Indeed, replacing $\mathcal{E}^\bullet$ and $\mathcal{F}^\bullet$ by complexes of locally free sheaves allows us to compute the derived tensor products and the derived pull-back as ordinary ones (the pull-back of a locally free sheaf is again locally free). But then the claim reduces to the classical statement for sheaves $f^*\mathcal{F} \otimes f^*\mathcal{E} \simeq f^*(\mathcal{F} \otimes \mathcal{E})$.

iii) Let $f : X \longrightarrow Y$ be a projective morphism. Then $Lf^* \dashv Rf_*$, i.e. there exist functorial morphisms

$$
\mathrm{Hom}(Lf^*\mathcal{F}^\bullet, \mathcal{E}^\bullet) \xrightarrow{\sim} \mathrm{Hom}(\mathcal{F}^\bullet, Rf_*\mathcal{E}^\bullet).
$$

Once more, one may suppose that $\mathcal{F}^\bullet$ is a complex of locally free sheaves and that $\mathcal{E}^\bullet$ is a complex of quasi-coherent injective sheaves. In this case, the derived functors are just the usual ones and the adjunction follows from the standard one $f^* \dashv f_*$ (cf. [45, II.5]).

iv) For simplicity, we will assume that X is smooth and projective over a field k. Then there are the following compatibilities of derived local Hom and derived tensor product. All complexes involved are supposed to be bounded complexes of coherent sheaves.

$$R\mathcal{H}om(\mathcal{F}^\bullet, \mathcal{E}^\bullet) \otimes^L \mathcal{G}^\bullet \simeq R\mathcal{H}om(\mathcal{F}^\bullet, \mathcal{E}^\bullet \otimes^L \mathcal{G}^\bullet) \tag{3.13}$$

$$R\mathcal{H}om(\mathcal{F}^\bullet, R\mathcal{H}om(\mathcal{E}^\bullet, \mathcal{G}^\bullet)) \simeq R\mathcal{H}om(\mathcal{F}^\bullet \otimes^L \mathcal{E}^\bullet, \mathcal{G}^\bullet) \tag{3.14}$$

$$R\mathcal{H}om(\mathcal{F}^\bullet, \mathcal{E}^\bullet \otimes^L \mathcal{G}^\bullet) \simeq R\mathcal{H}om(R\mathcal{H}om(\mathcal{E}^\bullet, \mathcal{F}^\bullet), \mathcal{G}^\bullet). \tag{3.15}$$

(Note that we need the smoothness, e.g. in order to ensure that $R\mathcal{H}om(\mathcal{F}^\bullet, \mathcal{E}^\bullet)$ is again bounded above so that the tensor product can be formed.) All these isomorphisms are rather obvious, once $\mathcal{F}^\bullet, \mathcal{E}^\bullet$, and $\mathcal{G}^\bullet$ are chosen to be bounded complexes of locally free coherent sheaves (cf. [45, II.5]).

Most important for us is the special case of the derived dual:

$$\mathcal{F}^{\bullet\vee} \otimes^L \mathcal{E}^\bullet \simeq R\mathcal{H}om(\mathcal{F}^\bullet, \mathcal{O}_X) \otimes^L \mathcal{E}^\bullet \simeq R\mathcal{H}om(\mathcal{F}^\bullet, \mathcal{E}^\bullet).$$

In other words, $R\mathcal{H}om(\mathcal{F}^\bullet, \quad)$ is isomorphic to the functor $\mathcal{F}^{\bullet\vee} \otimes^L (\quad) \simeq R\mathcal{H}om(\mathcal{F}^\bullet, \mathcal{O}_X) \otimes^L (\ )$.

If $\mathcal{F}^\bullet$ is a complex of locally free coherent sheaves, then $\mathcal{H}om(\mathcal{F}^\bullet, \mathcal{O}_X)$ is isomorphic to the complex that is obtained by genuinely dualizing the complex $\ldots \longrightarrow \mathcal{F}^{i-1} \longrightarrow \mathcal{F}^i \longrightarrow \mathcal{F}^{i+1} \longrightarrow \ldots$. and the tensor product need not be derived. Thus,

$$R\mathcal{H}om(\mathcal{F}^\bullet, \mathcal{E}^\bullet)^i \simeq \bigoplus_{q-p=i} \mathcal{H}om(\mathcal{F}^p, \mathcal{O}_X) \otimes \mathcal{E}^q.$$

We continue to assume that X is smooth and projective over a field k. Then the double dual of a complex $\mathcal{F}^\bullet \in \mathrm{D^b}(X)$ is canonically isomorphic to $\mathcal{F}^\bullet$. (Once more, we need the smoothness to ensure that the dual is again bounded above.) In other words,

$$\mathcal{F}^\bullet \simeq \mathcal{F}^{\bullet\vee\vee} \simeq R\mathcal{H}om(R\mathcal{H}om(\mathcal{F}^\bullet, \mathcal{O}_X), \mathcal{O}_X).$$

Indeed, we may assume that $\mathcal{F}^\bullet$ is a bounded complex of locally free coherent sheaves $\mathcal{F}^i$. Then the double dual is obtained by double dualizing the sheaves $\mathcal{F}^i$. But for a locally free sheaf the double dual is clearly naturally isomorphic to the original sheaf.

Let us also mention the following consequence.

Lemma 3.32 *For any $\mathcal{F}^\bullet \in \mathrm{D^b}(X)$ one has*

$$\mathrm{supp}(\mathcal{F}^\bullet) = \mathrm{supp}(\mathcal{F}^{\bullet\vee}).$$

Proof Consider the spectral sequence (see (3.8), p. 77)

$$E_2^{p,q} := \mathcal{E}xt^p(\mathcal{H}^{-q}(\mathcal{F}^\bullet), \mathcal{O}) \Rightarrow \mathcal{E}xt^{p+q}(\mathcal{F}^\bullet, \mathcal{O}) \simeq \mathcal{H}^{p+q}(\mathcal{F}^{\bullet\vee}).$$

From this one immediately concludes $\mathrm{supp}(\mathcal{F}^{\bullet\vee}) \subset \mathrm{supp}(\mathcal{F}^\bullet)$. Similarly, using $\mathcal{F}^{\bullet\vee\vee} \simeq \mathcal{F}^\bullet$ one shows the other inclusion and thus obtains equality. □

v) Let $\mathcal{F}^\bullet \in \mathrm{D}^-(X)$. Then by definition of the sheaf hom $\mathcal{H}om^\bullet$ one has $\Gamma \circ \mathcal{H}om^\bullet(\mathcal{F}^\bullet, \ ) = \mathrm{Hom}^\bullet(\mathcal{F}^\bullet, \ )$. Hence,

$$R\Gamma \circ R\mathcal{H}om(\mathcal{F}^\bullet, \ ) = R\mathrm{Hom}(\mathcal{F}^\bullet, \ ).$$

(Well, one has to verify that the image of a complex $\mathcal{E}^\bullet$ of injective sheaves under $\mathcal{H}om(\mathcal{F}^\bullet, \ )$ is Γ-acyclic. But this holds true [40].) An immediate consequence of this is the spectral sequence that relates local and global Ext:

$$E_2^{p,q} = H^p(X, \mathcal{E}xt^q(\mathcal{F}^\bullet, \mathcal{E}^\bullet)) \Rightarrow \mathrm{Ext}^{p+q}(\mathcal{F}^\bullet, \mathcal{E}^\bullet). \tag{3.16}$$

vi) Let $f : X \longrightarrow Y$ be a morphism of projective schemes and let $\mathcal{F}^\bullet \in \mathrm{D}^-(Y)$ and $\mathcal{E}^\bullet \in \mathrm{D}^{\mathrm{b}}(Y)$. Then there exists a natural isomorphism

$$Lf^* R\mathcal{H}om_Y(\mathcal{F}^\bullet, \mathcal{E}^\bullet) \xrightarrow{\sim} R\mathcal{H}om_X(Lf^*\mathcal{F}^\bullet, Lf^*\mathcal{E}^\bullet). \tag{3.17}$$

For the proof replace again all complexes by complexes of locally free sheaves.

vii) Consider a fibre product diagram

$$\begin{array}{ccc} X \times_Z Y & \xrightarrow{v} & Y \\ {\scriptstyle g}\downarrow & & \downarrow{\scriptstyle f} \\ X & \xrightarrow{u} & Z \end{array}$$

with $u : X \longrightarrow Z$ flat and $f : Y \longrightarrow Z$ proper. Then *flat base change* asserts the existence of a functorial isomorphism:

$$u^* Rf_* \mathcal{F}^\bullet \xrightarrow{\sim} Rg_* v^* \mathcal{F}^\bullet \tag{3.18}$$

for any $\mathcal{F}^\bullet \in \mathrm{D}(\mathbf{Qcoh}(Y))$.

As u and, therefore, v are flat, both functors u^* and v^* are exact and need not be derived. Furthermore, the formula also yields canonical isomorphisms $u^* R^i f_* \mathcal{F}^\bullet \simeq R^i g_* v^* \mathcal{F}^\bullet$ for any i.

Remark 3.33 i) Even without the flatness assumption one has a natural map $u^* Rf_* \mathcal{F}^\bullet \longrightarrow Rg_* v^* \mathcal{F}^\bullet$. Thus it suffices to prove that for u flat the induced cohomology maps are isomorphisms (see [45, III, 9.3]).

For a variant of the base change formula with f smooth and proper, but u arbitrary see [14, Lem.1.3].

ii) For completeness sake and later use we recall one of the fundamental results comparing higher direct images with fibrewise cohomology. For this, we let

$f : X \longrightarrow Y$ be an arbitrary proper morphism of varieties and $\mathcal{F}$ a coherent sheaf on X. Suppose

$$y \longmapsto \dim H^i(f^{-1}(y), \mathcal{F}|_{f^{-1}(y)})$$

is a constant function on the set of closed points of Y. If Y is integral, then $R^i f_* \mathcal{F}$ is locally free and the natural morphisms

$$R^i f_* \mathcal{F} \otimes k(y) \xrightarrow{\sim} H^i(f^{-1}(y), \mathcal{F}|_{f^{-1}(y)})$$

are bijective. See [45, III, 12.9].

Let us consider the special case of the product $X \times Y$, i.e. $Z = \mathrm{Spec}(k)$, with the two projections

$$\begin{array}{ccc} & X \times Y & \\ {}^{q}\swarrow & & \searrow^{p} \\ X & & Y. \end{array}$$

For $\mathcal{F}^\bullet \in \mathrm{D^b}(Y)$ flat base change yields

$$q_* p^* \mathcal{F}^\bullet \simeq R\Gamma(Y, \mathcal{F}^\bullet) \otimes \mathcal{O}_X.$$

From this and projection formula (3.11), one deduces the *Künneth formula*: for $\mathcal{F}^\bullet \in \mathrm{D^b}(Y)$ and $\mathcal{E}^\bullet \in \mathrm{D^b}(X)$

$$R\Gamma(X \times Y, q^*\mathcal{E}^\bullet \otimes^L p^*\mathcal{F}^\bullet) \simeq R\Gamma(X, \mathcal{E}^\bullet) \otimes R\Gamma(Y, \mathcal{F}^\bullet).$$

Note that the derived tensor product on the left hand side is in fact an ordinary tensor product.

3.4 Grothendieck–Verdier duality

In some sense, Grothendieck–Verdier duality is just another compatibility between the geometric derived functors, but it is not at all a formal consequence of the definitions as most of the ones discussed in the previous section.

For the most general version and the proof we have to refer to the literature, e.g. [32, 44].

Let $f : X \longrightarrow Y$ be a morphism of smooth schemes over a field k of relative dimension $\dim(f) := \dim(X) - \dim(Y)$. We introduce the *relative dualizing bundle*

$$\omega_f := \omega_X \otimes f^*\omega_Y^*.$$

Theorem 3.34 *For any $\mathcal{F}^\bullet \in \mathrm{D^b}(X)$ and $\mathcal{E}^\bullet \in \mathrm{D^b}(Y)$ there exists a functorial isomorphism*

$$Rf_* R\mathcal{H}om(\mathcal{F}^\bullet, Lf^*(\mathcal{E}^\bullet) \otimes \omega_f[\dim(f)]) \simeq R\mathcal{H}om(Rf_*\mathcal{F}^\bullet, \mathcal{E}^\bullet). \qquad (3.19)$$

Note that, since ω_f is locally free, the tensor product on the left hand side of (3.34) is indeed underived. Let us also introduce

$$f^! : \mathrm{D}^{\mathrm{b}}(Y) \longrightarrow \mathrm{D}^{\mathrm{b}}(X), \qquad \mathcal{E}^\bullet \longmapsto Lf^*(\mathcal{E}^\bullet) \otimes \omega_f[\dim(f)].$$

Corollary 3.35 *The functors $Lf^*, f^! : \mathrm{D}^{\mathrm{b}}(Y) \longrightarrow \mathrm{D}^{\mathrm{b}}(X)$ are left, respectively right adjoint to $Rf_* : \mathrm{D}^{\mathrm{b}}(X) \longrightarrow \mathrm{D}^{\mathrm{b}}(Y)$, i.e.*

$$Lf^* \dashv Rf_* \dashv f^!.$$

Proof One applies global sections to both sides of (3.19) and uses

$$R\Gamma \circ Rf_* \simeq R\Gamma \quad \text{and} \quad R\Gamma \circ R\mathcal{H}om \simeq R\mathrm{Hom}.$$

Taking cohomology in degree zero then yields

$$\mathrm{Hom}_{\mathrm{D}^{\mathrm{b}}(X)}(\mathcal{F}^\bullet, Lf^*(\mathcal{E}^\bullet) \otimes \omega_f[\dim(f)]) \simeq \mathrm{Hom}_{\mathrm{D}^{\mathrm{b}}(Y)}(Rf_*\mathcal{F}^\bullet, \mathcal{E})$$

as asserted. □

If we denote by $\mathbb{D}_X$ is the *dualizing functor*

$$\mathcal{F}^\bullet \longmapsto R\mathcal{H}om(\mathcal{F}^\bullet, \omega_X[\dim(X)]) = \mathcal{F}^{\bullet\vee} \otimes \omega_X[\dim(X)],$$

then

$$f^! \simeq \mathbb{D}_X \circ Lf^* \circ \mathbb{D}_Y^{-1}.$$

Moreover, the Grothendieck–Verdier duality can be equivalently expressed as

$$Rf_* \circ \mathbb{D}_X \simeq \mathbb{D}_Y \circ Rf_*. \tag{3.20}$$

Exercise 3.36 Prove that (3.20) is indeed equivalent to (3.19).

A special case of (3.20) is

$$Rf_*\omega_X[\dim(X)] \simeq (Rf_*\mathcal{O}_X)^\vee \otimes \omega_Y[\dim(Y)].$$

Remarks 3.37 i) Grothendieck–Verdier duality, e.g. in the form of (3.20), holds in much broader generality. What has to be changed is the definition of the dualizing functor $\mathbb{D}_X$. It has to be replaced by $\mathcal{F}^\bullet \longmapsto R\mathcal{H}om(\mathcal{F}^\bullet, K_X)$, where K_X is the *dualizing complex*. It turns out that K_X always exists (we are sloppy here by not specifying where it lives). Moreover, the variety X is Gorenstein if and only if K_X is a line bundle in degree $-n$. Also, X is Cohen–Macaulay if and only if K_X is (isomorphic to) a coherent sheaf. See [44, V.9].

ii) Classical Serre duality (see [45, II.7]) is a special case of Grothendieck–Verdier duality. Indeed, applied to $f : X \longrightarrow \mathrm{Spec}(k)$, the theorem yields canonical isomorphisms

$$\mathrm{Hom}_{\mathrm{D}^{\mathrm{b}}(X)}(\mathcal{F}^\bullet, \omega_X[\dim(X)]) \simeq \mathrm{Hom}_k(R\Gamma(\mathcal{F}^\bullet), k).$$

In particular, $\mathrm{Ext}^i(\mathcal{F}, \omega_X) = H^{n-i}(X, \mathcal{F})^*$ for any coherent sheaf $\mathcal{F}$ on X.

Also Serre duality for derived categories (see Theorem 3.12) can be deduced from it: Again we consider the projection $f : X \longrightarrow \mathrm{Spec}(k)$. If $\mathcal{E}^\bullet, \mathcal{F}^\bullet \in \mathrm{D}^\mathrm{b}(X)$, then

$$\begin{aligned}
&\mathrm{Hom}_{\mathrm{D^b}(X)}(\mathcal{F}^\bullet, \mathcal{E}^\bullet \otimes \omega_X[\dim(X)]) \\
&\simeq \mathrm{Hom}_{\mathrm{D^b}(X)}(R\mathcal{H}om(\mathcal{E}^\bullet, \mathcal{F}^\bullet), \omega_X[\dim(X)]) \quad \text{(use (3.15))} \\
&\simeq \mathrm{Hom}_{\mathrm{D^b}(\mathrm{Spec}(k))}(R\Gamma(R\mathcal{H}om(\mathcal{E}^\bullet, \mathcal{F}^\bullet)), k) \quad \text{(Cor. 3.35)} \\
&\simeq \mathrm{Hom}_{\mathrm{D^b}(X)}(\mathcal{E}^\bullet, \mathcal{F}^\bullet)^*.
\end{aligned}$$

iii) The derived categories $\mathrm{D^b}(X)$ and $\mathrm{D^b}(Y)$ of two smooth projective varieties X, respectively Y over k are endowed with Serre functors S_X, respectively S_Y (see Theorem 3.12). By definition,

$$f^! \simeq S_X \circ Lf^* \circ S_Y^{-1}$$

which yields yet another interpretation of $f^!$.

In fact, the existence of the left adjoint allows us to conclude immediately that $S_X \circ Lf^* \circ S_Y^{-1}$ is right adjoint of Rf_* (see Remark 1.31). Thus, strictly speaking Corollary 3.35 does not really require the full Grothendieck–Verdier duality; Serre duality in the form of Theorem 3.12 is enough.

Let us pass to the case of a closed embedding $i : X \hookrightarrow Y$ of codimension c of smooth varieties X and Y. If we keep the definition of the relative dualizing sheaf $\omega_i := \omega_X \otimes \omega_Y^*|_X$ as in Theorem 3.34, then (3.19) yields

Corollary 3.38 *Let $i : X \hookrightarrow Y$ be a smooth subvariety of codimension c of a smooth projective variety. For any $\mathcal{F}^\bullet \in \mathrm{D^b}(X)$ and any $\mathcal{E}^\bullet \in \mathrm{D^b}(Y)$ there exists a functorial isomorphism*

$$\mathrm{Hom}_X(\mathcal{F}^\bullet, Li^*(\mathcal{E}^\bullet) \otimes \omega_i[-c]) \simeq \mathrm{Hom}_Y(i_*\mathcal{F}^\bullet, \mathcal{E}^\bullet).$$

Proof Instead of viewing this as a special instance of the Grothendieck–Verdier duality, it can be proved directly by applying the derived version of Serre duality twice and using $Li^* \dashv i_*$. (Note that the direct image i_* need not be derived for a closed embedding.) Indeed,

$$\begin{aligned}
&\mathrm{Hom}_Y(i_*\mathcal{F}^\bullet, \mathcal{E}^\bullet) \\
&\simeq \mathrm{Hom}_Y(\mathcal{E}^\bullet, i_*\mathcal{F}^\bullet \otimes \omega_Y[\dim(Y)])^* \quad \text{(Serre duality on } Y\text{)} \\
&\simeq \mathrm{Hom}_X(Li^*(\mathcal{E}^\bullet), \mathcal{F}^\bullet \otimes i^*\omega_Y[\dim(Y)])^* \quad \text{(use } Li^* \dashv i_*\text{)} \\
&\simeq \mathrm{Hom}_X(\mathcal{F}^\bullet \otimes i^*\omega_Y[\dim(Y)], Li^*(\mathcal{E}^\bullet) \otimes \omega_X[\dim(X)]) \quad \text{(Serre duality on } X\text{)} \\
&\simeq \mathrm{Hom}_X(\mathcal{F}^\bullet, Li^*(\mathcal{E}^\bullet) \otimes \omega_X \otimes \omega_Y^*|_X[-c]).
\end{aligned}$$

□

Exercise 3.39 Give a proof that does not use Serre duality and, in particular, avoids the hypothesis Y projective.

In particular, this result allows one to compute the dual of the structure sheaf of a smooth closed subvariety.

Corollary 3.40 *Suppose $i : X \hookrightarrow Y$ is a smooth closed subvariety of codimension c of a smooth variety Y. The derived dual of $i_*\mathcal{O}_X$ is given by*

$$(i_*\mathcal{O}_X)^\vee \simeq i_*\omega_X \otimes \omega_Y^*[-c].$$

Proof It suffices to show that there are isomorphisms

$$\mathrm{Hom}_Y(\mathcal{G}^\bullet, (i_*\mathcal{O}_X)^\vee) \simeq \mathrm{Hom}_Y(\mathcal{G}^\bullet, i_*\omega_X \otimes \omega_Y^*[-c])),$$

which are functorial in $\mathcal{G}^\bullet \in \mathrm{D^b}(Y)$.

This follows from

$$\begin{aligned}\mathrm{Hom}_Y(\mathcal{G}^\bullet, (i_*\mathcal{O}_X)^\vee) &\simeq \mathrm{Hom}_Y(\mathcal{G}^\bullet \otimes^L i_*\mathcal{O}_X, \mathcal{O}_Y) \ \ \text{(use (3.14))}\\ &\simeq \mathrm{Hom}_Y(i_*Li^*(\mathcal{G}^\bullet), \mathcal{O}_Y) \ \ \text{(projection formula)}\\ &\simeq \mathrm{Hom}_X(Li^*(\mathcal{G}^\bullet), \omega_X \otimes \omega_Y^*|_X[-c]) \ \ \text{(Cor. 3.38)}\\ &\simeq \mathrm{Hom}_Y(\mathcal{G}^\bullet, i_*\omega_X \otimes \omega_Y^*[-c]) \ \ \text{(use } Li^* \dashv i_*).\end{aligned}$$

□

Examples 3.41 If $D \subset Y$ is a divisor, then together with the adjunction formula $\omega_D \simeq (\omega_Y \otimes \mathcal{O}(D))|_D$ the corollary says $(i_*\mathcal{O}_D)^\vee = i_*\mathcal{O}_D(D)[-1]$.

Exercise 3.42 Let $i : X \hookrightarrow Y$ be a smooth closed subvariety of codimension $c > 1$ of a smooth variety Y. Show that the derived dual $\mathcal{I}_X^\vee$ of its ideal sheaf $\mathcal{I}_X$ satisfies

$$\mathcal{H}^k(\mathcal{I}_X^\vee) \simeq \begin{cases} \mathcal{O}_Y & \text{if } k = 0\\ i_*\omega_X \otimes \omega_Y^* & \text{if } k = c-1\\ 0 & \text{otherwise.}\end{cases}$$

Notational convention As we will almost exclusively work on the level of derived categories and only the derived versions of the classical functors will make sense there, we shall in the sequel write $f_* : \mathrm{D^b}(X) \longrightarrow \mathrm{D^b}(Y)$ when Rf_* is meant. Similarly, we write $\mathcal{F}^\bullet \otimes \mathcal{E}^\bullet$ for the derived tensor product of two objects $\mathcal{E}^\bullet, \mathcal{F}^\bullet$ in the derived category $\mathrm{D^b}(X)$. Analogously, $R\mathcal{H}om$ becomes $\mathcal{H}om$ and Lf^* becomes f^*.

4

DERIVED CATEGORY AND CANONICAL BUNDLE – I

It turns out that the complexity of the derived category of a projective variety X depends very much on its geometry. This chapter is devoted to results by Bondal and Orlov which in particular show that for varieties with ample (anti-)canonical bundle the bounded derived category of coherent sheaves determines the variety. This will be proved in Section 4.1. Except for the case of elliptic curves, this settles completely the classification of derived categories of smooth projective varieties in dimension one.

In Section 4.2 we present a description of the group of autoequivalences of the bounded derived category of such varieties again due to Bondal and Orlov. The complexity of the triangulated category $\mathrm{D^b}(X)$ is reflected by the group of autoequivalences of $\mathrm{D^b}(X)$. Thus, these results say that derived categories of coherent sheaves on projective varieties with ample (anti-)canonical bundle tend to be easier than those for, e.g. abelian varieties or K3 surfaces. We will make this more explicit in later chapters.

We start out with the following general statement that applies to any smooth projective variety over a field k.

Proposition 4.1 *Let X and Y be smooth projective varieties over a field k. If there exists an exact equivalence*

$$\mathrm{D^b}(X) \xrightarrow{\sim} \mathrm{D^b}(Y)$$

of their derived categories, then

$$\dim(X) = \dim(Y).$$

Moreover, their canonical bundles ω_X and ω_Y are of the same order.

Proof The order of ω_X is the smallest positive integer $m \in \mathbb{Z}$ such that $\omega_X^{\otimes m}$ is isomorphic to the trivial bundle. The assertion of the proposition includes the case of infinite order.

Since both varieties are smooth projective, the derived categories $\mathrm{D^b}(X)$ and $\mathrm{D^b}(Y)$ come with natural Serre functors, e.g. $S_X(\mathcal{F}^\bullet) = \mathcal{F}^\bullet \otimes \omega_X[\dim(X)]$ for any $\mathcal{F}^\bullet \in \mathrm{D^b}(X)$ (cf. Theorem 3.12). Moreover, Lemma 1.30 tells us that any equivalence $F : \mathrm{D^b}(X) \longrightarrow \mathrm{D^b}(Y)$ commutes with S_X and S_Y.

Fix a closed point $x \in X$. Then $k(x) \simeq k(x) \otimes \omega_X \simeq S_X(k(x))[-\dim(X)]$ and, hence,

$$\begin{aligned} F(k(x)) &\simeq F(k(x) \otimes \omega_X) = F(S_X(k(x))[-\dim(X)]) \\ &\simeq F(S_X(k(x)))[-\dim(X)], \quad \text{since } F \text{ is exact} \\ &\simeq S_Y(F(k(x)))[-\dim(X)], \quad \text{since } F \circ S_X \simeq S_Y \circ F \\ &= F(k(x)) \otimes \omega_Y[\dim(Y) - \dim(X)]. \end{aligned}$$

Since F is an equivalence, $F(k(x))$ is a non-trivial bounded complex. If i is maximal (respectively minimal) with $\mathcal{H}^i(F(k(x))) \neq 0$, then we find (using that tensoring with the line bundle ω_Y commutes with cohomology):

$$\begin{aligned} 0 \neq \mathcal{H}^i(F(k(x))) &\simeq \mathcal{H}^i\left(F(k(x)) \otimes \omega_Y[\dim(Y) - \dim(X)]\right) \\ &\simeq \mathcal{H}^{i+\dim(Y)-\dim(X)}(F(k(x))) \otimes \omega_Y \end{aligned}$$

and hence $0 \neq \mathcal{H}^{i+\dim(Y)-\dim(X)}(F(k(x)))$, which contradicts the maximality (respectively minimality) if $\dim(Y) - \dim(X) > 0$ (< 0, respectively). Hence, $\dim(X) = \dim(Y) =: n$.

Suppose, $\omega_X^k \simeq \mathcal{O}_X$. Then $S_X^k[-kn] \simeq \mathrm{id}$ and hence

$$F^{-1} \circ S_Y^k[-kn] \circ F \simeq S_X^k[-kn] \simeq \mathrm{id}.$$

The latter clearly means $S_Y^k[-kn] \simeq \mathrm{id}$ and, therefore, $\omega_Y^k \simeq \mathcal{O}_Y$. □

Remark 4.2 Later we shall give another argument using the existence and uniqueness of the Fourier–Mukai kernel. See Corollary 5.21.

4.1 Ample (anti-)canonical bundle

It turns out that for smooth projective varieties with ample (anti-)canonical bundle ω_X the geometry of X is encoded by the derived category $\mathrm{D^b}(X)$ in a rather direct way. Before proving that $\mathrm{D^b}(X)$ actually determines X, let us show how to characterize certain geometric structures, like points or line bundles, intrinsically as objects in the derived category.

Definition 4.3 *Let $\mathcal{D}$ be a k-linear triangulated category with a Serre functor S. An object $P \in \mathcal{D}$ is called* point like of codimension d *if*

i) $S(P) \simeq P[d]$,
ii) $\mathrm{Hom}(P, P[i]) = 0$ *for* $i < 0$, *and*
iii) $k(P) := \mathrm{Hom}(P, P)$ *is a field.*

An object P satisfying iii) is called *simple*. As we continue to assume that all Hom's are finite-dimensional, the field $k(P)$ in iii) is automatically a finite field extension of k. So, if k is algebraically closed, it is just k.

Exercise 4.4 Suppose X is a smooth projective variety. Show that any point like object in $\mathrm{D^b}(X)$ is of codimension $d = \dim(X)$.

A sheaf $\mathcal{F}$ on X is *simple* if $\mathrm{Hom}(\mathcal{F}, \mathcal{F})$ is a field. Show that any simple sheaf $\mathcal{F}$ on a variety X with $\omega_X \simeq \mathcal{O}_X$ defines a point like object in $\mathrm{D^b}(X)$.

Lemma 4.5 *Suppose $\mathcal{F}^\bullet$ is a simple object in $\mathrm{D^b}(X)$ with with zero-dimensional support. If $\mathrm{Hom}(\mathcal{F}^\bullet, \mathcal{F}^\bullet[i]) = 0$ for $i < 0$, then*

$$\mathcal{F}^\bullet \simeq k(x)[m]$$

for some closed point $x \in X$ and some integer m.

Proof Let us first show that $\mathcal{F}^\bullet$ is concentrated in exactly one closed point. Otherwise $\mathcal{F}^\bullet$ could be written as a direct sum $\mathcal{F}^\bullet \simeq \mathcal{F}_1^\bullet \oplus \mathcal{F}_2^\bullet$ with $\mathcal{F}_j^\bullet \not\simeq 0$, $j = 1, 2$ (see Lemma 3.9). (The extra information that $\mathrm{supp}(\mathcal{F}_1^\bullet) \cap \mathrm{supp}(\mathcal{F}_2^\bullet) = \varnothing$ provided by the same lemma is not needed here.) However, a direct sum of two non-trivial complexes is never simple. Indeed, the projection to one of the two summands is not invertible.

Thus, we may assume that the support of all cohomology sheaves $\mathcal{H}^i$ of $\mathcal{F}^\bullet$ consists of the same closed point $x \in X$. Set

$$m_0 := \max\{i \mid \mathcal{H}^i \neq 0\} \quad \text{and} \quad m_1 := \min\{i \mid \mathcal{H}^i \neq 0\}.$$

Since both sheaves $\mathcal{H}^{m_0}$ and $\mathcal{H}^{m_1}$ are concentrated in $x \in X$, there exists a non-trivial homomorphism $\mathcal{H}^{m_0} \longrightarrow \mathcal{H}^{m_1}$.

(Indeed, if M is a finite module over a local noetherian ring $(A, \mathfrak{m})$ such that $\mathrm{supp}(M) = \{\mathfrak{m}\}$, then there exists a surjection $M \twoheadrightarrow A/\mathfrak{m}$ and an injection $A/\mathfrak{m} \hookrightarrow M$.)

The composition

$$\mathcal{F}^\bullet[m_0] \longrightarrow \mathcal{H}^{m_0} \longrightarrow \mathcal{H}^{m_1} \longrightarrow \mathcal{F}^\bullet[m_1],$$

where we apply Exercise 2.32, is non-trivial. By ii) this shows $m_0 = m_1$. Hence, $\mathcal{F}^\bullet \simeq \mathcal{F}[m]$ with $\mathcal{F}$ a coherent sheaf with support in x and $m := m_0 = m_1$.

The only such sheaf which is also simple is $k(x)$. Indeed, for any other sheaf, the homomorpism which is given by a non-trivial map from a quotient of $\mathcal{F}$ of the form $k(x)$ into the socle of $\mathcal{F}$ yields a non-invertible homomorphism. Hence, $\mathcal{F}^\bullet \simeq k(x)[m]$. □

Proposition 4.6 **(Bondal, Orlov)** *Let X be a smooth projective variety. Suppose that ω_X or ω_X^* is ample. Then the point like objects in $\mathrm{D^b}(X)$ are the objects which are isomorphic to $k(x)[m]$, where $x \in X$ is a closed point and $m \in \mathbb{Z}$. See [15].*

Proof The Serre functor on $\mathrm{D^b}(X)$ is given by $\mathcal{F}^\bullet \longmapsto \omega_X \otimes \mathcal{F}^\bullet[n]$, where n is the dimension of X. Thus, the skyscraper sheaf $k(x)$ of a closed point $x \in X$, as well as any shift $k(x)[m]$, does satisfy all three conditions in Definition 4.3. (In fact, ii) holds for any sheaf.)

Let us now assume that $P \in \mathrm{D^b}(X)$ satisfies i)–iii). We denote by $\mathcal{H}^i$ the cohomology sheaves of P, which are not all zero. Then condition i), which ensures $\mathcal{H}^i \otimes \omega_X[n] \simeq \mathcal{H}^i[d]$, yields $d = n$ and $\mathcal{H}^i \simeq \mathcal{H}^i \otimes \omega_X$.

Since ω_X or ω_X^* is ample, the latter condition shows that $\mathcal{H}^i$ is supported in dimension zero. Indeed, the Hilbert polynomial $P_{\mathcal{F}}(k) = \chi(\mathcal{F} \otimes \omega_X^k)$ of any coherent sheaf $\mathcal{F}$ is of degree $\dim \mathrm{supp}(\mathcal{F})$ (see [33]) and hence taking the tensor product of $\mathcal{F}$ with ω_X makes a difference if $\dim \mathrm{supp}(\mathcal{F}) > 0$.

The assertion now follows from Lemma 4.5. □

Remark 4.7 The condition on the canonical bundle is necessary. E.g. if ω_X is trivial (like for an abelian variety), then $\mathcal{O}_X$ (or any other simple sheaf) is a point like object.

Note also that the proof shows that, even without any positivity assumption on ω_X, any point like object in $\mathrm{D^b}(X)$ is of codimension $\dim(X)$.

One may also try to characterize line bundles as objects in the derived category. Let us first give the abstract definition.

Definition 4.8 *Let $\mathcal{D}$ be a triangulated category with a Serre functor S. An object $L \in \mathcal{D}$ is called* invertible *if for any point like object $P \in \mathcal{D}$ there exists $n_P \in \mathbb{Z}$ (depending also on L) such that*

$$\mathrm{Hom}(L, P[i]) = \begin{cases} k(P) & \text{if } i = n_P \\ 0 & \text{otherwise.} \end{cases}$$

Proposition 4.9 (Bondal, Orlov) *Let X be a smooth projective variety. Any invertible object in $\mathrm{D^b}(X)$ is of the form $L[m]$ with L a line bundle on X and $m \in \mathbb{Z}$. Conversely, if ω_X or ω_X^* is ample, then for any line bundle L and any $m \in \mathbb{Z}$ the object $L[m] \in \mathrm{D^b}(X)$ is invertible. See [15].*

Proof Let us suppose that L is an invertible object in $\mathrm{D^b}(X)$ and let m be maximal with $\mathcal{H}^m := \mathcal{H}^m(L) \not\simeq 0$. In particular, there exists the natural morphism

$$L \longrightarrow \mathcal{H}^m[-m]$$

inducing the identity on the m-th cohomology (see Exercise 2.32).

Pick a point $x_0 \in \mathrm{supp}(\mathcal{H}^m)$. Then there exists a non-trivial homomorphism $\mathcal{H}^m \longrightarrow k(x_0)$. Hence,

$$0 \neq \mathrm{Hom}(\mathcal{H}^m, k(x_0)) = \mathrm{Hom}(L, k(x_0)[-m])$$

and, therefore, $n_{k(x_0)} = -m$.

This could also have been deduced from the spectral sequence

$$E_2^{p,q} = \mathrm{Hom}(\mathcal{H}^{-q}(L), k(x_0)[p]) \Rightarrow \mathrm{Hom}(L, k(x_0)[p+q]). \tag{4.1}$$

A similar argument has been used before in the proof of Proposition 3.17.

Apply the same spectral sequence to deduce

$$E_2^{1,-m} = \mathrm{Hom}(\mathcal{H}^m, k(x_0)[1]) = \mathrm{Hom}(L, k(x_0)[1 + n_{k(x_0)}]) = 0.$$

Thus, as soon as $x_0 \in X$ is in the support of $\mathcal{H}^m$, we obtain $\mathrm{Ext}^1(\mathcal{H}^m, k(x_0)) = 0$.

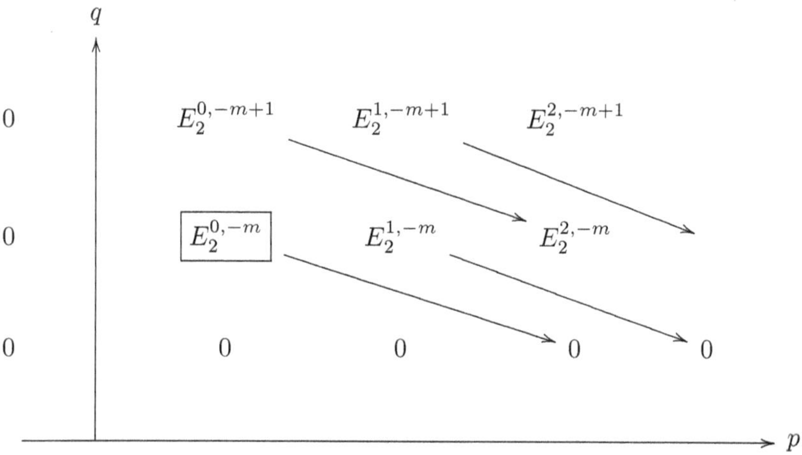

Next, we shall apply the following standard result in commutative algebra (cf. [17, X.3 Prop.4]): Any finite module M over an arbitrary noetherian local ring $(A, \mathfrak{m})$ with $\mathrm{Ext}^1_A(M, A/\mathfrak{m}) = 0$ is free.

The local-to-global spectral sequence (3.16), p. 85,

$$E_2^{p,q} = H^p(X, \mathcal{E}xt^q(\mathcal{H}^m, k(x_0))) \Rightarrow \mathrm{Ext}^{p+q}(\mathcal{H}^m, k(x_0))$$

allows us to pass from the global vanishing $\mathrm{Ext}^1(\mathcal{H}^m, k(x_0)) = 0$ to the local one $\mathcal{E}xt^1(\mathcal{H}^m, k(x_0)) = 0$. More precisely, as $\mathcal{E}xt^0(\mathcal{H}^m, k(x_0))$ is concentrated in $x_0 \in X$, one has

$$E_2^{2,0} = H^2(X, \mathcal{E}xt^0(\mathcal{H}^m, k(x_0))) = 0.$$

Therefore, $E_2^{0,1} = E_\infty^{0,1}$. Since $\mathcal{E}xt^1(\mathcal{H}^m, k(x_0))$ is also concentrated in $x_0 \in X$, it is a globally generated sheaf. Hence,

$$H^0(X, \mathcal{E}xt^1(\mathcal{H}^m, k(x_0))) = E_2^{0,1} = 0$$

implies $\mathcal{E}xt^1(\mathcal{H}^m, k(x_0)) = 0$. But then the aforementioned result from commutative algebra shows that $\mathcal{H}^m$ is free in $x_0 \in X$.

Since X is irreducible, we have in particular $\mathrm{supp}(\mathcal{H}^m) = X$. Thereby, there exists for any $x \in X$ a surjection $\mathcal{H}^m \twoheadrightarrow k(x)$. Hence,

$$\mathrm{Hom}(L, k(x)[-m]) = \mathrm{Hom}(\mathcal{H}^m, k(x)) \neq 0.$$

In particular, $n_{k(x)}$ does not depend on x. As by assumption,

$$k(x) = \mathrm{Hom}(L, k(x)[-m]) = \mathrm{Hom}(\mathcal{H}^m, k(x)),$$

the sheaf $\mathcal{H}^m$ has constant fibre dimension one. Hence, $\mathcal{H}^m$ is a line bundle.

It remains to show that $\mathcal{H}^i = 0$ for $i < m$. We use again the spectral sequence (4.1). Since $\mathcal{H}^m$ is locally free, the row $E_2^{q,-m}$ is trivial except for $q = 0$. Indeed,

$$E_2^{q,-m} = \mathrm{Ext}^q(\mathcal{H}^m, k(x)) = H^q(X, \mathcal{H}^{m*} \otimes k(x)) = 0$$

for $q > 0$.

The rest of the argument is by induction. Assume we have shown $\mathcal{H}^i = 0$ for $i_0 < i$. Then $E_2^{0,-i_0} = E_\infty^{0,-i_0}$. Since

$$E^{-i_0} = \mathrm{Hom}(L, k(x)[-i_0]) = 0,$$

this implies that $\mathrm{Hom}(\mathcal{H}^{i_0}, k(x)) = 0$ for any $x \in X$, i.e. $\mathcal{H}^{i_0} = 0$.

Let us now show that conversely any line bundle L on X, and hence any shift $L[m]$, defines an invertible object in $\mathrm{D^b}(X)$ whenever the (anti-)canonical bundle is ample.

By Proposition 4.6 the assumption on the canonical bundle implies that point like objects in $\mathrm{D^b}(X)$ are of the form $k(x)[m]$ for some closed point $x \in X$ and some $m \in \mathbb{Z}$. Hence,

$$\begin{aligned}\mathrm{Hom}(L[m], P[i]) &\simeq \mathrm{Hom}(L[m], k(x)[i+n])\\ &= H^0(X, L^*(x)[i+n-m])\\ &= H^{i+n-m}(X, L^*(x)) = 0\end{aligned}$$

except for $i = m - n$ when it is $k(x)$. We set $n_P := m - n$. □

Remark 4.10 If one naively defines the Picard group of a triangulated category endowed with a Serre functor as the set of invertible objects, then for smooth projective varieties with ample (anti-)canonical bundle ω_X one has $\mathrm{Pic}(\mathrm{D^b}(X)) = \mathrm{Pic}(X) \times \mathbb{Z}$.

Note however that varieties where such an easy description of all invertible objects does not hold are common.

Proposition 4.11 **(Bondal, Orlov)** *Let X and Y be smooth projective varieties and assume that the (anti-)canonical bundle of X is ample. If there exists an exact equivalence $\mathrm{D^b}(X) \simeq \mathrm{D^b}(Y)$, then X and Y are isomorphic. In particular, the (anti)-canonical bundle of Y is also ample. See [15].*

Proof Let us first sketch the idea of the proof which is strikingly simple. Assume that under an equivalence $F : \mathrm{D^b}(X) \longrightarrow \mathrm{D^b}(Y)$ the structure sheaf $\mathcal{O}_X$ is mapped to $\mathcal{O}_Y$. Since any equivalence is compatible with Serre functors (see Lemma 1.30) and $\dim(X) = \dim(Y) =: n$ (see Proposition 4.1), this proves

$$\begin{aligned}F(\omega_X^k) = F(S_X^k(\mathcal{O}_X))[-kn] &\simeq S_Y^k(F(\mathcal{O}_X))[-kn]\\ &\simeq S_Y^k(\mathcal{O}_Y)[-kn] = \omega_Y^k.\end{aligned}$$

Using that F is fully faithful, we conclude from this that

$$\begin{aligned} H^0(X, \omega_X^k) &= \mathrm{Hom}(\mathcal{O}_X, \omega_X^k) \simeq \mathrm{Hom}(F(\mathcal{O}_X), F(\omega_X^k)) \\ &= \mathrm{Hom}(\mathcal{O}_Y, \omega_Y^k) = H^0(Y, \omega_Y^k) \end{aligned}$$

for all k.

The product in $\bigoplus H^0(X, \omega_X^k)$ can be expressed in terms of compositions: namely, for $s_i \in H^0(X, \omega_X^{k_i}) = \mathrm{Hom}(\mathcal{O}_X, \omega_X^{k_i})$ one has

$$s_1 \cdot s_2 = S_X^{k_1}(s_2)[-k_1 n] \circ s_1$$

and similarly for sections on Y.

Hence, the induced bijection

$$\bigoplus H^0(X, \omega_X^k) \simeq \bigoplus H^0(Y, \omega_Y^k)$$

is a ring isomorphism. If the (anti-)canonical bundle of Y is also ample, then this shows

$$X \simeq \mathrm{Proj}\left(\bigoplus H^0(X, \omega_X^k)\right) \simeq \mathrm{Proj}\left(\bigoplus H^0(Y, \omega_Y^k)\right) \simeq Y.$$

Thus, under the two assumptions that $F(\mathcal{O}_X) \simeq \mathcal{O}_Y$ and that ω_Y (or ω_Y^*) is ample we have proved the assertion. Let us now explain how to reduce to this situation.

As the notions of point like and invertible objects in $\mathrm{D^b}$ are intrinsic, an exact equivalence $F : \mathrm{D^b}(X) \longrightarrow \mathrm{D^b}(Y)$ induces bijections

$$\begin{array}{ccc} \{\text{point like objects in } \mathrm{D^b}(X)\} & \stackrel{(*)}{\longleftrightarrow} & \{\text{point like objects in } \mathrm{D^b}(Y)\} \\ \Big\Vert & & \big\uparrow \\ \{k(x)[m] \mid x \in X, m \in \mathbb{Z}\} & & \{k(y)[m] \mid y \in Y, m \in \mathbb{Z}\} \end{array}$$

and

$$\begin{array}{ccc} \{\text{invertible objects in } \mathrm{D^b}(X)\} & \stackrel{(**)}{\longleftrightarrow} & \{\text{invertible objects in } \mathrm{D^b}(Y)\} \\ \Big\Vert & & \big\downarrow \\ \{L[m] \mid L \in \mathrm{Pic}(X)\} & & \{M[m] \mid M \in \mathrm{Pic}(Y)\}. \end{array}$$

As we have seen in Proposition 4.6, the point like objects in $\mathrm{D^b}(X)$ are all of the form $k(x)[m]$ for $x \in X$ a closed point and $m \in \mathbb{Z}$. By Proposition 4.9 any line bundle L, in particular $L = \mathcal{O}_X$, defines an invertible object in $\mathrm{D^b}(X)$. Thus, by $(**)$ also $F(\mathcal{O}_X)$ is an invertible object in $\mathrm{D^b}(Y)$ and hence, due to Proposition 4.9, of the form $M[m]$ for some line bundle M on Y.

Compose F with the two equivalences given by $M^* \otimes (\ \ )$, respectively by the shift T^{-m}. The new equivalence, which we continue to call F, satisfies $F(\mathcal{O}_X) \simeq \mathcal{O}_Y$.

In order to prove the ampleness of the (anti-)canonical bundle ω_Y, we shall first prove that point like objects in $\mathrm{D^b}(Y)$ are of the form $k(y)[m]$. We will conclude this, without assuming any positivity of ω_Y, simply from the existence of the equivalence F.

Due to $(*)$, one finds for any closed point $y \in Y$ a closed point $x_y \in X$ and an integer m_y such that $k(y) \simeq F(k(x_y)[m_y])$.

Suppose there exists a point like object $P \in \mathrm{D^b}(Y)$ which is not of the form $k(y)[m]$ and denote by $x_P \in X$ the closed point with $F(k(x_P)[m_P]) \simeq P$ for a certain $m_P \in \mathbb{Z}$.

Note that $x_P \neq x_y$ for all $y \in Y$. Hence we have for all $y \in Y$ and all $m \in \mathbb{Z}$

$$\begin{aligned}\mathrm{Hom}(P, k(y)[m]) &= \mathrm{Hom}(F(k(x_P))[m_p], F(k(x_y))[m_y + m]) \\ &= \mathrm{Hom}(k(x_P), k(x_y)[m_y + m - m_P]) = 0.\end{aligned}$$

Since the objects $k(y)[m]$ form a spanning class in $\mathrm{D^b}(Y)$ (see Proposition 3.17), this implies $P \simeq 0$, which is absurd. Hence, point like objects in $\mathrm{D^b}(Y)$ are exactly the objects of the form $k(y)[m]$.

Note that together with $F(\mathcal{O}_X) \simeq \mathcal{O}_Y$ this also implies that for any closed point $x \in X$ there exists a closed point $y \in Y$ such that $F(k(x)) \simeq k(y)$ (no shifts!). This is due to the easy observation that $m = 0$ if and only if $\mathrm{Hom}(\mathcal{O}_Y, k(y)[m]) \neq 0$.

It remains to prove that with ω_X^k (very) ample also ω_Y^k is (very) ample. Let us prove this in a very geometric way by showing that some power ω_Y^k separates points and tangents (cf. [45, II.7]). (Here we assume that k is algebraically closed, but see Remark 4.12.)

We continue to use that for any $k(y)$, with $y \in Y$ a closed point, there exists a closed point $x_y \in X$ with $F(k(x_y)) = k(y)$ and that $F(\omega_X^k) = \omega_Y^k$ for all $k \in \mathbb{Z}$.

The line bundle ω_Y^k separates points if for any two points $y_1 \neq y_2 \in Y$ the restriction map

$$r_{y_1,y_2} : \omega_Y^k \longrightarrow \omega_Y^k(y_1) \oplus \omega_Y^k(y_2) \simeq k(y_1) \oplus k(y_2)$$

induces a surjection $H^0(r_{y_1,y_2}) : H^0(Y, \omega_Y^k) \longrightarrow H^0(k(y_1) \oplus k(y_2))$. Let us denote $x_i := x_{y_i}$, $i = 1, 2$. Then

$$\begin{aligned}r_{y_1,y_2} \in \mathrm{Hom}(\omega_Y^k, k(y_1) \oplus k(y_2)) &\simeq \mathrm{Hom}(F(\omega_X^k), F(k(x_1) \oplus k(x_2))) \\ &\simeq \mathrm{Hom}(\omega_X^k, k(x_1) \oplus k(x_2)).\end{aligned}$$

It indeed corresponds to the restriction map $r_{x_1,x_2} : \omega_X^k \longrightarrow k(x_1) \oplus k(x_2)$ (up to isomorphism, which we will ignore), as there is only one non-trivial homomorphism $\omega_X^k \longrightarrow k(x_i)$ (up to scaling).

Altogether this yields the commutative diagram:

$$\begin{array}{ccc}
H^0(Y,\omega_Y^k) & \xrightarrow{H^0(r_{y_1,y_2})} & H^0(Y,k(y_1)\oplus k(y_2)) \\
\| & & \| \\
\mathrm{Hom}(\mathcal{O}_Y,\omega_Y^k) & \xrightarrow{r_{y_1,y_2}\circ} & \mathrm{Hom}(\mathcal{O}_Y,k(y_1)\oplus k(y_2)) \\
\| & & \| \\
\mathrm{Hom}(\mathcal{O}_X,\omega_X^k) & \xrightarrow{r_{x_1,x_2}\circ} & \mathrm{Hom}(\mathcal{O}_X,k(x_1)\oplus k(x_2)) \\
\| & & \| \\
H^0(X,\omega_X^k) & \xrightarrow{H^0(r_{x_1,x_2})} & H^0(X,k(x_1)\oplus k(x_2)).
\end{array}$$

As, by assumption, the line bundle ω_X^k is very ample for $k \gg 0$ (or $k \ll 0$) and, in particular, separates points, the map $H^0(r_{x_1,x_2})$ is surjective. The commutativity of the diagram allows us to conclude that also $H^0(r_{y_1,y_2})$ is surjective.

One proceeds in a similar fashion to prove that ω_Y^k separates tangent directions if ω_X^k does: Suppose $Z_y \subset Y$ is a subscheme of length two concentrated in $y \in Y$, i.e. Z_y is the point y endowed with a tangent direction. The exact sequence

$$0 \longrightarrow k(y) \longrightarrow \mathcal{O}_{Z_y} \longrightarrow k(y) \longrightarrow 0$$

shows that $\mathcal{O}_{Z_y}$ is given by a non-trivial extension class (with $x := x_y$):

$$\begin{aligned} e_Z \in \mathrm{Hom}(k(y),k(y)[1]) &= \mathrm{Hom}(F(k(x)),F(k(x))[1]) \\ &= \mathrm{Hom}(k(x),k(x)[1]). \end{aligned}$$

The latter, when viewed as a class in $\mathrm{Hom}(k(x),k(x)[1])$, defines a subscheme of length two $Z_x \subset X$ concentrated in x. Then $F(\mathcal{O}_{Z_x}) = \mathcal{O}_{Z_y}$. Moreover,

$$F\left(\omega_X^k \longrightarrow \mathcal{O}_{Z_x}\right) \simeq \omega_Y^k \longrightarrow \mathcal{O}_{Z_y},$$

where the homomorphisms on both sides are given by restriction (check this!).

As ω_X^k separates tangent directions, the restriction

$$H^0(X,\omega_X^k) \twoheadrightarrow H^0(X,\mathcal{O}_{Z_x})$$

is surjective. Now use $H^0(X,\omega_X^k) = H^0(Y,\omega_Y^k)$ and

$$\begin{aligned} H^0(Y,\mathcal{O}_{Z_y}) &\simeq \mathrm{Hom}(\mathcal{O}_Y,\mathcal{O}_{Z_y}) \simeq \mathrm{Hom}(F(\mathcal{O}_X),F(\mathcal{O}_{Z_x})) \\ &\simeq \mathrm{Hom}(\mathcal{O}_X,\mathcal{O}_{Z_x}) \simeq H^0(X,\mathcal{O}_{Z_x}), \end{aligned}$$

to deduce the surjectivity of $H^0(Y, \omega_Y^k) \twoheadrightarrow H^0(Y, \mathcal{O}_{Z_y})$, i.e. ω_Y^k separates the tangent direction in y given by $\mathcal{O}_{Z_y}$. □

Remarks 4.12 i) Bondal and Orlov give a different proof for the ampleness of ω_Y^k which is maybe less geometric, but has the advantage of working for fields that are not algebraically closed. They use rather directly the induced bijection between the sets of closed points of X, respectively Y.

ii) It is noteworthy that the above proof only uses that the equivalence is graded, i.e. that it commutes with the shift functor, but not that it maps a distinguished triangle to a distinguished triangle.

A different proof relying on the description that any equivalence is a Fourier–Mukai transform will be given in Section 6.1 (cf. Proposition 6.1 and Exercise 6.2) by proving that the (anti-)canonical rings of two smooth projective varieties with equivalent derived categories are isomorphic. This immediately yields the above proposition if we assume the same positivity for the two canonical bundles ω_X and ω_Y.

iii) Yet another proof of the above proposition, relying more on the original techniques of Gabriel [38] and of Thomason and Trobaugh [113] (see [38]), can be found in [101].

iv) Once the result is established, the reader might safely forget the notion of point like and invertible objects. They are not used any further and don't seen to appear anywhere else in the theory.

Corollary 4.13 *Let C be a curve of genus $g(C) \neq 1$ and let Y be a smooth projective variety. Then there exists an exact equivalence $\mathrm{D}^{\mathrm{b}}(C) \simeq \mathrm{D}^{\mathrm{b}}(Y)$ if and only if Y is a curve isomorphic to C.*

Proof If $g(C) = 0$, then $C \simeq \mathbb{P}^1$ and ω_C^* is ample. If $g(C) > 1$, then ω_C is ample. In both cases, the result of Bondal and Orlov applies. □

The remaining case of elliptic curves will be discussed in Section 5.2.

Remark 4.14 Kawamata in [63] has refined Proposition 4.11. He shows that also the nefness of the canonical bundle is preserved. Moreover, his more geometric approach allows him to construct birational correspondences between varieties of general type realizing the same derived category. This will be explained in Proposition 6.3.

4.2 Autoequivalences for ample (anti-)canonical bundle

After having discussed which projective varieties with ample (anti-)canonical bundle have equivalent derived categories, we now pass on to the question of how these equivalences arc realized. This immediately reduces to a description of the group of all autoequivalences of the bounded derived category of a smooth projective variety X. Here and in the sequel an *autoequivalence* means an exact k-linear equivalence $\mathrm{D}^{\mathrm{b}}(X) \xrightarrow{\sim} \mathrm{D}^{\mathrm{b}}(X)$. The set of all isomorphism classes of autoequivalences of $\mathrm{D}^{\mathrm{b}}(X)$ will be denoted $\mathrm{Aut}(\mathrm{D}^{\mathrm{b}}(X))$.

Examples 4.15 i) Any automorphism $f : X \longrightarrow X$ induces the autoequivalence

$$f_* : \mathrm{D^b}(X) \xrightarrow{\sim} \mathrm{D^b}(X).$$

Its inverse is given by $f^* : \mathrm{D^b}(X) \xrightarrow{\sim} \mathrm{D^b}(X)$.

ii) The shift functor generates a subgroup of $\mathrm{Aut}(\mathrm{D^b}(X))$ naturally isomorphic to $\mathbb{Z}$.

iii) If L is a line bundle on X, then

$$\otimes(\ ) : \mathrm{D^b}(X) \xrightarrow{\sim} \mathrm{D^b}(X)$$

is yet a third type of equivalence, which is isomorphic to the identity if and only if L is trivial. Hence, $\mathrm{Pic}(X) \hookrightarrow \mathrm{Aut}(\mathrm{D^b}(X))$.

Exercise 4.16 Show that the set $\mathrm{Aut}(\mathrm{D^b}(X))$ of all isomorphism classes of autoequivalences indeed forms a group.

As is shown by the next proposition, any autoequivalence of $\mathrm{D^b}(X)$, where X is a projective variety with ample (anti-)canonical bundle, is a composition of autoequivalences of type i)-iii).

Proposition 4.17 **(Bondal, Orlov)** *Let X be a smooth projective variety with ample (anti-)canonical bundle.*

The group of autoequivalences of $\mathrm{D^b}(X)$ *is generated by:* i) *automorphisms of* X, ii) *the shift functor T, and* iii) *twists by line bundles.*

In other words, one has

$$\mathrm{Aut}(\mathrm{D^b}(X)) \simeq \mathbb{Z} \times (\mathrm{Aut}(X) \ltimes \mathrm{Pic}(X)).$$

Proof In the proof of Proposition 4.11 we have seen that for any autoequivalence $F : \mathrm{D^b}(X) \longrightarrow \mathrm{D^b}(X)$ there exists a line bundle $M \in \mathrm{Pic}(X)$ and an integer m such that the composition of F with the twist functor $M \otimes (\ )$ and the shift functor T^{-m} maps line bundles to line bundles and, more specifically, $\mathcal{O}_X$ to $\mathcal{O}_X$. So we may assume that F already has this property.

We have furthermore seen that in this situation $F(\omega_X^{\otimes k}) = \omega_X^{\otimes k}$ for all $k \in \mathbb{Z}$ and that the induced isomorphism $\bigoplus H^0(X, \omega_X^k) = \bigoplus H^0(X, \omega_X^k)$ is compatible with the multiplicative structure.

Thus, F defines an automorphism of the graded ring $\bigoplus H^0(X, \omega_X^k)$ and thus an automorphism φ of $X \simeq \mathrm{Proj}\left(\bigoplus H^0(X, \omega_X^k)\right)$. Composing F with the equivalence φ^*, we obtain an autoequivalence which still maps line bundles to line bundles, $\mathcal{O}_X$ to $\mathcal{O}_X$, but which also acts as the identity on $\bigoplus H^0(X, \omega_X^k)$.

We conclude by proving that any autoequivalence with these properties is in fact isomorphic to the identity. This is an immediate consequence of Proposition 4.23, to be discussed in the next section, applied to the ample sequence $\omega_X^{\otimes k}$ (cf. Proposition 3.18).

As the shift functor commutes with line bundle twists and shifts and

$$\varphi^*(L \otimes \mathcal{E}^\bullet) \simeq \varphi^*(L) \otimes \varphi^*(\mathcal{E}^\bullet),$$

the fact that the autoequivalences i)-iii) generate $\mathrm{D^b}(X)$ suffices to conclude the asserted description of $\mathrm{Aut}(\mathrm{D^b}(X))$ as a group. □

Remark 4.18 Much more interesting is the group $\mathrm{Aut}(\mathrm{D^b}(X))$ for a smooth projective variety X with trivial canonical bundle. A complete description in the case of abelian varieties, due to Mukai and Orlov, will be given in Chapter 9. The case of K3 surfaces is still largely open, although a conjectural description has been put forward in [21].

4.3 Ample sequences in derived categories

Let us recall the definition of an ample sequence in an abelian category (see Section 3.1).

Definition 4.19 *Let $\mathcal{A}$ be a k-linear abelian category with finite-dimensional Hom's. A sequence of objects $L_i \in \mathcal{A}$, $i \in \mathbb{Z}$, is called* ample *if for any object $A \in \mathcal{A}$ there exists an integer $i_0(A)$ such that for $i < i_0(A)$ one has:*

i) *The natural morphism $\mathrm{Hom}(L_i, A) \otimes_k L_i \twoheadrightarrow A$ is surjective.*
ii) *If $j \neq 0$ then $\mathrm{Hom}(L_i, A[j]) = 0$.*
iii) $\mathrm{Hom}(A, L_i) = 0$.

Remark 4.20 Let us also recall the following crucial fact proved earlier in Lemma 2.73: Let L_i, $i \in \mathbb{Z}$, be an ample sequence in a k-linear abelian category $\mathcal{A}$ of finite homological dimension. Then, considered as objects in the derived category $\mathrm{D^b}(\mathcal{A})$, the L_i span $\mathrm{D^b}(\mathcal{A})$.

So, any ample sequence naturally defines a spanning class, but the notion of an ample sequence is indeed much stronger. E.g. Proposition 4.23 below is an assertion that could be formulated for any spanning class, but which can only be proven under the assumption that the spanning class is induced by an ample sequence.

Let us also mention the following variant of Proposition 1.49. See [92].

Corollary 4.21 **(Orlov)** *Let $\mathcal{A}$ be an abelian category of finite homological dimension with an ample sequence $L_i \in \mathcal{A}$, $i \in \mathbb{Z}$. An exact functor*

$$F : \mathrm{D^b}(\mathcal{A}) \longrightarrow \mathcal{D}$$

that admits adjoints $G \dashv F \dashv H$ is fully faithful if and only if for any $j \ll 0$, $i \ll j$, and all m one has:

$$\mathrm{Hom}(L_i, L_j[m]) \simeq \mathrm{Hom}(F(L_i), F(L_j)[m]).$$

Proof If all i and j are tested, the assertion follows directly from Proposition 1.49 and the fact that $\{L_i\}$ spans $\mathrm{D^b}(\mathcal{A})$. A direct proof, also giving this slightly stronger version, goes as follows.

Let $j \ll 0$, i.e. j smaller than a given (negative) number j_0, and consider the adjunction morphism $L_j \longrightarrow H(F(L_j))$. This may be completed to a distinguished triangle:

$$L_j \longrightarrow H(F(L_j)) \longrightarrow A^\bullet \longrightarrow L_j[1].$$

The long exact sequence obtained by applying $\mathrm{Hom}(L_i, \ )$, $i \ll j$, together with

$$\mathrm{Hom}(L_i, L_j) \simeq \mathrm{Hom}(F(L_i), F(L_j)) \simeq \mathrm{Hom}(L_i, H(F(L_j)))$$

shows that $\mathrm{Hom}(L_i, A^\bullet[k]) = 0$ for all k and all $i \ll j$. Proposition 2.73 (or rather Remark 2.75) then implies $A^\bullet = 0$ and, hence, $L_j \simeq H(F(L_j))$.

Consider for an arbitrary complex $A^\bullet \in \mathrm{D^b}(\mathcal{A})$ the adjunction morphism $G(F(A^\bullet)) \longrightarrow A^\bullet$. Again, it can be completed to a distinguished triangle

$$G(F(A^\bullet)) \longrightarrow A^\bullet \longrightarrow B^\bullet \longrightarrow G(F(A^\bullet))[1].$$

Applying $\mathrm{Hom}(\ , L_j)$ to it yields a long exact sequence. Since for $j < j_0$ the adjunction morphism $L_j \longrightarrow H(F(L_j))$ is an isomorphism, one finds

$$\begin{aligned}\mathrm{Hom}(G(F(A^\bullet)), L_j[k]) &\simeq \mathrm{Hom}(F(A^\bullet), F(L_j)[k])\\ &\simeq \mathrm{Hom}(A^\bullet, H(F(L_j))[k])\\ &\simeq \mathrm{Hom}(A^\bullet, L_j[k])\end{aligned}$$

for all k and all $j < j_0$. Thus, $\mathrm{Hom}(B^\bullet, L_j[k]) = 0$ for all k and all $j < j_0$.

Applying Proposition 2.73 once more yields $B^\bullet = 0$. Hence, $G \circ F \simeq \mathrm{id}$ which is enough to conclude that F is fully faithful (cf. Remark 1.24). □

Remark 4.22 If L_i, $i \in \mathbb{Z}$, is an ample sequence, then L_{ki}, $i \in \mathbb{Z}$, is an ample sequence for any $k \neq 0$. This roughly explains why testing the standard criterion only for $i \ll j \ll 0$ suffices.

Proposition 4.23 **(Bondal, Orlov)** *Let $F : \mathrm{D^b}(\mathcal{A}) \longrightarrow \mathrm{D^b}(\mathcal{A})$ be an exact autoequivalence. Suppose*

$$f : \mathrm{id}_{\{L_i\}} \xrightarrow{\sim} F|_{\{L_i\}}$$

is an isomorphism of functors on the full subcategory $\{L_i\}$ given by an ample sequence L_i in $\mathcal{A}$.

Then there exists a unique extension to an isomorphism

$$\tilde{f} : \mathrm{id} \xrightarrow{\sim} F.$$

See [15, 92].

Proof The proof of this statement is not really complicated, but somewhat lengthy. We will split it into several steps (following closely the presentation in [92]).

Step 1 We characterize objects in $\mathcal{A}$ in terms of the ample sequence as follows:

An objcct $A^\bullet \in \mathrm{D}^{\mathrm{b}}(\mathcal{A})$ is isomorphic to an object in $\mathcal{A}$ if and only if

$$\mathrm{Hom}_{\mathrm{D}^{\mathrm{b}}(\mathcal{A})}(L_i, A^\bullet[j]) = 0$$

for all $j \neq 0$ and $i \ll 0$.

One direction is immediate and the other one can easily be verified by using the spectral sequence (see Example 2.70):

$$E_2^{p,q} = \mathrm{Hom}_{\mathcal{A}}(L_i, H^q(A^\bullet)[p]) \Rightarrow \mathrm{Hom}_{\mathrm{D}^{\mathrm{b}}(\mathcal{A})}(L_i, A^\bullet[p+q]).$$

(We assume for simplicity that $\mathcal{A}$ has enough injectives. Later for $\mathcal{A} = \mathbf{Coh}(X)$ we will embed $\mathcal{A}$ into $\mathbf{Qcoh}(X)$, which has enough injectives, and consider the spectral sequence there.) Since $A^\bullet$ is a bounded complex, its cohomology is concentrated in, say, $[-k, k]$.

Hence, $E_2^{p,q} = 0$ for $|q| > k$. Due to condition ii) in the definition, one may find for any of the finitely many non-trivial cohomology objects $H^q(A^\bullet)$ an i_0 such that $\mathrm{Hom}(L_i, H^q(A^\bullet)[p]) = 0$ for $i < i_0$ and all $p \neq 0$. Let us fix one i_0 that works for all of them.

Thus, the spectral sequence is entirely supported on a finite segment of the vertical axis:

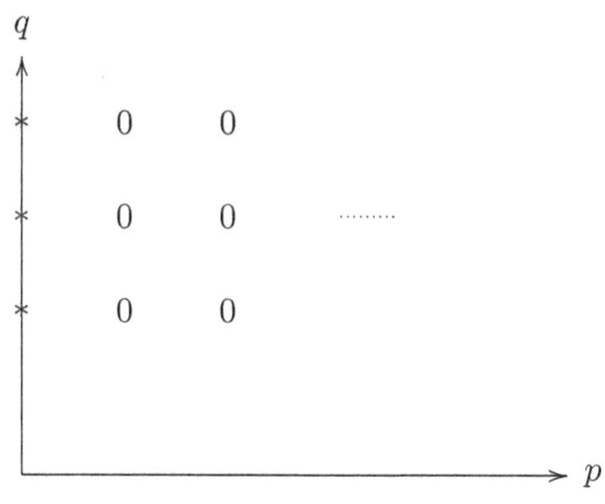

Thus, $\mathrm{Hom}(L_i, H^q(A^\bullet)) = \mathrm{Hom}(L_i, A^\bullet[q])$ for all q and all $i < i_0$. So, if $\mathrm{Hom}(L_i, A^\bullet[j]) = 0$ for all $j \neq 0$ and $i \ll 0$, then $\mathrm{Hom}(L_i, H^q(A^\bullet)) = 0$ for $i \ll 0$ and $q \neq 0$. Using property i) in the definition of an ample sequence applied to $H^q(A^\bullet)$ shows that $H^q(A^\bullet) = 0$ for $q \neq 0$. Thus, $A^\bullet$ is isomorphic to an object in $\mathcal{A}$.

Step 2 We show that for any $A \in \mathcal{A}$ also $F(A) \in \mathcal{A}$. Indeed, using the assumption that $\mathrm{id} \simeq F$ on $\{L_i\}$ yields

$$\mathrm{Hom}(L_i, F(A)[j]) \simeq \mathrm{Hom}(F(L_i), F(A)[j]) \simeq \mathrm{Hom}(L_i, A[j]) = 0$$

for all $j \neq 0$ and $i \ll 0$. Step one applies and yields $F(A) \in \mathcal{A}$.

Step 3 The aim of this step is to construct for any object $A \in \mathcal{A}$ an isomorphism $\tilde{f}_A : A \longrightarrow F(A)$ which is functorial in A and extends f.

Use property i) of $\{L_i\}$ applied to A to construct an exact sequence in $\mathcal{A}$ of the form

$$0 \longrightarrow B \longrightarrow L_i^k \longrightarrow A \longrightarrow 0$$

with $i \ll 0$. (Here and in the rest of this section L_i^k means $L_i^{\oplus k}$.) Its image under F, which is again a distinguished triangle, is a sequence in $\mathcal{A}$ and hence an exact sequence in $\mathcal{A}$ (see Exercise 2.27). We wish to complete the diagram

$$\begin{array}{ccccc} B & \hookrightarrow & L_i^k & \twoheadrightarrow & A \\ & & \downarrow{\scriptstyle f_{L_i}^k} & & \vdots\, {\scriptstyle \tilde{f}_A} \\ F(B) & \hookrightarrow & F(L_i)^k & \twoheadrightarrow & F(A) \end{array} \tag{4.2}$$

by a unique morphism $\tilde{f}_A : A \longrightarrow F(A)$. For the existence, it suffices to show that the composition $g : B \longrightarrow L_i^k \longrightarrow F(L_i)^k \longrightarrow F(A)$ is trivial.

In order to see this, choose a surjection $L_j^\ell \twoheadrightarrow B$ for $j \ll 0$ which yields the commutative diagram

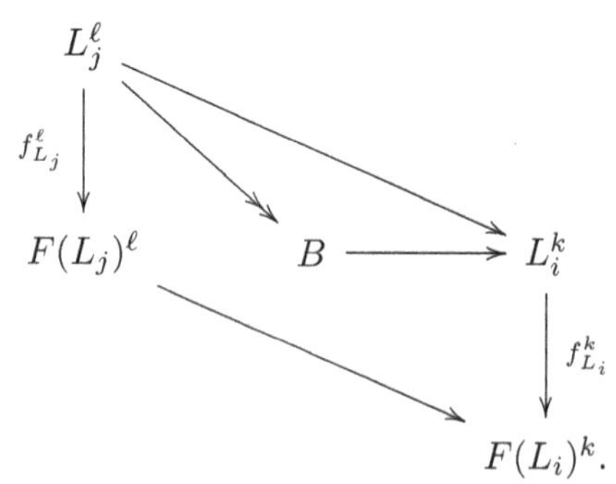

(We use here that $f : \mathrm{id}_{\{L_i\}} \simeq F_{\{L_i\}}$ is a functor morphism.)

Since the composition $L_j^\ell \longrightarrow L_i^k \longrightarrow A$ is trivial, the same is true for its image $F(L_j)^\ell \longrightarrow F(L_i)^k \longrightarrow F(A)$. Hence, the composition of the surjection $L_j^\ell \twoheadrightarrow B$ with $g : B \longrightarrow F(A)$ is trivial. Hence, $g : B \longrightarrow F(A)$ is trivial.

Thus, the desired $\tilde{f}_A : A \twoheadrightarrow F(A)$ exists. Its uniqueness follows from the injectivity of $\mathrm{Hom}(A, F(A)) \twoheadrightarrow \mathrm{Hom}(L_i^k, F(A))$ (we are still working within the abelian category $\mathcal{A}$).

Let us next show that the morphism $A \twoheadrightarrow F(A)$ we have constructed does not depend on the chosen surjection $L_i^k \twoheadrightarrow A$. As any two surjections can be dominated by a third one, it is enough to consider a situation of the type

$$\begin{array}{ccccc} L_j^\ell & \longrightarrow & L_i^k & \longrightarrow & A \\ \downarrow & \circlearrowleft & \downarrow & \circlearrowleft & \downarrow \tilde{f}_A \\ F(L_j)^\ell & \longrightarrow & F(L_i)^k & \longrightarrow & F(A) \end{array}$$

where we suppose that $\tilde{f}_A : A \twoheadrightarrow F(A)$ is induced as above by the surjection $L_i^k \twoheadrightarrow A$. Then also the outer rectangle is commutative, but we have seen that there is a unique choice for $A \twoheadrightarrow F(A)$ with this property. Hence, $\tilde{f}_A$ does not depend on the chosen surjection.

Finally, one proves that $\tilde{f}_A$ is functorial in A, i.e. that for any $\varphi : A_1 \twoheadrightarrow A_2$ the diagram

$$\begin{array}{ccc} A_1 & \xrightarrow{\tilde{f}_{A_1}} & F(A_1) \\ \varphi \downarrow & & \downarrow F(\varphi) \\ A_2 & \xrightarrow[\tilde{f}_{A_2}]{} & F(A_2) \end{array}$$

is commutative.

In order to compute $\tilde{f}_{A_1}$ and $\tilde{f}_{A_2}$, we may choose compatible surjections:

$$\begin{array}{ccc} L_i^k & \twoheadrightarrow & A_1 \\ \downarrow & & \downarrow \varphi \\ L_j^\ell & \twoheadrightarrow & A_2. \end{array}$$

Indeed, $\mathrm{Hom}(L_i, L_j^\ell) \twoheadrightarrow \mathrm{Hom}(L_i, A_2)$ is surjective for $i \ll 0$, as its cokernel is contained in $\mathrm{Hom}(L_i, B_2[1]) = 0$. This yields

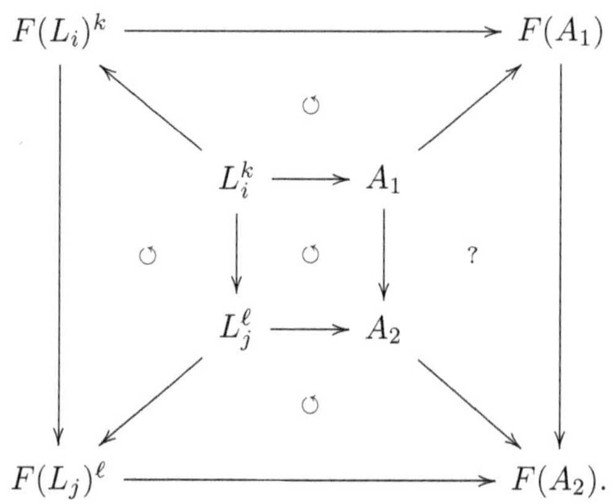

Using the commutativity of all the marked diagrams and of the exterior one, one finds that

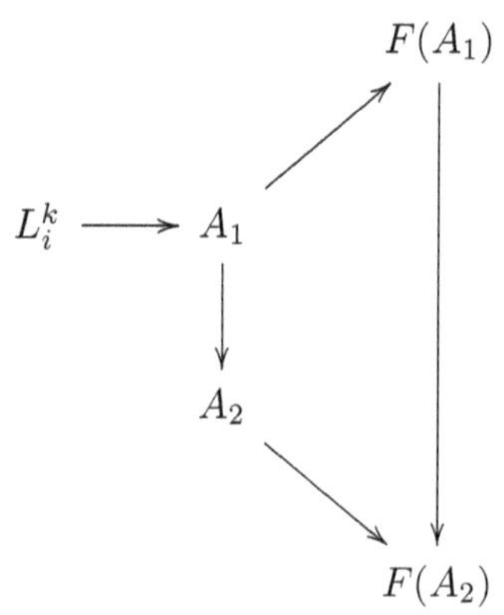

yields identical morphisms $L_i^k \longrightarrow F(A_2)$. Since $L_i^k \twoheadrightarrow A_1$ is surjective, this is enough to conclude. (Note that the functoriality of $\tilde{f}_A$ can also be seen as a generalization of the fact, proved earlier, that $\tilde{f}_A$ is independent of all the choices made.)

Finally, one verifies that the morphism $\mathrm{id} \longrightarrow F|_{\mathcal{A}}$ constructed in this way is in fact an isomorphism.

Here, we invoke the diagram (4.2) used to define $\tilde{f}_A : A \longrightarrow F(A)$:

$$\begin{array}{ccccc} B & \hookrightarrow & L_i^k & \twoheadrightarrow & A \\ \downarrow \tilde{f}_B & & \downarrow f_{L_i}^k & & \downarrow \tilde{f}_A \\ F(B) & \hookrightarrow & F(L_i)^k & \twoheadrightarrow & F(A). \end{array} \tag{4.3}$$

The morphism $B \longrightarrow F(B)$ on the left is indeed $\tilde{f}_B$, as it commutes with $f_{L_i^k}$, due to the functoriality shown above, and there is only one that does. From this diagram we immediately conclude that $\tilde{f}_A$ is surjective and that $\mathrm{Ker}(\tilde{f}_A) \simeq \mathrm{Coker}(\tilde{f}_B)$. Since a similar diagram with A replaced by B shows that $\tilde{f}_B$ is surjective, one finds that $\tilde{f}_A$ is in fact an isomorphism.

Step 4 In this last step we will define $\tilde{f}_{A^\bullet}$ for any $A^\bullet \in \mathrm{D^b}(\mathcal{A})$ recursively on the length of the complex $A^\bullet$. More precisely, we will assume that we have constructed an isomorphism $\tilde{f}_{A^\bullet} : A^\bullet \longrightarrow F(A^\bullet)$ for any complex $A^\bullet$ with

$$\mathrm{length}(A^\bullet) := \max\{q_1 - q_2 \mid H^{q_1}(A^\bullet) \neq 0 \neq H^{q_2}(A^\bullet)\} + 1 < N$$

such that it is functorial in $A^\bullet$. The case of complexes of length one has been dealt with above, so we assume $1 < N$.

Suppose $\mathrm{length}(A^\bullet) = N$. Let us write

$$A^\bullet : \ldots \longrightarrow A^{m-1} \longrightarrow A^m \longrightarrow 0.$$

For $i \ll 0$ we may assume that $\mathrm{Hom}(H^m(A^\bullet), L_i) = 0$ and that there exists a surjection $L_i^k \twoheadrightarrow A^m$. We pick one such surjection and consider it as a morphism $L_i^k[-m] \longrightarrow A^\bullet$, which then can be completed to a distinguished triangle

$$L_i^k[-m] \longrightarrow A^\bullet \longrightarrow B^\bullet \longrightarrow L_i^k[1-m]. \tag{4.4}$$

Since $H^m(L_i^k[-m]) = L_i^k \twoheadrightarrow A^m \twoheadrightarrow H^m(A^\bullet)$, the long cohomology sequence of this distinguished triangle shows that $H^i(B^\bullet) \simeq H^i(A^\bullet)$ for $i < m-1$ and $H^i(B^\bullet) = 0$ for $i \geq m$. As $1 < N$, this shows $\mathrm{length}(B^\bullet) < N$ and we might, therefore, use the induction hypothesis. We obtain the following diagram relating (4.4) and its image under F, which is also distinguished:

$$\begin{array}{ccccccc} L_i^k[-m] & \longrightarrow & A^\bullet & \longrightarrow & B^\bullet & \longrightarrow & L_i^k[1-m] \\ \downarrow f_{L_i}^k & & \vdots\, \tilde{f}_{A^\bullet} & & \downarrow \tilde{f}_{B^\bullet} & \circlearrowleft & \downarrow f_{L_i}^k \\ F(L_i^k)[-m] & \longrightarrow & F(A^\bullet) & \longrightarrow & F(B^\bullet) & \longrightarrow & F(L_i^k)[1-m]. \end{array}$$

Here, $\tilde{f}_{A^\bullet}$ exists due to TR3. Moreover, it is unique, because the kernel of

$$\mathrm{Hom}(A^\bullet, F(A^\bullet)) \longrightarrow \mathrm{Hom}(A^\bullet, F(B^\bullet))$$

is a quotient of

$$\begin{aligned}\mathrm{Hom}(A^\bullet, F(L_i^k)[-m]) &\simeq \mathrm{Hom}(A^\bullet, L_i^k[-m]) \\ &\simeq \mathrm{Hom}(H^m(A^\bullet), L_i^k) = 0\end{aligned}$$

(use the usual spectral sequence to prove the isomorphism). Since $\tilde{f}_{B^\bullet}$ and $f_{L_i^k}$ are isomorphisms by induction, the newly constructed morphism $\tilde{f}_{A^\bullet}$ is an isomorphism as well.

As in the case $N = 1$, we have to show that the morphism $\tilde{f}_{A^\bullet}$ is independent of the choices and that it is functorial in $A^\bullet$. Again, the first follows from the latter, but for clarity we prove them separately.

As before, in order to prove the independence of the surjection $L_i^k \twoheadrightarrow A^m$, we only have to deal with the situation $L_j^\ell \twoheadrightarrow L_i^k \twoheadrightarrow A^m$.

Consider the resulting diagram

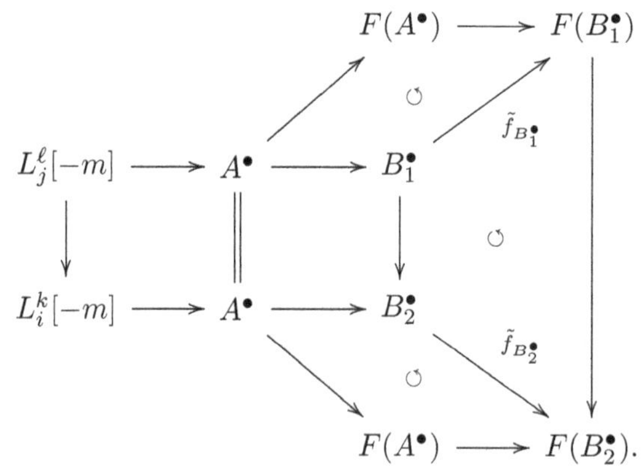

Here, the existence of $B_1^\bullet \longrightarrow B_2^\bullet$ is ensured by TR3 and the commutativity on the right follows from functoriality of $\tilde{f}$ for all complexes of length $< N$. Hence, the different ways to go from $A^\bullet$ to $F(B_2^\bullet)$ are identical. Using the injectivity of

$$\mathrm{Hom}(A^\bullet, F(A^\bullet)) \longrightarrow \mathrm{Hom}(A^\bullet, F(B_2^\bullet))$$

explained above, this shows that both morphisms $L_j^\ell \longrightarrow A^m$ and $L_i^k \longrightarrow A^m$ define the same $A^\bullet \longrightarrow F(A^\bullet)$.

Finally, we have to prove functoriality. Suppose that $\varphi : A^\bullet \longrightarrow C^\bullet$ is a morphism in $\mathrm{D^b}(\mathcal{A})$ of complexes $A^\bullet$ and $C^\bullet$ of length $\leq N$. In order to conclude, it will be enough to deduce functoriality of $\tilde{f}$ with respect to this morphism from the functoriality with respect to a morphism $\varphi_1 : A_1^\bullet \longrightarrow C_1^\bullet$ with

$$\operatorname{length}(A_1^\bullet) \leq \operatorname{length}(A^\bullet), \ \operatorname{length}(C_1^\bullet) \leq \operatorname{length}(C^\bullet),$$

and

$$\operatorname{length}(A_1^\bullet) + \operatorname{length}(C_1^\bullet) < \operatorname{length}(A^\bullet) + \operatorname{length}(C^\bullet).$$

Suppose $A^\bullet$ and $C^\bullet$ are of the form $\ldots \longrightarrow A^{n-1} \longrightarrow A^n \longrightarrow 0 \longrightarrow \ldots$, respectively $\ldots \longrightarrow C^{m-1} \longrightarrow C^m \longrightarrow 0 \longrightarrow \ldots$.

Suppose $m < n$. Then choose a surjection $L_i^k \longrightarrow A^n$ as before, which induces a distinguished triangle

$$L_i^k[-n] \longrightarrow A^\bullet \longrightarrow B^\bullet \longrightarrow L_i^k[1-n] \ .$$

Apply $\operatorname{Hom}(\ , C^\bullet)$ to it and use $\operatorname{Hom}(L_i[-n], C^\bullet) = 0$, as $m < n$. We find that $\operatorname{Hom}(B^\bullet, C^\bullet) \longrightarrow \operatorname{Hom}(A^\bullet, C^\bullet)$ is surjective.

Hence, $\varphi : A^\bullet \longrightarrow C^\bullet$ can be lifted to $\varphi_1 : B^\bullet \longrightarrow C^\bullet$. By construction, $\tilde{f}$ is functorial with respect to $A^\bullet \longrightarrow B^\bullet$, but we may also assume that $\tilde{f}$ is functorial with respect to $B^\bullet \longrightarrow C^\bullet$, for $\operatorname{length}(B^\bullet) < \operatorname{length}(A^\bullet)$ and $C^\bullet$ has not changed.

This shows the existence of the commutative diagram

$$\begin{array}{ccccc} & & \overset{\varphi}{\frown} & & \\ A^\bullet & \longrightarrow & B^\bullet & \xrightarrow[\varphi_1]{} & C^\bullet \\ \Big\downarrow \tilde{f}_{A^\bullet} & \circlearrowleft & \Big\downarrow \tilde{f}_{B^\bullet} & \circlearrowleft & \Big\downarrow \tilde{f}_{C^\bullet} \\ F(A^\bullet) & \longrightarrow & F(B^\bullet) & \xrightarrow{F(\varphi_1)} & F(C^\bullet), \\ & & \underset{F(\varphi)}{\smile} & & \end{array}$$

which is what had to be proven.

Suppose now that $n \leq m$. Choose a surjection $L_i^k \twoheadrightarrow C^m$ and construct a distinguished triangle

$$L_i^k[-m] \longrightarrow C^\bullet \xrightarrow{\psi} D^\bullet \longrightarrow L_i^k[1-m]$$

as before. Since $\operatorname{length}(D^\bullet) < \operatorname{length}(C^\bullet)$, the morphism $\tilde{f}$ is functorial with respect to the composition $\varphi_1 : A^\bullet \longrightarrow C^\bullet \longrightarrow D^\bullet$. It is also functorial with respect to $\psi : C^\bullet \longrightarrow D^\bullet$ by construction. Hence, one has commutative diagrams

$$\begin{array}{ccccc} & & \overset{\varphi_1}{\frown} & & \\ A^\bullet & \xrightarrow{\varphi} & C^\bullet & \xrightarrow{\psi} & D^\bullet \\ \downarrow & & \circlearrowleft & & \downarrow \\ F(A^\bullet) & \xrightarrow{F(\varphi)} & F(C^\bullet) & \xrightarrow{F(\psi)} & F(D^\bullet) \end{array} \quad \text{and} \quad \begin{array}{ccc} C^\bullet & \xrightarrow{\psi} & D^\bullet \\ \downarrow & \circlearrowleft & \downarrow \\ F(C^\bullet) & \xrightarrow{F(\psi)} & F(D^\bullet). \end{array}$$

The combination of both yields the commutative diagram

$$\begin{array}{ccccc} A^\bullet & \xrightarrow{\varphi} & C^\bullet & & \\ \downarrow & & & \searrow^{\tilde f_{C^\bullet}} & \\ F(A^\bullet) & & \circlearrowleft & & F(C^\bullet) \\ & \searrow_{F(\varphi)} & & & \downarrow{\scriptstyle F(\psi)} \\ & & F(C^\bullet) & \xrightarrow{F(\psi)} & F(D^\bullet). \end{array}$$

On the other hand, $F(\psi) \circ (\quad) : \mathrm{Hom}(A^\bullet, F(C^\bullet)) \longrightarrow \mathrm{Hom}(A^\bullet, F(D^\bullet))$ is injective, because $\mathrm{Hom}(A^\bullet, F(L_i^k)[-m]) = 0$ for $m > n$ and for $m = n$ we have $\mathrm{Hom}(H^m(A^\bullet), L_i^k) = 0$ for $i \ll 0$. Therefore, $F(\varphi) \circ \tilde f_{A^\bullet} = \tilde f_{C^\bullet} \circ \varphi$. □

Note that in the above proof we have really not used that $\{L_i\}$ is a spanning class of $\mathrm{D^b}(\mathcal{A})$, which stresses the fact that the assertion is about ample sequences and not about spanning classes. In particular, we need not assume that $\mathcal{A}$ is of finite homological dimension.

Kawamata successfully used the result of Bondal and Orlov to prove the following more general result.

As before we let $\mathrm{D^b}(\mathcal{A})$ be the derived category of an abelian category $\mathcal{A}$ of finite homological dimension with an ample sequence L_i, $i \in \mathbb{Z}$.

Proposition 4.24 (Kawamata) *Let $F : \mathrm{D^b}(\mathcal{A}) \longrightarrow \mathcal{D}$ be an exact fully faithful functor with left and right adjoint $G \dashv F \dashv H$ and let $F' : \mathrm{D^b}(\mathcal{A}) \longrightarrow \mathcal{D}$ be another exact functor admitting as well a left and a right adjoint $G' \dashv F' \dashv H'$. Furthermore, we suppose that $H \circ F'$ has a right adjoint and that $G' \circ F$ has a left adjoint.*

Then any isomorphism

$$\xi : F|_{\{L_i\}} \xrightarrow{\sim} F'|_{\{L_i\}}$$

of the restrictions of F and F' to the full subcategory $\{L_i\} \subset \mathrm{D^b}(\mathcal{A})$ can be extended uniquely to a functor isomorphism

$$\tilde{\xi} : F \xrightarrow{\sim} F'.$$

See [64].

Proof By Lemma 2.73 any ample sequence in $\mathcal{A}$ forms a spanning class in $\mathrm{D^b}(\mathcal{A})$. Since F' admits left and right adjoint and coincides with the fully faithful functor F on the spanning class, Proposition 1.49 tells us that also F' is fully faithful. In particular, the adjunction morphisms yield isomorphisms $\mathrm{id} \simeq H' \circ F'$ and $G' \circ F' \simeq \mathrm{id}$ (cf. Proposition 1.24).

The assumptions imply that $G' \circ F \dashv H \circ F'$. Indeed, one has functorial isomorphisms $\mathrm{Hom}(G' \circ F \;\;,\;\;) \simeq \mathrm{Hom}(F \;\;, F' \;\;) \simeq \mathrm{Hom}(\;\;, H \circ F')$. Similarly, one finds $G \circ F' \dashv H' \circ F$.

Since by hypothesis $H \circ F'$ has also a right adjoint, Proposition 1.49 applies and yields together with $H \circ F'|_{\{L_i\}} \simeq H \circ F|_{\{L_i\}} \simeq \mathrm{id}|_{\{L_i\}}$ that $H \circ F'$ is fully faithful. Similarly, one concludes that $G' \circ F$ is fully faithful.

The full faithfulness in turn implies that the natural adjunction morphisms $(G' \circ F) \circ (H \circ F') \longrightarrow \mathrm{id}$ and $\mathrm{id} \longrightarrow (H \circ F') \circ (G' \circ F)$ are isomorphisms, i.e. $G' \circ F$ and $H \circ F'$ are quasi-inverse to each other.

Due to Proposition 4.23, the natural isomorphism

$$f : \mathrm{id}|_{\{L_i\}} \xrightarrow{\sim} H \circ F|_{\{L_i\}} \xrightarrow{H(\xi)} H \circ F'|_{\{L_i\}}$$

can be extended uniquely to an isomorphism $\tilde{f} : \mathrm{id} \xrightarrow{\sim} H \circ F'$. The composition of $F(\tilde{f}) : F \xrightarrow{\sim} F \circ (H \circ F')$ with the adjunction $g_{F'(\;)} : (F \circ H) \circ F' \longrightarrow F'$ yields a canonical functor morphism $\tilde{\xi} : F \longrightarrow F'$. Note that restricted to the ample sequence $\{L_i\}$ this is nothing but ξ. Moreover, if F is an equivalence, then $g_{F'(\;)}$ is an isomorphism and thus so is $\tilde{\xi}$, which proves the assertion under the additional assumption that F is an equivalence.

Let us now show that for an arbitrary fully faithful functor F the morphism $\tilde{\xi}_{A^\bullet} : F(A^\bullet) \longrightarrow F'(A^\bullet)$ is an isomorphism for any $A^\bullet \in \mathrm{D^b}(\mathcal{A})$. Fix $A^\bullet$ and complete $\tilde{\xi}_{A^\bullet}$ to a distinguished triangle

$$F(A^\bullet) \xrightarrow{\tilde{\xi}_{A^\bullet}} F'(A^\bullet) \xrightarrow{\psi} B^\bullet \longrightarrow F(A^\bullet)[1].$$

Since the adjoint H is also exact, we obtain a distinguished triangle

$$H(F(A^\bullet)) \xrightarrow{H(\tilde{\xi}_{A^\bullet})} H(F'(A^\bullet)) \xrightarrow{H(\psi)} H(B^\bullet) \longrightarrow H(F(A^\bullet))[1].$$

By construction, the morphism $H(\tilde{\xi}_{A^\bullet})$ factorizes as

$$\begin{array}{ccc} H(F(A^\bullet)) & \xrightarrow{H(\tilde{\xi}_{A^\bullet})} & H(F'(A^\bullet)) \\ {\scriptstyle H(F(\tilde{f}))}\downarrow & & \uparrow{\scriptstyle H(g_{F'(A^\bullet)})} \\ H(F(H\circ F'))(A^\bullet) & \xrightarrow{\sim} & H((F\circ H)F'(A^\bullet)). \end{array}$$

We know that $H(F(\tilde{f}))$ is an isomorphism. On the other hand, $H(g) : H \circ F \circ H \longrightarrow H$ composed with the isomorphism $h_H : H \xrightarrow{\sim} H \circ F \circ H$ (by assumption F is fully faithful!) yields the identity (see Exercise 1.19). Hence, $H(g)$ is an isomorphism as well. Thus, $H(\tilde{\xi}_{A^\bullet})$ is an isomorphism and, therefore, $H(B^\bullet) \simeq 0$.

We show that this is enough to conclude that $B^\bullet \simeq 0$ and hence $F(A^\bullet) \simeq F'(A^\bullet)$. Indeed,

$$\begin{aligned} 0 = \mathrm{Hom}(L_i[j], H(B^\bullet)) &\simeq \mathrm{Hom}(F(L_i)[j], B^\bullet) \\ &\simeq \mathrm{Hom}(F'(L_i)[j], B^\bullet) \simeq \mathrm{Hom}(L_i[j], H'(B^\bullet)) \end{aligned}$$

for all i and j. Since $\{L_i\}$ is a spanning class, this shows $H'(B^\bullet) \simeq 0$. But then $\psi \in \mathrm{Hom}(F'(A^\bullet), B^\bullet) = \mathrm{Hom}(A^\bullet, H'(B^\bullet)) = 0$ and hence $\psi = 0$. Thus, $F(A^\bullet) \simeq F'(A^\bullet) \oplus B^\bullet[-1]$. But the projection to $B^\bullet[-1]$ must be trivial, because $\mathrm{Hom}(F(A^\bullet), B^\bullet[-1]) = \mathrm{Hom}(A^\bullet, H(B^\bullet)[-1]) = 0$. Therefore, $B \simeq 0$. □

Remarks 4.25 i) We leave it to the reader to modify the above proposition and its proof in the sense of Corollary 4.21, i.e. we have only to assume that the two functors coincide on the very negative part of the ample sequence.

ii) The most interesting special case of the above proposition is when both functors F and F' are equivalences and we have seen that the proof simplifies drastically in this case. But the more general case is needed when one wants to show that any fully faithful functor $F : \mathrm{D}^{\mathrm{b}}(X) \longrightarrow \mathrm{D}^{\mathrm{b}}(Y)$ (and not only any equivalence) is a Fourier–Mukai transform.

5

FOURIER–MUKAI TRANSFORMS

This chapter introduces the central notion of a Fourier–Mukai transform between derived categories. It is the derived version of the notion of a correspondence, which has been studied for all kinds of cohomology theories (e.g. Chow groups, singular cohomology, etc.) for many decades.

Functors that are of Fourier–Mukai type behave well in many respects. They are exact, admit left and right adjoints, can be composed, etc. In fact, Orlov's celebrated result, stated as Theorem 5.14 but not proved, says that any equivalence between derived categories of smooth projective varieties is of geometric origin, i.e. of Fourier–Mukai type.

Section 5.2 explains how to study Fourier–Mukai transforms by cohomological methods. We will show how the cohomological Fourier–Mukai transform behaves with respect to grading, Hodge structure, and Mukai pairing. This chapter concludes with an easy application to curves by showing that the derived category of a smooth curve determines the curve uniquely.

Objects in the derived category of coherent sheaves will sometimes be denoted by $\mathcal{E}$ and, when we want to stress that it is not simply a sheaf, by $\mathcal{E}^\bullet$. In particular, a Fourier–Mukai kernel, a notion to be introduced in Section 5.1, is often denoted $\mathcal{P}$, although it usually is a true complex.

5.1 What it is and Orlov's result

Let X and Y be smooth projective varieties and denote the two projections by

$$q : X \times Y \longrightarrow X \quad \text{and} \quad p : X \times Y \longrightarrow Y.$$

Definition 5.1 *Let $\mathcal{P} \in \mathrm{D}^{\mathrm{b}}(X \times Y)$. The induced* Fourier–Mukai transform *is the functor*

$$\Phi_{\mathcal{P}} : \mathrm{D}^{\mathrm{b}}(X) \longrightarrow \mathrm{D}^{\mathrm{b}}(Y) \ , \quad \mathcal{E}^\bullet \longmapsto p_*(q^*\mathcal{E}^\bullet \otimes \mathcal{P}) \ .$$

The object $\mathcal{P}$ is called the Fourier–Mukai kernel *of the Fourier–Mukai transform $\Phi_{\mathcal{P}}$.*

As before, we denote by p_*, q^*, and $\otimes$ the derived functors between the derived categories. Note, however, that q^* is the usual pull-back, as q is flat, and that $q^*\mathcal{E}^\bullet \otimes \mathcal{P}$ is the usual tensor product if $\mathcal{P}$ is a complex of locally free sheaves.

Remark 5.2 In the literature, $\Phi_{\mathcal{P}}$ is sometimes called an *integral functor* which is a Fourier–Mukai transform only when it is an equivalence. We call X and Y *Fourier–Mukai partners*, if there exists a Fourier–Mukai transform $\Phi_{\mathcal{P}}$ that is an equivalence.

The analogy to the classical Fourier transform is nicely explained in [82]. It is most striking in the case of abelian varieties as we shall see in Chapter 9. Roughly, L^2-functions are replaced by complexes of coherent sheaves and, in particular, the usual integral kernel by an object of the derived category in the product.

As the kernel $\mathcal{P}$ can also be used to define an exact functor $\mathrm{D^b}(Y) \longrightarrow \mathrm{D^b}(X)$ (in the opposite direction), the simplified notation we have chosen is sometimes ambiguous. To be more precise, one could write $\Phi_{\mathcal{P}}^{X\to Y}$ to indicate the direction $\mathrm{D^b}(X) \longrightarrow \mathrm{D^b}(Y)$ (which is of course useful only if $X \neq Y$).

Remark 5.3 Any Fourier–Mukai transform is the composition of three exact functors $q^* : \mathrm{D^b}(X) \longrightarrow \mathrm{D^b}(X \times Y)$, $(\ \ ) \otimes \mathcal{P} : \mathrm{D^b}(X \times Y) \longrightarrow \mathrm{D^b}(X \times Y)$, and $p_* : \mathrm{D^b}(X \times Y) \longrightarrow \mathrm{D^b}(Y)$. Thus, $\Phi_{\mathcal{P}}$ is itself exact.

Examples 5.4 Let us show that some of the equivalences already encountered in these notes are in fact Fourier–Mukai transforms. Geometrically more interesting ones will be studied in detail in later chapters.

i) The identity

$$\mathrm{id} : \mathrm{D^b}(X) \longrightarrow \mathrm{D^b}(X)$$

is naturally isomorphic to the Fourier–Mukai transform $\Phi_{\mathcal{O}_\Delta}$ with kernel the structure sheaf $\mathcal{O}_\Delta$ of the diagonal $\Delta \subset X \times X$. Indeed, with $\iota : X \xrightarrow{\sim} \Delta \subset X \times X$ denoting the diagonal embedding one has

$$\begin{aligned}\Phi_{\mathcal{O}_\Delta}(\mathcal{E}^\bullet) &= p_*(q^*\mathcal{E}^\bullet \otimes \mathcal{O}_\Delta) = p_*(q^*\mathcal{E}^\bullet \otimes \iota_*\mathcal{O}_X)\\ &\simeq p_*(\iota_*(\iota^*q^*\mathcal{E}^\bullet \otimes \mathcal{O}_X)) \quad \text{(projection formula(3.11))}\\ &\simeq (p\circ\iota)_*(q\circ\iota)^*\mathcal{E}^\bullet \simeq \mathcal{E}^\bullet \quad (\text{as } p\circ\iota = \mathrm{id} = q\circ\iota\ ).\end{aligned}$$

ii) Let $f : X \longrightarrow Y$ be a morphism. Then

$$f_* \simeq \Phi_{\mathcal{O}_{\Gamma_f}} : \mathrm{D^b}(X) \longrightarrow \mathrm{D^b}(Y),$$

where $\Gamma_f \subset X \times Y$ is the graph of f.

As a special instance, one may consider cohomology $H^*(X,\ \ )$ as the Fourier–Mukai transform $\Phi_{\mathcal{O}_X} : \mathrm{D^b}(X) \longrightarrow \mathrm{D^b}(\mathbf{Vec}_f(k))$, where $X \subset X \times \mathrm{Spec}(k)$ is considered as the graph of the projection $X \longrightarrow \mathrm{Spec}(k)$.

For arbitrary $f : X \longrightarrow Y$ one may also use $\mathcal{O}_{\Gamma_f}$ as the kernel for a Fourier–Mukai transform in the opposite direction which is nothing but the inverse image $f^* : \mathrm{D^b}(Y) \longrightarrow \mathrm{D^b}(X)$.

iii) Let L be a line bundle on X. Then $\mathcal{E}^\bullet \longmapsto \mathcal{E}^\bullet \otimes L$ defines an autoequivalence $\mathrm{D^b}(X) \longrightarrow \mathrm{D^b}(X)$ which is isomorphic to the Fourier–Mukai transform with kernel $\iota_*(L)$, where $\iota : X \xrightarrow{\sim} \Delta \subset X \times X$ is again the diagonal embedding of X.

iv) The shift functor $T : \mathrm{D^b}(X) \longrightarrow \mathrm{D^b}(X)$ can be described as the Fourier–Mukai transform with kernel $\mathcal{O}_\Delta[1]$.

v) Consider once more the diagonal embedding $\iota : X \xrightarrow{\sim} \Delta \subset X \times X$. Then

$$\Phi_{\iota_* \omega_X^k} \simeq S^k[-nk],$$

where S_X is the Serre functor $\mathcal{F}^\bullet \longmapsto \mathcal{F}^\bullet \otimes \omega_X[n]$ with $n = \dim(X)$ (see Definition 3.11).

vi) Suppose $\mathcal{P}$ is a coherent sheaf on $X \times Y$ flat over X. Consider the Fourier–Mukai transform $\Phi_\mathcal{P} : \mathrm{D^b}(X) \longrightarrow \mathrm{D^b}(Y)$. If $x \in X$ is a closed point with $k(x) \simeq k$, then

$$\Phi(k(x)) \simeq \mathcal{P}_x,$$

where $\mathcal{P}_x := \mathcal{P}|_{\{x\}\times Y}$ is considered as a sheaf on Y via the second projection $\{x\} \times Y \longrightarrow Y$.

vii) Let $\mathcal{P} \in \mathrm{D^b}(X \times Y)$ be a coherent sheaf on $X \times Y$ flat over X. This is commonly viewed as a family of coherent sheaves $\mathcal{P}_x$ on Y or as a deformation of the sheaf $\mathcal{P}_{x_0}$ for a distinguished closed point $x_0 \in X$. For simplicity we shall assume $k(x_0) \simeq k$. A tangent vector v at x_0 is determined by a subscheme $Z_v \subset X$ of length two concentrated in $x_0 \in X$. Pulling-back

$$0 \longrightarrow k(x) \longrightarrow \mathcal{O}_{Z_v} \longrightarrow k(x) \longrightarrow 0$$

and tensoring with $\mathcal{P}$ (remember, $\mathcal{P}$ is X-flat) yields

$$0 \longrightarrow \mathcal{P}_{x_0} \longrightarrow \mathcal{P}|_{Z_v \times Y} \longrightarrow \mathcal{P}_{x_0} \longrightarrow 0.$$

Viewed as a sequence on Y this gives rise to a class in $\mathrm{Ext}^1_Y(\mathcal{P}_{x_0}, \mathcal{P}_{x_0})$. In this way we obtain a linear map, the so-called *Kodaira–Spencer map*,

$$\kappa(x_0) : T_{x_0}X \longrightarrow \mathrm{Ext}^1_Y(\mathcal{P}_{x_0}, \mathcal{P}_{x_0}).$$

By construction, $\kappa(x_0)$ is compatible with $\Phi_{\mathcal{P}}$, i.e. one has the following commutative diagram:

$$\begin{array}{ccc}
T_{x_0}X \simeq \mathrm{Ext}^1_X(k(x), k(x)) & \xrightarrow{\kappa(x_0)} & \mathrm{Ext}^1(\mathcal{P}_{x_0}, \mathcal{P}_{x_0}) \\
\simeq \downarrow & & \downarrow \simeq \\
\mathrm{Hom}_{\mathrm{D^b}(X)}(k(x_0), k(x_0)[1]) & \xrightarrow[\Phi_{\mathcal{P}}]{} & \mathrm{Hom}_{\mathrm{D^b}(Y)}(\mathcal{P}_{x_0}, \mathcal{P}_{x_0}[1]).
\end{array}$$

Exercise 5.5 Spell out the details in iii) and vi).

Remark 5.6 We have seen and used already, that any equivalence is compatible with Serre duality (cf. Lemma 1.30). This is no longer true for arbitrary Fourier–Mukai transforms.

For example, if $f : X \longrightarrow \mathrm{Spec}(k)$ then the Fourier–Mukai transform f_* (see example ii) above) maps a sheaf $\mathcal{F}$ to its cohomology $H^*(X, \mathcal{F})$ and in general

$$\begin{aligned}
\mathrm{S}_{\mathrm{pt}} H^0(X, \mathcal{F}) = H^0(X, \mathcal{F}) &\not\simeq H^n(X, \mathcal{F} \otimes \omega_X) \\
&\simeq H^0(X, \mathcal{F} \otimes \omega_X[\dim(X)]) \\
&= H^0(X, \mathrm{S}_X(\mathcal{F})).
\end{aligned}$$

Clearly, any (exact) equivalence has a left and a right adjoint. This is in fact true for any Fourier–Mukai transform as we will explain now. More precisely, the left and the right adjoint of a Fourier–Mukai transform are again Fourier–Mukai transforms, the kernels of which can be described explicitly.

Definition 5.7 *For any object* $\mathcal{P} \in \mathrm{D^b}(X \times Y)$ *we let*

$$\mathcal{P}_{\mathrm{L}} := \mathcal{P}^{\vee} \otimes p^*\omega_Y[\dim(Y)] \quad \textit{and} \quad \mathcal{P}_{\mathrm{R}} := \mathcal{P}^{\vee} \otimes q^*\omega_X[\dim(X)],$$

both objects in $\mathrm{D^b}(X \times Y)$.

For the definition of the derived dual see p. 78.

Remark 5.8 The induced Fourier–Mukai transforms $\Phi_{\mathcal{P}_{\mathrm{R}}} : \mathrm{D^b}(Y) \longrightarrow \mathrm{D^b}(X)$ and $\Phi_{\mathcal{P}_{\mathrm{R}}} : \mathrm{D^b}(Y) \longrightarrow \mathrm{D^b}(X)$ can equivalently be described as

$$\Phi_{\mathcal{P}_{\mathrm{L}}} \simeq \Phi_{\mathcal{P}^{\vee}} \circ S_Y \quad \text{respectively} \quad \Phi_{\mathcal{P}_{\mathrm{R}}} \simeq S_X \circ \Phi_{\mathcal{P}^{\vee}}.$$

Proposition 5.9 **(Mukai)** *Let* $F = \Phi_{\mathcal{P}} : \mathrm{D^b}(X) \longrightarrow \mathrm{D^b}(Y)$ *be the Fourier–Mukai transform with Fourier–Mukai kernel* $\mathcal{P}$. *Then*

$$G := \Phi_{\mathcal{P}_{\mathrm{L}}} : \mathrm{D^b}(Y) \longrightarrow \mathrm{D^b}(X) \quad \textit{and} \quad H := \Phi_{\mathcal{P}_{\mathrm{R}}} : \mathrm{D^b}(Y) \longrightarrow \mathrm{D^b}(X)$$

are left, respectively right adjoint to F. *See [79].*

Proof The assertion is a direct consequence of the Grothendieck–Verdier duality (see Theorem 3.34 or rather Corollary 3.35). Indeed, for any $\mathcal{E}^\bullet \in \mathrm{D^b}(X)$ and $\mathcal{F}^\bullet \in \mathrm{D^b}(Y)$ one has a sequence of functorial isomorphisms:

$$\begin{aligned}
&\mathrm{Hom}_{\mathrm{D^b}(X)}(G(\mathcal{F}^\bullet), \mathcal{E}^\bullet)\\
&\quad = \mathrm{Hom}_{\mathrm{D^b}(X)}(q_*(\mathcal{P}_{\mathrm{L}} \otimes p^*\mathcal{F}^\bullet), \mathcal{E}^\bullet)\\
&\quad \simeq \mathrm{Hom}_{\mathrm{D^b}(X\times Y)}(\mathcal{P}_{\mathrm{L}} \otimes p^*\mathcal{F}^\bullet, q^*\mathcal{E}^\bullet \otimes p^*\omega_Y[\dim(Y)])\\
&\qquad\qquad\qquad\qquad\qquad\qquad \text{(Grothendieck – Verdier duality)}\\
&\quad \simeq \mathrm{Hom}_{\mathrm{D^b}(X\times Y)}(\mathcal{P}^\vee \otimes p^*\mathcal{F}^\bullet, q^*\mathcal{E}^\bullet)\\
&\quad \simeq \mathrm{Hom}_{\mathrm{D^b}(X\times Y)}(p^*\mathcal{F}^\bullet, \mathcal{P} \otimes q^*\mathcal{E}^\bullet) \qquad \text{(property of the dual, p. 84)}\\
&\quad \simeq \mathrm{Hom}_{\mathrm{D^b}(Y)}(\mathcal{F}^\bullet, p_*(\mathcal{P} \otimes q^*\mathcal{E}^\bullet)) \qquad \text{(since } p^* \dashv p_*)\\
&\quad = \mathrm{Hom}_{\mathrm{D^b}(Y)}(\mathcal{F}^\bullet, F(\mathcal{E}^\bullet)).
\end{aligned}$$

This shows $G \dashv F$. A similar calculation proves $F \dashv H$. The reader may write this down as an exercise. An alternative proof can be given using Remarks 5.8 and 1.31. By definition $H = S_X \circ G \circ S_Y^{-1}$ and, therefore, $G \dashv F$ yields $F \dashv H$ without any further work. □

This is certainly good news: the results of Section 1.3 apply to any Fourier–Mukai transform. In fact, due to a recent result of Bondal and van den Bergh (see [16]) any exact functor $F : \mathrm{D^b}(X) \longrightarrow \mathrm{D^b}(Y)$, whether it is of Fourier–Mukai type or not, admits left and right adjoints. Here, X and Y are supposed to be smooth projective in which case their derived categories are saturated.

In order to work out criteria that allow us to decide whether a given kernel defines a fully faithful functor or an equivalence, we have to consider the compositions $H \circ F$, $F \circ H$, etc. More generally, we will show that the composition of two arbitrary Fourier–Mukai transforms is again a Fourier–Mukai transform. We will give an explicit formula for the Fourier–Mukai kernel of the composition.

Let X, Y, and Z be smooth projective varieties over k a field. Consider objects $\mathcal{P} \in \mathrm{D^b}(X \times Y)$ and $\mathcal{Q} \in \mathrm{D^b}(Y \times Z)$. Then define the object $\mathcal{R} \in \mathrm{D^b}(X \times Z)$ by the formula

$$\mathcal{R} := \pi_{XZ*}(\pi_{XY}^*\mathcal{P} \otimes \pi_{YZ}^*\mathcal{Q}),$$

where π_{XZ}, π_{XY}, and π_{YZ} are the projections from $X \times Y \times Z$ to $X \times Z$, $X \times Y$, respectively $Y \times Z$.

Proposition 5.10 **(Mukai)** *The composition*

$$\mathrm{D^b}(X) \xrightarrow{\ \Phi_{\mathcal{P}}\ } \mathrm{D^b}(Y) \xrightarrow{\ \Phi_{\mathcal{Q}}\ } \mathrm{D^b}(Z)$$

is isomorphic to the Fourier–Mukai transform

$$\Phi_{\mathcal{R}} : \mathrm{D}^{\mathrm{b}}(X) \longrightarrow \mathrm{D}^{\mathrm{b}}(Z).$$

See [79].

Proof The proof is not difficult (e.g. Grothendieck–Verdier duality is not involved); it is the notation that causes most of the trouble. The following diagram, introducing the notation for all possible projections, might be helpful:

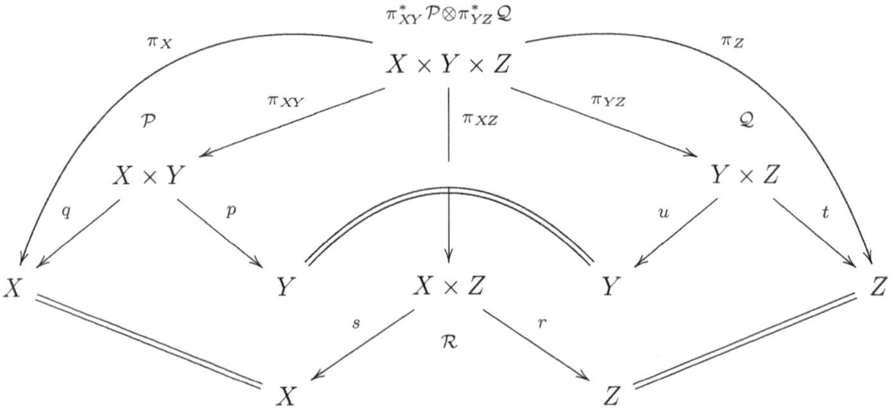

Then the proof consists of writing down the following functorial isomorphisms

$$\begin{aligned}
&\Phi_{\mathcal{R}}(\mathcal{E}^\bullet)\\
&\quad = r_*(s^*\mathcal{E}^\bullet \otimes \mathcal{R})\\
&\quad \simeq r_*\left(s^*\mathcal{E}^\bullet \otimes \pi_{XZ*}(\pi_{XY}^*\mathcal{P} \otimes \pi_{YZ}^*\mathcal{Q})\right)\\
&\quad \simeq r_*\left(\pi_{XZ*}(\pi_X^*\mathcal{E}^\bullet \otimes \pi_{XY}^*\mathcal{P} \otimes \pi_{YZ}^*\mathcal{Q})\right) \qquad \text{(projection formula)}\\
&\quad \simeq \pi_{Z*}(\pi_{XY}^*(q^*\mathcal{E}^\bullet \otimes \mathcal{P}) \otimes \pi_{YZ}^*\mathcal{Q}) \qquad \text{(use } r \circ \pi_{XZ} = \pi_Z)\\
&\quad \simeq t_*\pi_{YZ*}(\pi_{XY}^*(q^*\mathcal{E}^\bullet \otimes \mathcal{P}) \otimes \pi_{YZ}^*\mathcal{Q}) \qquad \text{(use } t \circ \pi_{YZ} = \pi_Z)\\
&\quad \simeq t_*(\pi_{YZ*}\pi_{XY}^*(q^*\mathcal{E}^\bullet \otimes \mathcal{P}) \otimes \mathcal{Q}) \qquad \text{(projection formula)}\\
&\quad \simeq t_*(u^*p_*(q^*\mathcal{E}^\bullet \otimes \mathcal{P}) \otimes \mathcal{Q}) \qquad (\pi_{YZ*} \circ \pi_{XY}^* = u^* \circ p_* \text{ see (3.18)})\\
&\quad = t_*(u^*\Phi_{\mathcal{P}}(\mathcal{E}^\bullet) \otimes \mathcal{Q}) = \Phi_{\mathcal{Q}}\left(\Phi_{\mathcal{P}}(\mathcal{E}^\bullet)\right).
\end{aligned}$$

□

Remark 5.11 If the composition $\Phi_{\mathcal{Q}} \circ \Phi_{\mathcal{P}}$ is not an equivalence, then the kernel $\mathcal{R}$ is in general not unique. The above choice of $\mathcal{R}$ is the natural one, e.g. with respect to the adjoint functors. More precisely, if $\mathcal{R}$ is given as above as $\pi_{XZ*}(\pi_{XY}^*\mathcal{P} \otimes \pi_{YZ}^*\mathcal{Q})$ then $\mathcal{R}_{\mathrm{R}} \simeq \pi_{XZ*}(\pi_{XY}^*\mathcal{P}_{\mathrm{R}} \otimes \pi_{YZ}^*\mathcal{Q}_{\mathrm{R}})$ and similarly for $\mathcal{R}_{\mathrm{L}}$.

Indeed, applying Grothendieck–Verdier duality yields

$$\begin{aligned}\mathcal{R}_{\mathrm{R}} &\simeq \mathcal{R}^{\vee} \otimes s^*\omega_X[\dim(X)] \simeq \mathcal{H}om(\mathcal{R}, \mathcal{O}_{X\times Z}) \otimes s^*\omega_X[\dim(X)] \\ &\simeq \pi_{XZ*}\mathcal{H}om(\pi_{XY}^*\mathcal{P} \otimes \pi_{YZ}^*\mathcal{Q}, \pi_Y^*\omega_Y[\dim(Y)]) \otimes s^*\omega_X[\dim(X)] \\ &\simeq \pi_{XZ*}(\pi_{XY}^*(\mathcal{P}^{\vee} \otimes q^*\omega_X[\dim(X)]) \otimes \pi_{YZ}^*(\mathcal{Q}^{\vee} \otimes u^*\omega_Y[\dim(Y)])) \\ &\simeq \pi_{XZ*}(\pi_{XY}^*\mathcal{P}_{\mathrm{R}} \otimes \pi_{YZ}^*\mathcal{Q}_{\mathrm{R}}).\end{aligned}$$

Exercise 5.12 Let $\mathcal{P} \in \mathrm{D}^{\mathrm{b}}(X \times Y)$ and $\Phi := \Phi_{\mathcal{P}} : \mathrm{D}^{\mathrm{b}}(X) \longrightarrow \mathrm{D}^{\mathrm{b}}(Y)$ be the associated Fourier–Mukai transform. Verify the following assertions:

i) For $f : Y \longrightarrow Z$ the composition $f_* \circ \Phi$ is isomorphic to the Fourier–Mukai transform with kernel $(\mathrm{id}_X \times f)_*\mathcal{P} \in \mathrm{D}^{\mathrm{b}}(X \times Z)$.
ii) For $f : Z \longrightarrow Y$ the composition $f^* \circ \Phi$ is isomorphic to the Fourier–Mukai transform with kernel $(\mathrm{id}_X \times f)^*\mathcal{P} \in \mathrm{D}^{\mathrm{b}}(X \times Z)$.
iii) For $g : W \longrightarrow X$ the composition $\Phi \circ g_*$ is isomorphic to the Fourier–Mukai transform with kernel $(g \times \mathrm{id}_Y)^*\mathcal{P} \in \mathrm{D}^{\mathrm{b}}(W \times Y)$.
iv) For $g : X \longrightarrow W$ the composition $\Phi \circ g^*$ is isomorphic to the Fourier–Mukai transform with kernel $(g \times \mathrm{id}_Y)_*\mathcal{P} \in \mathrm{D}^{\mathrm{b}}(W \times Y)$.

Exercise 5.13 Consider two kernels $\mathcal{P}_i \in \mathrm{D}^{\mathrm{b}}(X_i \times Y_i)$, $i = 1, 2$, and their exterior tensor product $\mathcal{P}_1 \boxtimes \mathcal{P}_2 \in \mathrm{D}^{\mathrm{b}}((X_1 \times X_2) \times (Y_1 \times Y_2))$.

i) Consider the induced Fourier–Mukai transforms $\Phi_{\mathcal{P}_i} : \mathrm{D}^{\mathrm{b}}(X_i) \longrightarrow \mathrm{D}^{\mathrm{b}}(Y_i)$, $i = 1, 2$, and $\Phi_{\mathcal{P}_1 \boxtimes \mathcal{P}_2} : \mathrm{D}^{\mathrm{b}}(X_1 \times X_2) \longrightarrow \mathrm{D}^{\mathrm{b}}(Y_1 \times Y_2)$. Show that there exist isomorphisms

$$\Phi_{\mathcal{P}_1 \boxtimes \mathcal{P}_2}(\mathcal{F}_1^{\bullet} \boxtimes \mathcal{F}_2^{\bullet}) \simeq \Phi_{\mathcal{P}_1}(\mathcal{F}_1^{\bullet}) \boxtimes \Phi_{\mathcal{P}_2}(\mathcal{F}_2^{\bullet}),$$

which are functorial in $\mathcal{F}_i^{\bullet} \in \mathrm{D}^{\mathrm{b}}(X_i)$, $i = 1, 2$.
ii) Show for $\mathcal{R} \in \mathrm{D}^{\mathrm{b}}(X_1 \times X_2)$ and its image $\mathcal{S} := \Phi_{\mathcal{P}_1 \boxtimes \mathcal{P}_2}(\mathcal{R}) \in \mathrm{D}^{\mathrm{b}}(Y_1 \times Y_2)$ the commutativity of the following diagram (see [94]):

$$\begin{array}{ccc} \mathrm{D}^{\mathrm{b}}(X_1) & \xleftarrow{\ \Phi_{\mathcal{P}_1}\ } & \mathrm{D}^{\mathrm{b}}(Y_1) \\ {\scriptstyle \Phi_{\mathcal{R}}}\downarrow & & \downarrow{\scriptstyle \Phi_{\mathcal{S}}} \\ \mathrm{D}^{\mathrm{b}}(X_2) & \xrightarrow[\ \Phi_{\mathcal{P}_2}\ ]{} & \mathrm{D}^{\mathrm{b}}(Y_2). \end{array}$$

Note that $\mathcal{P}_1$ is this time used to define a Fourier–Mukai transform in the opposite direction $\mathrm{D}^{\mathrm{b}}(Y_1) \longrightarrow \mathrm{D}^{\mathrm{b}}(X_1)$.

Let us next try to clarify the relation between arbitrary functors and those of Fourier–Mukai type. The answer is given by the following celebrated theorem of Orlov.

Theorem 5.14 **(Orlov)** *Let X and Y be two smooth projective varieties and let*

$$F : \mathrm{D}^{\mathrm{b}}(X) \longrightarrow \mathrm{D}^{\mathrm{b}}(Y)$$

be a fully faithful exact functor. If F admits right and left adjoint functors, then there exists an object $\mathcal{P} \in \mathrm{D}^{\mathrm{b}}(X \times Y)$ unique up to isomorphism such that F is isomorphic to $\Phi_{\mathcal{P}}$:

$$F \simeq \Phi_{\mathcal{P}}.$$

Proof We refrain from giving a proof of this highly non-trivial statement. There are two accounts of it in the literature: the original one due to Orlov in [92, 94] and another one due to Kawamata [64]. The proof uses Postnikov systems [39].

The assumption on the existence of the adjoint functor can be weakened or dropped altogether. Indeed, due to Remark 1.31 the existence of one of the two implies the existence of the other. The much deeper result in [16] ensures the existence of both adjoint functors at once. □

From a geometric point of view one might simply restrict one's attention to Fourier–Mukai transforms from the very beginning. This would avoid this difficult existence result altogether. (Note also that in more general situations, i.e. twisted derived categories, the existence of the kernel is not always known, not even for equivalences.)

In view of Orlov's result one might wonder whether any exact functor is a Fourier–Mukai transform. As a warning, that one might lose information when we pass from objects in the derived category of the product to Fourier–Mukai functors, we include the following example that was communicated to me independently by A. Căldăraru and D. Orlov, see [29].

Examples 5.15 Let E be an elliptic curve. Consider $\mathcal{O}_\Delta$ as an object of the derived category of $\mathrm{D}^{\mathrm{b}}(E\times E)$. Using Serre duality on the product, one finds that $\mathrm{Ext}^2(\mathcal{O}_\Delta, \mathcal{O}_\Delta)$ is one-dimensional. Thus, there exists a non-trivial morphism

$$\varphi : \mathcal{O}_\Delta \longrightarrow \mathcal{O}_\Delta[2]$$

in $\mathrm{D}^{\mathrm{b}}(E \times E)$.

As in general any morphism between objects on the product induces a morphism between their associated Fourier–Mukai transforms, this φ yields a morphism $\Phi_\varphi : \Phi_{\mathcal{O}_\Delta} \longrightarrow \Phi_{\mathcal{O}_\Delta[2]}$. Note that both Fourier–Mukai transforms are equivalences, in fact $\Phi_{\mathcal{O}_\Delta} = \mathrm{id}$ and $\Phi_{\mathcal{O}_\Delta[2]}$ is the double shift $\mathcal{F}^\bullet \longmapsto \mathcal{F}^\bullet[2]$.

Now, one proves that Φ_φ is zero, although φ is not. Indeed, for a sheaf $\mathcal{F}$ on E one has $\Phi_{\mathcal{O}_\Delta}(\mathcal{F}) = \mathcal{F}$ and $\Phi_{\mathcal{O}_\Delta[2]}(\mathcal{F}) = \mathcal{F}[2]$. As $\mathrm{Ext}^2(\mathcal{F}, \mathcal{F}) = 0$ (since E is one-dimensional), the map $\Phi_\varphi(\mathcal{F})$ must be trivial. To conclude, one uses the fact that any object in $\mathrm{D}^{\mathrm{b}}(E)$ is isomorphic to a direct sum of shifted sheaves (see Corollary 3.15) and, therefore, $\Phi_\varphi(\mathcal{F}^\bullet) : \mathcal{F}^\bullet \longrightarrow \mathcal{F}^\bullet[2]$ is trivial for any $\mathcal{F}^\bullet \in \mathrm{D}^{\mathrm{b}}(E)$.

Exercise 5.16 Verify the last conclusion.

Orlov's theorem is most often applied to equivalences:

Corollary 5.17 *Let* $F : \mathrm{D^b}(X) \xrightarrow{\sim} \mathrm{D^b}(Y)$ *be an equivalence between the derived categories of two smooth projective varieties. Then* F *is isomorphic to a Fourier–Mukai transform* $\Phi_{\mathcal{P}}$ *associated to a certain object* $\mathcal{P} \in \mathrm{D^b}(X \times Y)$*, which is unique up to isomorphism.* □

Exercise 5.18 Show that $\Phi_{\mathcal{P}}$ is an equivalence if and only if the following two conditions are satisfied:

i) $\pi_{13*}(\pi_{12}^*\mathcal{P} \otimes \pi_{23}^*\mathcal{P}_{\mathrm{L}}) \simeq \mathcal{O}_{\Delta_X}$ and
ii) $\pi_{13*}(\pi_{12}^*\mathcal{P}_{\mathrm{L}} \otimes \pi_{23}^*\mathcal{P}) \simeq \mathcal{O}_{\Delta_Y}$.

Here, we view $\mathcal{P}$ and $\mathcal{P}_{\mathrm{L}}$ as objects in $\mathrm{D^b}(X \times Y)$, respectively in $\mathrm{D^b}(Y \times X)$. Of course, a similar criterion works for $\mathcal{P}_{\mathrm{L}}$ replaced by $\mathcal{P}_{\mathrm{R}}$.

Exercise 5.19 Use the uniqueness statement of Orlov's result and the description of the right and left adjoint functors of a Fourier–Mukai transform, in order to show the following description of the derived dual of $\mathcal{O}_\Delta$:

$$\mathcal{O}_\Delta^\vee \simeq \mathcal{O}_\Delta[-n] \otimes p^*\omega_X^* \simeq \mathcal{O}_\Delta[-n] \otimes q^*\omega_X^*,$$

where $\Delta \subset X \times X$ is the diagonal of an n-dimensional smooth projective variety. Of course, alternative proofs of this statement exist (see Corollary 3.40).

Exercise 5.20 Let $\mathcal{P}_i \in \mathrm{D^b}(X_i \times X_i)$, $i = 1, 2$, be objects such that

$$\Phi_{\mathcal{P}_i} : \mathrm{D^b}(X_i) \longrightarrow \mathrm{D^b}(Y_i)$$

are equivalences.

Show that the exterior tensor product $\mathcal{P}_1 \boxtimes \mathcal{P}_2 \in \mathrm{D^b}((X_1 \times X_2) \times (Y_1 \times Y_2))$ defines an equivalence

$$\Phi_{\mathcal{P}_1 \boxtimes \mathcal{P}_2} : \mathrm{D^b}(X_1 \times X_2) \longrightarrow \mathrm{D^b}(Y_1 \times Y_2).$$

For an alternative proof, at least for the fact that this functor is fully faithful, see Corollary 7.4 and Exercise 7.14.

The following assertion had already been stated (and proven) as Proposition 4.1. We nevertheless outline another and more geometric proof here, which uses the existence and uniqueness of the Fourier–Mukai kernel.

Corollary 5.21 *Let* X *and* Y *be smooth projective varieties with equivalent derived categories* $\mathrm{D^b}(X)$ *and* $\mathrm{D^b}(Y)$*. Then* $\dim(X) = \dim(Y)$.

Proof The following argument is taken from [63].

By Orlov's result we know that any equivalence $F : \mathrm{D^b}(X) \xrightarrow{\sim} \mathrm{D^b}(Y)$ is of the form $\Phi_{\mathcal{P}}$ for some $\mathcal{P} \in \mathrm{D^b}(X \times Y)$. Moreover, F has a left adjoint given as the

Fourier–Mukai functor $\mathrm{D^b}(Y) \longrightarrow \mathrm{D^b}(X)$ with kernel $\mathcal{P}_\mathrm{L} = \mathcal{P}^\vee \otimes p^*\omega_Y[\dim(Y)]$ and a right adjoint $\mathrm{D^b}(Y) \longrightarrow \mathrm{D^b}(X)$ given as the Fourier–Mukai transform with kernel $\mathcal{P}_\mathrm{R} = \mathcal{P}^\vee \otimes q^*\omega_X[\dim(X)]$ (see Proposition 5.9).

Since F is an equivalence, its right and left adjoints are both quasi-inverse to F. Using the uniqueness of the Fourier–Mukai kernel, we conclude that $\mathcal{P}_\mathrm{L}$ and $\mathcal{P}_\mathrm{R}$ are isomorphic objects in $\mathrm{D^b}(X \times Y)$.

Hence,

$$\mathcal{P}^\vee \simeq \mathcal{P}^\vee \otimes (p^*\omega_X \otimes q^*\omega_Y^*[\dim(X) - \dim(Y)]).$$

With $\mathcal{P}^\vee$ an object of a bounded derived category, this immediately yields $\dim(X) = \dim(Y)$. □

Remark 5.22 In the proof we have tacitly deduced one of the standard facts that is used over and over again, namely that the kernel $\mathcal{P} \in \mathrm{D^b}(X \times Y)$ of a Fourier–Mukai transformation $\Phi_\mathcal{P}$ which is an equivalence satisfies

$$\mathcal{P} \otimes q^*\omega_X \simeq \mathcal{P} \otimes p^*\omega_Y.$$

We will come back to this necessary criterion in Proposition 7.6. There it will be turned into a sufficient criterion for a fully faithful functor to be an equivalence.

Here is another nice application of Orlov's existence result.

Corollary 5.23 *Suppose $\Phi : \mathrm{D^b}(X) \simeq \mathrm{D^b}(Y)$ is an equivalence such that for any closed point $x \in X$ there exists a closed point $f(x) \in Y$ with*

$$\Phi(k(x)) \simeq k(f(x)).$$

Then $f : X \longrightarrow Y$ defines an isomorphism and Φ is the composition of f_* with the twist by some line bundle $M \in \mathrm{Pic}(Y)$, i.e.

$$\Phi \simeq (M \otimes (\quad)) \circ f_*.$$

Proof In the first step one shows that there exists a morphism $X \longrightarrow Y$ which on the set of closed points induces the given map f.

If we think of Φ as a Fourier–Mukai transform $\Phi_\mathcal{P}$, then Lemma 3.31 implies that $\mathcal{P}$ is an X-flat sheaf on $X \times Y$. By assumption $\mathcal{P}|_{\{x\}\times Y} \simeq k(f(x))$. Choosing local sections of $\mathcal{P}$ shows that it indeed defines a morphism $X \longrightarrow Y$ inducing f on the closed points. By abuse of notation, the morphism will again be called f.

Next, one uses the assumption that Φ is an equivalence to prove that f is an isomorphism. Since the sheaves $k(x)$ span $\mathrm{D^b}(X)$, their images span $\mathrm{D^b}(Y)$. Thus, if $y \in Y$ is a closed point, then there exists a closed point $x \in X$ and an integer m with $\mathrm{Hom}(\Phi(k(x)), k(y)[m]) \neq 0$. This implies that any $k(y)$ is of the form $k(f(x))$ for some closed point $x \in X$, i.e. f is surjective on the set of closed points.

Similarly, two different points $x_1 \neq x_2 \in X$ give rise to two different points $f(x_1) \neq f(x_2)$, i.e. f is injective. In characteristic zero, this already suffices to conclude that f as a morphism between two smooth varieties is an isomorphism.

Without this assumptions, one argues by using a quasi-inverse Φ^{-1} to produce an honest inverse f^{-1}.

Eventually, $\mathcal{P}$ considered as a sheaf on its support, which is the graph of f, is a sheaf of constant fibre dimension one and hence a line bundle. Using $\mathrm{supp}(\mathcal{P}) \xrightarrow{\sim} Y$ given by the second projection allows us to view this line bundle as a line bundle M on Y. □

Orlov's existence result can also be used to give a somewhat round-about proof of the classical result of Gabriel saying that the abelian category of coherent sheaves on a scheme determines the scheme.

Corollary 5.24 **(Gabriel)** *Suppose X and Y are smooth projective varieties. If there exists an equivalence* $\mathbf{Coh}(X) \simeq \mathbf{Coh}(Y)$*, then X and Y are isomorphic.*

Proof Clearly, any equivalence

$$\Phi_0 : \mathbf{Coh}(X) \xrightarrow{\sim} \mathbf{Coh}(Y)$$

between the abelian categories can be extended to an equivalence

$$\Phi : \mathrm{D}^{\mathrm{b}}(X) \xrightarrow{\sim} \mathrm{D}^{\mathrm{b}}(Y)$$

of their derived categories.

A sheaf $\mathcal{F} \in \mathbf{Coh}(X)$ is called indecomposable if any non-trivial surjection $\mathcal{F} \twoheadrightarrow \mathcal{G}$ with $\mathcal{G} \in \mathbf{Coh}(X)$ is an isomorphism. It is not difficult to show that any indecomposable sheaf is of the form $k(x)$ with $x \in X$ a closed point.

The equivalence $\Phi_0 : \mathbf{Coh}(X) \xrightarrow{\sim} \mathbf{Coh}(Y)$ sends an indecomposable object to an indecomposable one. Hence, for any closed point $x \in X$ there exists a closed point $y \in Y$ with $\Phi_0(k(x)) \simeq k(y)$. This continues to hold for the extension $\Phi : \mathrm{D}^{\mathrm{b}}(X) \xrightarrow{\sim} \mathrm{D}^{\mathrm{b}}(Y)$. By Corollary 5.23 this implies that Φ is of the form $\mathcal{F}^\bullet \longmapsto M \otimes f_*\mathcal{F}^\bullet$ for some isomorphism $f : X \xrightarrow{\sim} Y$ and some line bundle M on Y.

Note that we have not only proved that X and Y are isomorphic, but that in fact any equivalence between their abelian categories is of the special form $\mathcal{F} \longmapsto M \otimes f_*\mathcal{F}$. □

5.2 Passage to cohomology

In this section we only consider smooth projective varieties over the complex numbers. We usually will not distinguish between a projective variety and the associated complex manifold. In particular, we will tacitly use the equivalence of the category of coherent sheaves on a projective variety and the category of analytic coherent sheaves on the associated complex manifold. When not mentioned otherwise, it is the Zariski topology that will be considered.

Let $\mathcal{F}^\bullet$ be a bounded complex of coherent sheaves $\mathcal{F}^i$, i.e. $\mathcal{F}^\bullet \in \mathrm{D^b}(X)$. To such a complex we associate the element

$$[\mathcal{F}^\bullet] := \sum (-1)^i [\mathcal{F}^i] \in K(X)$$

in the Grothendieck group $K(X)$. By definition $[\mathcal{E}^0] + [\mathcal{E}^2] = [\mathcal{E}^1]$ in $K(X)$ for any short exact sequence

$$0 \longrightarrow \mathcal{E}^0 \longrightarrow \mathcal{E}^1 \longrightarrow \mathcal{E}^2 \longrightarrow 0.$$

As any coherent sheaf on the smooth variety X admits a finite locally free resolution, the elements of $K(X)$ can be written as linear combinations $\sum a_i [\mathcal{E}^i]$ with $\mathcal{E}_i$ locally free sheaves. This allows us to define a ring structure on $K(X)$ by setting

$$[\mathcal{E}_1] \cdot [\mathcal{E}_2] := [\mathcal{E}_1 \otimes \mathcal{E}_2]$$

for locally free sheaves $\mathcal{E}_i$. With this definition, the trivial line bundle $\mathcal{O}_X$ becomes the identity element in $K(X)$.

To pass from the derived category $\mathrm{D^b}(X)$ to $K(X)$ one first defines the map

$$\mathrm{D^b}(X) \longrightarrow K(X), \qquad \mathcal{F}^\bullet \longmapsto [\mathcal{F}^\bullet] = \sum (-1)^i [\mathcal{F}^i].$$

Note that $[\mathcal{F}^\bullet[k]] = (-1)^k [\mathcal{F}^\bullet]$ and $[\mathcal{F}_1^\bullet \oplus \mathcal{F}_2^\bullet] = [\mathcal{F}_1^\bullet] + [\mathcal{F}_2^\bullet]$.

Observe that by the definition of $K(X)$ one has

$$[\mathcal{F}^\bullet] = \sum (-1)^i [\mathcal{H}^i(\mathcal{F}^\bullet)] \in K(X).$$

In particular, two isomorphic (in $\mathrm{D^b}(X)$) complexes $\mathcal{F}^\bullet \simeq \mathcal{E}^\bullet \in \mathrm{D^b}(X)$ define the same element in $K(X)$, i.e. $\mathcal{F}^\bullet \longmapsto [\mathcal{F}^\bullet]$ is defined on the set of isomorphism classes of objects in $\mathrm{D^b}(X)$.

The derived tensor product of complexes is the ordinary tensor product for complexes of locally free sheaves. Hence $[\mathcal{F}_1^\bullet \otimes \mathcal{F}_2^\bullet] = [\mathcal{F}_1^\bullet] \cdot [\mathcal{F}_2^\bullet]$. Thus, $\mathrm{D^b}(X) \longrightarrow K(X)$, $\mathcal{F}^\bullet \longmapsto [\mathcal{F}^\bullet]$ is compatible with the additive and the multiplicative structures given on both sides.

The Grothendieck group $K(X)$ is contravariant in the sense that for any morphism $f : X \longrightarrow Y$ the pull-back $\mathcal{F} \longmapsto f^*\mathcal{F}$ for locally free sheaves defines a ring homomorphism $f^* : K(Y) \longrightarrow K(X)$.

In order to view $K(X)$ covariantly, one defines for any coherent sheaf $\mathcal{F}$ on X the generalized direct image

$$f_![\mathcal{F}] := \sum (-1)^i [R^i f_*(\mathcal{F})]$$

(here we assume that f is projective or proper). This yields a group homomorphism:

$$f_! : K(X) \longrightarrow K(Y)$$

for any projective morphism $f : X \longrightarrow Y$.

Both maps are compatible with derived pull-back and derived direct image, i.e. for any $f : X \longrightarrow Y$ there are two commutative diagrams of the form

$$\begin{array}{ccc} \mathrm{D^b}(Y) & \xrightarrow{f^*} & \mathrm{D^b}(X) \\ {\scriptstyle [\,]}\downarrow & \circlearrowleft & \downarrow{\scriptstyle [\,]} \\ K(Y) & \xrightarrow[f^*]{} & K(X) \end{array} \qquad \begin{array}{ccc} \mathrm{D^b}(X) & \xrightarrow{f_*} & \mathrm{D^b}(Y) \\ {\scriptstyle [\,]}\downarrow & \circlearrowleft & \downarrow{\scriptstyle [\,]} \\ K(X) & \xrightarrow[f_!]{} & K(Y). \end{array}$$

(Recall that we write f^* and f_* for the derived functors Lf^*, respectively Rf_*.)

For the pull-back this is rather obvious, as we may represent any complex by a complex $\mathcal{F}^\bullet$ of locally free sheaves $\mathcal{F}^i$ and f^* can be computed by applying it to any $\mathcal{F}^i$.

In order to see the compatibility of the direct image, one has to show that $[Rf_*\mathcal{E}^\bullet] = \sum(-1)^i[R^if_*\mathcal{E}^\bullet]$ equals

$$f_![\mathcal{E}^\bullet] = \sum(-1)^i f_![\mathcal{H}^i(\mathcal{E}^\bullet)] = \sum(-1)^i \sum(-1)^j[R^jf_*\mathcal{H}^i(\mathcal{E}^\bullet)]$$

which is a consequence of the Leray spectral sequence (3.3)

$$E_2^{p,q} = R^pf_*\mathcal{H}^q(\mathcal{E}^\bullet) \Rightarrow R^{p+q}f_*(\mathcal{E}^\bullet)$$

and the observation

$$\sum(-1)^i[E_r^{p+ir,q-ir+i}] = \sum(-1)^i[E_{r+1}^{p+ir,q-ir+i}]$$

and hence

$$\sum(-1)^{p+q}[E_r^{p,q}] = \sum(-1)^{p+q}[E_{r+1}^{p,q}].$$

In complete analogy to the definition of the Fourier–Mukai functor $\Phi_{\mathcal{P}}$ one defines the *K-theoretic Fourier–Mukai transform*. Let $e \in K(X \times Y)$ be a given class in the Grothendieck group of the product of two projective varieties X and Y. Then, one defines

$$\Phi_e^K : K(X) \longrightarrow K(Y), \qquad f \longmapsto p_!(e \otimes q^*(f)).$$

Due to the aforementioned compatibilities, the two Fourier–Mukai maps commute:

$$\begin{array}{ccc} \mathrm{D^b}(X) & \xrightarrow{\Phi_{\mathcal{P}}} & \mathrm{D^b}(Y) \\ {\scriptstyle [\,]}\downarrow & \circlearrowleft & \downarrow{\scriptstyle [\,]} \\ K(X) & \xrightarrow[\Phi_{[\mathcal{P}]}^K]{} & K(Y). \end{array}$$

Remarks 5.25 i) In fact, the passage from a Fourier–Mukai transform between the derived categories of two varieties to a Fourier–Mukai transform of their K-groups does not really require the existence of a Fourier–Mukai kernel. Indeed, any exact functor $F : \mathrm{D^b}(X) \longrightarrow \mathrm{D^b}(Y)$ induces a group homomorphism $F^K : K(X) \longrightarrow K(Y)$ that commutes with the projections $[\ ] : \mathrm{D^b} \longrightarrow K$.

ii) So far, everything works for varieties over arbitrary fields. We could in fact go on without any further assumption on the field and consider the Fourier–Mukai transform on the level of the Chow groups $\mathrm{CH}^*(X)$, respectively $\mathrm{CH}^*(Y)$. As Chow groups and K-groups are actually isomorphic after tensoring with $\mathbb{Q}$ and we therefore would not gain much, we shall pass directly to cohomology where the assumption that the ground field is $\mathbb{C}$ comes in.

We next wish to descend further and consider a *cohomological Fourier–Mukai transform* for rational cohomology $H^*(X, \mathbb{Q})$. Here and in the sequel, $H^*(X, \mathbb{Q})$ denotes the cohomology of the constant sheaf $\mathbb{Q}$ on the associated complex manifold X. Recall that $H^*(X, \mathbb{Q})$ has a natural ring structure. The product of two classes $\alpha, \beta \in H^*(X, \mathbb{Q})$ will be written $\alpha.\beta$ or, simply, $\alpha\beta$. Any morphism (or continuous map) $f : X \longrightarrow Y$ induces a ring homomorphism

$$f^* : H^*(Y, \mathbb{Q}) \longrightarrow H^*(X, \mathbb{Q}).$$

If X and Y are compact and connected, e.g. X and Y projective varieties, we may use Poincaré duality $H^i(X, \mathbb{Q}) \simeq H^{2\dim(X)-i}(X, \mathbb{Q})^*$ and $H^i(Y, \mathbb{Q}) \simeq H^{2\dim(Y)-i}(Y, \mathbb{Q})^*$ to define

$$f_* : H^*(X, \mathbb{Q}) \longrightarrow H^{*+2\dim(Y)-2\dim(X)}(Y, \mathbb{Q}).$$

as the dual map. With this definition, the projection formula $f_*(f^*\alpha.\beta) = \alpha.f_*(\beta)$ holds.

For any cohomology class $\alpha \in H^*(X \times Y, \mathbb{Q})$ one introduces

$$\Phi^H_\alpha : H^*(X, \mathbb{Q}) \longrightarrow H^*(Y, \mathbb{Q}), \qquad \beta \longmapsto p_*(\alpha.q^*(\beta)).$$

The standard way to pass from the Grothendieck group $K(X)$ down to cohomology is via the Chern character

$$\mathrm{ch} : K(X) \longrightarrow H^*(X, \mathbb{Q}).$$

For the definition of the Chern character we refer, e.g. to [37, 45]. The underlying idea for its definition is that the Chern character is additive and that for a line bundle L one has

$$\mathrm{ch}(L) = \exp(\mathrm{c}_1(L)) = \sum \mathrm{c}_1(L)^i/i!.$$

The first Chern class c_1 can be defined as the image of $L \in \mathrm{Pic}(X) = H^1(X, \mathcal{O}_X^*)$ under the boundary map $H^1(X, \mathcal{O}_X^*) \longrightarrow H^2(X, \mathbb{Z})$ of the exponential sequence.

Reducing to line bundles is achieved by passing to certain fibre bundles over X where the pull-back of a given vector bundle can be described as a successive extension of line bundles.

However, the Chern character does not, in general, commute with the Fourier–Mukai transform on K-groups and cohomology. At this point the Todd class td has to be taken into account. By definition, the Todd class td is multiplicative, i.e. $\mathrm{td}(E_1 \oplus E_2) = \mathrm{td}(E_1).\mathrm{td}(E_2)$, and $\mathrm{td}(L)$ of a line bundle L is defined by the formal power series

$$\mathrm{td}(L) = \frac{\mathrm{c}_1(L)}{1 - \exp(-\mathrm{c}_1(L))}.$$

For a smooth variety one writes $\mathrm{td}(X)$ instead of $\mathrm{td}(\mathcal{T}_X)$. The key to the compatibility of the various Fourier–Mukai transforms is the Grothendieck–Riemann–Roch formula (see [37]):

Theorem 5.26 *Let $f : X \longrightarrow Y$ be a projective morphism of smooth projective varieties. Then for any $e \in K(X)$ one has*

$$\mathrm{ch}(f_!(e)).\mathrm{td}(Y) = f_*\left(\mathrm{ch}(e).\mathrm{td}(X)\right). \tag{5.1}$$

The Hirzebruch–Riemann–Roch formula can be viewed as the special case of the structure morphism $f : X \longrightarrow \mathrm{Spec}(k)$:

Corollary 5.27 *For any $e \in K(X)$ one has*

$$\chi(e) = \int_X (\mathrm{ch}(e).\mathrm{td}(X)).$$

More precisely, for $\mathcal{E}^\bullet \in \mathrm{D^b}(X)$ this reads

$$\sum_i (-1)^i \chi(\mathcal{E}^i) = \chi(\mathcal{E}^\bullet) = \int_X (\mathrm{ch}(\mathcal{E}^\bullet).\mathrm{td}(X)),$$

where by abuse of notation, we write $\mathrm{ch}(\mathcal{E}^\bullet)$ for $\mathrm{ch}([\mathcal{E}^\bullet])$.

Definition 5.28 *One defines the* Mukai vector *of a class $e \in K(X)$ or of an object $\mathcal{E}^\bullet \in \mathrm{D^b}(X)$ as the cohomology class*

$$v(e) := \mathrm{ch}(e).\sqrt{\mathrm{td}(X)} \quad \text{respectively} \quad v(\mathcal{E}^\bullet) := v([\mathcal{E}^\bullet]) = \mathrm{ch}(\mathcal{E}^\bullet).\sqrt{\mathrm{td}(X)}.$$

The square root $\sqrt{\mathrm{td}(X)}$ is a cohomology class whose square is $\mathrm{td}(X)$. Using the fact that the degree zero term of $\mathrm{td}(X)$ is $1 \in H^0(X, \mathbb{Q})$, its existence can be shown by a formal (but finite) power series calculation. Clearly, by definition the induced map

$$v : K(X) \longrightarrow H^*(X, \mathbb{Q})$$

is additive.

Corollary 5.29 *Let $e \in K(X \times Y)$. Then*

$$\Phi^H_{v(e)}\left(\mathrm{ch}(f).\sqrt{\mathrm{td}(X)}\right) = \mathrm{ch}\left(\Phi^K_e(f)\right).\sqrt{\mathrm{td}(Y)}$$

for any $f \in K(X)$. In other words, the following diagram commutes

$$\begin{array}{ccc} K(X) & \xrightarrow{\Phi^K_e} & K(Y) \\ \downarrow{\scriptstyle v} & & \downarrow{\scriptstyle v} \\ H^*(X,\mathbb{Q}) & \xrightarrow[\Phi^H_{v(e)}]{} & H^*(Y,\mathbb{Q}). \end{array}$$

Proof The assertion follows immediately from the commutativity of the following diagrams

$$\begin{array}{ccccccc} K(X) & \xrightarrow{q^*} & K(X \times Y) & \xrightarrow{.e} & K(X \times Y) & \xrightarrow{p_!} & K(Y) \\ \downarrow{\scriptstyle v} & & \downarrow{\scriptstyle v\sqrt{\mathrm{td}(Y)}^{-1}} & & \downarrow{\scriptstyle v\sqrt{\mathrm{td}(X)}} & & \downarrow{\scriptstyle v} \\ H^*(X) & \xrightarrow[q^*]{} & H^*(X \times Y) & \xrightarrow[.v(e)]{} & H^*(X \times Y) & \xrightarrow[p_*]{} & H^*(Y). \end{array}$$

The commutativity of the first two is easily deduced from the projection formula. The commutativity of the last one is a consequence of the Grothendieck–Riemann–Roch formula (5.1). □

Remark 5.30 The cohomological Fourier–Mukai transform Φ^H_α neither respects the grading of H^* nor the multiplicative structure (not even for $\alpha = v(e)$).

Let $\mathcal{P} \in \mathrm{D^b}(X \times Y)$ be the kernel of a Fourier–Mukai transform

$$\Phi_{\mathcal{P}} : \mathrm{D^b}(X) \longrightarrow \mathrm{D^b}(Y).$$

In the sequel, we will denote the induced cohomological Fourier–Mukai transform Φ^H_α with kernel $\alpha := v(\mathcal{P}) = \mathrm{ch}(\mathcal{P}).\sqrt{\mathrm{td}(X \times Y)}$ simply by

$$\Phi^H_{\mathcal{P}} : H^*(X,\mathbb{Q}) \longrightarrow H^*(Y,\mathbb{Q}).$$

With all characteristic classes (td, ch, etc.) being even, $\Phi^H_{\mathcal{P}}$ surely respects the parity, i.e.

$$\Phi^H_{\mathcal{P}}\left(H^{\mathrm{even}}(X)\right) \subset H^{\mathrm{even}}(Y) \text{ and } \Phi^H_{\mathcal{P}}\left(H^{\mathrm{odd}}(X)\right) \subset H^{\mathrm{odd}}(Y).$$

Remark 5.31 Note that one does not know how to associate a cohomological Fourier–Mukai transform to an equivalence $F : \mathrm{D^b}(X) \longrightarrow \mathrm{D^b}(Y)$ without using the existence of the kernel $\mathcal{P}$, the main problem being that in general the Chern character $\mathrm{ch} : K(X)_{\mathbb{Q}} \longrightarrow H^*(X, \mathbb{Q})$ is not surjective, i.e. often cohomology classes of objects in $\mathrm{D^b}(X)$ span a proper subspace of $H^*(X, \mathbb{Q})$.

Lemma 5.32 *Let* $\Phi_{\mathcal{P}} : \mathrm{D^b}(X) \longrightarrow \mathrm{D^b}(Y)$ *and* $\Phi_{\mathcal{Q}} : \mathrm{D^b}(Y) \longrightarrow \mathrm{D^b}(Z)$ *be two Fourier–Mukai transforms and let* $\Phi_{\mathcal{R}} : \mathrm{D^b}(X) \longrightarrow \mathrm{D^b}(Z)$ *their composition given as in Proposition 5.10. Then*

$$\Phi^H_{\mathcal{R}} = \Phi^H_{\mathcal{Q}} \circ \Phi^H_{\mathcal{P}}.$$

Proof As the proof is completely analogous to the proof of Proposition 5.10, we leave this to the reader. (Exercise!) □

Note that the analogous statement for the K-theoretic Fourier–Mukai transform, i.e. $\Phi^K_{[\mathcal{R}]} = \Phi^K_{[\mathcal{Q}]} \circ \Phi^K_{[\mathcal{P}]}$, is trivial due to the surjectivity of $\mathrm{D^b} \longrightarrow K$. Except for very special varieties we cannot expect that $K \longrightarrow H^*$ is surjective; the image of the Mukai vector

$$v : K(X) \longrightarrow H^*(X, \mathbb{Q})$$

might be very small compared to the full cohomology $H^*(X, \mathbb{Q})$. So, it is a nice surprise to have, nevertheless, the following

Proposition 5.33 *If* $\mathcal{P} \in \mathrm{D^b}(X \times Y)$ *defines an equivalence*

$$\Phi_{\mathcal{P}} : \mathrm{D^b}(X) \xrightarrow{\sim} \mathrm{D^b}(Y)$$

then the induced cohomological Fourier–Mukai transform

$$\Phi^H_{\mathcal{P}} : H^*(X, \mathbb{Q}) \xrightarrow{\sim} H^*(Y, \mathbb{Q})$$

is a bijection of rational vector spaces.

Proof If $\Phi_{\mathcal{P}}$ is an equivalence, then

$$\Phi_{\mathcal{P}_{\mathrm{R}}} \circ \Phi_{\mathcal{P}} \simeq \Phi_{\mathcal{O}_\Delta} (\simeq \mathrm{id}) \quad \text{and} \quad \Phi_{\mathcal{P}} \circ \Phi_{\mathcal{P}_{\mathrm{R}}} \simeq \Phi_{\mathcal{O}_\Delta} (\simeq \mathrm{id}).$$

Also recall that $\mathcal{O}_\Delta$ is the only object on the product that induces the identity Fourier–Mukai transform.

Due to the above lemma one has

$$\Phi_{\mathcal{R}} \simeq \Phi_{\mathcal{Q}} \circ \Phi_{\mathcal{P}} \;\Rightarrow\; \Phi^H_{\mathcal{R}} = \Phi^H_{\mathcal{Q}} \circ \Phi^H_{\mathcal{P}},$$

where $\mathcal{R} = \pi_{XZ*}(\pi^*_{XY}\mathcal{P} \otimes \pi^*_{YZ}\mathcal{Q})$ as in Proposition 5.10. Thus, we can conclude that

$$\Phi^H_{\mathcal{P}_{\mathrm{R}}} \circ \Phi^H_{\mathcal{P}} = \Phi^H_{\mathcal{O}_\Delta} \quad \text{and} \quad \Phi^H_{\mathcal{P}} \circ \Phi^H_{\mathcal{P}_{\mathrm{R}}} = \Phi^H_{\mathcal{O}_\Delta}.$$

Now, in order to ensure that $\Phi^H_{\mathcal{P}_R} : H^*(Y,\mathbb{Q}) \longrightarrow H^*(X,\mathbb{Q})$ is indeed inverse to $\Phi^H_{\mathcal{P}}$, it suffices to show that $\Phi^H_{\mathcal{O}_\Delta} = \mathrm{id}$.

Use the Grothendieck–Riemann–Roch formula (5.1) for the diagonal embedding $\iota : X \xrightarrow{\sim} \Delta \hookrightarrow X \times X$:

$$\mathrm{ch}(\mathcal{O}_\Delta).\mathrm{td}(X \times X) = \iota_* \left(\mathrm{ch}(\mathcal{O}_X).\mathrm{td}(X)\right) = \iota_* \mathrm{td}(X).$$

Dividing by $\sqrt{\mathrm{td}(X \times X)}$ and using $\iota^* \sqrt{\mathrm{td}(X \times X)} = \mathrm{td}(X)$ yields

$$\mathrm{ch}(\mathcal{O}_\Delta).\sqrt{\mathrm{td}(X \times X)} = \iota_*(1).$$

Hence,

$$\begin{aligned} &p_* \left(q^*(\beta).\mathrm{ch}(\mathcal{O}_\Delta).\sqrt{\mathrm{td}(X \times X)}\right) \\ &= p_* \left(q^*(\beta).\iota_*(1)\right) = p_* \left(\iota_*(\iota^* q^*(\beta))\right) = \beta, \end{aligned}$$

as $p \circ \iota = q \circ \iota = \mathrm{id}$. □

Exercise 5.34 Show that $v(\mathcal{O}_\Delta) = [\Delta]$.

Exercise 5.35 Show that $\Phi^K_{\mathcal{P}} : K(X) \longrightarrow K(Y)$ is an isomorphism of additive groups if $\Phi_{\mathcal{P}} : \mathrm{D^b}(X) \longrightarrow \mathrm{D^b}(Y)$ is an equivalence.

Exercise 5.36 Consider the shift functor $T : \mathrm{D^b}(X) \longrightarrow \mathrm{D^b}(X)$ (which is a Fourier–Mukai transform due to iv), Example 5.4). Show that the induced cohomological Fourier–Mukai transform T^H acts by multiplication with -1.

Exercise 5.37 Let $L \in \mathrm{Pic}(X)$ and $\Phi := L \otimes (\ \ ) : \mathrm{D^b}(X) \longrightarrow \mathrm{D^b}(X)$. Show that Φ^H is given by multiplication with $\mathrm{ch}(L) = \exp(\mathrm{c}_1(L))$. In particular, Φ^H does not respect the cohomological degree as long as $\mathrm{c}_1(L) \neq 0$.

Exercise 5.38 Use the fact that all characteristic classes of the kernel $\mathcal{P}$ of a Fourier–Mukai equivalence $\Phi_{\mathcal{P}} : \mathrm{D^b}(X) \xrightarrow{\sim} \mathrm{D^b}(Y)$ are even cohomology classes to deduce equality of the Euler numbers, i.e. $e(X) = e(Y)$.

Let us now consider in addition the *Hodge structure* on $H^*(X,\mathbb{Q})$. Since X is a smooth projective variety over $\mathbb{C}$, Hodge theory tells us that there is a natural direct sum decomposition

$$H^n(X,\mathbb{C}) = \bigoplus_{p+q=n} H^{p,q}(X)$$

with $\overline{H^{p,q}} = H^{q,p}$. Moreover, $H^{p,q}(X) \simeq H^q(X,\Omega^p)$.

The Chern classes, and hence all characteristic classes, are classes of type (p,p). Thus, the Mukai vector factorizes over the algebraic part of the cohomology

$$v(\ \ ) = \mathrm{ch}(\ \ ).\sqrt{\mathrm{td}(X)} : K(X) \longrightarrow \bigoplus H^{p,p}(X) \cap H^{2p}(X,\mathbb{Q}).$$

Proposition 5.39 *If* $\Phi_{\mathcal{P}} : \mathrm{D^b}(X) \xrightarrow{\sim} \mathrm{D^b}(Y)$ *is an equivalence, then the induced cohomological Fourier–Mukai transform* $\Phi_{\mathcal{P}}^H : H^*(X, \mathbb{Q}) \to H^*(Y, \mathbb{Q})$ *yields isomorphisms*

$$\bigoplus_{p-q=i} H^{p,q}(X) \simeq \bigoplus_{p-q=i} H^{p,q}(Y) \tag{5.2}$$

for all $i = -\dim(X), \ldots, 0, \ldots, \dim(X)$.

Proof As we have seen that $\Phi_{\mathcal{P}}^H$ defines an isomorphism of the rational cohomology groups, it suffices to show that its $\mathbb{C}$-linear extension satisfies

$$\Phi_{\mathcal{P}}^H(H^{p,q}(X)) \subset \bigoplus_{r-s=p-q} H^{r,s}(Y).$$

Consider the Künneth decomposition of $\mathrm{ch}(\mathcal{P}).\sqrt{\mathrm{td}(X \times Y)}$, which is of the form $\sum \alpha^{p',q'} \boxtimes \beta^{r,s}$ with $\alpha^{p',q'} \in H^{p',q'}(X)$ and $\beta^{r,s} \in H^{r,s}(Y)$. Moreover, only terms with $p'+r = q'+s$ contribute, for the class $\mathrm{ch}(\mathcal{P}).\sqrt{\mathrm{td}(X \times Y)}$ is algebraic, i.e. a sum of terms of type (t,t).

For $\alpha \in H^{p,q}(X)$ only those terms in $\sum \alpha^{p',q'} \boxtimes \beta^{r,s}$ with

$$(p,q) + (p',q') = (\dim(X), \dim(X))$$

contribute to $\Phi_{\mathcal{P}}^H(\alpha)$. In fact,

$$\Phi_{\mathcal{P}}^H(\alpha) = \sum \left(\int_X \alpha \wedge \alpha^{p',q'} \right) \beta^{r,s} \in \bigoplus H^{r,s}(Y).$$

Hence, $p - q = q' - p' = r - s$. □

Remark 5.40 There is a construction that works over any field, which uses Hochschild cohomology in order to associate to any derived equivalence a vector space isomorphism

$$\bigoplus_{p+q=i} H^p(X, \bigwedge^q \mathcal{T}_X \otimes \omega_X) \simeq \bigoplus_{p+q=i} H^p(Y, \bigwedge^q \mathcal{T}_Y \otimes \omega_Y).$$

As $H^p(X, \bigwedge^q \mathcal{T}_X \otimes \omega_X) \simeq H^p(X, \Omega_X^{n-q})$ with $n = \dim(X)$, this can also be interpreted as an isomorphism

$$\bigoplus_{p-q=i-n} H^p(X, \Omega_X^q) \simeq \bigoplus_{p-q=i-n} H^p(Y, \Omega_X^q). \tag{5.3}$$

which is of the same form as the one in (5.2).

Note however that (5.2) and (5.3) are not supposed to commute; an extra factor $\sqrt{\mathrm{td}}$ has to be put in on both sides. For more details see Remark 6.3.

It turns out that $\Phi_{\mathcal{P}}^H : H^*(X, \mathbb{Q}) \xrightarrow{\sim} H^*(Y, \mathbb{Q})$ associated to an equivalence $\Phi_{\mathcal{P}} : \mathrm{D^b}(X) \xrightarrow{\sim} \mathrm{D^b}(Y)$ is also compatible with a natural quadratic form that can be defined on $H^*(X,\mathbb{Q})$. This was first observed by Mukai in the case of K3 surfaces. The general definition was recently given by Căldăraru in [28].

The basic idea is the following. If $\Phi_{\mathcal{P}} : \mathrm{D}^{\mathrm{b}}(X) \xrightarrow{\sim} \mathrm{D}^{\mathrm{b}}(Y)$ is an equivalence, then for any $\mathcal{E}^\bullet, \mathcal{F}^\bullet \in \mathrm{D}^{\mathrm{b}}(X)$ one finds, using the induced isomorphism $\mathrm{Ext}^i_X(\mathcal{E}^\bullet, \mathcal{F}^\bullet) \simeq \mathrm{Ext}^i_Y(\Phi_{\mathcal{P}}(\mathcal{E}^\bullet), \Phi_{\mathcal{P}}(\mathcal{F}^\bullet))$, the equality

$$\chi(\mathcal{E}^\bullet, \mathcal{F}^\bullet) = \chi(\Phi_{\mathcal{P}}(\mathcal{E}^\bullet), \Phi_{\mathcal{P}}(\mathcal{F}^\bullet)). \tag{5.4}$$

(Here, by definition, $\chi(\mathcal{E}^\bullet, \mathcal{F}^\bullet) := \sum (-1)^i \dim \mathrm{Ext}^i(\mathcal{E}^\bullet, \mathcal{F}^\bullet)$.) If both sides can be understood as a bilinear pairing of the Mukai vectors, then one might expect that $\Phi^H_{\mathcal{P}}$ respects this pairing also for classes that are not in the image of the Mukai vector.

Using the Hirzebruch–Riemann–Roch formula, $\chi(\mathcal{E}^\bullet, \mathcal{F}^\bullet)$ can be expressed as

$$\begin{aligned}\chi(\mathcal{E}^\bullet, \mathcal{F}^\bullet) = \chi(X, \mathcal{E}^{\bullet\vee} \otimes \mathcal{F}^\bullet) &= \int_X \mathrm{ch}(\mathcal{E}^{\bullet\vee}).\mathrm{ch}(\mathcal{F}^\bullet).\mathrm{td}(X) \\ &= \int_X \left(\mathrm{ch}(\mathcal{E}^{\bullet\vee}).\sqrt{\mathrm{td}(X)}\right).\left(\mathrm{ch}(\mathcal{F}^\bullet).\sqrt{\mathrm{td}(X)}\right).\end{aligned}$$

Now, $\mathrm{ch}(\mathcal{F}^\bullet).\sqrt{\mathrm{td}(X)} = v(\mathcal{F}^\bullet)$. But how can $\mathrm{ch}(\mathcal{E}^{\bullet\vee}).\sqrt{\mathrm{td}(X)}$ be expressed in terms of $v(\mathcal{E}^\bullet)$? The answer is given by the following lemma. To formulate it, let us introduce $v^\vee := \sum (-1)^k v_k$ for any $v = \sum v_k \in \bigoplus H^{2k}(X,\mathbb{Q})$. This operation is easily checked to be multiplicative (see ii), Exercise 5.43).

Lemma 5.41 *With this notation one has*

$$v(\mathcal{E}^{\bullet\vee}) = \mathrm{ch}(\mathcal{E}^{\bullet\vee}).\sqrt{\mathrm{td}(X)} = v(\mathcal{E}^\bullet)^\vee.\exp(\mathrm{c}_1(X)/2).$$

Proof Since $\mathrm{c}_k(\mathcal{E}^\vee) = (-1)^k \mathrm{c}_k(\mathcal{E})$ for any locally free sheaf $\mathcal{E}$, the Chern character satisfies $\mathrm{ch}(\mathcal{E}^{\bullet\vee}) = \mathrm{ch}(\mathcal{E}^\bullet)^\vee$ and hence

$$v(\mathcal{E}^{\bullet\vee}) = \mathrm{ch}(\mathcal{E}^{\bullet\vee}).\sqrt{\mathrm{td}(X)} = v(\mathcal{E}^\bullet)^\vee.\left(\frac{\sqrt{\mathrm{td}(X)}}{\sqrt{\mathrm{td}(X)}^\vee}\right).$$

It therefore suffices to prove $\sqrt{\mathrm{td}(X)} = \sqrt{\mathrm{td}(X)}^\vee.\exp(\mathrm{c}_1(X)/2)$ or, equivalently, $\mathrm{td}(X) = \mathrm{td}(X)^\vee.\exp(\mathrm{c}_1(X))$. The latter can easily be deduced from the splitting principle by writing $\mathrm{td}(X) = \prod \frac{\gamma_i}{1-\exp(-\gamma_i)}$ and

$$\mathrm{td}(X)^\vee.\exp(\mathrm{c}_1(X)) = \prod \frac{(-\gamma_i)}{1-\exp(\gamma_i)}.\prod \exp(\gamma_i).$$

□

With this observation in mind, the following definition seems very natural.

Definition 5.42 *Let $v = \sum v_j \in \bigoplus H^j(X,\mathbb{C})$. Then one defines the* dual of v *by*

$$v^\vee := \sum \sqrt{-1}^j v_j \in H^*(X,\mathbb{C}).$$

The Mukai pairing *on* $H^*(X,\mathbb{C})$ *is the quadratic form*

$$\langle v, v'\rangle_X := \int_X \exp(\mathrm{c}_1(X)/2).(v^\vee.v').$$

Clearly, both definitions of the dual $v^\vee$ coincide for even cohomology classes. Moreover, by the very construction, one has for all $\mathcal{E}^\bullet, \mathcal{F}^\bullet \in \mathrm{D^b}(X)$:

$$\chi(\mathcal{E}^\bullet, \mathcal{F}^\bullet) = \langle v(\mathcal{E}^\bullet), v(\mathcal{F}^\bullet)\rangle. \tag{5.5}$$

Exercise 5.43 Prove the following assertions.

i) Suppose $\mathrm{c}_1(X) = 0$. Then the form $\langle\ ,\ \rangle_X$ is symmetric for X even dimensional and alternating otherwise.
ii) Taking duals is multiplicative, i.e. $v^\vee.w^\vee = (v.w)^\vee$.
iii) If $p : X \times Y \longrightarrow Y$ is the second projection, then

$$p_*(v)^\vee = (-1)^{\dim(X)} p_*(v^\vee)$$

for any $v \in H^*(X \times Y, \mathbb{C})$.

The following proposition is Căldăraru's generalization of the original result of Mukai for K3 surfaces.

Proposition 5.44 *Let* $\Phi_{\mathcal{P}} : \mathrm{D^b}(X) \xrightarrow{\sim} \mathrm{D^b}(Y)$ *be an equivalence. Then*

$$\Phi^H_{\mathcal{P}} : H^*(X,\mathbb{Q}) \xrightarrow{\ \sim\ } H^*(Y,\mathbb{Q})$$

is isometric with respect to the Mukai pairing, i.e. for all $v, v' \in H^*(X,\mathbb{Q})$ *one has*

$$\langle v, v'\rangle_X = \left\langle \Phi^H_{\mathcal{P}}(v), \Phi^H_{\mathcal{P}}(v')\right\rangle_Y.$$

Proof It suffices to show that $\langle \Phi^H_{\mathcal{P}}(v), w\rangle_Y = \langle v, \Phi^{H^{-1}}_{\mathcal{P}}(w)\rangle_X$ for all $v \in H^*(X,\mathbb{Q})$ and $w \in H^*(Y,\mathbb{Q})$. To see this, note first that the inverse functor of $\Phi_{\mathcal{P}}$ is isomorphic to $\Phi_{\mathcal{P}_{\mathrm{L}}}$, where $\mathcal{P}_{\mathrm{L}} = \mathcal{P}^\vee \otimes p^*\omega_Y[n]$ with $n := \dim(X) = \dim(Y)$ (see Proposition 5.9). By Lemma 5.32 this then shows $\Phi^{H^{-1}}_{\mathcal{P}} = \Phi^H_{\mathcal{P}_{\mathrm{L}}}$.

Using Exercise 5.43 and Lemma 5.41 one computes

$$\begin{aligned}
&\langle \Phi^H_{\mathcal{P}}(v), w\rangle_Y \\
&= \int_Y \exp(\mathrm{c}_1(Y)/2).p_*(q^*v.v(\mathcal{P}))^\vee.w \\
&= (-1)^n \int_{X\times Y} p^*\exp(\mathrm{c}_1(Y)/2).(q^*v.v(\mathcal{P}))^\vee.p^*w \\
&= (-1)^n \int_{X\times Y} p^*\exp(\mathrm{c}_1(Y)/2).q^*v^\vee.v(\mathcal{P})^\vee.p^*w \\
&= (-1)^n \int_{X\times Y} p^*\exp(\mathrm{c}_1(Y)/2).q^*v^\vee.v(\mathcal{P}^\vee).(\exp(\mathrm{c}_1(X\times Y)/2))^{-1}.p^*w \\
&= \int_{X\times Y} q^*v^\vee.v(\mathcal{P}_{\mathrm{L}}).q^*\exp(\mathrm{c}_1(X)/2).p^*w \\
&= \langle v, \Phi^H_{\mathcal{P}_{\mathrm{L}}}(w)\rangle_X .
\end{aligned}$$

(We also used the fact that the shift $[n]$ acts by $(-1)^n$. See Exercise 5.36.) □

As an application of the above observations, we will prove that two elliptic curves E and E' have equivalent derived categories, i.e.

$$\mathrm{D^b}(E) \simeq \mathrm{D^b}(E') \iff E \simeq E'.$$

Suppose $\Phi_{\mathcal{P}} : \mathrm{D^b}(E) \xrightarrow{\sim} \mathrm{D^b}(E')$ is an equivalence. Then the induced cohomological Fourier–Mukai transform $\Phi^H_{\mathcal{P}} : H^*(E,\mathbb{Q}) \longrightarrow H^*(E',\mathbb{Q})$ is a direct sum of isomorphisms

$$H^1(E,\mathbb{Q}) \simeq H^1(E',\mathbb{Q}) \ \text{ and } \ (H^0\oplus H^2)(E,\mathbb{Q}) \simeq (H^0\oplus H^2)(E',\mathbb{Q})$$

(cf. Remark 5.30).

Moreover, by Proposition 5.39 one knows that $\Phi^H_{\mathcal{P}} : H^1(E,\mathbb{Q}) \longrightarrow H^1(E',\mathbb{Q})$ respects the Hodge decomposition $H^1 = H^{1,0}\oplus H^{0,1}$. On the other hand, the weight-one Hodge structure determines the elliptic curve. More precisely, $E \simeq H^{1,0}(E)^*/H_1(E,\mathbb{Z}) \simeq H^{0,1}(E)/H^1(E,\mathbb{Z})$.

Hence, it suffices to show that for a derived equivalence

$$\Phi_{\mathcal{P}} : \mathrm{D^b}(E) \xrightarrow{\ \sim\ } \mathrm{D^b}(E')$$

of two elliptic curves the cohomological Fourier–Mukai transform is defined over $\mathbb{Z}$, i.e.

$$\Phi^H_{\mathcal{P}} : H^1(E,\mathbb{Z}) \longrightarrow H^1(E',\mathbb{Z}).$$

This follows from the observation that $\mathrm{td}(E \times E') = 1$ and $\mathrm{ch}(\mathcal{P}) = r + c_1(\mathcal{P}) + (1/2)(c_1^2 - 2c_2)(\mathcal{P})$, where the degree four term, which might a priori be non-integral, does not contribute to $H^1(E) \longrightarrow H^1(E')$. (As a matter of fact, it can be shown that also $\mathrm{ch}_2(\mathcal{P})$ is integral.)

Remark 5.45 The situation is more complicated and more interesting for higher dimensional abelian varieties, as shall be explained in Chapter 9.

Combined with Corollary 4.13 we thus have proven the following folklore result.

Corollary 5.46 *Let C be a smooth complex projective curve and let Y be a smooth complex projective variety. Then*

$$\mathrm{D^b}(C) \simeq \mathrm{D^b}(Y) \quad \Longleftrightarrow \quad C \simeq Y.$$

□

With the exception of the group of autoequivalences of the derived category of an elliptic curve, treated in broader generality in Chapter 9, we thus have achieved a complete understanding of the derived category of smooth projective curves.

6

DERIVED CATEGORY AND CANONICAL BUNDLE – II

With this chapter we return to questions already dealt with in Chapter 4. More precisely, we address the question of how much of the positivity of the canonical bundle of a variety is preserved under derived equivalence. The main difference to the treatment of Chapter 4 is that we now make extensive use of Orlov's existence result (Theorem 5.14). The discussion will thus be more geometric, as we use the description of derived equivalences as Fourier–Mukai transforms, and the results will be finer.

In Section 6.1 we present Orlov's refinement of his joint result with Bondal, that was presented as Proposition 4.11, by showing that Kodaira dimension and canonical ring are preserved under derived equivalence. The original result can in fact be seen as a corollary to this. As the same techniques can be used to derive the invariance of Hochschild cohomology under derived equivalence, a fact alluded to before, this is included here.

Kawamata went one step further and showed that also nefness of the canonical bundle and the numerical Kodaira dimension are preserved under derived equivalence. Proofs of these results can be found in Section 6.3.

Section 6.4 studies the relation between derived and birational equivalence. We will come back to this in later chapters. This section concludes with a conjecture, put forward by Bondal, Orlov, and Kawamata, that clarifies the relation between these two equivalence relations. The most famous special case of it is the conjecture that two birational Calabi–Yau varieties have equivalent derived categories (proved in dimension three by Bridgeland, see Section 11.4).

Section 6.2 contains technical results on the geometry of the support of the Fourier–Mukai kernel of an equivalence. They are crucial for the proofs in Sections 6.3 and 6.4, but also of independent interest. The last Section 6.5 collects definitions and standard facts on (numerical) Kodaira dimension, nef line bundles, and the like.

In this chapter, all varieties are defined over an algebraically closed field of characteristic zero. This assumption simplifies some of the arguments in Section 6.2.

6.1 Kodaira dimension under derived equivalence

Recall that the result of Bondal and Orlov in particular shows that for two smooth projective varieties with equivalent derived categories the canonical bundle ω_Y is ample if and only if ω_X is ample. In fact, they prove $X \simeq Y$ in this case. Roughly, this is achieved by identifying the canonical rings which, under the ampleness assumption, is enough to deduce isomorphy of the varieties. We will see that the existence of the Fourier–Mukai kernel (provided by Theorem 5.14) not only allows us to prove finer results, but that it also provides a more geometric proof of the original one.

A formal consequence of Orlov's theorem 5.14 and the fact that any equivalence commutes with Serre functors is the following result. For the definition of the canonical ring and the Kodaira dimension of a variety see Section 6.5.

Proposition 6.1 (Orlov) *Suppose X and Y are smooth projective varieties with equivalent derived categories $\mathrm{D^b}(X) \simeq \mathrm{D^b}(Y)$.*

Then there exists a ring isomorphism $R(X) \simeq R(Y)$ and, in particular, $\mathrm{kod}(X) = \mathrm{kod}(Y)$. *See [94].*

Proof Every equivalence is a Fourier–Mukai transform, i.e. there exists a complex $\mathcal{P} \in \mathrm{D^b}(X \times Y)$ such that $\Phi_{\mathcal{P}} : \mathrm{D^b}(X) \longrightarrow \mathrm{D^b}(Y)$ is isomorphic to the given equivalence. In particular, its right and left adjoint are isomorphic. Since the kernel of a Fourier–Mukai equivalence is uniquely determined, this yields $\mathcal{Q} := \mathcal{P}^{\vee} \otimes q^*\omega_X[n] \simeq \mathcal{P}^{\vee} \otimes p^*\omega_Y[n]$, where $n = \dim(X) = \dim(Y)$ (see Proposition 4.1 or Corollary 5.21).

Clearly, $\Phi_{\mathcal{Q}} : \mathrm{D^b}(Y) \longrightarrow \mathrm{D^b}(X)$ as a quasi-inverse of $\Phi_{\mathcal{P}} : \mathrm{D^b}(X) \longrightarrow \mathrm{D^b}(Y)$ is an equivalence, but one can also show that $\Phi_{\mathcal{Q}} : \mathrm{D^b}(X) \longrightarrow \mathrm{D^b}(Y)$ (note the change of the direction) is an equivalence. This will be done first. (For an alternative proof of this fact see Remark 7.7.)

Consider the composition

$$\mathrm{D^b}(X) \xrightarrow{\Phi_{\mathcal{P}}} \mathrm{D^b}(Y) \xrightarrow{\Phi_{\mathcal{Q}}} \mathrm{D^b}(X),$$

which is isomorphic to the identity, for $\Phi_{\mathcal{P}}$ is fully faithful and $\Phi_{\mathcal{Q}} \dashv \Phi_{\mathcal{P}}$. On the other hand, we have computed in Proposition 5.10 the kernel of this equivalence as $\mathcal{R} = \pi_{13*}(\pi_{12}^*\mathcal{P} \otimes \pi_{23}^*\mathcal{Q})$. Due to the uniqueness, this yields $\mathcal{R} \simeq \mathcal{O}_\Delta \in \mathrm{D^b}(X \times X)$.

Applying the automorphism $\tau_{12} : X \times X \longrightarrow X \times X$ that interchanges the two factors, one finds

$$\begin{aligned}\mathcal{O}_\Delta \simeq \tau_{12}^*\mathcal{O}_\Delta \simeq \tau_{12}^*\mathcal{R} &\simeq \pi_{13*}\tau_{13}^*(\pi_{12}^*\mathcal{P} \otimes \pi_{23}^*\mathcal{Q}) \\ &\simeq \pi_{13*}(\pi_{12}^*\mathcal{Q} \otimes \pi_{23}^*\mathcal{P}).\end{aligned}$$

Hence, the composition

$$\mathrm{D^b}(X) \xrightarrow{\Phi_{\mathcal{Q}}} \mathrm{D^b}(Y) \xrightarrow{\Phi_{\mathcal{P}}} \mathrm{D^b}(X) \tag{6.1}$$

is also isomorphic to the identity. As this is the composition of $\Phi_{\mathcal{Q}}$ with its adjoint functor, this proves that $\Phi_{\mathcal{Q}} : \mathrm{D^b}(X) \longrightarrow \mathrm{D^b}(Y)$ is fully faithful (cf. Corollary 1.23). Note that so far, we have only used that $\Phi_{\mathcal{P}}$ is fully faithful and that its adjoints are given by $\Phi_{\mathcal{Q}}$.

Now, interchanging the rôle of $\mathcal{P}$ and $\mathcal{Q}$ and using that $\Phi_{\mathcal{Q}} : \mathrm{D^b}(Y) \longrightarrow \mathrm{D^b}(X)$ as a quasi-inverse of $\Phi_{\mathcal{P}} : \mathrm{D^b}(X) \longrightarrow \mathrm{D^b}(Y)$ is fully faithful, the same arguments prove that

$$\mathrm{D^b}(Y) \xrightarrow{\ \Phi_{\mathcal{P}}\ } \mathrm{D^b}(X) \xrightarrow{\ \Phi_{\mathcal{Q}}\ } \mathrm{D^b}(Y) \tag{6.2}$$

is isomorphic to the identity.

The two facts, that both compositions (6.1) and (6.2) are isomorphic to the identity, yield the assertion that with $\Phi_{\mathcal{P}}$ also $\Phi_{\mathcal{Q}} : \mathrm{D^b}(X) \longrightarrow \mathrm{D^b}(Y)$ is an equivalence.

Next, use the kernel $\mathcal{Q} \boxtimes \mathcal{P} \in \mathrm{D^b}((X \times X) \times (Y \times Y))$ to define the Fourier–Mukai equivalence

$$\Phi_{\mathcal{Q}\boxtimes\mathcal{P}} : \mathrm{D^b}(X \times X) \xrightarrow{\ \sim\ } \mathrm{D^b}(Y \times Y).$$

Denote $\Phi_{\mathcal{Q}\boxtimes\mathcal{P}}(\iota_*\omega_X^k)$ by $\mathcal{S} \in \mathrm{D^b}(Y \times Y)$. (We use the same notation ι for both diagonal inclusions $X \hookrightarrow X \times X$ and $Y \hookrightarrow Y \times Y$.)

Then $\Phi_{\mathcal{S}} : \mathrm{D^b}(Y) \longrightarrow \mathrm{D^b}(Y)$ is an equivalence that can be computed as the composition (see Exercise 5.13):

$$\mathrm{D^b}(Y) \xrightarrow{\ \Phi_{\mathcal{Q}}\ } \mathrm{D^b}(X) \xrightarrow{\ \Phi_{\iota_*\omega_X^k}\ } \mathrm{D^b}(X) \xrightarrow{\ \Phi_{\mathcal{P}}\ } \mathrm{D^b}(Y).$$

Since $\Phi_{\iota_*\omega_X^k}$ is isomorphic to $S_X^k[-kn]$ (see Example 5.4) and since any equivalence commutes with the Serre functors S_X and S_Y, we obtain $\Phi_{\mathcal{S}} \simeq S_Y^k[-kn]$. Hence, due to the uniqueness of the Fourier–Mukai kernel $\mathcal{S} \simeq \iota_*\omega_Y^k$.

Thus, for all $k \in \mathbb{Z}$ we have $\Phi_{\mathcal{Q}\boxtimes\mathcal{P}}(\iota_*\omega_X^k) \simeq \iota_*\omega_Y^k$. Since $\Phi_{\mathcal{Q}\boxtimes\mathcal{P}}$ is an equivalence, we obtain isomorphisms

$$\mathrm{Hom}_{X\times X}(\iota_*\omega_X^k, \iota_*\omega_X^\ell) \simeq \mathrm{Hom}_{Y\times Y}(\iota_*\omega_Y^k, \iota_*\omega_Y^\ell)$$

for all $k, \ell \in \mathbb{Z}$. The case $k = 0$ and $\ell \geq 0$ induces the claimed bijection

$$\begin{aligned} H^0(X, \omega_X^\ell) &= \mathrm{Hom}_{X\times X}(\iota_*\mathcal{O}_X, \iota_*\omega_X^\ell) \\ &\simeq \mathrm{Hom}_{Y\times Y}(\iota_*\mathcal{O}_Y, \iota_*\omega_Y^\ell) = H^0(Y, \omega_Y^\ell). \end{aligned}$$

As in the proof of Proposition 4.11, one shows that the multiplicative structure of the canonical ring $R(X) = \bigoplus_{\ell\geq 0} H^0(X, \omega_X^\ell)$ is given by composition and hence compatible with any functor. In other words, the induced bijection

$$R(X) = \bigoplus_{\ell\geq 0} H^0(X, \omega_X^\ell) \simeq \bigoplus_{\ell\geq 0} H^0(Y, \omega_Y^\ell) = R(Y)$$

is indeed a ring isomorphism. □

Exercise 6.2 Show that the same arguments also provide a ring isomorphism of the anti-canonical rings, i.e. $R(X, \omega_X^*) \simeq R(Y, \omega_Y^*)$ and hence $\mathrm{kod}(X, \omega_X^*) = \mathrm{kod}(Y, \omega_Y^*)$.

Note that both cases together provide an alternative proof of the original result of Bondal and Orlov (see Proposition 4.11) for the case that both canonical bundles, ω_X and ω_Y, are (anti-)ample.

Remark 6.3 The techniques of the above proof can be used to compare other invariants of X and Y, which are not directly relevant to the birational geometry of derived equivalent varieties treated in this chapter. In this sense, the following is a digression which may be skipped.

In the discussion we follow Orlov's presentation in [94], but we also recommend [29, 75].

For any smooth projective variety X one introduces the bigraded ring

$$HH(X) := \bigoplus_{i,\ell} HA_{i,\ell}(X) \quad \text{with} \quad HA_{i,\ell}(X) := \mathrm{Ext}^i_{X\times X}(\iota_*\mathcal{O}_X, \iota_*\omega_X^\ell).$$

The algebra structure is defined by composition in $\mathrm{D^b}(X \times X)$.

This bigraded ring contains several interesting substructures. We have encountered the canonical ring $R(X)$ which can be identified with the subring $\bigoplus_{\ell\geq 0} HA_{0,\ell}(X)$.

In another direction, one may look at the *Hochschild cohomology* of X, i.e. at the subring

$$HH^*(X) := \bigoplus_i HA_{i,0}(X) \simeq \bigoplus_i \mathrm{Ext}^i_{X\times X}(\iota_*\mathcal{O}_X, \iota_*\mathcal{O}_X)$$

or at the *Hochschild homology*

$$HH_*(X) := \bigoplus_i HA_{i,1}(X) \simeq \bigoplus_i \mathrm{Ext}^i_{X\times X}(\iota_*\mathcal{O}_X, \iota_*\omega_X),$$

which can be viewed as a graded module over $HH^*(X)$.

As was shown in the proof of Proposition 6.1, any equivalence

$$\Phi : \mathrm{D^b}(X) \xrightarrow{\sim} \mathrm{D^b}(Y)$$

induces an isomorphism

$$R(X) = \bigoplus_\ell HA_{0,\ell}(X) \simeq \bigoplus_\ell HA_{0,\ell}(Y) = R(Y)$$

of graded rings. As should be clear from a quick look at that proof, this isomorphism extends to an isomorphism

$$\Phi^{HH} : HH(X) \xrightarrow{\sim} HH(Y)$$

which respects the bigrading and the multiplicative structure.

Let us study more closely the two induced isomorphisms

$$\Phi^{HH^*} : HH^*(X) \xrightarrow{\sim} HH^*(Y) \quad \text{and} \quad \Phi^{HH_*} : HH_*(X) \longrightarrow HH_*(Y).$$

The composition in $HH(X)$ endows the Hochschild homology $HH_*(X)$ with the structure of a module over the Hochschild cohomology $HH^*(X)$. Derived equivalences preserve this module structure.

To make the isomorphisms between the Hochschild cohomology of X and Y more transparent, we invoke a result usually attributed to Swan [109] and put in the geometric context by Kontsevich. It says that the spectral sequence (see (3.16), p. 85)

$$E_2^{p,q} = H^p(X \times X, \mathcal{E}xt^q(\iota_*\mathcal{O}_X, \iota_*\mathcal{O}_X)) \Rightarrow \mathrm{Ext}^{p+q}(\iota_*\mathcal{O}_X, \iota_*\mathcal{O}_X)$$

degenerates, i.e.

$$\mathrm{Ext}^i(\iota_*\mathcal{O}_X, \iota_*\mathcal{O}_X) \simeq \bigoplus_{p+q=i} H^p(X \times X, \mathcal{E}xt^q(\iota_*\mathcal{O}_X, \iota_*\mathcal{O}_X)).$$

Using $\mathcal{E}xt^q(\iota_*\mathcal{O}_X, \iota_*\mathcal{O}_X) \simeq \bigwedge^q \mathcal{T}_X$ (see Example 11.9), this yields

$$HH^i(X) = \mathrm{Ext}^i(\iota_*\mathcal{O}_X, \iota_*\mathcal{O}_X) \simeq \bigoplus_{p+q=i} H^p(X, \bigwedge^q \mathcal{T}_X).$$

The isomorphism that appears naturally in the argument is called the *Hochschild–Kostant–Rosenberg isomorphism*. Be aware that this isomorphism does not respect the multiplicative structures given on the two sides. Conjecturally, multiplying by $\sqrt{\mathrm{td}(X)}^{-1}$ is needed to make it multiplicative.

For $i = 0$ we do not find anything interesting, but already the case $i = 1$ provides us with the highly non-trivial isomorphism

$$\Phi^{HH^1} : H^0(X, \mathcal{T}_X) \oplus H^1(X, \mathcal{O}_X) \xrightarrow{\sim} H^0(Y, \mathcal{T}_Y) \oplus H^1(Y, \mathcal{O}_Y)$$

for any equivalence $\Phi : \mathrm{D^b}(X) \xrightarrow{\sim} \mathrm{D^b}(Y)$. We will provide a geometric interpretation for this in Section 9.4 (see Proposition 9.45). Roughly, $H^1(X, \mathcal{O}_X)$, respectively, $H^0(X, \mathcal{T}_X)$, are the tangent spaces of the Picard group, respectively the group of automorphisms, of X.

Similarly, the induced isomorphisms of Hochschild homology can be better understood if combined with the degenerate spectral sequence

$$E_2^{p,q} = H^p(X \times X, \mathcal{E}xt^q(\iota_*\mathcal{O}_X, \iota_*\omega_X)) \Rightarrow \mathrm{Ext}^{p+q}(\iota_*\mathcal{O}_X, \iota_*\omega_X),$$

which yields the Hochschild–Kostant–Rosenberg isomorphism for Hochschild homology

$$HH_i(X) = \mathrm{Ext}^i(\iota_*\mathcal{O}_X, \iota_*\omega_X) \simeq \bigoplus_{p+q=i} H^p(X, \bigwedge^q \mathcal{T}_X \otimes \omega_X).$$

For a derived equivalence Φ the induced isomorphism Φ^{HH_*} yields the isomorphism alluded to before (see Remark 5.40)

$$\bigoplus_{p+q=i} H^p(X, \bigwedge^q \mathcal{T}_X \otimes \omega_X) \simeq \bigoplus_{p+q=i} H^p(Y, \bigwedge^q \mathcal{T}_Y \otimes \omega_Y).$$

As was remarked in Remark 5.40, one can identify the direct sum with the $(i-n)$-th column of the Hodge diamond and one thus obtains isomorphisms

$$\bigoplus_{p-q=i-n} H^p(X, \Omega_X^q) \simeq \bigoplus_{p-q=i-n} H^p(Y, \Omega_Y^q).$$

It is believed that for an equivalence

$$\Phi : \mathrm{D^b}(X) \xrightarrow{\sim} \mathrm{D^b}(Y)$$

the two isomorphisms Φ^{HH_*} and Φ^H commute under these identifications up to twisting with $\sqrt{\mathrm{td}}$. More precisely, the following diagram should commute:

$$\begin{array}{ccc} HH_*(X) & \xrightarrow{\Phi^{HH_*}} & HH_*(Y) \\ \downarrow{\scriptstyle \sqrt{\mathrm{td}(X)}} & & \downarrow{\scriptstyle \sqrt{\mathrm{td}(Y)}} \\ H^*(X, \mathbb{C}) & \xrightarrow{\Phi^H} & H^*(Y, \mathbb{C}). \end{array}$$

The evidence for this conjecture is manifold. E.g. Căldăraru shows that it holds true on the image of the Mukai vector, which itself is contained in HH_0.

6.2 Geometrical aspects of the Fourier–Mukai kernel

In this section we prove a series of technical but useful facts that shed light on the geometry of the support of the Fourier–Mukai kernel $\mathcal{P}$ of an equivalence

$$\Phi_{\mathcal{P}} : \mathrm{D^b}(X) \xrightarrow{\sim} \mathrm{D^b}(Y).$$

Sometimes, $\mathcal{P}$ is a locally free sheaf on $X \times Y$, e.g. the Poincaré sheaf on the product of an abelian variety and its dual (see Chapter 9), and then nothing interesting can be said about $\mathrm{supp}(\mathcal{P})$, which is just all $X \times Y$. However, often the kernel $\mathcal{P}$ is concentrated on a smaller subvariety, e.g. on the graph of a morphism or a correspondence, and then it encodes information about the geometric relationship between X and Y. This usually happens if the canonical bundles of the varieties enjoy some kind of positivity.

The results, which will be presented as a series of lemmas, are in the original literature often implicitly contained in the proofs of deeper results, some of which shall be discussed later. Most of the material is taken from [63].

Throughout this section we shall consider a Fourier–Mukai equivalence

$$\Phi_{\mathcal{P}} : \mathrm{D^b}(X) \xrightarrow{\sim} \mathrm{D^b}(Y)$$

with Fourier–Mukai kernel $\mathcal{P} \in \mathrm{D^b}(X \times Y)$. Its support (see Definition 3.8)

$$\mathrm{supp}(\mathcal{P}) = \bigcup \mathrm{supp}(\mathcal{H}^i(\mathcal{P})) \subset X \times Y,$$

is a closed subset with possibly many irreducible components. We also recall that by Lemma 3.32 and the fact that taking the tensor product with a line bundle does not change the support of a complex one has:

$$\mathrm{supp}(\mathcal{P}) = \mathrm{supp}(\mathcal{P}^\vee) = \mathrm{supp}(\mathcal{P}_\mathrm{R}) = \mathrm{supp}(\mathcal{P}_\mathrm{L}).$$

This is in fact true without $\Phi_{\mathcal{P}}$ being an equivalence. For the stronger statement $\mathcal{P}_\mathrm{R} \simeq \mathcal{P}_\mathrm{L}$ one needs $\Phi_{\mathcal{P}}$ to be an equivalence. Also note that $\mathcal{P} \otimes q^*\omega_X \simeq \mathcal{P} \otimes p^*\omega_Y$ in this case (see Remark 5.22).

We can hope to extract geometrically meaningful information from the support $\mathrm{supp}(\mathcal{P})$ of the kernel of a Fourier–Mukai equivalence only if $\mathrm{supp}(\mathcal{P})$ is a proper subset. This is, however, not always the case. E.g. the Poincaré bundle is a line bundle and hence has support on the whole product (see Chapter 9). In fact, if $\mathrm{supp}(\mathcal{P}) = X \times Y$, then the two canonical bundles ω_X and ω_Y are both of finite order, see Exercise 6.10.

In the sequel we will often abbreviate

$$\mathcal{H}^i := \mathcal{H}^i(\mathcal{P})$$

and use $\mathcal{H}^i \otimes q^*\omega_X \simeq \mathcal{H}^i \otimes p^*\omega_Y$.

Lemma 6.4 *The natural projection* $\mathrm{supp}(\mathcal{P}) \twoheadrightarrow X$ *is surjective.*

Proof We shall use the spectral sequence (see (3.9), p. 80)

$$E_2^{r,s} = \mathcal{T}or_{-r}(\mathcal{H}^s, q^*k(x)) \Rightarrow \mathcal{T}or_{-(r+s)}(\mathcal{P}, q^*k(x))$$

and the fact that $\mathcal{T}or_i(\mathcal{F}, \mathcal{E})$ is local (and hence trivial for coherent sheaves $\mathcal{F}, \mathcal{E}$ with disjoint support).

Thus, for a closed point $x \in X$ in the complement of $q(\mathrm{supp}(\mathcal{P}))$, the derived tensor product $\mathcal{P} \otimes q^*k(x)$ is trivial. Therefore, $\Phi_{\mathcal{P}}(k(x)) \simeq 0$, which is absurd for the equivalence $\Phi_{\mathcal{P}} : \mathrm{D^b}(X) \xrightarrow{\sim} \mathrm{D^b}(Y)$. □

As the situation is completely symmetric and $\mathrm{supp}(\mathcal{P}) = \mathrm{supp}(\mathcal{P}_\mathrm{R})$, one immediately derives from the lemma also the surjectivity of the other projection $\mathrm{supp}(\mathcal{P}) \twoheadrightarrow Y$.

Corollary 6.5 *There exists an integer $i \in \mathbb{Z}$ and an irreducible component Z of* $\mathrm{supp}(\mathcal{H}^i)$ *that projects onto X.* □

Again, the corollary applies also to the projection to Y, but a priori the integer i and the irreducible component Z might have to be chosen differently then.

Lemma 6.6 *Let C be a complete reduced curve and let $\varphi : C \longrightarrow X \times Y$ be a morphism with image in* supp($\mathcal{P}$). *Then*

$$\deg(\varphi^* q^* \omega_X) = \deg(\varphi^* p^* \omega_Y).$$

In other words, the pull-backs $q^*\omega_X|_{\mathrm{supp}(\mathcal{P})}$ and $p^*\omega_Y|_{\mathrm{supp}(\mathcal{P})}$ are numerically equivalent.

Proof We may assume that the curve C is irreducible and smooth. Then there exists an integer i with $\varphi(C) \subset \mathrm{supp}(\mathcal{H}^i)$, i.e. the underived pull-back $\varphi^*\mathcal{H}^i$ is a sheaf with a possibly non-trivial torsion part $T(\varphi^*\mathcal{H}^i)$, but such that its locally free part $\mathcal{F} := \varphi^*\mathcal{H}^i/T(\varphi^*\mathcal{H}^i)$ is non-trivial. In other words, $\mathcal{F}$ is a locally free sheaf of positive rank, say r.

On the other hand, as $\Phi_{\mathcal{P}}$ is an equivalence, one has $\mathcal{P} \otimes q^*\omega_X \simeq \mathcal{P} \otimes p^*\omega_Y$ and thus $\mathcal{H}^i \otimes q^*\omega_X \simeq \mathcal{H}^i \otimes p^*\omega_Y$. Pulled-back to C it yields $\mathcal{F} \otimes \varphi^*q^*\omega_X \simeq \mathcal{F} \otimes \varphi^*p^*\omega_Y$ and after taking determinants $\varphi^*q^*\omega_X^r \simeq \varphi^*p^*\omega_Y^r$. This suffices to conclude. □

Corollary 6.7 *The canonical bundle ω_X is numerically trivial if and only if the canonical bundle ω_Y is.*

Proof Suppose ω_X is numerically trivial. Then in particular $\deg \varphi^*q^*\omega_X = 0$ for any curve $\varphi : C \longrightarrow X \times Y$. Thus, the lemma shows that $p^*\omega_Y|_{\mathrm{supp}(\mathcal{P})}$ is numerically trivial.

A line bundle is numerically trivial if and only if the line bundle and its dual are both nef (see Definition 6.26). Hence, Lemma 6.27 applied to the surjective morphism $\mathrm{supp}(\mathcal{P}) \twoheadrightarrow Y$ (see Lemma 6.4) shows that ω_Y and ω_Y^* are also both nef. Hence, ω_Y is numerically trivial as well. □

Corollary 6.8 *Suppose $Z \subset$* supp($\mathcal{P}$) *is a closed subvariety such that the restriction of ω_X (or its dual ω_X^*) to the image of $q : Z \longrightarrow X$ is ample. Then $p : Z \longrightarrow Y$ is a finite morphism.*

Proof Suppose $p : Z \longrightarrow Y$ is not finite. Then there exists an irreducible curve $\varphi : C \hookrightarrow Z$ such that $p \circ \varphi : C \longrightarrow Y$ is constant. Thus, $\varphi^*p^*\omega_Y$ is a (numerically) trivial line bundle on C. Lemma 6.6 shows that $\varphi^*q^*\omega_X$ is also numerically trivial. As $p \circ \varphi$ is constant, the composition $q \circ \varphi$ is necessarily non-trivial. Since ω_X (or its dual ω_X^*) is ample on $q(Z)$ and hence on $q(\varphi(C))$, this yields a contradiction. □

Here is a refined version of the same principle.

Lemma 6.9 *Let $Z \subset$* supp($\mathcal{P}$) *be a closed irreducible subvariety with normalization $\mu : \widetilde{Z} \longrightarrow Z$. Then there exists an integer $r > 0$ such that*

$$\pi_X^*\omega_X^r \simeq \pi_Y^*\omega_Y^r,$$

where $\pi_X := q \circ \mu$ and $\pi_Y := p \circ \mu$.

Proof Let us first prove the following general fact:

- Let Z be a normal variety over a field k and let $\mathcal{F}$ be a coherent sheaf on Z generically of rank r. If $L_1, L_2 \in \mathrm{Pic}(Z)$ are two line bundles such that $\mathcal{F} \otimes L_1 \simeq \mathcal{F} \otimes L_2$, then $L_1^r \simeq L_2^r$.

Proof. Clearly, we may divide out by the torsion of $\mathcal{F}$ and can, therefore, assume that $\mathcal{F}$ is torsion free to begin with (the generic rank is unchanged while doing this). Since Z is normal, this means that $\mathcal{F}$ is locally free on the open complement U of a codimension two subset.

As $\det(\mathcal{F} \otimes L_i|_U) \simeq (\det(\mathcal{F}) \otimes L_i^r)|_U$, one has $L_1^r|_U \simeq L_2^r|_U$. The induced trivializing section $s \in H^0(U, L_1^r \otimes L_2^{-r})$ extends to a section $\tilde{s} \in H^0(Z, L_1^r \otimes L_2^{-r})$, which automatically is trivializing and, hence, induces an isomorphism $L_1^r \simeq L_2^r$. For the last two statements one uses $\mathrm{codim}(Z \setminus U) \geq 2$ and the normality of Z. □

Now let $Z \subset \mathrm{supp}(\mathcal{P})$ be a closed irreducible subvariety and let $\mu : \widetilde{Z} \longrightarrow Z$ be its normalization. Then there exists an integer i with $Z \subset \mathrm{supp}(\mathcal{H}^i)$, i.e. $\mu^*\mathcal{H}^i$ is a coherent sheaf on $\widetilde{Z}$ of generically positive rank, say $r > 0$. Pulling-back $\mathcal{H}^i \otimes q^*\omega_X \simeq \mathcal{H}^i \otimes p^*\omega_Y$ via μ to the normal variety $\widetilde{Z}$ allows one to conclude by using the above general fact. □

Exercise 6.10 Suppose $\Phi_{\mathcal{P}} : \mathrm{D^b}(X) \xrightarrow{\sim} \mathrm{D^b}(Y)$ is a Fourier–Mukai equivalence with kernel $\mathcal{P} \in \mathrm{D^b}(X \times Y)$ such that $\mathrm{supp}(\mathcal{P}) = X \times Y$. Show that ω_X and ω_Y are both of finite order. (In fact, by Proposition 4.1 of the same finite order.)

Lemma 6.11 *The fibres of the projection* $\mathrm{supp}(\mathcal{P}) \longrightarrow X$ *are connected.*

Proof Suppose there exists a point $x \in X$ over which the fibre is not connected. Write $\mathrm{supp}(\mathcal{P}) \cap (\{x\} \times Y) = Y_1 \sqcup Y_2$ as a disjoint union of two non-empty closed subsets $Y_1, Y_2 \subset Y$.

Recall that by Lemma 3.29 we have $\mathrm{supp}(\mathcal{P}) \cap (\{x\} \times Y) = \mathrm{supp}(\mathcal{P}|_{\{x\} \times Y})$. Hence, $\Phi_{\mathcal{P}}(k(x))$ has a disconnected support and can, therefore, be written as a direct sum $\mathcal{F}_1^\bullet \oplus \mathcal{F}_2^\bullet$ with $\mathrm{supp}(\mathcal{F}_i^\bullet) = Y_i$, $i = 1, 2$ (cf. Lemma 3.9).

In particular, $\mathrm{End}(\Phi(k(x))) = \mathrm{End}(\mathcal{F}_1^\bullet \oplus \mathcal{F}_2^\bullet)$ is not a field. This contradicts $k = \mathrm{End}(k(x)) \simeq \mathrm{End}(\Phi(k(x)))$. □

Corollary 6.12 *Let* $Z \subset \mathrm{supp}(\mathcal{P})$ *be an irreducible component that surjects onto* X. *If* $\dim(Z) = \dim(X)$, *then* $q : Z \longrightarrow X$ *is a birational morphism. Moreover, if such a component exists, then no other component of* $\mathrm{supp}(\mathcal{P})$ *dominates* X.

Proof Let us prove the last assertion first. Recall that due to Lemma 6.11 every fibre of $\mathrm{supp}(\mathcal{P}) \longrightarrow X$ is connected. Consider the generic fibre of $\bigcup Z_i \longrightarrow X$, where the Z_i are the irreducible components of $\mathrm{supp}(\mathcal{P})$ different from Z. It is either empty or contains the corresponding (zero-dimensional!) fibre of $Z \longrightarrow X$. The latter would imply $Z \subset \bigcup Z_i$ which is absurd.

In order to prove that $q : Z \twoheadrightarrow X$ is birational, we pick a generic point $x \in X$. The intersection

$$\{y_1, \ldots, y_\ell\} := Z \cap (\{x\} \times Y)$$

is finite and disjoint from any other irreducible component of $\mathrm{supp}(\mathcal{P})$. Applying the lemma proves $\ell = 1$, i.e. $Z \twoheadrightarrow X$ is birational. □

Remark 6.13 So far we have considered the irreducible components of $\mathrm{supp}(\mathcal{P})$ with their reduced scheme structure, which is not very natural but usually sufficient. If a component Z as in the corollary exists, then the assertion is in fact still valid even when Z is considered with its natural scheme structure, which a priori might be non-reduced.

More precisely, under the same assumptions one shows that for a generic point $x \in X$ the image $\Phi_{\mathcal{P}}(k(x))$ is of the form $k(y)[m]$. Indeed, $\mathcal{F}^\bullet := \Phi_{\mathcal{P}}(k(x))$ is concentrated in some point $y \in Y$ and $\mathrm{Hom}(\mathcal{F}^\bullet, \mathcal{F}^\bullet[i]) = \mathrm{Hom}(k(x), k(x)[i]) = 0$ for $i < 0$. Then conclude by Lemma 4.5.

The following is a refinement of Corollary 5.23.

Corollary 6.14 *Suppose there exists a closed point $x_0 \in X$ such that*

$$\Phi_{\mathcal{P}}(k(x_0)) \simeq k(y_0)$$

for a certain closed point $y_0 \in Y$. Then one finds an open neighbourhood $x_0 \in U \subset X$ and a morphism $f : U \twoheadrightarrow Y_0$ with $f(x_0) = y_0$ and such that

$$\Phi_{\mathcal{P}}(k(x)) \simeq k(f(x))$$

for all closed points $x \in U$.

Proof The assumption says in particular that the fibre over x_0 of the morphism $\mathrm{supp}(\mathcal{P}) \twoheadrightarrow X$ is zero-dimensional. This clearly holds true then for all points in an open neighbourhood $U \subset X$ of x_0. In other words, for any $x \in U$ the complex $\Phi_{\mathcal{P}}(k(x))$ is concentrated in points. As before, $\mathrm{Hom}(\Phi_{\mathcal{P}}(k(x)), \Phi_{\mathcal{P}}(k(x))[i]) = 0$ for $i < 0$. Thus, Lemma 4.5 applies and shows that $\Phi_{\mathcal{P}}(k(x))$ is of the form $k(y)[m]$. Due to semi-continuity the shift m needs to be constant locally around $x_0 \in U$.

To conclude, one imitates the proof of Corollary 5.23 in order to verify that the induced map $U \twoheadrightarrow Y$ (of sets of closed points) is induced by an honest morphism. □

6.3 Nefness under derived equivalence

After the technical preparations in the last section, the following result is proven easily. (For the definition of nef and the numerical Kodaira dimension see Section 6.5.)

Proposition 6.15 (Kawamata) *Let X and Y be smooth projective varieties with equivalent derived categories $\mathrm{D^b}(X)$ and $\mathrm{D^b}(Y)$. Then the (anti-)canonical bundle of X is nef if and only if the (anti-)canonical bundle of Y is nef. See [63].*

Proof Due to Theorem 5.14 we know that any equivalence

$$F : \mathrm{D^b}(X) \xrightarrow{\sim} \mathrm{D^b}(Y)$$

is of the form $\Phi_{\mathcal{P}}$ with a uniquely determined kernel $\mathcal{P} \in \mathrm{D^b}(X \times Y)$.

Consider the projection $q : \mathrm{supp}(\mathcal{P}) \twoheadrightarrow X$ which is surjective due to Lemma 6.4. Following Lemma 6.27 we know that ω_X is nef if and only if $q^*\omega_X$ is a nef line bundle on $\mathrm{supp}(\mathcal{P})$.

Suppose ω_Y is nef. Then again due to Lemma 6.27 the line bundle $p^*\omega_Y$ is nef on $\mathrm{supp}(\mathcal{P})$. In other words, $\deg(\varphi^*p^*\omega_Y) \geq 0$ for any curve $\varphi : C \longrightarrow \mathrm{supp}(\mathcal{P})$. By Lemma 6.6, $\deg(\varphi^*p^*\omega_Y) = \deg(\varphi^*q^*\omega_X)$. Thus, $\deg(\varphi^*q^*\omega_X) \geq 0$ for any curve $\varphi : C \longrightarrow \mathrm{supp}(\mathcal{P})$, i.e. $q^*\omega_X$ is a nef line bundle on $\mathrm{supp}(\mathcal{P})$.

Similarly, one proves that if ω_Y^* is nef, then so is ω_X^*. Of course, interchanging the rôle of X and Y and repeating the arguments also shows that ω_X (or ω_X^*) nef implies ω_Y (respectively ω_Y^*) nef. □

An immediate consequence is the following result, which has been stated already earlier as Corollary 6.7.

Corollary 6.16 *If X and Y are smooth projective varieties with equivalent derived categories, then ω_X is numerically trivial if and only if ω_Y is numerically trivial.*

Proof Just note that a line bundle is numerically trivial if and only if the line bundle and its dual are both nef. □

Remark 6.17 The corollary complements nicely Proposition 4.1 which in particular shows that ω_X is trivial if and only if ω_Y is trivial.

Proposition 6.18 (Kawamata) *Let X and Y be smooth projective varieties with equivalent derived categories $\mathrm{D^b}(X)$ and $\mathrm{D^b}(Y)$. Then equality of numerical Kodaira dimensions holds: $\nu(X) = \nu(Y)$. See [63].*

Proof This time we apply the stronger Lemma 6.9.

We denote by $\mathcal{H}^i$ the cohomology sheaves $\mathcal{H}^i(\mathcal{P})$. Now apply Corollary 6.5 which shows that there exists at least one cohomology $\mathcal{H}^i$ and an irreducible component Z of $\mathrm{supp}(\mathcal{H}^i)$ such that $p : Z \twoheadrightarrow Y$ is surjective.

Denote the normalization of Z by $\mu : \widetilde{Z} \longrightarrow Z$ and the two projections to X and Y by $\pi_X = q \circ \mu$, respectively $\pi_Y = p \circ \mu$. Due to Lemma 6.9 one finds an integer $r > 0$ with $\pi_X^*\omega_X^r \simeq \pi_Y^*\omega_Y^r$.

Now use the general fact that $\nu(L) = \nu(L^r)$ for any line bundle L and any $r \neq 0$ and Lemma 6.30 to prove $\nu(X, \omega_X) \geq \nu(Y, \omega_Y)$. But due to the symmetry of the situation, this is enough to conclude $\nu(X, \omega_X) = \nu(Y, \omega_Y)$. □

6.4 Derived equivalence versus birationality

As has been proved in Sections 6.1 and 6.3, the (numerical) Kodaira dimension of a smooth projective variety is an invariant of its derived category. But it is also a birational invariant. So one might, and should, wonder whether birational varieties have equivalent derived categories and, conversely, whether derived equivalent varieties are birational. In this generality, the answer to both questions is negative. If, however, the Kodaira dimension or the Kodaira dimension of the anti-canonical bundle is maximal, then an affirmative answer to the second question has been obtained by Kawamata. Moreover, a very precise conjecture concerning the first one has been formulated by Bondal, Orlov, and Kawamata.

This section presents a few results clarifying some of these questions. More can be found in the later chapters. In particular, we shall study examples of varieties with equivalent categories which are not birational (e.g. certain abelian varieties or K3 surfaces, see Chapters 9 and 10) and of birationally equivalent varieties which realize inequivalent derived categories (e.g. a simple blow-up, see Chapter 11).

Proposition 6.19 (Kawamata) *As above, let X and Y be two smooth projective varieties over an algebraically closed field. Suppose there exists an exact equivalence*

$$\mathrm{D}^{\mathrm{b}}(X) \xrightarrow{\sim} \mathrm{D}^{\mathrm{b}}(Y).$$

If $\mathrm{kod}(X,\omega_X) = \dim(X)$ or $\mathrm{kod}(X,\omega_X^*) = \dim(X)$, then X and Y are birational and, more precisely, there exists a birational correspondence

$$\begin{array}{ccc} & Z & \\ {}^{\pi_X}\swarrow & & \searrow^{\pi_Y} \\ X & & Y \end{array}$$

with $\pi_X^*\omega_X \simeq \pi_Y^*\omega_Y$. See [63].

Proof We shall only treat the case $\mathrm{kod}(X,\omega_X) = \dim(X)$, the other being completely analogous.

Let $H \subset X$ be a smooth ample hypersurface. The exact sequence

$$0 \longrightarrow \mathcal{O}(-H) \longrightarrow \mathcal{O} \longrightarrow \mathcal{O}_H \longrightarrow 0$$

induces an exact sequence

$$0 \longrightarrow H^0(\omega_X^\ell(-H)) \longrightarrow H^0(\omega_X^\ell) \longrightarrow H^0(\omega_X^\ell|_H)$$

for any ℓ. If $\mathrm{kod}(X,\omega_X) = \dim(X)$, then $\dim H^0(X,\omega_X^\ell)$ grows like $\ell^{\dim(X)}$. On the other hand, as $\dim(H) < \dim(X)$, the dimension of $H^0(\omega_X^\ell|_H)$ has smaller growth. Thus, for $\ell \gg 0$ the line bundle $\omega_X^\ell(-H)$ has a section. In other words, $\omega_X^\ell(-H) \simeq \mathcal{O}(D)$ for some effective divisor D or, equivalently,

$$\omega_X^\ell \simeq \mathcal{O}(H) \otimes \mathcal{O}(D)$$

with H ample and D effective. (This fact is called Kodaira's lemma.)

Due to Lemma 6.4 there exists an irreducible component Z of $\operatorname{supp}(\mathcal{P})$ that surjects onto X. Moreover, the pull-backs of (some power of) ω_X and ω_Y under $\pi_X : \widetilde{Z} \longrightarrow X$, respectively $\pi_Y : \widetilde{Z} \longrightarrow Y$ coincide, where $\widetilde{Z} \longrightarrow Z$ is the normalization (see Lemma 6.9).

Let us show that

$$\pi_Y : \widetilde{Z} \setminus \pi_X^{-1}(D) \longrightarrow Y$$

is quasi-finite, i.e. has finite fibres. In other words, at most curves completely mapped into D via π_X are contracted by the projection to Y.

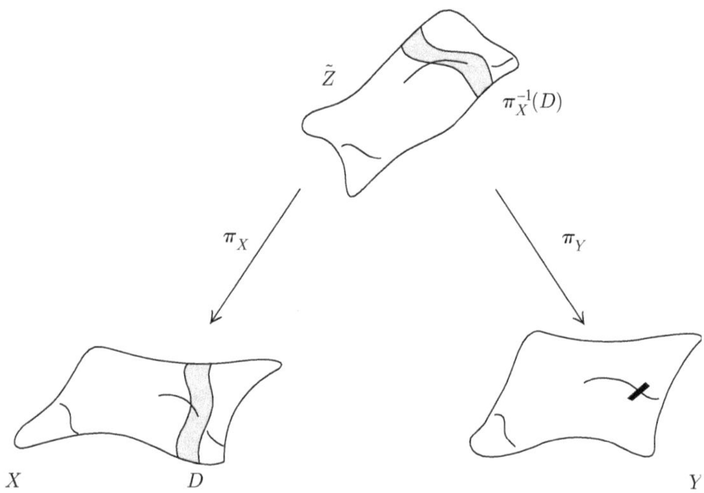

Suppose that there exists an irreducible curve $C \subset \widetilde{Z}$ contracted by π_Y and such that $C \not\subseteq \pi_X^{-1}(D)$. Then, $\deg \pi_Y^*(\omega_Y)|_C = 0$. On the other hand,

$$\deg \pi_X^*(\omega_X)|_C \geq (1/\ell) \deg \pi_X^* \mathcal{O}(H)|_C,$$

as the intersection of $\pi_X(C)$ and D consists of at most finitely many points. Moreover, since C is contracted by π_Y, the other projection $\pi_X : C \longrightarrow X$ must be finite. As H is ample this implies $\deg \pi_X^* \mathcal{O}(H)|_C > 0$. Altogether, this contradicts $\pi_X^* \omega_X^r|_C \simeq \pi_Y^* \omega_Y^r|_C$ implied by Lemma 6.9 or the more elementary $\deg(\pi_X^* \omega_X) = \deg(\pi_Y^* \omega_Y)$ of Lemma 6.6.

Hence, $Z \longrightarrow Y$ is generically finite and thus $\dim(Z) \leq \dim(Y)$. On the other hand, Z dominates X and therefore $\dim(X) \leq \dim(Z)$. As $\dim(X) = \dim(Y)$, this shows that the correspondence $X \xleftarrow{\pi_X} \widetilde{Z} \xrightarrow{\pi_Y} Y$ maps generically finitely onto X and onto Y.

Now apply Corollary 6.12 to conclude that we have in fact constructed a birational correspondence

$$\begin{array}{ccc} & \widetilde{Z} & \\ {}^{\pi_X}\swarrow & & \searrow^{\pi_Y} \\ X & & Y. \end{array}$$

Moreover, by construction $\pi_X^*\omega_X^r \simeq \pi_Y^*\omega_Y^r$ for some $r > 0$.

On the other hand,

$$\pi_X^*\omega_X + \mathcal{O}\left(\sum a_i E_i\right) \simeq \pi_Y^*\omega_Y + \mathcal{O}\left(\sum a_i' E_i\right),$$

where the E_i are exceptional with respect to π_X or π_Y. (Well, this can only be ensured if, e.g. $\widetilde{Z}$ is smooth, but we may actually replace $\widetilde{Z}$ by a desingularization. In fact, if the isomorphism exists on a desingularization, it also exists on the normal variety $\widetilde{Z}$.)

Passing to the r-th power shows $\mathcal{O}(\sum r(a_i - a_i')E_i) \simeq \mathcal{O}$. Thus, it suffices to show that whenever a linear combination $\sum \alpha_i E_i$ is linearly equivalent to zero, then all α_i are trivial. In our case, this would yield $r(a_i - a_i') = 0$ and, hence, $a_i = a_i'$.

Here is the sketch of the argument. Away from the pairwise intersections of the different exceptional divisors, they can all be contracted at once. So we suppose for simplicity that there is a single contraction $\widetilde{Z} \twoheadrightarrow X$ contracting all E_i. Suppose $\sum \alpha_i E_i$ is linearly equivalent to zero with $\alpha_i < 0$ for $i \leq k$ and $\alpha_i \geq 0$ for $i > k$. We may assume $k > 0$, otherwise change the global sign.

Now, let $s \in H^0(\mathcal{O}(-\sum_1^k \alpha_i E_i))$ be the unique section vanishing to order $-\alpha_i$ along the divisors E_i, $i = 1, \ldots, k$. A trivializing section of $\mathcal{O}(\sum \alpha_i E_i)$ multiplied by s would yield a section of $\mathcal{O}(\sum_{i \geq k+1} \alpha_i E_i)$ vanishing along the divisors E_i with $i \leq k$. However, $\mathcal{O}(\sum_{i \geq k+1} \alpha_i E_i)$ admits only one global section up to scaling, namely the one that vanishes only along E_i, $i > k$ (of order α_i).

(Indeed, by contracting the exceptional divisors E_i, $i \geq k+1$, two sections of $\mathcal{O}(\sum_{i \geq k+1} \alpha_i E_i)$ give rise to two functions on the complement of a closed subset of X of codimension ≥ 2 which by Hartogs differ by a scalar factor.)

This yields a contradiction.

If we don't want to assume the existence of the single contraction, we have to work with a morphism $\widetilde{Z} \supset U \twoheadrightarrow V$ onto a quasi-projective variety that can be dominated by open subsets in X or Y whose complement is of codimension two. □

Let us show how to use the arguments of the last proof for yet another alternative proof of Proposition 4.11 (cf. Exercise 6.2).

Corollary 6.20 *If X and Y have derived equivalent categories and ω_X or ω_X^* is ample, then $X \simeq Y$.*

Proof Consider the birational correspondence $Z \subset X \times Y$ constructed above. If $C \subset Z$ is a curve contracted by the projection $\pi_Y : Z \longrightarrow Y$, then $\pi_Y^* \omega_Y|_C \simeq \mathcal{O}_C$, but $\pi_X^* \omega_X|_C$ is ample. This contradicts $\pi_X^* \omega_X \simeq \pi_Y^* \omega_Y$. To be more precise we have to pass to the normalization $\widetilde{Z} \longrightarrow Z$, but this is a finite map.

Hence, $Z \longrightarrow Y$ is an isomorphism. So, there exists a birational morphism $\pi_X : Y \simeq Z \longrightarrow X$ with $\pi_X^* \omega_X \simeq \omega_Y$. The determinant of the differential of π_X can be seen as a section of $\pi_X^* \omega_X \otimes \omega_Y^* \simeq \mathcal{O}_Y$, which is either trivial or non-vanishing everywhere. Thus, since $\pi_X : Y \simeq Z \longrightarrow X$ is birational, it is in fact smooth and hence an isomorphism.

The argument for ω_X^* ample is identical. □

Definition 6.21 *Two varieties X and Y are called* K-equivalent *if there exists a birational correspondence*

$$\begin{array}{ccc} & Z & \\ {}^{\pi_X}\swarrow & & \searrow^{\pi_Y} \\ X & & Y \end{array}$$

with $\pi_X^* \omega_X \simeq \pi_Y^* \omega_Y$.

Corollary 6.22 *Two D-equivalent varieties X and Y with X of maximal (anti-canonical) Kodaira dimension are K-equivalent.* □

Remark 6.23 Without this additional assumption the statement is false. E.g. we will present work of Mukai showing that there exist non-isomorphic and hence non-birational abelian varieties with equivalent derived categories (see Chapter 9). But even if one adds the assumption that X and Y are birational, D-equivalence does not in general imply K-equivalence (see [114] and Section 12.2).

The converse of the corollary is the following.

Conjecture 6.24 *Let X and Y be two smooth projective varieties. If X and Y are K-equivalent, then they are also D-equivalent.*

Thus, for birationally equivalent varieties of maximal (anti-canonical) Kodaira dimension one expects:

$$\text{D} - \text{equivalent} \quad \Longleftrightarrow \quad \text{K} - \text{equivalent}.$$

The conjecture also predicts that two birational Calabi–Yau varieties, i.e. varieties with trivial canonical bundle, are derived equivalent. Some progress has been made in low dimensions (see Chapter 11 for more details), but the general question, even for Calabi–Yau manifolds, is still wide open.

As is probably clear from a closer inspection of the above proof, there is no reason to hope that the birational correspondence induces the sought for equivalence. In fact, there are explicit examples known where this is not true (cf. Section 11.4).

6.5 Recap: Kodaira dimension, canonical ring, etc.

This section recalls a few definitions and facts from higher-dimensional algebraic geometry.

Definition 6.25 *Let X be a smooth projective variety and let $L \in \mathrm{Pic}(X)$. The* Kodaira dimension $\mathrm{kod}(X, L)$ *of L on X is the integer m such that*

$$h^0(X, L^\ell) := \dim H^0(X, L^\ell)$$

grows like a polynomial of degree m for $\ell \gg 0$. By definition, $\mathrm{kod}(X, L) = -\infty$ *if $h^0(X, L^\ell) = 0$ for all $\ell > 0$.*

There are equivalent descriptions of the Kodaira dimension (cf. [115]): E.g. under the assumption that $\mathrm{kod}(X, L) \geq 0$, one has

$$\begin{aligned} \mathrm{kod}(X, L) &= \max\{\dim(\mathrm{Im}(\varphi_{L^\ell})) \mid \ell \geq 0\} && (6.3) \\ &= \mathrm{trdeg}_k Q(R(X, L)) - 1. && (6.4) \end{aligned}$$

Here, $\varphi_{L^\ell} : X \dashrightarrow \mathbb{P}^{h^0(L^\ell)-1}$ is the rational map defined by the linear system $|L^\ell|$, $R(X, L)$ is the *canonical ring of* L, i.e.

$$R(X, L) := \bigoplus_{\ell \geq 0} H^0(X, L^\ell)$$

and $Q(R(X, L))$ denotes its field of fractions. Note that $\mathrm{kod}(X, L) \leq \dim(X)$ for any line bundle L.

The case that interests us most here is when $L \simeq \omega_X$. Then one calls

$$\mathrm{kod}(X) := \mathrm{kod}(X, \omega_X)$$

the *Kodaira dimension* of X and

$$R(X) := R(X, \omega_X)$$

the *canonical ring* of X.

A standard fact in higher dimensional algebraic geometry says that the Kodaira dimension is a birational invariant, i.e. if X and Y are two birational smooth projective varieties, then $\mathrm{kod}(X) = \mathrm{kod}(Y)$ (see [115]).

Definition 6.26 *A line bundle L on a proper scheme X over a field k is called* nef *if for any morphism $\varphi : C \longrightarrow X$ from a complete reduced curve C one has*

$$\deg(\varphi^* L) \geq 0.$$

Of course, it suffices to test curves that are embedded into X, as one might replace $\varphi : C \longrightarrow X$ by the image $C' = \varphi(C)$ (use $\deg \varphi^* L = \deg(\varphi) \cdot \deg(L|_{C'})$). In another direction, it suffices to test $\varphi : C \longrightarrow X$ with C smooth and irreducible, as we can always pass to the normalization of C.

The degree of a line bundle M on a curve C over a field k is defined by the Riemann–Roch formula

$$\chi(C, M^\ell) = \deg(M) \cdot \ell + \chi(C, \mathcal{O}_C).$$

Clearly, a line bundle L is nef if and only if some positive power L^i, $i > 0$, is nef.

Here are a few simple facts for nef line bundles:

Lemma 6.27 *Let $\pi : Z \longrightarrow X$ be a projective morphism of proper schemes and $L \in \mathrm{Pic}(X)$.*

i) *If L is a nef line bundle on X then $\pi^*(L)$ is nef.*
ii) *If π is surjective, then L is nef if and only if $\pi^*(L)$ is nef.*

Proof Let $\varphi : C \longrightarrow Z$ be a given curve. Then the composition with π yields $\pi \circ \varphi : C \longrightarrow X$. This immediately shows i).

To see ii), one constructs for any irreducible curve $\varphi : C \longrightarrow X$ a ramified cover $\psi : \widetilde{C} \longrightarrow C$ by an irreducible curve $\widetilde{C}$ such that $\varphi \circ \psi : \widetilde{C} \longrightarrow X$ factorizes over $Z \longrightarrow X$. Using $\deg(\psi^*\varphi^*L) = \deg(\psi) \cdot \deg(\varphi^*L)$ this finishes the proof.

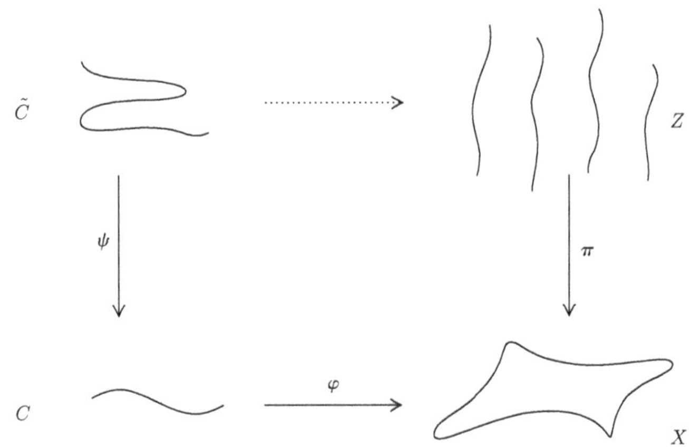

The construction of $\psi : \widetilde{C} \longrightarrow C$ is standard algebraic geometry: By working with the fibre product $C \times_X Z$, we may reduce to the claim that any dominant projective morphism $Z \longrightarrow C$ onto a curve admits a multisection. By embedding Z into some $\mathbb{P}^N \times C$ this may be achieved by intersecting with a generic linear subspace in $\mathbb{P}^N$ of the appropriate dimension. □

By definition, the intersection number $([M]^m.W)$ of a line bundle M on a proper scheme W of dimension m is the degree m coefficient of the polynomial $\chi(W, M^\ell)$ (cf. [33]).

Definition 6.28 *The* numerical Kodaira dimension $\nu(X, L)$ *of a line bundle L on a projective scheme X is the maximal integer m such that there exists a proper morphism $\varphi : W \longrightarrow X$ with W of dimension m with*

$$([\varphi^*(L)]^m.W) \neq 0.$$

As in the definition of nefness, it suffices to test closed subschemes $W \subset X$.

Remark 6.29 A line bundle L is called *numerically trivial* if $\nu(X,L) = 0$ or, equivalently, if for any curve $\varphi : C \longrightarrow X$ one has $\deg \varphi^* L = 0$. Clearly, L is numerically trivial if and only if L and L^* are both nef.

In general, there is no relation between the Kodaira dimension and the numerical Kodaira dimension. Only if L is nef, then $\mathrm{kod}(X,L) \leq \nu(X,L)$ (Exercise!).

For the canonical bundle, one writes

$$\nu(X) := \nu(X, \omega_X)$$

and calls it the *numerical Kodaira dimension of* X.

Lemma 6.30 *Let $\pi : Z \longrightarrow X$ be a projective morphism of projective schemes and $L \in \mathrm{Pic}(X)$.*

i) *Then $\nu(X,L) \geq \nu(Z, \pi^* L)$.*
ii) *If $\pi : Z \longrightarrow X$ is surjective, then $\nu(X,L) = \nu(Z, \pi^* L)$.*

Proof The first assertion follows from the definition, as any proper $\varphi : W \longrightarrow Z$ can be composed with π.

To see ii), consider a proper morphism $\varphi : W \longrightarrow X$. Then there exists a generically finite surjective morphism $\psi : \widetilde{W} \longrightarrow W$ and a morphism $\widetilde{\varphi} : \widetilde{W} \longrightarrow Z$ such that $\pi \circ \widetilde{\varphi} = \varphi \circ \psi$. The existence is ensured by arguments similar to those in the proof of Lemma 6.27. Since

$$([\widetilde{\varphi}^* \pi^* L]^m . \widetilde{W}) = \deg(\psi) \cdot ([\varphi^* L]^m . W),$$

this shows $\nu(X,L) \leq \nu(Z, \pi^* L)$. □

7

EQUIVALENCE CRITERIA FOR FOURIER–MUKAI TRANSFORMS

In the preceding chapters we have studied equivalences between derived categories of smooth projective varieties and how they are reflected by the geometry, cohomology, etc. The time is ripe to develop criteria that allow us to decide whether a given Fourier–Mukai transform is in fact an equivalence. In order to do this, we shall follow the procedure outlined in Chapter 1. So we will first try to understand full faithfulness of a Fourier–Mukai transform. This will be discussed in Section 7.1. Then, in Section 7.2, we will address the question under which circumstances a fully faithful Fourier–Mukai transform does define an equivalence. As it turns out, this is often the easier part of the programme. Section 7.3, where varieties with torsion canonical bundle and their canonical cover are investigated, is logically independent and can also be read later.

We consider smooth projective varieties over an algebraically closed field k of characteristic zero. Earlier, this was imposed in order to simplify the arguments, but here it is crucial and we will point out where it comes in.

7.1 Fully faithful

Consider the Fourier–Mukai transform $\Phi_{\mathcal{P}} : \mathrm{D^b}(X) \longrightarrow \mathrm{D^b}(Y)$ between the derived categories of two smooth projective varieties X and Y given by an object $\mathcal{P} \in \mathrm{D^b}(X \times Y)$. For the following proposition compare the references [14] and [18].

Proposition 7.1 **(Bondal, Orlov)** *The functor $\Phi_{\mathcal{P}}$ is fully faithful if and only if for any two closed points $x, y \in X$ one has*

$$\mathrm{Hom}(\Phi_{\mathcal{P}}(k(x)), \Phi_{\mathcal{P}}(k(y))[i]) = \begin{cases} k & \textit{if } \; x = y \; \textit{ and } \; i = 0 \\ 0 & \textit{if } \; x \neq y \; \textit{ or } \; i < 0 \; \textit{ or } \; i > \dim(X). \end{cases}$$

Proof The proof is an application of Proposition 1.49. The verification of all the hypotheses is rather long and we will split the proof into several steps. We closely follow Bridgeland's account of the proof (cf. [18]).

Step 1. Points are spanning Here we just recall Proposition 3.17 which says that objects of the form $k(x)[i]$ with $x \in X$ a closed point and $i \in \mathbb{Z}$ form a spanning class in $\mathrm{D^b}(X)$.

Since the Fourier–Mukai transform $F := \Phi_{\mathcal{P}}$ admits a left adjoint $G := \Phi_{\mathcal{P}_{\mathrm{L}}}$ and a right adjoint $H := \Phi_{\mathcal{P}_{\mathrm{R}}}$ (cf. Proposition 5.9), we may apply Proposition 1.49. Thus, F is fully faithful if and only if the natural homomorphisms

$$\mathrm{Hom}(k(x), k(y)[i]) \longrightarrow \mathrm{Hom}(F(k(x)), F(k(y))[i])$$

are bijective for arbitrary closed points $x, y \in X$ and any integer i. For $x \neq y$ this holds true by assumption. Thus, it remains to discuss the case $x = y$. In this case, the assumption a priori yields the bijectivity only for $i \notin [1, \dim(X)]$.

Step 2. Reduction to $\mathbf{G(F(k(x))) \simeq k(x)}$ By Lemma 1.21 we know that the bijectivity of

$$\mathrm{Hom}(k(x), k(x)[i]) \longrightarrow \mathrm{Hom}(F(k(x)), F(k(x))[i]) \tag{7.1}$$

is equivalent to the bijectivity of

$$\mathrm{Hom}(k(x), k(x)[i]) \xrightarrow{\circ g_{k(x)}} \mathrm{Hom}(G(F(k(x))), k(x)[i]), \tag{7.2}$$

which is induced by the adjunction morphism $g : G \circ F \longrightarrow \mathrm{id}_{\mathrm{D^b}(X)}$.

If we can show that $G(F(k(x))) \simeq k(x)$, then the adjunction morphism

$$g_{k(x)} : G(F(k(x))) \longrightarrow k(x)$$

is either an isomorphism, which immediately yields bijectivity for all i in (7.2), or $g_{k(x)}$ is zero.

(Recall that we assumed that k is algebraically closed and, hence, that $k(x)$ is isomorphic to k concentrated in x.)

We can actually exclude that $g_{k(x)}$ is zero, since the composition of

$$F(g_{k(x)}) : F(G(F(k(x)))) \longrightarrow F(k(x))$$

with the adjunction morphism

$$h_{F(k(x))} : F(k(x)) \longrightarrow F(G(F(k(x))))$$

yields the identity (see Exercise 1.19) and $F(k(x)) \neq 0$ due to the assumption that $\mathrm{End}(F(k(x))) = k$.

Step 3. Proof of $\mathbf{G(F(k(x))) \simeq k(x)}$ under additional hypothesis Let us fix a closed point $x \in X$. We shall first show $G(F(k(x))) \simeq k(x)$ under two additional assumptions:

i) $G(F(k(x)))$ is a sheaf and
ii) The homomorphism (7.1) is at least injective for $i = 1$ (which is equivalent to the injectivity of (7.2) for $i = 1$).

Let us denote $G(F(k(x)))$ by $\mathcal{F}$, which is a sheaf due to i). Then by adjunction and assumption one has $\mathrm{Hom}(\mathcal{F}, k(y)) = \mathrm{Hom}(F(k(x)), F(k(y))) = 0$ for any closed point $y \neq x$. Hence, $\mathcal{F}$ is concentrated in x. As explained earlier, the adjunction morphism $\delta := g_{k(x)} : \mathcal{F} \twoheadrightarrow k(x)$ is not trivial and hence surjective. We have to show that δ is in fact bijective. Consider the short exact sequence

$$0 \longrightarrow \mathrm{Ker}(\delta) \longrightarrow \mathcal{F} \xrightarrow{\delta} k(x) \longrightarrow 0. \tag{7.3}$$

Clearly, $\mathrm{Ker}(\delta)$ is also concentrated in $k(x)$ and in order to show $\mathrm{Ker}(\delta) \simeq 0$ it suffices to prove $\mathrm{Hom}(\mathrm{Ker}(\delta), k(x)) = 0$. Applying $\mathrm{Hom}(\ \ , k(x))$ to (7.3) and using $\mathrm{Hom}(\mathcal{F}, k(x)) = k$, yields the exact sequence

$$0 \longrightarrow \mathrm{Hom}(\mathrm{Ker}(\delta), k(x)) \longrightarrow \mathrm{Hom}(k(x), k(x)[1]) \xrightarrow{\circ\delta} \mathrm{Hom}(\mathcal{F}, k(x)[1]).$$

The last map is injective due to ii) and hence $\mathrm{Ker}(\delta) = 0$.

Step 4. Verification of the additional hypothesis i) We shall use the following general lemma, which is a variation on the fact that sheaves of the form $k(y)$ are spanning (cf. Proposition 3.17).

Lemma 7.2 *Let X be a smooth projective variety, $x \in X$ a closed point, and $\mathcal{F}^\bullet \in \mathrm{D^b}(X)$. Suppose* $\mathrm{Hom}(\mathcal{F}^\bullet, k(y)[i]) = 0$ *for any closed point $y \neq x$ and any $i \in \mathbb{Z}$ and* $\mathrm{Hom}(\mathcal{F}^\bullet, k(x)[i]) = 0$ *for $i < 0$ or $i > \dim(X)$.*

Then $\mathcal{F}^\bullet$ is isomorphic to a sheaf concentrated in $x \in X$.

Proof We will abbreviate the cohomology sheaves of $\mathcal{F}^\bullet$ by $\mathcal{H}^q$. For a fixed point $y \in X$ we consider the spectral sequence (see (2.8) p. 58)

$$E_2^{p,q} := \mathrm{Hom}(\mathcal{H}^{-q}, k(y)[p]) \Rightarrow \mathrm{Hom}(\mathcal{F}^\bullet, k(y)[p+q]). \tag{7.4}$$

Let m_0 be maximal with $y \in \mathrm{supp}(\mathcal{H}^{m_0})$. Then $E_2^{0,-m_0} \neq 0$ and $E_2^{p,q} = 0$ for $q < -m_0$. Hence,

$$0 \neq E_2^{0,-m_0} = E_\infty^{0,-m_0} = E^{-m_0} = \mathrm{Hom}(\mathcal{F}^\bullet, k(y)[-m_0]).$$

Hence, $y = x$ and $-d \leq m_0 \leq 0$, where $d := \dim(X)$. In other words, all cohomology sheaves of $\mathcal{F}^\bullet$ are concentrated in $x \in X$ and in degree $-d \leq i \leq 0$.

On the other hand, $\mathrm{Hom}(\mathcal{H}^{-q}, k(x)[p]) = 0$ for $p \notin [0, \dim(X)]$ and, therefore, $E_2^{p,q} = 0$ for $p \notin [0, \dim(X)]$ in the spectral sequence (7.4) with $y = x$.

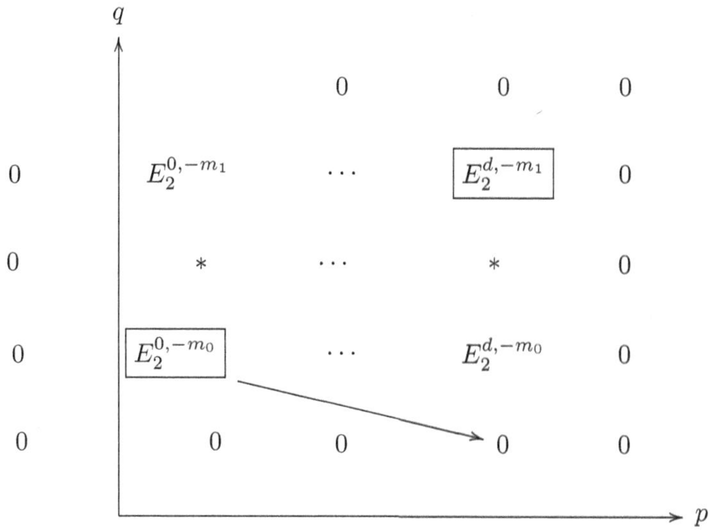

Let now m_1 be minimal with $\mathcal{H}^{m_1} \neq 0$. By what has been shown, we know $m_1 \leq m_0 \leq 0$. Since the sheaf $\mathcal{H}^{m_1}$ is concentrated in x, one finds, by applying Serre duality, that $\mathrm{Hom}(\mathcal{H}^{m_1}, k(x)[d]) \simeq \mathrm{Hom}(k(x), \mathcal{H}^{m_1})^* \neq 0$.

A quick look at the spectral sequence above reveals that a non-trivial element in $E_2^{d,-m_1} = \mathrm{Hom}(\mathcal{H}^{m_1}, k(x)[d])$ survives and yields a non-trivial element in $E^{d-m_1} = \mathrm{Hom}(\mathcal{F}^\bullet, k(x)[d-m_1])$. By assumption the latter group is zero if $d - m_1 > d$, which thus only leaves the possibility $m_1 = m_0 = 0$. This proves that $\mathcal{F}^\bullet$ is isomorphic to a sheaf concentrated in x. □

Thus, we have proved the first of our two additional assumptions in Step 3, namely that $G(F(k(x)))$ is a sheaf for any $x \in X$. Indeed, $\mathcal{F}^\bullet := G(F(k(x)))$ satisfies the assumption of the lemma, because

$$\begin{aligned} \mathrm{Hom}(\mathcal{F}^\bullet, k(y)[i]) &\simeq \mathrm{Hom}(G(F(k(x))), k(y)[i]) \\ &\simeq \mathrm{Hom}(F(k(x)), F(k(y))[i]) = 0 \end{aligned}$$

for $i \notin [0, \dim(X)]$ or $x \neq y$ by assumption.

Step 5. Verification of the additional hypothesis ii) for generic x The composition $G \circ F$ is a Fourier–Mukai transform. We denote its kernel by $\mathcal{Q}$. As we have just seen, $i_x^* \mathcal{Q} = G(F(k(x)))$ is a sheaf (concentrated in x), for any point $x \in X$. Here, i_x^* is the derived pull-back of the inclusion $i_x : \{x\} \times X \hookrightarrow X \times X$. Now Lemma 3.31 applies and shows that $\mathcal{Q}$ is a sheaf on $X \times X$ flat over the first factor.

Note that applying this lemma we use for the first time that the functor F is in fact a Fourier–Mukai transform.

Let us now prove that $\mathrm{Hom}(k(x), k(x)[1]) \longrightarrow \mathrm{Hom}(F(k(x)), F(k(x))[1])$ is injective for generic $x \in X$. The composition with the functor G yields the map

$$\kappa(x) : \mathrm{Hom}(k(x), k(x)[1]) \longrightarrow \mathrm{Hom}(G(F(k(x))), G(F(k(x)))[1])$$

and we will rather show the injectivity of this map. This is clearly sufficient to ensure assertion ii) in step 3.

Using the flatness of $\mathcal{Q}$ and the explanations in vii), Example 5.4, we know that $\kappa(x)$ is the Kodaira–Spencer map of the flat family $\mathcal{Q}$ over $X \times X$ defining $G \circ F$.

On the other hand, the map $f : x \longmapsto \mathcal{Q}_x$ is injective, since for any x the sheaf $\mathcal{Q}_x = G(F(k(x)))$ is concentrated in x (see the arguments at the beginning of step 3). Hence, the tangent map $\kappa(x) := df(x)$ is injective for $x \in X$ generic. (Note that here we definitely use the assumption that the characteristic is zero![3])

Step 6. End of proof. Why generic is enough We may apply step 3 to a generic $x \in X$. Thus, $\mathcal{Q}_x \simeq k(x)$ for generic $x \in X$. On the other hand, $\mathcal{Q}$ is flat over X and hence the Hilbert polynomial of $\mathcal{Q}_x$ is independent of $x \in X$ (cf. [45, III.9]). As we know in addition that $\mathcal{Q}_x$ is concentrated in x for any $x \in X$, we find $\mathcal{Q}_x \simeq k(x)$ for all $x \in X$. □

Remark 7.3 At first sight, the proposition looks like a translation of the general method to test full faithfulness to the case of the spanning class given by closed points. However, it is much stronger than this, as it asserts that the difficult cohomology groups $\mathrm{Ext}^i(k(x), k(x))$ with $0 < i \leq \dim(X)$ need not be tested. To give an idea what they look like, one can show that $\mathrm{Ext}^i(k(x), k(x)) \simeq \bigwedge^i T_x$ with $T_x \simeq \mathrm{Ext}^1(k(x), k(x))$ the Zariski tangent space at $x \in X$ (cf. Section 11.1)

Here is one immediate consequence:

Corollary 7.4 *Consider two fully faithful Fourier–Mukai transforms* $\Phi_{\mathcal{P}} : \mathrm{D}^{\mathrm{b}}(X) \longrightarrow \mathrm{D}^{\mathrm{b}}(Y)$ *and* $\Phi_{\mathcal{P}'} : \mathrm{D}^{\mathrm{b}}(X') \longrightarrow \mathrm{D}^{\mathrm{b}}(Y')$. *Then the product*

$$\Phi_{\mathcal{P} \boxtimes \mathcal{P}'} : \mathrm{D}^{\mathrm{b}}(X \times X') \longrightarrow \mathrm{D}^{\mathrm{b}}(Y \times Y')$$

is again fully faithful.

Proof First note that $\Phi_{\mathcal{P} \boxtimes \mathcal{P}'}(k(x) \boxtimes k(x')) \simeq \Phi_{\mathcal{P}}(k(x)) \boxtimes \Phi_{\mathcal{P}'}(k(x'))$ for any closed point $(x, x') \in X \times X'$ (see Exercise 5.13).

To conclude, apply the Künneth formula

$$\mathrm{Hom}(\mathcal{F}^\bullet \boxtimes \mathcal{F}'^\bullet, \mathcal{G}^\bullet \boxtimes \mathcal{G}'^\bullet) = \bigoplus_{i+j=0} \mathrm{Hom}(\mathcal{F}^\bullet, \mathcal{G}^\bullet[i]) \otimes \mathrm{Hom}(\mathcal{F}'^\bullet, \mathcal{G}'^\bullet[j]).$$

□

[3] To make this rigorous one has to use a little deformation theory. Bridgeland avoids this by arguing with the Hilbert scheme, but I wanted to stick to this geometrically intuitive argument.

Let us mention one special case of the above proposition, where the kernel $\mathcal{P}$ is an actual sheaf. In this case we denote by $\mathcal{P}_x$ the restriction of $\mathcal{P}$ to $\{x\}\times Y \subset X \times Y$, which is naturally isomorphic to Y.

Corollary 7.5 *Let $\mathcal{P}$ be a coherent sheaf on $X \times Y$ flat over X. Then $\Phi_{\mathcal{P}}$ is fully faithful if and only if the following two conditions are satisfied:*

i) *For any point $x \in X$ one has $\mathrm{Hom}(\mathcal{P}_x, \mathcal{P}_x) \simeq k$.*
ii) *If $x \neq y$, then $\mathrm{Ext}^i(\mathcal{P}_x, \mathcal{P}_y) = 0$ for all i.*

Proof This is an immediate consequence of the proposition, for the flatness of $\mathcal{P}$ over X ensures that $\Phi(k(x)) = \mathcal{P}_x$. □

We will come back to this special situation on various occasions. Most often, we will consider the case of a vector bundle $\mathcal{P}$, e.g. the Poincaré bundle in Chapter 9. Clearly, conditions i) and ii) say that all induced bundles $\mathcal{P}_x$ are simple and pairwise orthogonal in the sense that $\mathrm{Ext}^i(\mathcal{P}_x, \mathcal{P}_y) =$ for all i whenever $x \neq y$.

7.2 Equivalences

Suppose $\Phi_{\mathcal{P}} : \mathrm{D^b}(X) \longrightarrow \mathrm{D^b}(Y)$ is a Fourier–Mukai transform between smooth projective varieties of the same dimension. If $\Phi_{\mathcal{P}}$ is an equivalence, then $\Phi_{\mathcal{P}_{\mathrm{R}}} \simeq \Phi_{\mathcal{P}_{\mathrm{L}}}$ and hence $\mathcal{P}^{\vee} \otimes q^*\omega_X \simeq \mathcal{P}^{\vee} \otimes p^*\omega_Y$ (cf. the arguments in the proof of Corollary 5.21). Dualizing once more yields $\mathcal{P} \otimes q^*\omega_X \simeq \mathcal{P} \otimes p^*\omega_Y$. One might wonder whether this condition in itself is sufficient to ensure that a given $\Phi_{\mathcal{P}}$ is an equivalence. Unfortunately, this is not true, as an example of any sheaf on the product on two non-isomorphic elliptic curves reveals. However, it holds true whenever the Fourier–Mukai transform is already known to be fully faithful, but this is a rather weak result:

Proposition 7.6 *Suppose $\Phi_{\mathcal{P}} : \mathrm{D^b}(X) \longrightarrow \mathrm{D^b}(Y)$ is a fully faithful Fourier–Mukai transform between smooth projective varieties. Then $\Phi_{\mathcal{P}}$ is an equivalence if and only if*

$$\dim(X) = \dim(Y) \quad \textit{and} \quad \mathcal{P} \otimes q^*\omega_X \simeq \mathcal{P} \otimes p^*\omega_Y.$$

Proof Any equivalence $\Phi_{\mathcal{P}}$ satisfies these two conditions due to Proposition 4.1 (or Corollary 5.21) and Remark 5.22.

For the converse we want to apply Proposition 1.54. So, we need to ensure that under the assumptions the adjoint functors $G \dashv F := \Phi_{\mathcal{P}} \dashv H$, which exist due to Proposition 5.9, satisfy the condition:

- If $H(\mathcal{F}^\bullet) \simeq 0$ for $\mathcal{F}^\bullet \in \mathrm{D^b}(Y)$, then $G(\mathcal{F}^\bullet) \simeq 0$.
  But this is obvious, as dualizing our hypothesis yields $G = \Phi_{\mathcal{P}_{\mathrm{L}}} \simeq \Phi_{\mathcal{P}_{\mathrm{R}}} = H$.

□

Remark 7.7 We shall give an alternative proof of an assertion encountered earlier in the proof of Proposition 6.1 saying that $\Phi_{\mathcal{P}_{\mathrm{L}}} : \mathrm{D^b}(X) \longrightarrow \mathrm{D^b}(Y)$ is an equivalence if $\Phi_{\mathcal{P}} : \mathrm{D^b}(X) \longrightarrow \mathrm{D^b}(Y)$ is one (both functors go in the same direction). One first shows that $\Phi_{\mathcal{P}_{\rm L}}$ is fully faithful by applying Proposition 7.1. Indeed,

$$\begin{aligned}\mathrm{Hom}(\Phi_{\mathcal{P}_{\rm L}}(k(x)),\Phi_{\mathcal{P}_{\rm L}}(k(y))[i]) &\simeq \mathrm{Hom}(i_x^*\mathcal{P}^\vee, i_y^*\mathcal{P}^\vee[i])\\ &\simeq \mathrm{Hom}((i_x^*\mathcal{P})^\vee,(i_y^*\mathcal{P})^\vee[i])\\ &\simeq \mathrm{Hom}(i_y^*\mathcal{P}, i_x^*\mathcal{P}[i])\\ &\simeq \mathrm{Hom}(\Phi_{\mathcal{P}}(k(y)),\Phi_{\mathcal{P}}(k(x))[i]).\end{aligned}$$

For the second isomorphism use that dualizing and pull-back commute (cf. (3.17), p. 85). In order to see that $\Phi_{\mathcal{P}_{\rm L}}:\mathrm{D^b}(X)\longrightarrow\mathrm{D^b}(Y)$ is in fact an equivalence, apply the proposition. Indeed, as $\Phi_{\mathcal{P}}$ is an equivalence, we have $\mathcal{P}_{\rm L}\otimes q^*\omega_X\simeq\mathcal{P}_{\rm L}\otimes p^*\omega_Y$.

Corollary 7.8 *Suppose X and Y are smooth projective varieties of the same dimension and with trivial canonical bundle ω_X, respectively ω_Y. Then any fully faithful exact functor $\mathrm{D^b}(X)\longrightarrow\mathrm{D^b}(Y)$ is an equivalence.* □

Exercise 7.9 Find a proof of this corollary that only uses the existence of a left (or right) adjoint and not Orlov's existence result. (Hint: Use Remark 1.31.)

Here is another application of the same techniques.

Proposition 7.10 *Consider a fully faithful Fourier–Mukai transform*

$$\Phi_{\mathcal{P}}:\mathrm{D^b}(X)\longrightarrow\mathrm{D^b}(Y)$$

and suppose that its right adjoint $H=\Phi_{\mathcal{P}_{\rm R}}$ satisfies

$$H(\mathcal{F}^\bullet\otimes\omega_Y)\simeq H(\mathcal{F}^\bullet)\otimes\omega_X.$$

Then, $\Phi_{\mathcal{P}}$ is an equivalence.

Proof We shall again apply Proposition 1.54. Let $G\dashv F:=\Phi_{\mathcal{P}}\dashv H$. Suppose $H(\mathcal{F}^\bullet)=0$. Then for any $\mathcal{F}^\bullet\in\mathrm{D^b}(Y)$

$$\begin{aligned}\mathrm{Hom}(G(\mathcal{F}^\bullet),\mathcal{E}^\bullet) &\simeq \mathrm{Hom}(\mathcal{F}^\bullet,F(\mathcal{E}^\bullet))\\ &\simeq \mathrm{Hom}(F(\mathcal{E}^\bullet),\mathcal{F}^\bullet\otimes\omega_Y[\dim(Y)])^*\\ &\simeq \mathrm{Hom}(\mathcal{E}^\bullet,H(\mathcal{F}^\bullet\otimes\omega_Y)[\dim(Y)])^*=0,\end{aligned}$$

where we use that $H(\mathcal{F}^\bullet\otimes\omega_Y)\simeq H(\mathcal{F}^\bullet)\otimes\omega_X\simeq 0$.

Thus, $\mathrm{Hom}(G(\mathcal{F}^\bullet),\mathcal{E}^\bullet)=0$ for any $\mathcal{E}^\bullet\in\mathrm{D^b}(X)$ and hence also $G(\mathcal{F}^\bullet)\simeq 0$. Note that only '$H(\mathcal{F}^\bullet)\simeq 0$ implies $H(\mathcal{F}^\bullet\otimes\omega_Y)=0$' has been used. □

A similar argument shows that another sufficient assumption would be that H commutes with Serre functors. Of course, once we assume that both varieties have the same dimension, these two assumptions are equivalent.

It is not difficult to modify the above proof in order to get the original

Proposition 7.11 **(Bridgeland)** *Suppose* $\Phi_{\mathcal{P}} : \mathrm{D^b}(X) \longrightarrow \mathrm{D^b}(Y)$ *is fully faithful. Then* $\Phi_{\mathcal{P}}$ *is an equivalence if and only if*

$$\Phi_{\mathcal{P}}(k(x)) \otimes \omega_Y \simeq \Phi_{\mathcal{P}}(k(x))$$

for all closed points $x \in X$. *See [18].*

Proof Note that the assertion follows directly from Corollary 1.56 if we assume in addition that $\dim(X) = \dim(Y)$. In this case $\Phi_{\mathcal{P}}(k(x)) \otimes \omega_Y \simeq \Phi_{\mathcal{P}}(k(x))$ says that $\Phi_{\mathcal{P}}$ commutes with Serre functors on the spanning class $\{k(x)\}$.

If we don't know yet that $\dim(X) = \dim(Y)$, then we argue as above. Assume $H(\mathcal{F}^\bullet) \simeq 0$, where as before $G \dashv F \dashv H$ with $F := \Phi_{\mathcal{P}}$. Then

$$\begin{aligned} \mathrm{Hom}(G(\mathcal{F}^\bullet), k(x)[i]) &\simeq \mathrm{Hom}(\mathcal{F}^\bullet, F(k(x))[i]) \\ &\simeq \mathrm{Hom}(\mathcal{F}^\bullet, F(k(x)) \otimes \omega_Y[i]) && \text{(by assumption)} \\ &\simeq \mathrm{Hom}(F(k(x)), \mathcal{F}^\bullet[\dim(Y) - i])^* && \text{(Serre duality)} \\ &\simeq \mathrm{Hom}(k(x), H(\mathcal{F}^\bullet)[\dim(Y) - i])^* = 0. \end{aligned}$$

Since by Proposition 3.17 the objects of the form $k(x)$ span $\mathrm{D^b}(X)$, this suffices to conclude that $G(\mathcal{F}^\bullet) \simeq 0$.

The other direction, namely that $F(k(x)) \simeq F(k(x)) \otimes \omega_Y$ for any equivalence F, is easier. Either, one applies $\mathcal{P}_{\mathrm{R}} \simeq \mathcal{P}_{\mathrm{L}}$ or one argues as follows. Since F is an equivalence, it commutes with Serre functors and, furthermore, $\dim(X) = \dim(Y)$. Hence

$$\begin{aligned} F(k(x)) &\simeq F(k(x) \otimes \omega_X) = F(S_X(k(x))[-\dim(X)]) \\ &\simeq S_Y(F(k(y)))[-\dim(Y)] \simeq F(k(x)) \otimes \omega_Y. \end{aligned}$$

(See the proof of Proposition 4.1 for similar arguments.) □

We again mention explicitly the special case when the kernel is a sheaf.

Corollary 7.12 *Let* $\mathcal{P}$ *be a sheaf on* $X \times Y$ *flat over* X. *Assume that* $\Phi_{\mathcal{P}}$ *is fully faithful. Then* $\Phi_{\mathcal{P}}$ *is an equivalence if and only if* $\mathcal{P}_x \simeq \mathcal{P}_x \otimes \omega_Y$ *for all* $x \in X$. □

This corollary is often combined with Corollary 7.5, see Chapter 9. The proof of the following version of Proposition 7.11 is left to the reader. It uses the spanning class given by an ample line bundle.

Proposition 7.13 *Let* $F : \mathrm{D^b}(X) \longrightarrow \mathrm{D^b}(Y)$ *be a fully faithful exact functor with* X *and* Y *smooth projective varieties. Suppose that* $F(L^i \otimes \omega_X) \simeq F(L^i) \otimes \omega_Y$ *for all powers* L^i, $i \in \mathbb{Z}$, *of an ample line bundle* L *on* X.

Then F *is an equivalence.* □

Exercise 7.14 Prove the analogue of Corollary 7.4 for equivalences, i.e. show that $\Phi_{\mathcal{P} \boxtimes \mathcal{P}'}$ is an equivalence for any two equivalences $\Phi_{\mathcal{P}}$ and $\Phi_{\mathcal{P}'}$.

7.3 Canonical quotients

This section is a digression and you might as well skip it at first reading. It will be used later when working through the Enriques classification of algebraic surfaces from the derived category point of view (see Section 12.3). It is included here as an illustration of some of the techniques encountered so far.

There is a standard construction that trivializes a line bundle of finite order by passing to an étale cover $\widetilde{X} \to X$. Let us recall some of the details. Consider an arbitrary line bundle L on X and let

$$\pi : |L| \to X$$

be the associated affine bundle over X. The pull-back π^*L admits a canonical section $t \in H^0(|L|, \pi^*L)$ which in a closed point $\ell \in L(x)$ over a closed point $x = \pi(\ell) \in X$ takes the value $\ell \in (\pi^*L)(\ell) = L(x)$. In fact, t trivializes π^*L away from the zero section.

Suppose now that $L^n \simeq \mathcal{O}_X$ for some finite $n > 0$ and choose a trivializing section $s \in H^0(X, L^n)$. The equation $t^n - \pi^*s \in H^0(|L|, \pi^*L^n)$ defines a subscheme $\widetilde{X} \subset |L|$. A local calculation shows that the induced projection $\pi : \widetilde{X} \to X$ is étale. Moreover, $\pi_*\mathcal{O}_{\widetilde{X}} \simeq \bigoplus_{k=0}^{n-1} L^{-k}$ (see [5, I.17]). If the order of L is exactly n, then $\widetilde{X}$ is connected. Moreover, the cyclic group $G := \mathbb{Z}/n\mathbb{Z}$ acts freely by covering maps such that $\widetilde{X}/G = X$.

Suppose X is a smooth projective variety with a canonical bundle ω_X of finite order n. The above construction applied to $L = \omega_X^*$ yields the *canonical cover* $\pi : \widetilde{X} \to X$. Note that the canonical cover behaves well with respect to products, i.e. if $\pi_X : \widetilde{X} \to X$ and $\pi_Y : \widetilde{Y} \to Y$ are the canonical covers of X, respectively Y, then their product $\pi_X \times \pi_Y : \widetilde{X} \times \widetilde{Y} \to X \times Y$ is the canonical cover of $X \times Y$. Similarly, $\widetilde{X} \times \widetilde{Y} \to \widetilde{X} \times Y$ is the canonical cover of $\widetilde{X} \times Y$, etc.

If $\mathrm{D^b}(X) \simeq \mathrm{D^b}(Y)$ then by Proposition 4.1 with ω_X of order n, ω_Y is also of order n. So one might ask whether the derived equivalence of X and Y implies derived equivalence of their canonical covers $\widetilde{X}$, respectively $\widetilde{Y}$. This turns out to be true and will be proved here. The result of Bridgeland and Maciocia is more precise than this. In order to phrase it, we need the following

Definition 7.15 *Suppose X and Y are smooth projective varieties with canonical bundles of order n. Let $\pi_X : \widetilde{X} \to X$ and $\pi_Y : \widetilde{Y} \to Y$ denote the canonical covers. We say that an equivalence $\Phi : \mathrm{D^b}(X) \simeq \mathrm{D^b}(Y)$* lifts *to an equivalence $\widetilde{\Phi} : \mathrm{D^b}(\widetilde{X}) \simeq \mathrm{D^b}(\widetilde{Y})$ if the two diagrams*

$$\begin{array}{ccc} \mathrm{D^b}(\widetilde{X}) & \xrightarrow{\widetilde{\Phi}} & \mathrm{D^b}(\widetilde{Y}) \\ \pi_{X*}\downarrow & (*) & \downarrow\pi_{Y*} \\ \mathrm{D^b}(X) & \xrightarrow[\Phi]{} & \mathrm{D^b}(Y) \end{array} \qquad \begin{array}{ccc} \mathrm{D^b}(\widetilde{X}) & \xrightarrow{\widetilde{\Phi}} & \mathrm{D^b}(\widetilde{Y}) \\ \pi_X^*\uparrow & (**) & \uparrow\pi_Y^* \\ \mathrm{D^b}(X) & \xrightarrow[\Phi]{} & \mathrm{D^b}(Y) \end{array}$$

commute.

Eventually, one is interested in $\mathbb{Z}/n\mathbb{Z}$-equivariant lifts $\widetilde{\Phi}$. This means that there exists an automorphism $\mu : \mathbb{Z}/n\mathbb{Z} \xrightarrow{\sim} \mathbb{Z}/n\mathbb{Z}$ such that

$$\widetilde{\Phi} \circ g^* \simeq \mu(g)^* \circ \widetilde{\Phi}$$

for any $g \in \mathbb{Z}/n\mathbb{Z}$. Of course, it suffices to check this condition for a generator of $\mathbb{Z}/n\mathbb{Z}$, which in the sequel will be denoted $g \in \mathbb{Z}/n\mathbb{Z}$.

In order to construct a lift $\widetilde{\Phi}$ for a given Φ, one needs the following standard (at least for sheaves) fact.

Lemma 7.16 *Suppose $\pi : \widetilde{X} \longrightarrow X$ is the canonical cover of X and $\mathcal{P} \in \mathrm{D^b}(X)$. Then $\mathcal{P} \simeq \mathcal{P} \otimes \omega_X$ if and only if there exists an object $\widetilde{\mathcal{P}} \in \mathrm{D^b}(\widetilde{X})$ with $\pi_*\widetilde{\mathcal{P}} \simeq \mathcal{P}$. See [23].* □

We shall apply this to the product situation:

Corollary 7.17 *Suppose $\mathcal{P} \in \mathrm{D^b}(X \times Y)$ satisfies $\mathcal{P} \otimes q^*\omega_X \simeq \mathcal{P} \otimes p^*\omega_Y$. Then there exists an object $\widetilde{\mathcal{P}} \in \mathrm{D^b}(\widetilde{X} \times \widetilde{Y})$ with*

$$(\mathrm{id}_X \times \pi_Y)^*\mathcal{P} \simeq (\pi_X \times \mathrm{id}_{\widetilde{Y}})_*\widetilde{\mathcal{P}} \in \mathrm{D^b}(X \times \widetilde{Y}). \tag{7.5}$$

Proof In order to apply Lemma 7.16, it suffices to check $(\mathrm{id}_X \times \pi_Y)^*\mathcal{P} \simeq (\mathrm{id}_X \times \pi_Y)^*\mathcal{P} \otimes \omega_{X \times \widetilde{Y}}$. As $\pi_Y^*\omega_Y \simeq \omega_{\widetilde{Y}} \simeq \mathcal{O}_{\widetilde{Y}}$, this follows from

$$\begin{aligned}(\mathrm{id}_X \times \pi_Y)^*\mathcal{P} \otimes \omega_{X \times \widetilde{Y}} &\simeq (\mathrm{id}_X \times \pi_Y)^*(\mathcal{P} \otimes \omega_{X \times Y}) \\ &\simeq (\mathrm{id}_X \times \pi_Y)^*(\mathcal{P} \otimes (p^*\omega_Y^2)) \simeq (\mathrm{id}_X \times \pi_Y)^*\mathcal{P}.\end{aligned}$$

□

Proposition 7.18 (Bridgeland, Maciocia) *Suppose X and Y are smooth projective varieties with canonical bundles of finite order. Then any equivalence $\Phi : \mathrm{D^b}(X) \simeq \mathrm{D^b}(Y)$ admits an equivariant lift $\widetilde{\Phi} : \mathrm{D^b}(\widetilde{X}) \simeq \mathrm{D^b}(\widetilde{Y})$. See [23].*

Proof Equivalences are always of Fourier–Mukai type (see Proposition 5.14). So, we can work with the corresponding kernels. Suppose $\Phi = \Phi_{\mathcal{P}}$. As Φ is an equivalence, its kernel satisfies $\mathcal{P} \otimes q^*\omega_X \simeq \mathcal{P} \otimes p^*\omega_Y$. Hence, the corollary applies and yields an object $\widetilde{\mathcal{P}}$ satisfying (7.5). The induced Fourier–Mukai transform will be denoted $\widetilde{\Phi} := \Phi_{\widetilde{\mathcal{P}}}$.

Let us first check the commutativity of (∗∗) in Definition 7.15.

Exercise 5.12 and (7.5) yield an isomorphism between

$$\mathrm{D^b}(X) \xrightarrow{\ \Phi\ } \mathrm{D^b}(Y) \xrightarrow{\ \pi_Y^*\ } \mathrm{D^b}(\widetilde{Y})$$

and

$$\mathrm{D^b}(X) \xrightarrow{\ \pi_X^*\ } \mathrm{D^b}(\widetilde{X}) \xrightarrow{\ \widetilde{\Phi}\ } \mathrm{D^b}(\widetilde{Y}),$$

i.e. $\pi_Y^* \circ \Phi \simeq \widetilde{\Phi} \circ \pi_X^*$.

Before proving that also $(*)$ commutes, let us prove that $\widetilde{\Phi}$ is an equivalence. As $\widetilde{X}$ and $\widetilde{Y}$ both have trivial canonical bundle, it suffices to prove full faithfulness. For this purpose choose an ample line bundle L on X which naturally induces ample sequences L^i in $\mathbf{Coh}(X)$ and $\pi_X^* L^i$ in $\mathbf{Coh}(\widetilde{X})$. Indeed, since π_X is finite, $\pi_X^* L$ is again ample.

Invoke Proposition 1.49 and Corollary 3.19 to see that full faithfulness follows from the bijectivity of the natural maps

$$\widetilde{\Phi}_{\pi_X^* L^i, \pi_X^* L^j} : \mathrm{Hom}_{\mathrm{D^b}(\widetilde{X})}(\pi_X^* L^i, \pi_X^* L^j) \longrightarrow \mathrm{Hom}_{\mathrm{D^b}(\widetilde{Y})}(\widetilde{\Phi}(\pi_X^* L^i), \widetilde{\Phi}(\pi_X^* L^j)).$$

By adjunction and projection formula (3.11)

$$\begin{aligned}\mathrm{Hom}_{\mathrm{D^b}(\widetilde{X})}(\pi_X^* L^i, \pi_X^* L^j) &\simeq \mathrm{Hom}_{\mathrm{D^b}(X)}(L^i, L^j \otimes \pi_{X*}\mathcal{O}_{\widetilde{X}}) \\ &\simeq \bigoplus_{k=0}^{n-1} \mathrm{Hom}_{\mathrm{D^b}(X)}(L^i, L^j \otimes \omega_X^k)\end{aligned}$$

and, similarly,

$$\begin{aligned}\mathrm{Hom}_{\mathrm{D^b}(\widetilde{Y})}(\widetilde{\Phi}(\pi_X^* L^i), \widetilde{\Phi}(\pi_X^* L^j)) &\simeq \mathrm{Hom}_{\mathrm{D^b}(\widetilde{Y})}(\pi_Y^* \Phi(L^i), \pi_Y^* \Phi(L^j)) \\ &\simeq \bigoplus_{k=0}^{n-1} \mathrm{Hom}_{\mathrm{D^b}(Y)}(\Phi(L^i), \Phi(L^j) \otimes \omega_Y^k).\end{aligned}$$

For the latter we use $\widetilde{\Phi} \circ \pi_X^* \simeq \pi_Y^* \circ \Phi$, which also ensures that $\widetilde{\Phi}_{\pi_X^* L^i, \pi_X^* L^j}$ respects the direct sum decomposition, i.e. it is the direct sum of the natural bijections

$$\Phi_{L^i, L^j \otimes \omega_X^k} : \mathrm{Hom}_{\mathrm{D^b}(X)}(L^i, L^j \otimes \omega_X^k) \longrightarrow \mathrm{Hom}_{\mathrm{D^b}(Y)}(\Phi(L^i), \Phi(L^j) \otimes \omega_Y^k).$$

Note that $\Phi(L^j \otimes \omega_X^k) \simeq \Phi(L^j) \otimes \omega_Y^k$, for Φ is an equivalence and as such commutes with Serre functors.

Once $\pi_Y^* \circ \Phi \simeq \widetilde{\Phi} \circ \pi_X^*$ and the fact that $\widetilde{\Phi}$ is an equivalence are established, the commutativity of $(*)$ can be proved as follows: take right adjoints to obtain $\Phi_{\mathcal{P}_\mathrm{R}} \circ \pi_{Y*} \simeq \pi_{X*} \circ \Phi_{\widetilde{\mathcal{P}}_\mathrm{R}}$, and then compose with Φ from the left and with $\widetilde{\Phi}$ from the right.

It remains to prove that $\widetilde{\Phi}$ is equivariant. As before, we pick a generator $g \in \mathbb{Z}/n\mathbb{Z}$. Then consider the autoequivalence

$$\widetilde{\Psi} := \widetilde{\Phi} \circ g^* \circ \widetilde{\Phi}^{-1} : \mathrm{D^b}(\widetilde{Y}) \xrightarrow{\sim} \mathrm{D^b}(Y).$$

Clearly, $\widetilde{\Psi}$ lifts the identity on $\mathrm{D^b}(Y)$. Thus, it suffices that any lift of the identity is of the form $h^* : \mathrm{D^b}(\widetilde{Y}) \longrightarrow \mathrm{D^b}(\widetilde{Y})$ for some $h \in \mathbb{Z}/n\mathbb{Z}$; the isomorphism $\mu : \mathbb{Z}/n\mathbb{Z} \xrightarrow{\sim} \mathbb{Z}/n\mathbb{Z}$ is then defined by $g \longmapsto h$.

Since $\widetilde{\Psi}$ lifts the identity, there exists for any closed point $y \in \widetilde{Y}$ a closed point $y' \in Y$ such that $\pi_{Y*}\widetilde{\Psi}(k(y)) \simeq k(y')$. Hence, $\widetilde{\Psi}(k(y))$ is also isomorphic to $k(f(y))$ for some closed point $f(y) \in \widetilde{Y}$. Thus Corollary 5.23 applies and shows that f describes an automorphism of $\widetilde{Y}$ and $\widetilde{\Psi} \simeq (M \otimes (\ \ )) \circ f_*$ for some line bundle M on $\widetilde{Y}$. Clearly, f covers the identity on Y and is thus induced by some $h \in \mathbb{Z}/n\mathbb{Z}$. Evaluating $\widetilde{\Psi}$ on $\mathcal{O}_{\widetilde{Y}} = \pi_{\widetilde{Y}}^*\mathcal{O}_Y$ yields $M \simeq \mathcal{O}_{\widetilde{Y}}$. $\square$

8

SPHERICAL AND EXCEPTIONAL OBJECTS

It is clearly a difficult task to construct interesting autoequivalences of a given derived category or to uncover the complete structure of the derived category itself. Only very few general principles are known and this chapter is devoted to the presentation of those that are related to the existence of special objects in the derived category.

In the first section we shall introduce spherical objects, a notion that has been motivated by considerations in the context of mirror symmetry. Spherical objects naturally induce autoequivalences and their action on cohomology can be described precisely. In particular, we will be in the position to construct interesting non-trivial autoequivalences that act trivially on cohomology.

Considering more than one spherical object yields more autoequivalences. For certain configurations of spherical objects this construction gives rise to an action of the braid group. These results, due to Seidel and Thomas, are the topic of Section 8.2.

The results of Section 8.3 are almost classical. We give an account of the Beilinson spectral sequence and how it is used to deduce a complete description of the derived category of the projective space and, more generally, of the derived category of a projective bundle. This will use the language of exceptional sequences and semi-orthogonal decompositions encountered in Section 1.4.

The final section gives a simplified account of work of Horja which extends the theory of spherical objects and their associated twists to a broader geometric context.

8.1 Autoequivalences induced by spherical objects

In this section X denotes a smooth projective variety over a field k. As we will not use Proposition 7.1 the field does not necessarily need to be algebraically closed or of characteristic zero.

Definition 8.1 *An object* $\mathcal{E}^\bullet \in \mathrm{D^b}(X)$ *is called* spherical *if*

i) $\mathcal{E}^\bullet \otimes \omega_X \simeq \mathcal{E}^\bullet$ *and*

ii) $\mathrm{Hom}(\mathcal{E}^\bullet, \mathcal{E}^\bullet[i]) = \begin{cases} k & \text{if } i = 0, \dim(X) \\ 0 & \text{otherwise.} \end{cases}$

Condition ii) can equivalently be expressed as

$$\mathrm{Hom}(\mathcal{E}^\bullet, \mathcal{E}^\bullet[*]) \simeq H^*(S^{\dim(X)}, k),$$

where $S^{\dim(X)}$ is the real sphere of dimension $\dim(X)$. This explains the name.

In the following we shall denote spherical objects (and in fact any complex) simply by $\mathcal{E}$. Whether they are sheaves, as they indeed are in most of the examples, is of no importance.

Note that choosing an isomorphism in i) and applying Serre duality yields a canonical isomorphism $\mathrm{Hom}(\mathcal{E},\mathcal{E}) \simeq \mathrm{Hom}(\mathcal{E},\mathcal{E}[\dim(X)])^*$.

Exercise 8.2 Let $\mathcal{E}$ be a spherical object. Show that $\mathcal{E}^\vee$, $\mathcal{E}[i]$ for any $i \in \mathbb{Z}$, and $\mathcal{E} \otimes L$ for any $L \in \mathrm{Pic}(X)$ are again spherical objects.

Using the cone construction (cf. Definition 2.15), one associates to any object $\mathcal{E} \in \mathrm{D^b}(X)$ the following object $\mathcal{P} := \mathcal{P}_\mathcal{E}$ in the derived category $\mathrm{D^b}(X \times X)$ of the product:

$$\mathcal{P}_\mathcal{E} := \mathrm{C}\left(q^*\mathcal{E}^\vee \otimes p^*\mathcal{E} \longrightarrow \mathcal{O}_\Delta\right). \tag{8.1}$$

Here, $\mathcal{O}_\Delta$ is the structure sheaf of the diagonal $\Delta \subset X \times X$ viewed as a sheaf on $X \times X$. So, more accurately $\mathcal{O}_\Delta = \iota_*\mathcal{O}_X$ with ι the diagonal embedding $\iota : X \xrightarrow{\sim} \Delta \subset X \times X$. The homomorphism in (8.1) is given as the composition of the restriction

$$q^*\mathcal{E}^\vee \otimes p^*\mathcal{E} \longrightarrow \iota_*\iota^*((q^*\mathcal{E}^\vee \otimes p^*\mathcal{E})) = \iota_*(\mathcal{E}^\vee \otimes \mathcal{E}),$$

and of the direct image $\iota_*\mathrm{tr}$ of the trace map (see p. 77)

$$\mathrm{tr} : \mathcal{E}^\vee \otimes \mathcal{E} \longrightarrow \mathcal{O}_X.$$

In other words and more accurately, $\mathcal{P}_\mathcal{E}$ is an object that completes the natural morphism $q^*\mathcal{E}^\vee \otimes p^*\mathcal{E} \longrightarrow \mathcal{O}_\Delta$ to a distinguished triangle

$$q^*\mathcal{E}^\vee \otimes p^*\mathcal{E} \longrightarrow \mathcal{O}_\Delta \longrightarrow \mathcal{P}_\mathcal{E} \longrightarrow q^*\mathcal{E}^\vee \otimes p^*\mathcal{E}[1].$$

(The cone cannot be defined in general, as the trace is not a true morphism of complexes, but only defined as a morphism in the derived category.)

In fact, the cone construction is not functorial due to the non-uniqueness in the axiom TR3 (see Definition 1.32). The object $\mathcal{P}_\mathcal{E}$ exists, but it is defined only up to non-unique isomorphism.

Definition 8.3 *The* spherical twist *associated to a spherical object $\mathcal{E} \in \mathrm{D^b}(X)$ is by definition the Fourier–Mukai transform*

$$T_\mathcal{E} := \Phi_{\mathcal{P}_\mathcal{E}} : \mathrm{D^b}(X) \longrightarrow \mathrm{D^b}(X)$$

with kernel $\mathcal{P}_\mathcal{E}$.

Exercise 8.4 Let $\mathcal{E}$ be a spherical object. Show that there exists a natural isomorphism $T_\mathcal{E} \simeq T_{\mathcal{E}[1]}$ (see Exercise 8.2).

Exercise 8.5 Let $\mathcal{E}$ be a spherical object.

i) Show that for any object $\mathcal{F} \in \mathrm{D^b}(X)$ there exists an isomorphism

$$T_{\mathcal{E}}(\mathcal{F}) \simeq \mathrm{C}\left(\mathrm{Hom}(\mathcal{E}, \mathcal{F}[*]) \otimes \mathcal{E} \longrightarrow \mathcal{F} \right).$$

More precisely,

$$T_{\mathcal{E}}(\mathcal{F}) \simeq \mathrm{C}\left(\bigoplus_i (\mathrm{Hom}(\mathcal{E}, \mathcal{F}[i]) \otimes \mathcal{E}[-i]) \longrightarrow \mathcal{F} \right)$$

with the first morphism given by evaluation. (Use that trace and evaluation yield the same homomorphism $V^\vee \otimes V \longrightarrow k$.)

(Again, we are sloppy here. The cone does not make sense. What is meant, of course, is that the image under the spherical twist completes the evaluation morphism (in the derived category) to a distinguished triangle.)

ii) Use i) to prove

$$T_{\mathcal{E}}(\mathcal{E}) \simeq \mathcal{E}[1 - \dim(X)] \quad \text{and} \quad T_{\mathcal{E}}(\mathcal{F}) \simeq \mathcal{F} \tag{8.2}$$

for any $\mathcal{F} \in \mathrm{D^b}(X)$ with $\mathrm{Hom}(\mathcal{E}, \mathcal{F}[i]) = 0$ for all $i \in \mathbb{Z}$ (i.e. $\mathcal{F} \in \mathcal{E}^\perp$).

The following result was suggested by Kontsevich. A complete proof was given by Seidel and Thomas in [106].

Proposition 8.6 *Let $\mathcal{E}$ be a spherical object in the derived category $\mathrm{D^b}(X)$ of a smooth projective variety X. Then the induced spherical twist*

$$T_{\mathcal{E}} : \mathrm{D^b}(X) \xrightarrow{\sim} \mathrm{D^b}(X)$$

is an autoequivalence.

Proof As often, the difficult part of the proof is to show that the functor is fully faithful. Indeed, once this has been established, one easily deduces that $T_{\mathcal{E}}$ is an equivalence from $\mathcal{E} \otimes \omega \simeq \mathcal{E}$,

$$\begin{aligned} \mathcal{P}_{\mathcal{E}} \otimes q^*\omega_X &= \mathrm{C}\left(q^*(\mathcal{E}^\vee \otimes \omega_X) \otimes p^*\mathcal{E} \longrightarrow \mathcal{O}_\Delta \otimes q^*\omega_X \right) \\ &\simeq \mathrm{C}\left(q^*\mathcal{E}^\vee \otimes p^*\mathcal{E} \longrightarrow \mathcal{O}_\Delta \otimes p^*\omega_X \right) \\ &\simeq \mathcal{P}_{\mathcal{E}} \otimes p^*\omega_X, \end{aligned}$$

and Proposition 7.6.

(One could also argue that $T_{\mathcal{E}}$ commutes with Serre functors on objects $\mathcal{F} = \mathcal{E}$ and $\mathcal{F} \in \mathcal{E}^\perp$, which follows directly from (8.2). Corollary 1.56 then yields the assertion as soon as one knows that these objects span $\mathrm{D^b}(X)$, which will be shown next.)

In order to show that $T_{\mathcal{E}}$ is fully faithful, we apply Proposition 1.49. The spanning class Ω we wish to consider in the present situation consists of $\mathcal{E}$ and all $\mathcal{F} \in \mathrm{D^b}(X)$ such that $\mathrm{Hom}(\mathcal{E}, \mathcal{F}[i]) = 0$ for all $i \in \mathbb{Z}$. In other words, $\Omega := \{\mathcal{E}\} \cup \mathcal{E}^{\perp}$.

Let us first verify that Ω is indeed spanning. Suppose $\mathcal{F} \in \mathrm{D^b}(X)$ is an object such that $\mathrm{Hom}(\mathcal{G}, \mathcal{F}[i]) = 0$ for all $\mathcal{G} \in \Omega$ and all $i \in \mathbb{Z}$. In particular, $\mathrm{Hom}(\mathcal{E}, \mathcal{F}[i]) = 0$ for all $i \in \mathbb{Z}$ and, therefore, $\mathcal{F} \in \mathcal{E}^{\perp} \subset \Omega$. Thus, $\mathrm{id} \in \mathrm{Hom}(\mathcal{F}, \mathcal{F}) = 0$, which yields $\mathcal{F} \simeq 0$.

If $\mathcal{F} \in \mathrm{D^b}(X)$ is such that $\mathrm{Hom}(\mathcal{F}, \mathcal{G}[i]) = 0$ for all $\mathcal{G} \in \Omega$ and all $i \in \mathbb{Z}$, then Serre duality shows $\mathrm{Hom}(\mathcal{G}, \mathcal{F} \otimes \omega_X[i]) = 0$ for all $\mathcal{G} \in \Omega$ and all $i \in \mathbb{Z}$. As was shown before, this implies $\mathcal{F} \otimes \omega_X \simeq 0$ and hence $\mathcal{F} \simeq 0$.

In order to verify that

$$T_{\mathcal{E}} : \mathrm{Hom}(\mathcal{G}_1, \mathcal{G}_2[i]) \longrightarrow \mathrm{Hom}(T_{\mathcal{E}}(\mathcal{G}_1), T_{\mathcal{E}}(\mathcal{G}_2))[i]) \tag{8.3}$$

is an isomorphism for all $i \in \mathbb{Z}$ and all $\mathcal{G}_1, \mathcal{G}_2 \in \Omega$ we shall use the description of the image $T_{\mathcal{E}}(\mathcal{G})$ for $\mathcal{G} \in \Omega$ given in Exercise 8.5.

Thus (8.3) is an isomorphism for $\mathcal{G}_1 = \mathcal{E}$ and $\mathcal{G}_2 \in \mathcal{E}^{\perp}$ or vice versa, because in these cases both sides are simply trivial. (Serre duality and the assumption $\mathcal{E} \otimes \omega_X \simeq \mathcal{E}$ come in here.)

Next one considers the case $\mathcal{G}_1 \simeq \mathcal{G}_2 \simeq \mathcal{E}$. The image of $\mathrm{id} \in \mathrm{Hom}(\mathcal{E}, \mathcal{E})$ is again the identity $\mathrm{id} = T_{\mathcal{E}}(\mathrm{id}) : \mathcal{E}[1 - \dim(X)] \longrightarrow \mathcal{E}[1 - \dim(X)]$. It is similarly straightforward to check that (8.3) is the identity for $i = \dim(X)$.

Eventually, one deals with the case $\mathcal{G}_1, \mathcal{G}_2 \in \mathcal{E}^{\perp}$. Here, one finds that (8.3) composed with the isomorphisms $T_{\mathcal{E}}(\mathcal{G}_i) \simeq \mathcal{G}_i$ obtained in Exercise 8.5 yields a bijection $\mathrm{Hom}(\mathcal{G}_1, \mathcal{G}_2) = \mathrm{Hom}(\mathcal{G}_1, \mathcal{G}_2)$. Hence (8.3) is bijective as well. □

Exercise 8.7 In view of the distinguished triangle

$$T(\mathcal{F})[-1] \longrightarrow \bigoplus \mathrm{Hom}(\mathcal{E}, \mathcal{F}[i]) \otimes \mathcal{E}[-i] \longrightarrow \mathcal{F} \longrightarrow T(\mathcal{F})$$

one might wonder whether the two triangulated subcategories $\langle \mathcal{E} \rangle$ and $\mathcal{E}^{\perp}$, which span $\mathrm{D^b}(X)$ (see the proof), actually define a semi-orthogonal decomposition of $\mathrm{D^b}(X)$ (see Definition 1.59). Why don't they?

Spherical objects are almost exclusively studied on Calabi–Yau manifolds, i.e. varieties with trivial canonical bundle $\omega_X \simeq \mathcal{O}_X$. In this case, the condition i) in Definition 8.1 is automatically satisfied. For Calabi–Yau manifolds the preceding exercise can be generalized to the following one.

Exercise 8.8 Show that the derived category $\mathrm{D^b}(X)$ of a Calabi–Yau manifold X does not admit any non-trivial semi-orthogonal decomposition.

Remark 8.9 The original proof of Seidel and Thomas is phrased purely in terms of homological algebra, i.e. not using the description of the spherical twist as a Fourier–Mukai transform. The above short proof is taken from [95].

Examples 8.10 i) Consider a smooth projective curve C and a closed point $x \in C$. Then $k(x)$ is a spherical object. The spherical twist $T_{k(x)}$ turns out to be isomorphic to the functor given by $\mathcal{F} \longmapsto \mathcal{F} \otimes \mathcal{O}(x)$, where $\mathcal{O}(x)$ is the line bundle associated to $x \in C$.

One way to see this is to show that there exists a functorial isomorphism $T_{k(x)} \simeq \mathcal{O}(x) \otimes (\ \ )$ on the subcategory spanned by all line bundles and then to apply Proposition 4.23 or 4.24.

Now, for a line bundle L one has $\mathrm{Hom}(k(x), L) = 0$ and $\mathrm{Ext}^1(k(x), L) = \mathrm{Hom}(k(x), L[1])$ is spanned by the unique extension provided by

$$0 \longrightarrow L \longrightarrow L(x) \longrightarrow k(x) \longrightarrow 0.$$

Hence, there exists a functorial isomorphism

$$T_{k(x)}(L) \simeq \mathrm{C}\left(\mathrm{Ext}^1(k(x), L)[-1] \otimes k(x) \longrightarrow L\right),$$

i.e. $T_{k(x)}(L) \simeq L(x)$.

ii) If X is a true Calabi–Yau manifold, i.e. $\omega_X \simeq \mathcal{O}_X$ and $H^i(X, \mathcal{O}_X) = 0$ for $0 < i < \dim(X)$, then any line bundle L on X is a spherical object. For $L = \mathcal{O}_X$ the associated Fourier–Mukai kernel is the shifted ideal sheaf $\mathcal{I}_\Delta$ of the diagonal.

iii) Let X be a smooth projective surface and $C \subset X$ a smooth irreducible rational curve with $C^2 = -2$. Then $\mathcal{O}_C$ is a spherical object. More generally, any $\mathcal{O}_C(k)$ is a spherical object. Here, $\mathcal{O}_C(k)$ is the pull-back of $\mathcal{O}_{\mathbb{P}^1}(k)$ under an isomorphism $C \simeq \mathbb{P}^1$. Any smooth irreducible rational curve in a K3 surface satisfies the hypothesis $C^2 = -2$.

iv) Let now C be a smooth rational curve contained in a true Calabi–Yau variety X of dimension three. Assume that the normal bundle of C is isomorphic to $\mathcal{N}_{C/X} \simeq \mathcal{O}(-1) \oplus \mathcal{O}(-1)$. Then $\mathcal{O}_C$ is a spherical object. Indeed, by Serre duality

$$\mathrm{Hom}(\mathcal{O}_C, \mathcal{O}_C[3]) \simeq \mathrm{Hom}(\mathcal{O}_C, \mathcal{O}_C)^* \simeq k$$

and

$$\mathrm{Hom}(\mathcal{O}_C, \mathcal{O}_C[2]) \simeq \mathrm{Hom}(\mathcal{O}_C, \mathcal{O}_C[1])^* \simeq \mathrm{Ext}^1_X(\mathcal{O}_C, \mathcal{O}_C)^*.$$

The latter group is trivial, as it measures infinitesimal deformations of $\mathcal{O}_C$ and there are none under the assumption $H^0(C, \mathcal{N}_{C/X}) = 0$. (The relation between the Ext-groups and the normal bundle will be discussed in broader generality in Section 11.2.)

v) The previous example can be generalized to cover a situation that will interest us in Chapter 11. Consider a smooth subvariety $P \subset X$ in a true Calabi–Yau variety X of dimension $2n+1$. Suppose

$$P \simeq \mathbb{P}^n \quad \text{and} \quad \mathcal{N}_{P/X} \simeq \mathcal{O}(-1)^{\oplus n+1}.$$

By using the isomorphism $\mathcal{E}xt^q_X(\mathcal{O}_P, \mathcal{O}_P) \simeq \bigwedge^q \mathcal{N}_{P/X}$ proved in Proposition 11.8 the spectral sequence (3.16)

$$E_2^{p,q} = H^p(X, \mathcal{E}xt^q_X(\mathcal{O}_P, \mathcal{O}_P)) \Rightarrow \mathrm{Ext}^{p+q}_X(\mathcal{O}_P, \mathcal{O}_P)$$

becomes

$$E_2^{p,q} = H^p(P, \bigwedge^q \mathcal{N}_{P/X}) \Rightarrow \mathrm{Ext}_X^{p+q}(\mathcal{O}_P, \mathcal{O}_P).$$

The standard Bott formulae yield $E_2^{p,q} = 0$ except for $(p,q) = (0,0)$ or $(n, n+1)$. Moreover, $E_2^{0,0} = E_2^{n,n+1} = k$. This shows

$$\mathrm{Ext}_X^*(\mathcal{O}_P, \mathcal{O}_P) \simeq H^*(S^{2n+1}, k)$$

and, hence, since X has trivial canonical bundle, $\mathcal{O}_P \in \mathrm{D^b}(X)$ is spherical.

vi) Abelian varieties (see Chapter 9), true Calabi–Yau varieties (see iii) above) and algebraic symplectic varieties form the building blocks for all varieties with trivial canonical bundle. By definition an algebraic symplectic variety is a variety that possesses a global section of Ω_X^2 which is non-degenerate at every point. The Pfaffian of such a section trivializes the canonical bundle ω_X. In particular Hodge theory tells us that $H^2(X, \mathcal{O}_X) = \overline{H^0(X, \Omega_X^2)}$ is non-trivial. Of course, this also holds true for abelian varieties of dimension at least two.

In characteristic zero, a variety of dimension at least three with non-trivial $H^2(X, \mathcal{O}_X)$ admits no spherical objects of non-trivial rank. This is shown by means of the trace map which induces a surjection $\mathrm{Ext}^2(\mathcal{E}, \mathcal{E}) \longrightarrow H^2(X, \mathcal{O}_X)$, whenever $\mathrm{rk}(\mathcal{E}) \neq 0$. (The rank of a complex is of course defined as the alternating sum of the ranks of all the participating sheaves.)

One can in fact show that abelian varieties of dimension at least two do not admit any spherical objects. Most likely, the same holds true for algebraic symplectic manifolds from dimension four on. Thus, in higher dimensions the theory of spherical twists is only of interest for genuine Calabi–Yau manifolds.

Remark 8.11 The kernel of the inverse T^{-1} of a spherical twist $T := T_{\mathcal{E}}$ is not so easily described. However, as in Exercise 8.5, i) its effect on objects can be described in terms of distinguished triangles.

More precisely, for any object $\mathcal{G} \in \mathrm{D^b}(X)$ there exists a distinguished triangle of the form

$$T^{-1}\mathcal{G} \longrightarrow \mathcal{G} \longrightarrow \mathcal{E}[d] \otimes \mathrm{Hom}(\mathcal{E}, \mathcal{G}[*]) \longrightarrow T^{-1}\mathcal{G}[1]. \qquad (8.4)$$

Indeed, applying the exact functor T^{-1} to the original distinguished triangle in i), Exercise 8.5 yields

$$T^{-1}\mathcal{G} \longrightarrow \mathcal{G} \longrightarrow T^{-1}\mathcal{E}[1] \otimes \mathrm{Hom}(\mathcal{E}, \mathcal{G}[*]) \longrightarrow T^{-1}\mathcal{G}[1].$$

Using $T^{-1}\mathcal{E} \simeq \mathcal{E}[d-1]$ (see Exercise 8.5, ii)) we obtain (8.4).

When studying any kind of Fourier–Mukai transform the natural reflex is to first exhibit its action on cohomology. This will be done now for spherical twists. More precisely, for any spherical object $\mathcal{E} \in \mathrm{D^b}(X)$ and its associated spherical twist $T_{\mathcal{E}} : \mathrm{D^b}(X) \longrightarrow \mathrm{D^b}(X)$ we shall describe

$$T_{\mathcal{E}}^H : H^*(X, \mathbb{Q}) \longrightarrow H^*(X, \mathbb{Q}).$$

To $\mathcal{E}$, as to any other object in $\mathrm{D}^{\mathrm{b}}(X)$, one associates its Mukai vector

$$v(\mathcal{E}) := \mathrm{ch}(\mathcal{E})\sqrt{\mathrm{td}(X)} \in H^*(X,\mathbb{Q}).$$

If $\langle\ ,\ \rangle$ denotes the Mukai pairing on $H^*(X,\mathbb{Q})$ introduced in Section 5.2, then

$$\langle v(\mathcal{E}), v(\mathcal{E})\rangle = \chi(\mathcal{E},\mathcal{E}) = \sum_i (-1)^i \dim \mathrm{Hom}(\mathcal{E},\mathcal{E}[i]).$$

Thus, for a spherical object $\mathcal{E}$ we have

$$\langle v(\mathcal{E}), v(\mathcal{E})\rangle = \begin{cases} 2 & \text{if } \dim(X) \equiv 0\ (2) \\ 0 & \text{if } \dim(X) \equiv 1\ (2). \end{cases}$$

In fact, the second equality holds true for any object $\mathcal{E}$, as the Mukai pairing on an odd dimensional variety is alternating (see Exercise 5.43).

Lemma 8.12 *Let $\mathcal{E} \in \mathrm{D}^{\mathrm{b}}(X)$ be a spherical object. Then the induced spherical twist $T_{\mathcal{E}}$ acts on $H^*(X,\mathbb{Q})$ by*

$$T_{\mathcal{E}}^H : v \longmapsto v - \langle v(\mathcal{E}), v\rangle \cdot v(\mathcal{E}).$$

Proof By definition $T_{\mathcal{E}}$ is the Fourier–Mukai transform with kernel

$$\mathcal{P}_{\mathcal{E}} = \mathrm{C}(q^*\mathcal{E}^\vee \otimes p^*\mathcal{E} \longrightarrow \mathcal{O}_\Delta)$$

whose Mukai vector is of the form $v(\mathcal{O}_\Delta) - q^*v(\mathcal{E}^\vee).p^*v(\mathcal{E}) = [\Delta] - q^*v(\mathcal{E}^\vee)p^*v(\mathcal{E})$ (see Exercise 5.34). Hence,

$$\begin{aligned} T_{\mathcal{E}}(v) &= v - \left(\int_X v.v(\mathcal{E}^\vee)\right) \cdot v(\mathcal{E}) \\ &= v - \left(\int_X \exp(\mathrm{c}_1(X)/2).v.v(\mathcal{E})^\vee\right) \cdot v(\mathcal{E}) \\ &= v - \langle v(\mathcal{E}), v\rangle \cdot v(\mathcal{E}), \end{aligned}$$

where we use Lemma 5.41 and the fact that $v(\mathcal{E})^\vee$ is an even cohomology class so that $\int_X v(\mathcal{E})^\vee.v = \int_X v.v(\mathcal{E})^\vee$ for all v. □

In particular,

$$T_{\mathcal{E}}^H(v(\mathcal{E})) = \begin{cases} -v(\mathcal{E}) & \text{if } \dim(X) \equiv 0\ (2) \\ v(\mathcal{E}) & \text{if } \dim(X) \equiv 1\ (2), \end{cases}$$

which can also be seen as a consequence of $T_{\mathcal{E}}(\mathcal{E}) \simeq \mathcal{E}[1-\dim(X)]$ (cf. Exercise 8.5, ii)).

Corollary 8.13 *If $\mathcal{E}$ is a spherical object on an even dimensional variety X, then the spherical twist $T_{\mathcal{E}}$ acts on $H^*(X,\mathbb{Q})$ by reflection in the hyperplane orthogonal to $v(\mathcal{E})$.* □

Remark 8.14 In particular, in the even dimensional case $(T_{\mathcal{E}}^H)^2 = \mathrm{id}_{H^*}$. On the other hand, $T_{\mathcal{E}}^2$ sends $\mathcal{E}$ to $\mathcal{E}[2-2\dim(X)]$ and any $\mathcal{F} \in \langle\mathcal{E}\rangle^\perp$ to $\mathcal{F}$ again. Hence, $T_{\mathcal{E}}^2$ is neither the identity nor a pure shift functor.

In other words, on an even dimensional variety any spherical object $\mathcal{E}$ gives rise to an element

$$T_{\mathcal{E}}^2 \in \mathrm{Ker}\big(\,\mathrm{Aut}(\mathrm{D^b}(X)) \longrightarrow \mathrm{Aut}(H^*(X,\mathbb{Q}))\,\big)$$

no non-trivial power of which is contained in the subgroup generated by the double shift $\mathcal{F} \longmapsto \mathcal{F}[2]$.

Exercise 8.15 Suppose $\mathcal{E}$ is a spherical object on a variety X of odd dimension. Show that $(T_{\mathcal{E}}^H)^k(v) = v - k \cdot \langle v, v(\mathcal{E})\rangle \cdot v(\mathcal{E})$ for any $k \in \mathbb{Z}$. Thus, if $v(\mathcal{E}) \neq 0$ then $T_{\mathcal{E}}^H$ (and hence $T_{\mathcal{E}}$) is of infinite order. For a geometric instance see Examples 8.10, iv).

Besides abelian varieties, to be treated in Chapter 9, there are essentially two other types of smooth projective varieties with trivial canonical bundle. Those with trivial $H^0(X, \Omega_X^i)$ for $0 < i < \dim(X)$, which were called true Calabi–Yau varieties above, and those with an everywhere non-degenerate two-form σ spanning $H^0(X, \Omega^2)$. The latter are called *(irreducible) algebraic symplectic varieties*, for σ can be viewed as the algebraic analogue of a symplectic form. On a true Calabi–Yau variety, any line bundle provides an example of a spherical object, whereas on algebraic symplectic varieties of dimension at least four we do not expect any spherical object to exist at all.

In this sense, algebraic symplectic varieties are not covered by the previous discussion (see however the discussion of $\mathbb{EZ}$-spherical objects in Section 8.4 and in particular Examples 8.49, iv)). There is however another type of object on algebraic symplectic varieties that give rise to autoequivalences; these are the so-called $\mathbb{P}$-objects and their associated $\mathbb{P}$-twists.

There is a striking analogy between the theory of spherical twists and of $\mathbb{P}$-twists and an amusing interplay between them, when the symplectic variety can be put in a certain family. This has been explained in [56]. Here, we shall just outline the construction of a $\mathbb{P}$-twist and state that it is an equivalence. The techniques are similar, but slightly more involved than in the spherical case.

In the following discussion we will not explicitly require X to be an algebraic symplectic variety, but this is the type of variety we have in mind.

Definition 8.16 *An object $\mathcal{E}^\bullet \in \mathrm{D^b}(X)$ in the derived category of a smooth projective variety X is a* $\mathbb{P}^n$-object *if*

$$\text{i) } \mathcal{E}^\bullet \otimes \omega_X \simeq \mathcal{E}^\bullet \quad \textit{and}$$

$$\text{ii) } \mathrm{Hom}(\mathcal{E}^\bullet, \mathcal{E}^\bullet[*]) \simeq H^*(\mathbb{P}^n, k).$$

The isomorphism in ii) is supposed to be an isomorphism of graded algebras, where the multiplication in $\mathrm{Hom}(\mathcal{E}^\bullet, \mathcal{E}^\bullet[*])$ is given by composition. Note that Serre duality implies $\dim(X) = 2n$.

Examples 8.17 i) Suppose X is an algebraic symplectic variety of dimension $2n$ and $P := \mathbb{P}^n \subset X$. One can show that $\mathcal{N}_{P/X} \simeq \Omega_P$ and hence $\mathcal{E}xt^q(\mathcal{O}_P, \mathcal{O}_P) \simeq \Omega_P^q$ (cf. Proposition 11.8). Thus the spectral sequence

$$E_2^{p,q} = H^p(X, \mathcal{E}xt^q(\mathcal{O}_P, \mathcal{O}_P)) \Rightarrow \mathrm{Ext}_X^{p+q}(\mathcal{O}_P, \mathcal{O}_P)$$

yields a ring isomorphism $\mathrm{Ext}_X^*(\mathcal{O}_P, \mathcal{O}_P) \simeq H^*(P, \Omega_P^*) = H^*(\mathbb{P}^n, \mathbb{C})$. Hence, $\mathcal{O}_P \in \mathrm{D^b}(X)$ is a $\mathbb{P}^n$-object.

ii) If X is an algebraic symplectic variety, then $H^*(X, \mathcal{O}_X) \simeq H^*(\mathbb{P}^n, \mathbb{C})$. Hence, any line bundle L on X is a $\mathbb{P}^n$-object. Indeed, $\mathrm{Ext}^*(L, L) \simeq H^*(X, \mathcal{O}_X)$.

Let $\mathcal{E} \in \mathrm{D^b}(X)$ be a $\mathbb{P}^n$-object and $\bar{h} \in \mathrm{Hom}(\mathcal{E}, \mathcal{E}[2])$ a generator thought of as a morphism $h : \mathcal{E}[-2] \longrightarrow \mathcal{E}$. The image of $\bar{h}$ under the natural isomorphism $\mathrm{Hom}(\mathcal{E}, \mathcal{E}[2]) \simeq \mathrm{Hom}(\mathcal{E}^\vee, \mathcal{E}^\vee[2])$ will be denoted $\bar{h}^\vee$, which represents a morphism $h^\vee : \mathcal{E}^\vee[-2] \longrightarrow \mathcal{E}^\vee$.

Then introduce $H := h^\vee \boxtimes \mathrm{id} - \mathrm{id} \boxtimes h$ on $X \times X$ which is a morphism

$$H : (\mathcal{E}^\vee \boxtimes \mathcal{E})[-2] \longrightarrow \mathcal{E}^\vee \boxtimes \mathcal{E}.$$

The cone $\mathcal{H} := \mathrm{C}(H)$ of this morphism fits in a distinguished triangle

$$(\mathcal{E}^\vee \boxtimes \mathcal{E})[-2] \xrightarrow{H} \mathcal{E}^\vee \boxtimes \mathcal{E} \longrightarrow \mathcal{H} \longrightarrow (\mathcal{E}^\vee \boxtimes \mathcal{E})[-1]. \tag{8.5}$$

Recall that the kernel of the spherical twist associated to a spherical object is by definition the cone of the trace morphism $\mathrm{tr} : \mathcal{E}^\vee \boxtimes \mathcal{E} \longrightarrow \mathcal{O}_\Delta$.

As it turns out, the trace factorizes over a morphism $\mathrm{t} : \mathcal{H} \longrightarrow \mathcal{O}_\Delta$. Indeed, applying $\mathrm{Hom}(\ \ , \mathcal{O}_\Delta)$ to (8.5) yields a long exact sequence.

By definition of H, the boundary morphisms

$$\begin{array}{ccc} \mathrm{Hom}_{X\times X}(\mathcal{E}^\vee \boxtimes \mathcal{E}, \mathcal{O}_\Delta[i]) & \longrightarrow & \mathrm{Hom}_{X\times X}(\mathcal{E}^\vee \boxtimes \mathcal{E}, \mathcal{O}_\Delta[i+2]) \\ \simeq \mathrm{Hom}_X(\mathcal{E}, \mathcal{E}[i]) & & \simeq \mathrm{Hom}_X(\mathcal{E}, \mathcal{E}[i+2]) \end{array}$$

are given by $\bar{h} - \bar{h} = 0$. Hence $\mathrm{Hom}(\mathcal{H}, \mathcal{O}_\Delta) \longrightarrow \mathrm{Hom}(\mathcal{E}^\vee \boxtimes \mathcal{E}, \mathcal{O}_\Delta)$ is an isomorphism, giving the unique lift t of the trace map.

To any $\mathbb{P}^n$-object $\mathcal{E} \in \mathrm{D^b}(X)$ one associates the cone

$$\mathcal{Q}_\mathcal{E} := \mathrm{C}\left(\mathcal{H} \xrightarrow{\mathrm{t}} \mathcal{O}_\Delta\right) \in \mathrm{D^b}(X).$$

Definition 8.18 *Let $\mathcal{E} \in \mathrm{D^b}(X)$ be a $\mathbb{P}^n$-object. The associated $\mathbb{P}^n$-*twist *$P_\mathcal{E}$ is the Fourier–Mukai transform*

$$P_\mathcal{E} := \Phi_{\mathcal{Q}_\mathcal{E}} : \mathrm{D^b}(X) \longrightarrow \mathrm{D^b}(X)$$

with kernel $\mathcal{Q}_\mathcal{E}$.

Proposition 8.19 (Huybrechts, Thomas) *For any $\mathbb{P}^n$-object $\mathcal{E} \in \mathrm{D^b}(X)$ the associated $\mathbb{P}^n$-twist is an autoequivalence*

$$P_{\mathcal{E}} : \mathrm{D^b}(X) \xrightarrow{\sim} \mathrm{D^b}(X).$$

See [56].

8.2 Braid group actions

Let us start with a few recollections on the braid group.

The *braid group* B_{m+1} on $(m+1)$-strands is the group that is generated by elements $\beta_1, \ldots, \beta_m$ subject to the relations

$$\beta_i \cdot \beta_{i+1} \cdot \beta_i = \beta_{i+1} \cdot \beta_i \cdot \beta_{i+1} \quad \text{for all } i = 1, \ldots, m-1$$
$$\beta_i \cdot \beta_j = \beta_j \cdot \beta_i \quad \text{if } |i-j| \geq 2.$$

The first relation is best pictured by the geometric realization:

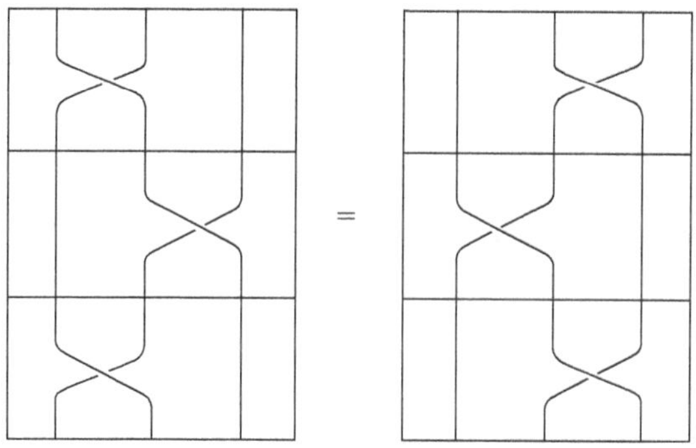

Definition 8.20 *An* A_m-configuration of spherical objects *in* $\mathrm{D^b}(X)$ *consists of spherical objects* $\mathcal{E}_1, \ldots, \mathcal{E}_m \in \mathrm{D^b}(X)$ *such that*

$$\bigoplus_k \dim \mathrm{Hom}(\mathcal{E}_i, \mathcal{E}_j[k]) = \begin{cases} 1 & \text{if } |i-j| = 1 \\ 0 & \text{if } |i-j| > 1. \end{cases}$$

The following technical lemma is the key to the understanding of the simultaneous action of all the twist functors induced by an A_m-configuration.

Lemma 8.21 *Let $\mathcal{E} \in \mathrm{D^b}(X)$ be a spherical object. Then for any autoequivalence $\Phi : \mathrm{D^b}(X) \xrightarrow{\sim} \mathrm{D^b}(X)$ there exists an isomorphism*

$$\Phi \circ T_{\mathcal{E}} \simeq T_{\Phi(\mathcal{E})} \circ \Phi.$$

Proof To make the assertion plausible we will apply an ad hoc argument to produce an isomorphism $\Phi(T_{\mathcal{E}}(\mathcal{G})) \simeq T_{\Phi(\mathcal{E})}(\Phi(\mathcal{G}))$ for any object $\mathcal{G}$.

Exercise 8.5 applied to $T := T_{\Phi(\mathcal{E})}$ and the object $\Phi(\mathcal{G})$ provides a distinguished triangle

$$\Phi(\mathcal{E}) \otimes \mathrm{Hom}(\Phi(\mathcal{E}), \Phi(\mathcal{G})[*]) \longrightarrow \Phi(\mathcal{G}) \longrightarrow T(\Phi(\mathcal{G})). \tag{8.6}$$

As Φ is an equivalence and hence $\mathrm{Hom}(\Phi(\mathcal{E}), \Phi(\mathcal{G})[*]) \simeq \mathrm{Hom}(\mathcal{E}, \mathcal{G}[*])$, (8.6) can be written as a distinguished triangle

$$\Phi(\mathcal{E}) \otimes \mathrm{Hom}(\mathcal{E}, \mathcal{G}[*]) \xrightarrow{\ \varphi\ } \Phi(\mathcal{G}) \longrightarrow T(\Phi(\mathcal{G})). \tag{8.7}$$

On the other hand, Exercise 8.5 applied to $T_{\mathcal{E}}$ and the object $\mathcal{G}$ yields a distinguished triangle

$$\mathcal{E} \otimes \mathrm{Hom}(\mathcal{E}, \mathcal{G}[*]) \xrightarrow{\ \psi_0\ } \mathcal{G} \longrightarrow T_{\mathcal{E}}(\mathcal{G}), \tag{8.8}$$

which after applying Φ provides a distinguished triangle

$$\Phi(\mathcal{E}) \otimes \mathrm{Hom}(\mathcal{E}, \mathcal{G}[*]) \xrightarrow{\ \psi\ } \Phi(\mathcal{G}) \longrightarrow \Phi(T_{\mathcal{E}}(\mathcal{G}))$$

with $\psi := \Phi(\psi_0)$. A moment's thought reveals $\varphi = \psi$. Hence, by TR3 there exists a, not necessarily unique, isomorphism $\Phi(T_{\mathcal{E}}(\mathcal{G})) \simeq T_{\Phi(\mathcal{E})}(\Phi(\mathcal{G}))$.

The problem that remains is to show that these individual isomorphisms constructed in this way for each given $\mathcal{G}$ indeed glue to an honest functor isomorphism $\Phi \circ T_{\mathcal{E}} \simeq T_{\Phi(\mathcal{E})} \circ \Phi$, the difficulty being caused by the non-uniqueness of the completing morphism in TR3.

The existence of an isomorphism is equivalent to the commutativity of the diagram

$$\begin{array}{ccc} \mathrm{D^b}(X) & \xleftarrow{\ \Phi^{-1}\ } & \mathrm{D^b}(X) \\ {\scriptstyle T_{\mathcal{E}}}\downarrow & & \downarrow{\scriptstyle T_{\Phi(\mathcal{E})}} \\ \mathrm{D^b}(X) & \xrightarrow[\ \Phi\ ]{} & \mathrm{D^b}(X). \end{array}$$

Denote the Fourier–Mukai kernel of Φ by $\mathcal{S} \in \mathrm{D^b}(X \times X)$, i.e. $\Phi \simeq \Phi_{\mathcal{S}}$. Then Exercise 5.13, ii) shows that it suffices to prove $\Phi_{\mathcal{S}_{\mathrm{R}} \boxtimes \mathcal{S}}(\mathcal{P}_{\mathcal{E}}) \simeq \mathcal{P}_{\Phi(\mathcal{E})}$. (Recall that $\Phi_{\mathcal{S}_{\mathrm{R}}} \simeq \Phi^{-1}$.)

To this end, apply $\Phi_{\mathcal{S}_{\mathrm{R}} \boxtimes \mathcal{S}}$ to the distinguished triangle

$$\mathcal{E}^{\vee} \boxtimes \mathcal{E} \longrightarrow \mathcal{O}_{\Delta} \longrightarrow \mathcal{P}_{\mathcal{E}}$$

which yields a distinguished triangle

$$\Phi_{\mathcal{S}_{\mathrm{R}}}(\mathcal{E}^{\vee}) \boxtimes \Phi_{\mathcal{S}}(\mathcal{E}) \longrightarrow \Phi_{\mathcal{S}_{\mathrm{R}} \boxtimes \mathcal{S}}(\mathcal{O}_{\Delta}) \longrightarrow \Phi_{\mathcal{S}_{\mathrm{R}} \boxtimes \mathcal{S}}(\mathcal{P}_{\mathcal{E}}) \tag{8.9}$$

(cf. Exercise 5.13, i)).

Invoking the commutative diagram of Exercise 5.13, i)

$$\begin{array}{ccc} \mathrm{D^b}(X) & \xleftarrow{\Phi^{-1}} & \mathrm{D^b}(X) \\ \Big\downarrow{\scriptstyle \mathrm{id}} & & \Big\downarrow{\scriptstyle \Phi_{\Phi_{\mathcal{S}_{\mathrm{R}}\boxtimes\mathcal{S}}(\mathcal{O}_\Delta)}} \\ \mathrm{D^b}(X) & \xrightarrow[\Phi]{} & \mathrm{D^b}(X) \end{array}$$

shows $\Phi_{\mathcal{S}_{\mathrm{R}}\boxtimes\mathcal{S}}(\mathcal{O}_\Delta) \simeq \mathcal{O}_\Delta$.

Using Grothendieck–Verdier duality (see Theorem 3.34) one finds

$$\begin{aligned} \Phi_{\mathcal{S}_{\mathrm{R}}}(\mathcal{E}^\vee) &= p_*(q^*\mathcal{E}^\vee \otimes \mathcal{S}^\vee \otimes q^*\omega_X[\dim(X)]) \\ &\simeq \mathcal{H}om(p_*(q^*\mathcal{E}\otimes\mathcal{S}), \mathcal{O}_X) \\ &\simeq p_*\mathcal{H}om(q^*\mathcal{E}\otimes\mathcal{S}, \omega_p[\dim(p)]) \simeq \Phi_{\mathcal{S}}(\mathcal{E})^\vee. \end{aligned}$$

Hence, (8.9) becomes

$$\Phi_{\mathcal{S}}(\mathcal{E})^\vee \boxtimes \Phi_{\mathcal{S}}(\mathcal{E}) \longrightarrow \mathcal{O}_\Delta \longrightarrow \Phi_{\mathcal{S}_{\mathrm{R}}\boxtimes\mathcal{S}}(\mathcal{P}_{\mathcal{E}}).$$

All identifications are sufficiently canonical to ensure that the first morphism in this triangle is again just the trace map. Thus, the two objects $\Phi_{\mathcal{S}_{\mathrm{R}}\boxtimes\mathcal{S}}(\mathcal{P}_{\mathcal{E}})$ and $\mathcal{P}_{\Phi(\mathcal{E})}$ completing it to a distinguished triangle are necessarily isomorphic. □

In the following, the proposition shall be applied to the case that also Φ is a spherical twist.

Proposition 8.22 (Seidel, Thomas) *Suppose $\mathcal{E}_1,\ldots,\mathcal{E}_m \in \mathrm{D^b}(X)$ is an A_m-configuration of spherical objects. Then, for the induced spherical twists $T_i := T_{\mathcal{E}_i}$ one finds*

$$\begin{aligned} T_i\circ T_j &\simeq T_j\circ T_i && \text{for } |i-j|>1 \\ T_i\circ T_{i+1}\circ T_i &\simeq T_{i+1}\circ T_i\circ T_{i+1} && \text{for } i=1,\ldots,m-1. \end{aligned}$$

Proof The first isomorphism follows directly from the previous lemma and the assumption $\mathrm{Hom}(\mathcal{E}_j,\mathcal{E}_i[*])=0$ which implies $T_j(\mathcal{E}_i)\simeq\mathcal{E}_i$ (see Exercise 8.5).

For the second assertion one first applies the lemma to conclude

$$\begin{aligned} T_i\circ T_{i+1}\circ T_i &\simeq T_i\circ T_{T_{i+1}(\mathcal{E}_i)}\circ T_{i+1} \\ &\simeq T_{T_i(T_{i+1}(\mathcal{E}_i))}\circ T_i\circ T_{i+1}. \end{aligned}$$

Hence, it suffices to prove $T_i(T_{i+1}(\mathcal{E}_i))\simeq\mathcal{E}_{i+1}[\ell]$ for some ℓ (cf. Exercise 8.4).

If necessary, we shift $\mathcal{E}_{i+1}$ such that $\dim\mathrm{Hom}(\mathcal{E}_{i+1},\mathcal{E}_i)=1$. Thus, applying Exercise 8.5 to T_{i+1} produces a distinguished triangle

$$\mathcal{E}_{i+1} \longrightarrow \mathcal{E}_i \longrightarrow T_{i+1}(\mathcal{E}_i) \longrightarrow \mathcal{E}_{i+1}[1]$$

and by applying the exact functor T_i to it yet another one

$$T_i(\mathcal{E}_{i+1}) \xrightarrow{\varphi_1} T_i(\mathcal{E}_i) \longrightarrow T_i(T_{i+1}(\mathcal{E}_i)) \longrightarrow T_i(\mathcal{E}_{i+1})[1].$$

For the term in the middle we use the isomorphism $T_i(\mathcal{E}_i) \simeq \mathcal{E}_i[1-d]$, where $d = \dim(X)$ (cf. Exercise 8.5).

Similarly, $T_i(\mathcal{E}_{i+1})$ is described by the distinguished triangle

$$\mathcal{E}_i[-d] \longrightarrow \mathcal{E}_{i+1} \longrightarrow T_i(\mathcal{E}_{i+1}) \longrightarrow \mathcal{E}_i[1-d]$$

or, equivalently,

$$T_i(\mathcal{E}_{i+1}) \xrightarrow{\varphi_2} \mathcal{E}_i[1-d] \longrightarrow \mathcal{E}_{i+1}[1] \longrightarrow T_i(\mathcal{E}_{i+1})[1].$$

Since $\mathrm{Hom}(\mathcal{E}_{i+1}, \mathcal{E}_i) \simeq \mathrm{Hom}(T_i(\mathcal{E}_{i+1}), T_i(\mathcal{E}_i))$ is of dimension one, the two morphisms φ_1 and φ_2 coincide up to scaling. Hence, $T_i(T_{i+1}(\mathcal{E}_i)) \simeq \mathcal{E}_{i+1}[1]$ by TR3. □

Remark 8.23 The proposition can be rephrased by saying that any A_m-configuration of spherical objects in $\mathrm{D^b}(X)$ induces a group homomorphism

$$B_{m+1} \longrightarrow \mathrm{Aut}(\mathrm{D^b}(X)),$$

i.e. a 'representation' of the braid group on $\mathrm{D^b}(X)$.

Theorem 8.24 (Seidel, Thomas) *Suppose $\mathcal{E}_1, \ldots, \mathcal{E}_m \in \mathrm{D^b}(X)$ is an A_m-configuration of spherical objects on a smooth projective variety X of dimension at least two. Then the induced representation*

$$B_{m+1} \hookrightarrow \mathrm{Aut}(\mathrm{D^b}(X))$$

is injective. See [106].

The proof of this deep theorem is far beyond the scope of these notes. Roughly, one tries to extract a 'smaller' representation of the braid group for which faithfulness can be shown more easily. In fact, it is shown that any non-trivial element g of the braid group acts non-trivially on at least one of the spherical objects $\mathcal{E}_i$, i.e. $g(\mathcal{E}_i) \not\simeq \mathcal{E}_i$. Clearly, by the very nature of the braid group some fair amount of topology has to come in at some point.

If X is even-dimensional and the Mukai vectors $v(\mathcal{E}_i)$ are linearly independent, then the braid group action

$$B_{m+1} \longrightarrow \mathrm{Aut}(\mathrm{D^b}(X))$$

covers the Weyl group action

$$W_m \longrightarrow \mathrm{Aut}(H^*(X, \mathbb{Q}))$$

given by the reflections in the hyperplanes orthogonal to the $v(\mathcal{E}_i)$. Hence, in order to prove injectivity of $B_{m+1} \longrightarrow \mathrm{Aut}(\mathrm{D^b}(X))$ it suffices to verify the faithfulness of the induced action of the pure braid group, i.e. that the kernel of $B_{m+1} \longrightarrow W_m$ injects into $\mathrm{Aut}(\mathrm{D^b}(X))$. To have an example in mind, consider T_i^2, which acts trivially on cohomology, i.e. represents an element in the pure braid group. The action on $\mathrm{D^b}(X)$ is non-trivial as has been shown by Remark 8.14.

Note that linearly dependent Mukai vectors are frequent, e.g. in the example below, where a counterexample to the assertion without the condition on the dimension of X is explained.

Examples 8.25 That the assumption on the dimension is indeed needed can be seen by the following example taken from [106]. Let C be a smooth elliptic curve. Choose two closed points $x_1, x_2 \in C$ such that $\mathcal{O}(x_1 - x_2)$ is a non-trivial line bundle of order two, e.g. let x_1 be the origin and x_2 a point of order two. Clearly, $k(x_1), \mathcal{O}_C, k(x_2)$ is an A_3-configuration, but the induced representation

$$B_4 \longrightarrow \mathrm{Aut}(\mathrm{D^b}(X))$$

is not injective. E.g. consider $T_{k(x_i)} \simeq \mathcal{O}(x_i) \otimes (\ \ )$, $i = 1, 2$, (see Example 8.10, i)) for which one computes that

$$T_{k(x_1)} \circ T_{k(x_2)}^{-1} \simeq (\mathcal{O}(x_1 - x_2) \otimes (\ \ ))$$

is of order two, a relation that does not hold in B_4.

That the braid group comes up in the context of autoequivalences of derived categories seems rather mysterious. It is however something that is clearly foreseen by the principles of mirror symmetry. We refrain from making any comments in this direction, but the reader is urged to consult [69, 106, 112].

Remark 8.26 Szendrői constructs in [111] braid group actions for more general configurations, i.e. of Artin groups of more complicated Dynkin diagrams. His construction covers other geometrically interesting situations incorporating certain phenomena coming from deforming Calabi–Yau varieties. In certain cases he shows faithfulness of the action by reducing it to the original result of Seidel and Thomas.

8.3 Beilinson spectral sequence

Clearly, the Fourier–Mukai transform $\Phi_{\mathcal{O}_\Delta} : \mathrm{D^b}(\mathbb{P}^n) \longrightarrow \mathrm{D^b}(\mathbb{P}^n)$ whose kernel is the structure sheaf of the diagonal $\mathcal{O}_\Delta$ in $\mathbb{P}^n \times \mathbb{P}^n$ is nothing but the identity and, as such, not very interesting. However, due to the existence of a very special resolution of $\mathcal{O}_\Delta$ as a sheaf on the product, there is nevertheless a highly intriguing structure that emerges, the *Beilinson spectral sequence.*

We continue to use the notation $\mathcal{F} \boxtimes \mathcal{G} := q^*\mathcal{F} \otimes p^*\mathcal{G}$. As in the previous section, we denote complexes simply by $\mathcal{E}$, $\mathcal{F}$, $\mathcal{G}$, etc.

Lemma 8.27 *There exists a natural locally free resolution of $\mathcal{O}_\Delta$ of the form*

$$0 \longrightarrow \bigwedge^n(\mathcal{O}(-1)\boxtimes\Omega(1)) \longrightarrow \bigwedge^{n-1}(\mathcal{O}(-1)\boxtimes\Omega(1)) \longrightarrow \cdots$$

$$\cdots \longrightarrow \mathcal{O}(-1)\boxtimes\Omega(1) \longrightarrow \mathcal{O}_{\mathbb{P}^n\times\mathbb{P}^n} \longrightarrow \mathcal{O}_\Delta \longrightarrow 0.$$

Proof To make the argument more transparent, let us write $\mathbb{P}^n$ as $\mathbb{P}(V)$. The Euler sequence can then be written as

$$0 \longrightarrow \Omega(1) \longrightarrow V^*\otimes\mathcal{O} \longrightarrow \mathcal{O}(1) \longrightarrow 0.$$

Recall that the fibre of $\mathcal{O}(-1)$ at a point $\ell\in\mathbb{P}(V)$ is by definition identified with the line $\ell\subset V$. Also, the fibre of $\Omega(1)$ in ℓ' is the subspace of those linear maps $\varphi: V\longrightarrow k$ that are trivial on $\ell'\subset V$. Thus, the Euler sequence at a point $\ell\in\mathbb{P}(V)$ is

$$0 \longrightarrow \ell^\perp \longrightarrow V^* \longrightarrow \ell^* \longrightarrow 0.$$

The homomorphism

$$\mathcal{O}(-1)\boxtimes\Omega(1) \longrightarrow \mathcal{O}_{\mathbb{P}(V)\times\mathbb{P}(V)}$$

at a point $(\ell,\ell')\in\mathbb{P}(V)\times\mathbb{P}(V)$ is by definition given by $(x,\varphi)\longmapsto\varphi(x)$, where $x\in\ell$ and $\varphi|_{\ell'}=0$. Clearly, the image of this map is the ideal sheaf of the diagonal $\Delta\subset\mathbb{P}(V)\times\mathbb{P}(V)$.

Now, using the standard construction of the *Koszul complex* yields the claimed resolution: This is a very general principle that associates to any section $s\in H^0(X,\mathcal{E})$ of a locally free sheaf $\mathcal{E}$ of rank r on a smooth variety with zero locus $Z:=Z(s)$ of codimension r, the Koszul complex

$$0 \longrightarrow \bigwedge^r\mathcal{E}^* \longrightarrow \cdots \longrightarrow \mathcal{E}^* \longrightarrow \mathcal{O}_X \longrightarrow \mathcal{O}_Z \longrightarrow 0,$$

which is a locally free resolution of $\mathcal{O}_Z$. The maps are given by contraction with the section s, i.e. $\varphi\longmapsto i_s\varphi$. □

Note that the structure sheaf of the diagonal $\Delta\subset X\times X$ for more complicated varieties (e.g. K3 surfaces) cannot be resolved by sheaves of the form $\mathcal{F}\boxtimes\mathcal{G}$.

In the following, we denote the complex

$$0 \longrightarrow \bigwedge^n(\mathcal{O}(-1)\boxtimes\Omega(1)) \longrightarrow \cdots \longrightarrow \mathcal{O}(-1)\boxtimes\Omega(1) \longrightarrow \mathcal{O}_{\mathbb{P}^n\times\mathbb{P}^n}$$

by $L^\bullet$. So, $L^0=\mathcal{O}_{\mathbb{P}^n\times\mathbb{P}^n}$ and $L^{-n}=\bigwedge^n(\mathcal{O}(-1)\boxtimes\Omega(1))$. As an object in $\mathrm{D^b}(\mathbb{P}^n\times\mathbb{P}^n)$ the complex $L^\bullet$ is isomorphic to the sheaf $\mathcal{O}_\Delta$.

Proposition 8.28 (Beilinson) *For any coherent sheaf $\mathcal{F}$ on $\mathbb{P}^n$ there exist two natural spectral sequences:*

$$E_1^{r,s} := H^s(\mathbb{P}^n, \mathcal{F}(r)) \otimes \Omega^{-r}(-r) \Rightarrow E^{r+s} = \begin{cases} \mathcal{F} & \textit{if } r+s=0 \\ 0 & \textit{otherwise} \end{cases} \tag{8.10}$$

and

$$E_1^{r,s} := H^s(\mathbb{P}^n, \mathcal{F} \otimes \Omega^{-r}(-r)) \otimes \mathcal{O}(r) \Rightarrow E^{r+s} = \begin{cases} \mathcal{F} & \textit{if } r+s=0 \\ 0 & \textit{otherwise.} \end{cases} \tag{8.11}$$

Proof The proof is a consequence of the spectral sequence

$$E_1^{r,s} = R^sF(A^r) \Rightarrow R^{r+s}F(A)$$

for any complex $A^\bullet \in \mathrm{D^b}(\mathcal{A})$ and the right derived functor RF of a left exact functor F (cf. Remark 2.67).

In our situation, we consider $A^\bullet := q^*(\mathcal{F}) \otimes L^\bullet$ and the derived functor of the direct image with respect to the second projection $p : \mathbb{P}^n \times \mathbb{P}^n \longrightarrow \mathbb{P}^n$. (Note that the tensor product need not to be derived, as $L^\bullet$ is a complex of locally free sheaves.) Thus, $A^r = \mathcal{F}(r) \boxtimes \Omega^{-r}(-r)$ and hence

$$R^sF(A^r) \simeq H^s(\mathbb{P}^n, \mathcal{F}(r)) \otimes \Omega^{-r}(-r)$$

due to the projection formula and base change (see p. 83).

On the other hand, $A^\bullet$ is quasi-isomorphic to $q^*\mathcal{F} \otimes \mathcal{O}_\Delta \simeq \iota_*\mathcal{F}$, with the diagonal embedding $\iota : X \xrightarrow{\sim} \Delta \subset X \times X$. Therefore, $p_*(A^\bullet) \simeq p_*(\iota_*\mathcal{F}) \simeq \mathcal{F}$. This proves (8.10).

Interchanging the rôle of p and q, the same argument proves (8.11). □

Since $\Omega^{-r}(-r)$ is only non-trivial for $r \in [-n, 0]$, the $E_1^{r,s}$ in (8.10) and (8.11) are trivial for $r < -n$ or $r > 0$ independently of $\mathcal{F}$. On the other hand, $E_1^{r,s} = 0$ for $s < 0$ and $s > n$. Thus, both spectral sequences are concentrated in the second quadrant.

Corollary 8.29 *Any sequence of line bundles of the form*

$$\mathcal{O}(a), \mathcal{O}(a+1), \ldots, \mathcal{O}(a+n)$$

on $\mathbb{P}^n$ defines a full exceptional sequence in $\mathrm{D^b}(\mathbb{P}^n)$.

Proof For the definition of a full exceptional sequence see Section 1.4.

Firstly, any line bundle $\mathcal{O}(i)$ on $\mathbb{P}^n$ is an exceptional object in $\mathrm{D^b}(\mathbb{P}^n)$, for

$$\mathrm{Hom}(\mathcal{O}(i), \mathcal{O}(i)[\ell]) \simeq H^\ell(\mathbb{P}^n, \mathcal{O}) = \begin{cases} k & \text{if } \ell = 0 \\ 0 & \text{otherwise.} \end{cases}$$

Secondly, for $a \le j < i \le a+n$ and hence $-n \le j-i < 0$ and all ℓ one has

$$\mathrm{Hom}(\mathcal{O}(i), \mathcal{O}(j)[\ell]) \simeq H^\ell(\mathbb{P}^n, \mathcal{O}(j-i)) = 0.$$

Thus, $\mathcal{O}(a), \ldots, \mathcal{O}(a+n)$ is an exceptional sequence.

In order to show that the exceptional sequence is full, i.e. that they generate $\mathrm{D}^{\mathrm{b}}(\mathbb{P}^n)$, it suffices to prove that any object $\mathcal{F}$ in the orthogonal complement $\langle \mathcal{O}(a), \ldots, \mathcal{O}(a+n)\rangle^{\perp}$ is trivial. We shall explain that this follows directly from the Beilinson spectral sequence if the object $\mathcal{F}$ is simply a sheaf (or a shifted sheaf). For genuine complexes, however, we have to go back to the proof of the Beilinson spectral sequence.

As it is a nice application of the Beilinson spectral sequence, we prove the special case of a sheaf first and then give the argument for the general case.

Applying (8.10) to a coherent sheaf $\mathcal{F}$, or rather its twist $\mathcal{F}(-a)$, yields a spectral sequence with

$$E_1^{r,s} = H^s(\mathbb{P}^n, \mathcal{F}(-a)(r)) \otimes \Omega^{-r}(-r) \simeq \mathrm{Hom}(\mathcal{O}(i), \mathcal{F}[s]) \otimes \Omega^{-r}(-r).$$

Here, $i = a - r$. As $\Omega^{-r}(-r)$ is non-trivial only for $-n \leq r \leq 0$; the same holds true for $E_1^{r,s}$.

If now $\mathcal{F}$ in addition is orthogonal to any $\mathcal{O}(i)$ for $i = a, \ldots, a+n$, i.e. $\mathrm{Hom}(\mathcal{O}(i), \mathcal{F}[s]) = 0$ for all s and $i = a, \ldots, a+n$ (or, in other words, $-n \leq r \leq 0$), then all $E_1^{r,s}$ are in fact trivial and hence also the object the spectral sequence converges to, i.e. $\mathcal{F}(-a)$, is trivial. This proves $\mathcal{F} \simeq 0$.

For the general case, we shall split the resolution $L^\bullet \longrightarrow \mathcal{O}_\Delta$ introduced before into short exact sequences:

$$0 \longrightarrow \bigwedge^n(\mathcal{O}(-1) \boxtimes \Omega(1)) \longrightarrow \bigwedge^{n-1}(\mathcal{O}(-1) \boxtimes \Omega(1)) \longrightarrow M_{n-1} \longrightarrow 0$$

$$0 \longrightarrow M_{n-1} \longrightarrow \bigwedge^{n-2}(\mathcal{O}(-1) \boxtimes \Omega(1)) \longrightarrow M_{n-2} \longrightarrow 0$$

$$\cdots$$

$$0 \longrightarrow M_1 \longrightarrow \mathcal{O}_{\mathbb{P}^n \times \mathbb{P}^n} \longrightarrow \mathcal{O}_\Delta \longrightarrow 0.$$

Each of these short exact sequences can be regarded as a distinguished triangle in $\mathrm{D}^{\mathrm{b}}(\mathbb{P}^n \times \mathbb{P}^n)$. Tensor product with $p^*\mathcal{F}$ and direct image under the first projection q yields distinguished triangles on the first factor:

$$\Phi_{M_{i+1}}(\mathcal{F}) \longrightarrow \Phi_{\mathcal{O}(-i)\boxtimes\Omega^i(i)}(\mathcal{F}) \longrightarrow \Phi_{M_i}(\mathcal{F}) \longrightarrow \Phi_{M_{i+1}}(\mathcal{F})[1].$$

(Note that compared to the special case treated earlier we changed the order of the two projections here. So, morally we use (8.11) this time.)

Clearly, $\Phi_{\mathcal{O}(-i)\boxtimes\Omega^i(i)}(\mathcal{F}) \simeq H^*(\mathbb{P}^n, \mathcal{F} \otimes \Omega^i(i)) \otimes \mathcal{O}(-i)$ is contained in $\langle \mathcal{O}(-i)\rangle$. By induction this proves that $\Phi_{M_i}(\mathcal{F}) \in \langle \mathcal{O}(-n), \ldots, \mathcal{O}(-i)\rangle$ for all i and eventually $\mathcal{F} \simeq \Phi_{\mathcal{O}_\Delta}(\mathcal{F}) \in \langle \mathcal{O}(-n), \ldots, \mathcal{O}\rangle$.

Thus, $\mathcal{O}(-n), \ldots, \mathcal{O}(-1), \mathcal{O}$ is a full exceptional sequence. Since tensor product with $\mathcal{O}(a+n)$ defines an equivalence and the image of $\mathcal{O}(-n), \ldots, \mathcal{O}(-1), \mathcal{O}$

under this equivalence is the exceptional sequence $\mathcal{O}(a), \ldots, \mathcal{O}(a+n-1), \mathcal{O}(a+n)$, the latter is full as well. □

Exercise 8.30 Show that also $\Omega^p(p)$ with $p = 0, \ldots, n$ form a full exceptional sequence in $\mathrm{D}^{\mathrm{b}}(\mathbb{P}^n)$.

Both full exceptional sequences on $\mathbb{P}^n$ are in fact strong.

Definition 8.31 *An exceptional collection* $E_1, \ldots, E_m \in \mathrm{D}^{\mathrm{b}}(X)$ *is* strong *if in addition* $\mathrm{Hom}(E_i, E_j[\ell]) = 0$ *for all* i, j *and* $\ell \neq 0$.

Exercise 8.32 Prove that $\mathcal{O}, \ldots, \mathcal{O}(n)$ and $\mathcal{O}, \Omega(1), \ldots, \Omega^n(n)$ are both strong full exceptional collections in $\mathrm{D}^{\mathrm{b}}(X)$.

Exercise 8.33 Prove that $\mathcal{O}, \mathcal{O}(1,0), \mathcal{O}(0,1), \mathcal{O}(1,1) \in \mathrm{D}^{\mathrm{b}}(\mathbb{P}^1 \times \mathbb{P}^1)$ is a strong full exceptional sequence.

The reason why the notion of strong full exceptional sequences is interesting is the following result, which we only state.

Theorem 8.34 (Bondal) *Let* X *be a smooth projective variety. Suppose there is a strong full exceptional sequence* $E_1, \ldots, E_m \in \mathrm{D}^{\mathrm{b}}(X)$. *If* $E := \bigoplus E_i$ *and* $A := \mathrm{End}(E)$, *then*

$$R\mathrm{Hom}(E, \ \ ) : \mathrm{D}^{\mathrm{b}}(X) \xrightarrow{\sim} \mathrm{D}^{\mathrm{b}}(\mathrm{mod}\text{--}A)$$

is an exact equivalence. See [12].

Here, mod–A is the abelian category of right A-modules. The A-module structure of $R\mathrm{Hom}(E, \ \ )$ is given by composition. The theorem had first been proved for the two exceptional sequences on $\mathbb{P}^n$ discussed above.

Remark 8.35 It is a straightforward exercise to generalize the Beilinson spectral sequence to the relative setting. More precisely, if

$$\pi : \mathbb{P}(\mathcal{N}) \longrightarrow Y$$

is a projective bundle over a smooth projective variety Y with relative tautological line bundle $\mathcal{O}_\pi(1)$ and relative cotangent bundle Ω_π, then for any coherent sheaf $\mathcal{F}$ there exists a spectral sequence

$$E_1^{r,s} = \pi^* R^s \pi_*(\mathcal{F} \otimes \mathcal{O}_\pi(r)) \otimes \Omega_\pi^{-r}(-r) \Rightarrow \begin{cases} \mathcal{F} & \text{if } r+s=0 \\ 0 & \text{otherwise.} \end{cases} \tag{8.12}$$

In order to prove this, one constructs a locally free resolution

$$0 \longrightarrow \bigwedge^n(q^*\mathcal{O}_\pi(-1)\otimes p^*\Omega_\pi(1)) \longrightarrow \cdots \longrightarrow q^*\mathcal{O}_\pi(-1)\otimes p^*\Omega_\pi(1)$$

$$\longrightarrow \mathcal{O}_{\mathbb{P}(\mathcal{N})\times_Y\mathbb{P}(\mathcal{N})} \longrightarrow \mathcal{O}_\Delta \longrightarrow 0.$$

Here, $p,q:\mathbb{P}(\mathcal{N})\times_Y\mathbb{P}(\mathcal{N})\longrightarrow\mathbb{P}(\mathcal{N})$ are the two projections. As before, $E_1^{r,s}=0$ for $r\notin[-\mathrm{rk}(\mathcal{N})+1,0]$. Moreover, if $R\pi_*(\mathcal{F}\otimes\mathcal{O}_\pi(r))=0$ for $r\in[-\mathrm{rk}(\mathcal{N})+1,0]$, then $\mathcal{F}\simeq 0$.

The relative version of the above corollary produces a semi-orthogonal decomposition of $\mathrm{D^b}(\mathbb{P}(\mathcal{N}))$ (cf. Definition 1.59). The proof, being similar to that of Corollary 8.29, is left to the reader.

Corollary 8.36 *Let $\mathcal{N}$ be a vector bundle of rank r. Then for any $a\in\mathbb{Z}$ the sequence of full subcategories*

$$\pi^*\mathrm{D^b}(Y)\otimes\mathcal{O}(a),\ldots,\pi^*\mathrm{D^b}(Y)\otimes\mathcal{O}(a+r-1)\subset\mathrm{D^b}(\mathbb{P}(\mathcal{N}))$$

defines a semi-orthogonal decomposition of $\mathrm{D^b}(\mathbb{P}(\mathcal{N}))$. *See [91].* □

Exercise 8.37 Generalize Exercise 8.30 to the relative setting.

Remark 8.38 There is a wealth of other results about exceptional objects especially on Fano varieties, e.g. $\mathbb{P}^n$, $\mathbb{P}^n\times\mathbb{P}^m$. The above are only the most classical ones. More concrete examples can be found in [103].

Very recently Kawamata has proved the long standing conjecture saying that any toric variety admits a full exceptional collection of sheaves (see [65]). The most amazing aspect of his proof is that one has to pass via mildly singular toric varieties. In doing so, he actually proves the same result for the derived categories of those. Note, however, that the stronger question, whether on a toric variety one always finds full exceptional sequences of line bundles, remains open for the time being.

There is plenty of literature on the subject, so we content ourselves with this glimpse on these fascinating topics. The interested reader might start further reading with [103].

Spherical and exceptional objects serve different purposes: the former are studied on Calabi–Yau varieties, whereas the latter are found for on varieties with negative canonical bundle. Nonetheless, there are striking similarities between these two types of special objects.

Not only is being spherical or exceptional phrased in terms of the Ext-groups of the object, but both types give rise to certain actions: the spherical twists studied in Section 8.1 and so called *mutations*, defined completely analogously. (Note however that mutations do not act on the whole derived category, but rather on collections of exceptional objects.) Moreover, braid groups appear in both contexts.

8.4 They go together

The striking similarity between spherical and exceptional objects can be partially explained. In this section, we will first present two results, due to Seidel and Thomas, that shed some light on this mysterious relationship. Work of Horja makes clear that a mixture of both conditions leads to a new kind of spherical object, so called $\mathbb{EZ}$-spherical objects, that give rise to generalized twist functors. The second part of this section will be devoted to some of his results. Note that $\mathbb{EZ}$-spherical objects are not just an idle generalization of what has been seen in Section 8.1, but this notion can indeed successfully be applied to many interesting geometric situations. Some of these will be presented on the way.

Proposition 8.39 (Seidel, Thomas) *Suppose $f : X \longrightarrow Y$ is a morphism of smooth projective varieties such that $f_*\mathcal{O}_X$ sits in a distinguished triangle of the form*

$$\mathcal{O}_Y \longrightarrow f_*\mathcal{O}_X \longrightarrow \omega_Y[-c] \longrightarrow \mathcal{O}_Y[1]$$

with $c := \dim(X) - \dim(Y)$. Suppose in addition that $\omega_X \simeq \mathcal{O}_X$.

If $\mathcal{E} \in \mathrm{D^b}(Y)$ is an exceptional object, then $f^*\mathcal{E}$ is spherical. See [106].

Proof First note that the assumed distinguished triangle excludes $\dim(X) = 0$.

By adjunction and the projection formula, one has

$$\mathrm{Hom}(f^*\mathcal{E}, f^*\mathcal{E}[i]) \simeq \mathrm{Hom}(\mathcal{E}, \mathcal{E} \otimes f_*\mathcal{O}_X[i]).$$

Applying $\mathrm{Hom}(\mathcal{E},\ \ )$ to the distinguished triangle

$$\mathcal{E} \longrightarrow \mathcal{E} \otimes f_*\mathcal{O}_X \longrightarrow \mathcal{E} \otimes \omega_Y[-c] \longrightarrow \mathcal{E}[1]$$

yields a long exact sequence on which the following conclusions are based.

By using Serre duality, one finds

$$\mathrm{Hom}(\mathcal{E}, \mathcal{E} \otimes \omega_Y[-c-i]) = \mathrm{Hom}(\mathcal{E}, \mathcal{E}[\dim(X) + i])^* = 0$$

for $i = 0, 1$ ($> -\dim(X)$), which allows us to conclude that

$$\mathrm{Hom}(\mathcal{E}, \mathcal{E} \otimes f_*\mathcal{O}_X) = \mathrm{Hom}(\mathcal{E}, \mathcal{E}) = k.$$

Similarly, $\mathrm{Hom}(\mathcal{E}, \mathcal{E}[\dim(X) + i]) = 0$ for $i = 0, 1$ yields

$$\begin{aligned}\mathrm{Hom}(\mathcal{E}, \mathcal{E} \otimes f_*\mathcal{O}_X[\dim(X)]) &\simeq \mathrm{Hom}(\mathcal{E}, \mathcal{E} \otimes \omega_Y[\dim(X) - c]) \\ &\simeq \mathrm{Hom}(\mathcal{E}, \mathcal{E})^* = k.\end{aligned}$$

All the other groups are trivial. This proves the assertion. □

Remark 8.40 i) For $c > 0$ the condition of the proposition is satisfied if

$$R^i f_*\mathcal{O}_X = \begin{cases} \mathcal{O}_Y & \text{if } i = 0 \\ \omega_Y & \text{if } i = c \\ 0 & \text{otherwise.} \end{cases}$$

ii) The analogy ends here. In general, the pull-back of an exceptional collection $\mathcal{E}_1, \ldots, \mathcal{E}_m$ does not give rise to an A_m-configuration on X (not even for $m = 2$).

Examples 8.41 The following three typical examples are taken from [106].

i) Suppose $f : X \longrightarrow Y$ is a smooth morphism with honest Calabi–Yau manifolds as fibres F, i.e. $\omega_F \simeq \mathcal{O}_F$ and $H^i(F, \mathcal{O}_F) = 0$ for $0 < i < c$. E.g. the fibres F could be elliptic curves or K3 surfaces. The smoothness assumptions can be weakened.
ii) Consider a K3 surface X realized as a degree two cover $f : X \longrightarrow \mathbb{P}^2$ ramified along a sextic. Here, $c = 0$.
iii) Suppose $f : X \hookrightarrow Y$ is the inclusion of a smooth hypersurface with $\mathcal{O}(X) \simeq \omega_Y^*$. The inclusion does indeed satisfy the assumptions of the proposition with $c = -1$. Concrete examples are provided by a quartic in $\mathbb{P}^3$ or a quintic in $\mathbb{P}^4$.

In Proposition 8.4 an exceptional object becomes spherical under pull-back. In the following result we study the direct image of an exceptional object.

Proposition 8.42 (Seidel, Thomas) *Suppose $i : Y \hookrightarrow X$ is a smooth hypersurface with $i^*\omega_X \simeq \mathcal{O}_Y$. If $\mathcal{E} \in \mathrm{D^b}(Y)$ is exceptional, then $i_*\mathcal{E}$ is a spherical object in $\mathrm{D^b}(X)$. See [106].*

We omit the proof as it is very close in spirit to the arguments of the subsequent discussion. (In fact, the argument will be given in Examples 8.49, ii) but it will use a result of one of the later chapters.)

Propositions 8.4 and 8.4 can be seen from a broader perspective, as was explained by Horja in [46]. His main result shall be presented here, although under a few simplifying assumptions (e.g. smoothness of all the participating varieties). Also note that the main result of Section 8.1 can be viewed as a special case of what follows. For methodological and historical reasons we postponed the general case until now.

For the rest of this section we fix the following notation. Let X, E, and Z be smooth projective varieties, such that E comes with an embedding i and a smooth surjective projection:

$$i : E \hookrightarrow X, \qquad q : E \twoheadrightarrow Z.$$

The various dimension will be denoted $d := \mathrm{codim}(E \hookrightarrow X)$, $n := \dim(X)$, and $k := \dim(q) = \dim(E) - \dim(Z)$.

In the following we shall be interested in objects $\mathcal{E} \in \mathrm{D^b}(E)$ that behave like exceptional objects on the fibres of q and more like spherical objects with respect to the ambient variety X. Before stating the precise condition, we need to introduce for any $\mathcal{G} \in \mathrm{D^b}(Z)$ the natural morphism

$$\varphi_{\mathcal{G}} : \mathcal{G} \longrightarrow q_* \left(\mathcal{E}^\vee \otimes i^! i_*(\mathcal{E} \otimes q^*\mathcal{G}) \right).$$

The most important case is that of $\mathcal{G} = \mathcal{O}_Z$.

Here, $i^! : \mathrm{D^b}(X) \longrightarrow \mathrm{D^b}(E)$ is the functor

$$\mathcal{F} \longmapsto i^*\mathcal{F} \otimes \omega_i[\dim(i)] = i^*\mathcal{F} \otimes \omega_E \otimes \omega_X^*|_E[-d],$$

which is right adjoint to i_* (see p. 86). More precisely, for all $\mathcal{F}' \in \mathrm{D^b}(E)$ one has $\mathcal{H}om_X(i_*\mathcal{F}', \mathcal{F}) \simeq i_*\mathcal{H}om_E(\mathcal{F}', i^!\mathcal{F})$ and in particular $i_*i^!\mathcal{F} \simeq \mathcal{H}om_X(i_*\mathcal{O}_E, \mathcal{F})$.

By definition $\varphi_\mathcal{G}$ corresponds to the identity $i_*(\mathcal{E} \otimes q^*\mathcal{G}) \longrightarrow i_*(\mathcal{E} \otimes q^*\mathcal{G})$ under the chain of functorial isomorphisms

$$\begin{aligned}
&\mathrm{Hom}(\mathcal{G}, q_*(\mathcal{E}^\vee \otimes i^!i_*(\mathcal{E} \otimes q^*\mathcal{G})))\\
&\quad = \mathrm{Hom}(q^*\mathcal{G}, \mathcal{E}^\vee \otimes i^!i_*(\mathcal{E} \otimes q^*\mathcal{G}))\\
&\quad \simeq \mathrm{Hom}(q^*\mathcal{G} \otimes \mathcal{E}, i^!i_*(\mathcal{E} \otimes q^*\mathcal{G}))\\
&\quad \simeq \mathrm{Hom}(i_*(\mathcal{E} \otimes q^*\mathcal{G}), i_*(\mathcal{E} \otimes q^*\mathcal{G})).
\end{aligned}$$

Definition 8.43 *An object $\mathcal{E} \in \mathrm{D^b}(E)$ is called* $E\!\!\!Z$-spherical *if the following two conditions hold true:*

i) *For any $\mathcal{G} \in \mathrm{D^b}(Z)$, the natural morphism $\varphi_\mathcal{G}$ introduced before can be completed to a distinguished triangle of the form*

$$\mathcal{G} \xrightarrow{\varphi_\mathcal{G}} q_*(\mathcal{E}^\vee \otimes i^!i_*(\mathcal{E} \otimes q^*\mathcal{G})) \longrightarrow \mathcal{L} \otimes \mathcal{G}[-d-k] \longrightarrow \mathcal{G}[1] \quad (8.13)$$

with $\mathcal{L}$ isomorphic to a line bundle on Z (independent of $\mathcal{G}$) with the property that $q^*\mathcal{L}$ is isomorphic to the restriction of a line bundle on X.

ii) *There is an isomorphism $\mathcal{E} \otimes i^*\omega_X \simeq \mathcal{E}$.*

Remark 8.44 In many situations $\varphi_\mathcal{G}$ will simply be $\varphi_{\mathcal{O}_Z} \otimes \mathrm{id}_\mathcal{G}$, e.g. if $d = 0$, and it then suffices to verify condition i) just for $\mathcal{G} = \mathcal{O}_Z$. Of course, the problem at this point is that in general there is no reason to believe that $q_*(\mathcal{E}^\vee \otimes i^!i_*(\mathcal{E} \otimes q^*\mathcal{G})) \simeq q_*(\mathcal{E}^\vee \otimes i^!i_*(\mathcal{E})) \otimes \mathcal{G}$ for all $\mathcal{G}$. However, as we will see, this is needed for the proof of the main result.

In fact, in a first version of his paper Horja only required i) for $\mathcal{G} = \mathcal{O}_Z$. The definition he eventually adopted is different from ours. Further comments can be found in Remark 8.50.

Exercise 8.45 Show that $\mathcal{L} \simeq \omega_Z$. (Dualize (8.13) for $\mathcal{G} = \mathcal{O}_Z$ and apply Grothendieck–Verdier duality.)

The following discussion is modeled on the arguments in Section 8.1.

Spanning class Consider $\Omega := \Omega_1 \bigcup \Omega_2$ with

$$\begin{aligned}
\Omega_1 &:= \{i_*(\mathcal{E} \otimes q^*\mathcal{G}) \mid \mathcal{G} \in \mathrm{D^b}(Z)\}\\
\Omega_2 &:= \{\mathcal{F} \in \mathrm{D^b}(X) \mid q_*(\mathcal{E}^\vee \otimes i^!\mathcal{F}) = 0\}.
\end{aligned}$$

Note that this Ω is actually invariant under shift.

In order to show that Ω is spanning we first consider an object $\mathcal{F} \in \Omega_1^{\perp}$, i.e. $\mathrm{Hom}(i_*(\mathcal{E} \otimes q^*\mathcal{G}), \mathcal{F}) = 0$ for all $\mathcal{G} \in \mathrm{D}^{\mathrm{b}}(Z)$. Using $i_* \dashv i^!$ and $q^* \dashv q_*$, this implies $\mathrm{Hom}(\mathcal{G}, q_*(\mathcal{E}^\vee \otimes i^!\mathcal{F})) = 0$ for all $\mathcal{G} \in \mathrm{D}^{\mathrm{b}}(Z)$ and thus $q_*(\mathcal{E}^\vee \otimes i^!\mathcal{F}) \simeq 0$. In other words, $\mathcal{F} \in \Omega_2$. If also $\mathcal{F} \in \Omega_2^{\perp}$, then it would be orthogonal to itself and hence trivial. Thus, any object $\mathcal{F} \in \Omega^{\perp}$ is trivial.

The triviality of any $\mathcal{F}$ with $\mathrm{Hom}(\mathcal{F}, \mathcal{H}) = 0$ for all $\mathcal{H} \in \Omega$ can be reduced to the argument above. Indeed, combining with Serre duality $\mathrm{Hom}(\mathcal{F}, \mathcal{H}) \simeq \mathrm{Hom}(\mathcal{H}, \mathcal{F} \otimes \omega_X[n])^*$ it yields $\mathcal{F} \otimes \omega_X \simeq 0$, which is equivalent to $\mathcal{F} \simeq 0$.

Exercise 8.46 Prove the vanishing $\mathrm{Hom}(\mathcal{F}_1, \mathcal{F}_2) = 0 = \mathrm{Hom}(\mathcal{F}_2, \mathcal{F}_1)$ for any $\mathcal{F}_i \in \Omega_i$, $i = 1, 2$. (Condition ii) is used for the second one.)

***EZ*-spherical twist** The Fourier–Mukai kernel for the *EZ*-spherical twist we are interested in will again be obtained by a cone construction. We will use the following notation:

$$\begin{array}{ccc} \Delta_E & \overset{i}{\hookrightarrow} & \Delta_X \\ {\scriptstyle k}\downarrow & & \downarrow{\scriptstyle \iota} \\ E \times_Z E & \underset{j}{\hookrightarrow} & X \times X. \end{array}$$

Here $\Delta_E \subset E \times E$ and $\Delta_X \subset X \times X$ are the two diagonals, which will tacitly be identified with E, respectively X.

First note that adjunction induces a natural morphism $i_*i^!\mathcal{O}_{\Delta_X} \longrightarrow \mathcal{O}_{\Delta_X}$. Its direct image under ι yields a morphism $\iota_*i_*i^!\mathcal{O}_{\Delta_X} \longrightarrow \iota_*\mathcal{O}_{\Delta_X}$.

Next, the trace map induces a natural morphism

$$k_*(\mathcal{E}^\vee \otimes \mathcal{E} \otimes i^!\mathcal{O}_{\Delta_X}) \longrightarrow k_*i^!\mathcal{O}_{\Delta_X}$$

which can be composed with the restriction map

$$(\mathcal{E}^\vee \otimes i^!\mathcal{O}_X) \boxtimes_Z \mathcal{E} \longrightarrow k_*\left(k^*\left((\mathcal{E}^\vee \otimes i^!\mathcal{O}_X) \boxtimes_Z \mathcal{E}\right)\right) \simeq k_*(\mathcal{E}^\vee \otimes \mathcal{E} \otimes i^!\mathcal{O}_{\Delta_X}).$$

(As the notation suggests $\boxtimes_Z$ denotes the tensor product of the two sheaves that are obtained by pull-back under the two projections $E \times_Z E \longrightarrow E$.)

Since $j_*k_*i^!\mathcal{O}_{\Delta_X} \simeq \iota_*i_*i^!\mathcal{O}_{\Delta_X}$, the composition of the two morphisms leads to the natural morphism

$$j_*\left((\mathcal{E}^\vee \otimes i^!\mathcal{O}_X) \boxtimes_Z \mathcal{E}\right) \longrightarrow \mathcal{O}_{\Delta_X}.$$

The cone of this morphism, i.e. the object completing it to a distinguished triangle, shall be denoted

$$\mathcal{P}_{\mathcal{E}} := \mathrm{C}\left(j_*\left((\mathcal{E}^\vee \otimes i^!\mathcal{O}_X) \boxtimes_Z \mathcal{E}\right) \longrightarrow \mathcal{O}_{\Delta_X} \right).$$

Definition 8.47 *For an EZ-spherical object $\mathcal{E} \in \mathrm{D^b}(E)$ one defines the induced EZ-spherical twist $T_{\mathcal{E}}$ as the Fourier–Mukai transform*

$$T_{\mathcal{E}} := \Phi_{\mathcal{P}_{\mathcal{E}}} : \mathrm{D^b}(X) \longrightarrow \mathrm{D^b}(X).$$

Cone description of the twist Similarly to the standard twist, the image of a complex under the EZ-spherical twist $T := T_{\mathcal{E}}$ can be described by a distinguished triangle (cf. Exercise 8.5). More precisely, for any $\mathcal{F} \in \mathrm{D^b}(X)$ there exists a distinguished triangle

$$T(\mathcal{F})[-1] \longrightarrow i_*\left(q^*q_*(\mathcal{E}^\vee \otimes i^!\mathcal{F}) \otimes \mathcal{E}\right) \xrightarrow{\ \eta\ } \mathcal{F} \longrightarrow T(\mathcal{F}). \qquad (8.14)$$

This is evident from the description of the Fourier–Mukai kernel of T and $i^!\mathcal{F} \simeq i^!\mathcal{O}_X \otimes i^*\mathcal{F}$.

Moreover, by construction, the morphism η corresponds to the identity $\mathrm{id} : q_*(\mathcal{E}^\vee \otimes i^!\mathcal{F}) \to q_*(\mathcal{E}^\vee \otimes i^!\mathcal{F})$ under the functorial isomorphism

$$\begin{aligned} &\mathrm{Hom}(i_*(q^*q_*(\mathcal{E}^\vee \otimes i^!\mathcal{F}) \otimes \mathcal{E}), \mathcal{F}) \\ &\simeq \mathrm{Hom}(q^*q_*(\mathcal{E}^\vee \otimes i^!\mathcal{F}) \otimes \mathcal{E}, i^!\mathcal{F}) \\ &\simeq \mathrm{Hom}(q^*q_*(\mathcal{E}^\vee \otimes i^!\mathcal{F}), \mathcal{E}^\vee \otimes i^!\mathcal{F}) \\ &\simeq \mathrm{Hom}(q_*(\mathcal{E}^\vee \otimes i^!\mathcal{F}), q_*(\mathcal{E}^\vee \otimes i^!\mathcal{F})). \end{aligned}$$

Now consider $\mathcal{F} \in \Omega_2$. Then $i_*(q^*q_*(\mathcal{E}^\vee \otimes i^!\mathcal{F}) \otimes \mathcal{E}) = 0$ and hence

$$T(\mathcal{F}) \simeq \mathcal{F}.$$

If, on the other hand, $\mathcal{F} \in \Omega_1$, i.e. $\mathcal{F} = i_*(\mathcal{E} \otimes q^*\mathcal{G})$ for some $\mathcal{G} \in \mathrm{D^b}(Z)$, then

$$i_*(q^*q_*(\mathcal{E}^\vee \otimes i^!\mathcal{F}) \otimes \mathcal{E}) \simeq i_*(q^*q_*(\mathcal{E}^\vee \otimes i^!i_*(\mathcal{E} \otimes q^*\mathcal{G})) \otimes \mathcal{E}).$$

Now use the condition i) on an EZ-spherical object. Applying the exact functors q^*, $\mathcal{E} \otimes (\ \ )$, and i_* to (8.13), we obtain the distinguished triangle

$$\begin{aligned} \mathcal{F} &\xrightarrow{\ \psi\ } i_*\left(q^*\left(q_*(\mathcal{E}^\vee \otimes i^!i_*(\mathcal{E} \otimes q^*\mathcal{G}))\right) \otimes \mathcal{E}\right) \\ &\longrightarrow i_*\left((\mathcal{E} \otimes q^*\mathcal{G}) \otimes q^*\mathcal{L}\right)[-d-k] \longrightarrow \mathcal{F}[1]. \end{aligned} \qquad (8.15)$$

One checks that $\eta \circ \psi$ is the identity. Combining the distinguished triangles (8.14) and (8.15) and applying the axioms TR in the usual way

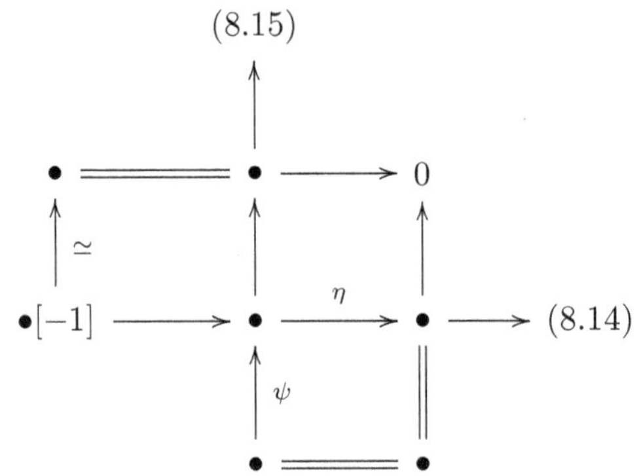

allows us to deduce an isomorphism

$$T(\mathcal{F}) = T(i_*(\mathcal{E} \otimes q^*\mathcal{G})) \simeq i_*((\mathcal{E} \otimes q^*\mathcal{G}) \otimes q^*\mathcal{L})[1 - d - k],$$

which describes $T(\mathcal{F})$ for $\mathcal{F} \in \Omega_1$. In particular, we have shown that with $\mathcal{F} \in \Omega_i$ $(i = 1, 2)$ also $T(\mathcal{F}) \in \Omega_i$.

Fully faithful The criterion for full faithfulness via a spanning class applies. Indeed, if $\mathcal{F}_1, \mathcal{F}_2 \in \Omega_2$, then $T(\mathcal{F}_i) \simeq \mathcal{F}_i$ yields isomorphisms $\mathrm{Hom}(\mathcal{F}_1, \mathcal{F}_2) \simeq \mathrm{Hom}(T(\mathcal{F}_1), T(\mathcal{F}_2))$. If $\mathcal{F}_1, \mathcal{F}_2 \in \Omega_1$, i.e. $\mathcal{F}_i = i_*(\mathcal{E} \otimes q^*\mathcal{G}_i)$ for certain $\mathcal{G}_i$, $i = 1, 2$, then

$$\begin{aligned}\mathrm{Hom}(T(\mathcal{F}_1), T(\mathcal{F}_2)) &\simeq \mathrm{Hom}(\mathcal{E} \otimes q^*(\mathcal{G}_1 \otimes \mathcal{L}), i^! i_*(\mathcal{E} \otimes q^*(\mathcal{G}_2 \otimes \mathcal{L})))\\ &\simeq \mathrm{Hom}(\mathcal{E} \otimes q^*\mathcal{G}_1, i^! i_*(\mathcal{E} \otimes q^*\mathcal{G}_2)).\end{aligned}$$

Note that for the last isomorphism we use the assumption that $q^*\mathcal{L}$ is the restriction of a line bundle on X.

As Ω_1 and Ω_2 are orthogonal to each other, the case $\mathcal{F}_1 \in \Omega_1$ and $\mathcal{F}_2 \in \Omega_2$ need not be tested. In order to prove $\Omega_1 \subset \Omega_2^\perp$ one uses condition ii).

Equivalence We have seen that $T = T_{\mathcal{E}}$ is fully faithful. So, writing it as the Fourier–Mukai transform $T = \Phi_{\mathcal{P}_{\mathcal{E}}}$ it suffices to check that $\mathcal{P}_{\mathcal{E}} \otimes q^*\omega_X \simeq \mathcal{P}_{\mathcal{E}} \otimes p^*\omega_X$. But this follows easily from the explicit description of $\mathcal{P}_{\mathcal{E}}$ and the assumption $\mathcal{E} \otimes i^*\omega_X \simeq \mathcal{E}$.

We summarize the discussion by the following

Proposition 8.48 (Horja) *The EZ-spherical twist $T_{\mathcal{E}}$ associated to an EZ-spherical object $\mathcal{E} \in \mathrm{D^b}(E)$ induces an autoequivalence*

$$T_{\mathcal{E}} : \mathrm{D^b}(X) \xrightarrow{\sim} \mathrm{D^b}(X).$$

See [46]. □

Examples 8.49 i) Let $Z = \mathrm{Spec}(k)$. Then $\mathcal{E}$ is EZ-spherical if and only if $i_*\mathcal{E}$ is spherical. In this case, the spherical twist $T_{i_*\mathcal{E}}$ is isomorphic to the EZ-spherical twist $T_{\mathcal{E}}$.

Indeed, in this case adjunction allows us to interpret $R^jq_*(\mathcal{E}^\vee \otimes i^!i_*\mathcal{E})$ as $\mathrm{Hom}(\mathcal{E}, i^!i_*\mathcal{E}[j]) \simeq \mathrm{Hom}(i_*\mathcal{E}, i_*\mathcal{E}[j])$. The assumption $\mathcal{E} \otimes i^*\omega_X \simeq \mathcal{E}$ shows that $\mathrm{Hom}(i_*\mathcal{E}, i_*\mathcal{E}[n])$ is at least one-dimensional. This suffices to conclude the equivalence. In order to verify $T_{i_*\mathcal{E}} \simeq T_{\mathcal{E}}$ one may use the cone description (8.14) or identify the kernels directly (use Grothendieck–Verdier duality to show $i_*(\mathcal{E}^\vee \otimes i^!\mathcal{O}_X) \simeq (i_*(\mathcal{E}))^\vee$).

ii) Let $E \subset X$ be a smooth divisor and $Z = \mathrm{Spec}(k)$. For the discussion of this case we invoke a few things that will only be explained in Chapter 11, e.g. Corollary 11.4. In particular, we shall make use of the distinguished triangle

$$\mathcal{E} \otimes \mathcal{O}_E(-E)[1] \longrightarrow i^*i_*\mathcal{E} \longrightarrow \mathcal{E} \longrightarrow \mathcal{E} \otimes \mathcal{O}_E(-E)[2].$$

Tensoring with $\mathcal{E}^\vee \otimes \mathcal{O}_E(E)[-1]$ turns this into a distinguished triangle of the form

$$\mathcal{E}^\vee \otimes \mathcal{E} \longrightarrow \mathcal{E}^\vee \otimes i^!i_*\mathcal{E} \longrightarrow \mathcal{E}^\vee \otimes \mathcal{E} \otimes \mathcal{O}_E(E)[-1] \longrightarrow \mathcal{E}^\vee \otimes \mathcal{E}[1].$$

Now apply q_* to it. If $\mathcal{E} \otimes i^*\omega_X \simeq \mathcal{E}$, then $\mathcal{E}$ is EZ-spherical if and only if $\mathcal{E}$ is an exceptional object in $\mathrm{D^b}(E)$. Similar arguments also prove Proposition 8.4, but this is of no importance as EZ-spherical is enough to ensure that the twist $T_{\mathcal{E}}$, which is isomorphic to $T_{i_*\mathcal{E}}$, is an equivalence.

iii) This example generalizes the approach of Proposition 8.4. Consider the variety $E := F \times Z$ which is assumed to be of codimension one in X and suppose there exists an exceptional object $\mathcal{E} \in \mathrm{D^b}(F)$. Then the pull-back $\pi_1^*\mathcal{E}$ via the first projection describes an EZ-spherical object. Of course, $q : E \longrightarrow Z$ is chosen to be the second projection. We leave the details to the reader.

iv) Let $q : E \longrightarrow Z$ be a $\mathbb{P}^1$-bundle and $E \subset X$ a smooth divisor in a variety X with trivial ω_X. Then $\mathcal{O}_E$ is EZ-spherical. More generally, any line bundle on E is EZ-spherical.

The concrete geometric situation we have in mind here is the exceptional divisor $E = \mathbb{P}(\Omega_S) \subset X$ inside the Hilbert scheme $X = \mathrm{Hilb}^2(S)$ of length-two subschemes in a K3 or abelian surface S.

Remark 8.50 For $d > 0$ Horja calls $\mathcal{E} \in \mathrm{D^b}(E)$ EZ-spherical if $q_*(\mathcal{E}^\vee \otimes \mathcal{E}) \simeq \mathcal{O}_Z$ and $q_*(\mathcal{E}^\vee \otimes \mathcal{E} \otimes \bigwedge^k \mathcal{N}) = 0$ for $0 < k < d$. Here $\mathcal{N}$ is the normal bundle of $E \subset X$. This definition has of course the advantage that it does not involve arbitrary objects $\mathcal{G} \in \mathrm{D^b}(Z)$.

It can be shown, though the proof is not completely trivial, that his definition is stronger than the one we worked with here.

9

ABELIAN VARIETIES

From a historical point of view, this chapter should have come first. Mukai's starting point for introducing what is nowadays called the Fourier–Mukai transform was the Poincaré bundle on the product of an abelian variety and its dual. It yields an important instance of a derived equivalence between a projective variety and another, in general non-isomorphic, one. As K3 surfaces treated in Chapter 10, abelian varieties occupy a distinguished place between varieties with ample canonical and those with ample anti-canonical bundle.

The first section is meant as a reminder of those facts from the rich theory of abelian varieties that are relevant from our derived point of view. The story begins in Section 9.2 where we prove Mukai's result saying that the Poincaré bundle taken as a Fourier–Mukai kernel does indeed define an equivalence. For principally polarized abelian varieties this equivalence can be viewed as an autoequivalence and, as Mukai showed, extended to a certain Sl_2-action.

Sections 9.4 and 9.5 present a general investigation of derived equivalences between abelian varieties and derived autoequivalences of a single abelian variety. Most of these results are due to Orlov. The situation is much more interesting than for varieties with ample (anti-)canonical bundle, but, contrary to the case of K3 surfaces, everything one wants to know can in principle be computed.

Throughout, we will work over an algebraically closed field k of characteristic zero, so that we can freely use all results of the preceding chapters. Sometimes, more direct arguments can be given by actually working over the complex numbers, e.g. when we want to use singular cohomology. However, most of the results of this chapter hold true for abelian varieties over arbitrary fields.

9.1 Basic definitions and facts

Let us begin with a few recollections from the theory of abelian varieties necessary for the understanding of the later sections. For a thorough treatment see any of the many text books on abelian varieties, e.g. [11, 34, 85].

Let us begin with the algebraic definition of an abelian variety:

Definition 9.1 *An* abelian variety A *is a projective connected algebraic group over* k.

In particular, A comes with morphisms

$$m : A \times A \longrightarrow A, \quad \iota : A \longrightarrow A, \quad \text{and } e : \mathrm{Spec}(k) \longrightarrow A$$

satisfying the usual axioms of a group. In the sequel we often write $a+b$ for $m(a,b)$, $-a$ for $\iota(a)$, and $0 \in A$ for $e \in A$.

Any closed point $a \in A$ gives rise to the translation

$$t_a : A \xrightarrow{\sim} A, \quad b \longmapsto m(a,b).$$

Remark 9.2 Here is a list of some basic facts:

i) Any abelian variety is smooth and the underlying group is commutative.

ii) If $k = \mathbb{C}$, then the associated complex manifold, by abuse of notation also denoted by A, is a compact complex Lie group.

iii) More precisely, the associated complex manifold is isomorphic (as a complex Lie group) to a complex torus $\mathbb{C}^g/\Gamma$. This can be made concrete in various ways. Here are two:

Consider the exponential map $\exp : T_eA \longrightarrow A$. It indeed induces an isomorphism of T_eA/Γ for some discrete subgroup $\Gamma \subset T_eA$.

Alternatively, consider the Albanese morphism

$$A \longrightarrow \mathrm{Alb}(A) = H^0(A,\Omega)^*/H_1(A,\mathbb{Z})$$

given by $a \longmapsto \int_e^a$. Note that the quotient is indeed a torus, due to standard Hodge theory.

Once (the induced complex manifold) A is written as $\mathbb{C}^g/\Gamma$, one has

$$H^i(A,\mathbb{Z}) \simeq \bigwedge^i \Gamma^*.$$

iv) The cotangent bundle Ω_A of an abelian variety A is trivial. In particular, $\omega_A \simeq \mathcal{O}_A$ and $\mathrm{c}_i(A) = 0$, $i > 0$.

v) The second description in iii) allows us to give a quick proof of the following assertion: A morphism $\varphi : A_1 \longrightarrow A_2$ between two abelian varieties is a homomorphism, i.e. also compatible with the group structure, if and only if $\varphi(e_1) = e_2$.

Indeed, any morphism $\varphi : A_1 \longrightarrow A_2$ defines a natural linear map

$$\varphi^* : H^0(A_1,\Omega_{A_1})^* \longrightarrow H^0(A_2,\Omega_{A_2})^*,$$

which is compatible with the Albanese map if and only if $\varphi(e_1) = e_2$. Clearly, dividing out by the lattices $H_1(A_i,\mathbb{Z})$, $i = 1,2$, defines a homomorphism.

An important example of a homomorphism is the morphism 'multiplication by n':

$$n : A \longrightarrow A, \qquad a \longmapsto n \cdot a,$$

which is defined for any $n \in \mathbb{Z}$. Another description for $n > 0$ is given by $m \circ \Delta_n$, where $\Delta_n : A \longrightarrow A^n$ is the n-fold diagonal embedding and $m : A^n \longrightarrow A$ is the sum. For $n < 0$ one has $n = \iota \circ (-n)$. (As we used ι to denote the inversion on an abelian variety, we had to switch to Δ for the diagonal embedding.)

Multiplication by $n \neq 0$ is a typical case of a very interesting class of homomorphisms:

Definition 9.3 *An* isogeny *between two abelian varieties A_i, $i = 1, 2$, is a finite surjective homomorphism $A_1 \twoheadrightarrow A_2$. The* degree *of an isogeny $\varphi : A_1 \twoheadrightarrow A_2$ is the order of the kernel $K_\varphi := \varphi^{-1}(e_2)$.*

It is easy to determine the degree of the multiplication by $n : A \twoheadrightarrow A$, namely $\deg(n) = n^{2g}$, where $g = \dim(A)$ as before.

Let us next discuss line bundles on abelian varieties. We start out with an important and completely general result.

Proposition 9.4 **(See-saw principle)** *Let X be an irreducible complete variety, T an integral scheme, and $L \in \mathrm{Pic}(X \times T)$. Suppose that $L_t := L|_{X \times \{t\}}$ is trivial for all (closed) point $t \in T$.*

Then there exists a line bundle M on T with $L \simeq p^*M$. See [85, I.5, Cor.6].

Sketch of proof First note that a line bundle L on X is trivial if and only if $H^0(X, L) \neq 0$ and $H^0(X, L^*) \neq 0$. Semi-continuity of $h^0(L_t)$ shows that being trivial is a closed condition. Thus, testing closed points is enough.

Standard results on semi-continuity and direct images also show that the assumption $h^0(L_t) \equiv 1$, $t \in T$, implies that $M := p_*L$ is a line bundle on T. The adjunction morphism $p^*M = p^*p_*L \longrightarrow L$ is an isomorphism on each fibre $X \times \{t\}$ and hence an isomorphism on $X \times T$. □

Remark 9.5 Here are useful additions to the above:

i) If moreover L is trivial on at least one fibre of the projection $X \times T \twoheadrightarrow X$, then L is trivial.
ii) Suppose L and L' are two line bundles on $X \times T$ such that $L_t \simeq L'_t$ for all closed points $t \in T$. Then $L \simeq L' \otimes p^*M$ for some line bundle M on T.

Examples 9.6 Suppose L is a line bundle on an abelian variety A. Then

$$m^*L \simeq q^*L \otimes p^*L \iff t_a^*L \simeq L \text{ for all } a \in A.$$

As usual, $p, q : A \times A \twoheadrightarrow A$ denote the projections. Indeed, $(m^*L)|_{p^{-1}(a)} \simeq t_a^*L$ and $(q^*L \otimes p^*L)|_{p^{-1}(a)} \simeq L$ and, therefore, the above criteria apply.

The following is a very useful consequence of the see-saw principle:

Theorem 9.7 **(of the cube)** *Let $X \times Y \times Z$ be the product of three irreducible complete varieties with chosen closed points $x_0 \in X$, $y_0 \in Y$, and $z_0 \in Z$.*

Then, a line bundle L on $X \times Y \times Z$ is trivial if and only if the three restrictions $L|_{\{x_0\} \times Y \times Z}$, $L|_{X \times \{y_0\} \times Z}$, and $L|_{X \times Y \times \{z_0\}}$ are trivial.

The assumptions ensure that L is trivial on the fibre of the projection $X \times Y \times Z \twoheadrightarrow X$ over $x_0 \in X$ and one has to show that this is actually true for any fibre. Connecting any other point by a complete curve with x_0, one reduces to the case that X is a curve. For the rest of the argument, see e.g. [85].

If all three varieties are smooth projective with $H^1(\ ,\mathbb{Z}) = 0$, then the result follows from the inclusion $\mathrm{Pic}(\ ) \hookrightarrow H^2(\ ,\mathbb{Z})$ and the Künneth formula $H^2(X \times Y \times Z, \mathbb{Z}) = H^2(X,\mathbb{Z}) \oplus H^2(Y,\mathbb{Z}) \oplus H^2(Z,\mathbb{Z})$. If e.g. $H^1(X,\mathbb{Z}) \neq 0$, then one can argue with the exponential sequence.

Let us mention a few immediate consequences of the theorem of the cube for line bundles on abelian varieties:

i) Consider the three projections $\pi_i : A \times A \times A \longrightarrow A$ together with the morphisms $m_{ij} : A \times A \times A \longrightarrow A$ given as $m \circ (\pi_i, \pi_j)$, and the triple sum $m : A \times A \times A \longrightarrow A$, $(a,b,c) \longmapsto a+b+c$. Then

$$m^*L \otimes \pi_1^*L \otimes \pi_2^*L \otimes \pi_3^*L \simeq m_{12}^*L \otimes m_{13}^*L \otimes m_{23}^*L. \tag{9.1}$$

This follows directly from the general statement, where the distinguished closed points in all three factors are chosen to be the origin $e \in A$.

ii) Consider the multiplication $n : A \longrightarrow A$. Then for any line bundle L on A one has

$$n^*L \simeq L^{(n^2+n)/2} \otimes \iota^*L^{(n^2-n)/2}. \tag{9.2}$$

Pull-back the equation (9.1) via $(n, \mathrm{id}, \iota) : A \longrightarrow A \times A \times A$, $a \longmapsto (na, a, -a)$. This does not give (9.2) right away, but can be used to express $(n+1)^*L$ in terms of n^*L and $(n-1)^*L$. Then one argues by recursion, the cases $n = 0, \pm 1$ being trivial.

iii) The last application we want to mention is the *theorem of the square.* It says

$$t_{a+b}^*L \otimes L \simeq t_a^*L \otimes t_b^*L \tag{9.3}$$

for all line bundles L on A and all closed points $a, b \in A$. This isomorphism is obtained directly as the pull-back of (9.1) via $A \longrightarrow A \times A \times A$, $c \longmapsto (c, a, b)$.

Another way to express the same fact is to say that

$$\varphi_L : A(k) \longrightarrow \mathrm{Pic}(A), \quad a \longmapsto t_a^*L \otimes L^* \tag{9.4}$$

is a group homomorphism. Here, $A(k)$ denotes the set of closed points of A.

Definition 9.8 *Let A be an abelian variety. Then*

$$\mathrm{Pic}^0(A) := \{L \in \mathrm{Pic}(A) \mid t_a^*L \simeq L \text{ for all } a \in A\},$$

which is a subgroup of the Picard group Pic(A).

Translation invariant line bundles enjoy many interesting properties. E.g. if $L \in \mathrm{Pic}^0(A)$, then for all $n \in \mathbb{Z}$ one has

$$n^*L \simeq L^n. \tag{9.5}$$

Indeed, using (9.2) one reduces to the case $n = -1$, which can be proved by pulling-back $m^*L \simeq q^*L \otimes p^*L$ under the morphism $A \longrightarrow A \times A$, $a \longmapsto (a, -a)$ (use Example 9.6).

Slightly more difficult to prove is the fact that non-trivial translation invariant line bundles have trivial cohomology:

Lemma 9.9 *Let $\mathcal{O} \not\simeq L \in \mathrm{Pic}^0(A)$. Then $H^i(A, L) = 0$ for all i.*

Proof The first step is to show that $H^0(A, L) = 0$. Indeed, if not then there exists a non-trivial section s of L, which in turn induces a non-trivial section ι^*s of ι^*L. If L is not trivial, both vanish along a non-trivial effective divisor and so does their tensor product, which is a section of $L \otimes \iota^*L$. The latter is trivial due to (9.5). Contradiction.

Suppose k is minimal with $H^k(A, L) \neq 0$. Then use $m^*L \simeq q^*L \otimes p^*L$ and the Künneth formula to write

$$H^k(A \times A, m^*L) = \bigoplus_{i+j=k} H^i(A, L) \otimes H^j(A, L).$$

As $A \longrightarrow A \times A$, $a \longmapsto (a, e)$ composed with m is the identity, the pull-back $H^k(A, L) \longrightarrow H^k(A \times A, m^*L)$ is injective. However, this yields a contradiction using that k is minimal and $H^0(A, L) = 0$. $\square$

Until now, we have studied $\mathrm{Pic}^0(A)$ as a group. Let us next try to endow it with a geometric structure. Using the exponential sequence, one defines the *dual variety*

$$\widehat{A} := H^1(A, \mathcal{O})/H^1(A, \mathbb{Z}) \longrightarrow \mathrm{Pic}(A) = H^1(A, \mathcal{O}^*) \xrightarrow{\mathrm{c}_1} H^2(A, \mathbb{Z}).$$

Again, Hodge theory (see, e.g. [42, 51]) tells us that $\widehat{A}$ has the structure of a complex torus. The group structure induces the usual morphisms, which shall be called $\widehat{m}$, $\hat{\iota}$, respectively $\hat{e}$.

It is not difficult to show that the two subgroups $\widehat{A}$ and $\mathrm{Pic}^0(A)$ of $\mathrm{Pic}(A)$ actually coincide. One inclusion is easy: Since the induced action of t_a on $H^1(A, \mathcal{O})$ is trivial, one has $\widehat{A} \subset \mathrm{Pic}^0(A)$. On the other hand, for $L \in \mathrm{Pic}^0(A)$ one uses $\iota^*L \simeq L^*$ and hence $\iota^*\mathrm{c}_1(L) = -\mathrm{c}_1(L)$, but $\iota^*|_{H^2(A,\mathbb{Z})} = \bigwedge^2 \iota^*|_{H^1(A,\mathbb{Z})} = \bigwedge^2(-1) = \mathrm{id}$. Since $H^2(A, \mathbb{Z})$ is torsion free, one finds $\mathrm{c}_1(L) = 0$ and, therefore, $L \in \widehat{A}$.

Over the complex numbers, line bundles on an abelian variety (or, more generally, on a complex torus) can be described in terms of *Appell–Humbert* data. This part is slightly technical and we encourage the reader to consult [67, 85] for more details.

Let us construct line bundles on a complex torus $A = V/\Lambda$ as quotients of the trivial line bundle $\mathbb{C} \times V$ on V by a lifted action of Λ. More precisely, one defines $\lambda \cdot (z, v) = (A_\lambda(v) \cdot z, v + \lambda)$, with $A : \Lambda \times V \longrightarrow \mathbb{C}^*$ satisfying

$$A_{\lambda_1}(v + \lambda_2) \cdot A_{\lambda_2}(v) = A_{\lambda_1+\lambda_2}(v).$$

An Appell–Humbert datum (AH-datum for short) is a pair (α, H), where $\alpha : \Lambda \longrightarrow \mathrm{U}(1)$ and H an hermitian form on V such that:

$$\mathrm{Im}(H)(\Lambda \times \Lambda) \subset \mathbb{Z} \quad \text{and} \quad \alpha(\lambda_1 + \lambda_2) = (-1)^{\mathrm{Im}(H)(\lambda_1,\lambda_2)} \cdot \alpha(\lambda_1) \cdot \alpha(\lambda_2).$$

To every AH-datum (α, H) one associates

$$A_\lambda(v) = \alpha(\lambda) \cdot e^{\pi H(v,\lambda) + (\pi/2) H(\lambda,\lambda)},$$

which satisfies the above cocycle condition. Thus, to any AH-datum (α, H) one can associate a line bundle $L_{(\alpha,H)}$ on A.

Theorem 9.10 *The map $(\alpha, H) \longmapsto L_{(\alpha,H)}$ defines an isomorphism of groups*

$$\{(\alpha, H) \mid \mathrm{AH} - \mathrm{data}\} \longleftrightarrow \mathrm{Pic}(A).$$

Here, the group structure for AH-data is given by $(\alpha, H) + (\alpha', H') = (\alpha \cdot \alpha', H + H')$. See [85, Ch.I.2].

The first Chern class of the line bundle $L_{(\alpha,H)}$ can be described in terms of the AH-datum (α, H) as

$$\mathrm{c}_1(L_{(\alpha,H)}) = \mathrm{Im}(H) \in \bigwedge^2 \Lambda^* = H^2(A, \mathbb{Z}).$$

In particular, $L_{(\alpha,H)} \in \mathrm{Pic}^0(A)$ if and only if $H = 0$. Conversely, if $L \in \mathrm{Pic}^0(A)$ corresponds to a point $[v] \in H^1(A, \mathcal{O})/H^1(A, \mathbb{Z}) = \widehat{A}$ then $L = L_{(\alpha,0)}$ with $\alpha(\lambda) = e^{(2\pi i) v(\lambda)}$.

We next wish to construct the *Poincaré bundle* $\mathcal{P}$ on $\widehat{A} \times A$ by means of the AH-construction.

We want the Poincaré bundle to have the following two characteristic properties:

i) If $\alpha \in \widehat{A}$ corresponds to a line bundle $L \in \mathrm{Pic}(A)$ on A, then $\mathcal{P}|_{\{\alpha\} \times A}$ is isomorphic to L.
ii) The restriction $\mathcal{P}|_{\widehat{A} \times \{e\}}$ is trivial.

From the see-saw principle it immediately follows that the Poincaré bundle, if it exists, is unique.

To describe an AH-datum for the product $\widehat{A} \times A$ write

$$\widehat{A} \times A = (V^* \times V) \,/\, (\Lambda^* \times \Lambda) = \big(H^1(A, \mathbb{R}) \times H_1(A, \mathbb{R})\big) \,/\, \big(H^1(A, \mathbb{Z}) \times H_1(A, \mathbb{Z})\big)$$

and define

$$\alpha_{\mathcal{P}} : H^1(A, \mathbb{Z}) \times H_1(A, \mathbb{Z}) \longrightarrow \mathrm{U}(1), \quad (\lambda, \mu) \longmapsto (-1)^{\lambda(\mu)}$$

and

$$H_{\mathcal{P}}\left((u_1, v_1), (u_2, v_2)\right) := -v_1(iu_2) - v_2(iu_1) + i(v_1(u_2) - v_2(u_1)).$$

Then one verifies that the associated line bundle $\mathcal{P}$ satisfies i) and ii). Also note that the definition is completely symmetric, i.e. the line bundle $\mathcal{P}$ can also be seen as the Poincaré line bundle on $\widehat{\widehat{A}} \times \widehat{A} = A \times \widehat{A}$.

Exercise 9.11 Show that $(\hat{\iota} \times \mathrm{id})^*\mathcal{P} \simeq (\mathrm{id} \times \iota)^*\mathcal{P}$.

The first Chern class $c_1(\mathcal{P}) \in H^2(A \times \widehat{A}, \mathbb{Z})$ of the Poincaré bundle $\mathcal{P}$ can be described as follows.

Using the Künneth formula, one writes

$$H^2(A \times \widehat{A}, \mathbb{Z}) = H^2(A, \mathbb{Z}) \oplus H^1(A, \mathbb{Z}) \otimes H^1(\widehat{A}, \mathbb{Z}) \oplus H^2(\widehat{A}, \mathbb{Z}).$$

By construction of the dual abelian variety, $H^1(\widehat{A}, \mathbb{Z}) = H^1(A, \mathbb{Z})^*$. Then $c_1(\mathcal{P})$ is contained in

$$H^1(A, \mathbb{Z}) \otimes H^1(\widehat{A}, \mathbb{Z}) = H^1(A, \mathbb{Z}) \otimes H^1(A, \mathbb{Z})^*$$

and corresponds to the identity there. If we choose a basis $\{e_i\}$, $i = 1, \ldots, 2g$, of $H^1(A, \mathbb{Z})$ and denote the dual basis by e_i^*, then

$$c_1(\mathcal{P}) = \sum e_i \wedge e_i^*.$$

An easy calculation yields

$$c_1(\mathcal{P})^n = \sum_{\ell_1 < \ldots < \ell_n} (-1)^{n(n-1)/2} \cdot n! \cdot (e_{\ell_1} \wedge \ldots \wedge e_{\ell_n}) \wedge (e_{\ell_1}^* \wedge \ldots \wedge e_{\ell_n}^*) \qquad (9.6)$$

and, in particular,

$$c_1(\mathcal{P})^{2g}/(2g)! = (-1)^g (e_1 \wedge \ldots \wedge e_{2g}) \wedge (e_1^* \wedge \ldots \wedge e_{2g}^*). \qquad (9.7)$$

Remark 9.12 Let us be a little more specific about the identification $\rho : A \longrightarrow \widehat{\widehat{A}}$. By definition this is the isomorphism such that under

$$A \times \widehat{A} \xrightarrow{\rho \times \mathrm{id}} \widehat{\widehat{A}} \times \widehat{A} \xrightarrow{\sim} \widehat{A} \times \widehat{\widehat{A}},$$

where the second isomorphism is given by interchanging the two factors, the Poincaré bundle $\mathcal{P}$ for the abelian variety A corresponds to the Poincaré bundle $\widehat{\mathcal{P}}$ for the dual abelian variety $\widehat{A}$.

In the above notation, $c_1(\mathcal{P}) = \sum e_i \wedge e_i^*$ and $c_1(\widehat{\mathcal{P}}) = \sum e_i^* \wedge e_i^{**}$. Here we use the identification $H^1(\widehat{\widehat{A}}, \mathbb{Z}) \simeq H^1(\widehat{A}, \mathbb{Z})^* \simeq H^1(A, \mathbb{Z})^{**}$.

Thus, due to $e_i \wedge e_i^* = -e_i^* \wedge e_i$ the isomorphism $\rho : A \xrightarrow{\sim} \widehat{\widehat{A}}$ induces on cohomology the homomorphism

$$H^1(A, \mathbb{Z}) \xrightarrow{\sim} H^1(\widehat{\widehat{A}}, \mathbb{Z}) \xrightarrow{\sim} H^1(A, \mathbb{Z})^{**}, \qquad e_i \longmapsto -e_i^{**},$$

which is the standard one up to the sign.

Consider the projections $\pi_{12} : \widehat{A} \times A \times \widehat{A} \longrightarrow \widehat{A} \times A$, $\pi_{23} : \widehat{A} \times A \times \widehat{A} \longrightarrow A \times \widehat{A}$, $\pi_2 : \widehat{A} \times A \times \widehat{A} \longrightarrow A$, and $\widehat{m}_{13} : \widehat{A} \times A \times \widehat{A} \longrightarrow \widehat{A} \times A$ given as the product of $\widehat{m} : \widehat{A} \times \widehat{A} \longrightarrow \widehat{A}$ and π_2.

Lemma 9.13 $\pi_{12}^*\mathcal{P} \otimes \pi_{23}^*\mathcal{P} \simeq \widehat{m}_{13}^*\mathcal{P}$.

Proof This can be seen by using the see-saw principle with respect to π_2. Restricting $\pi_{12}^*\mathcal{P} \otimes \pi_{23}^*\mathcal{P}$ and $\widehat{m}_{13}^*\mathcal{P}$ to a fibre $\pi_2^{-1}(a)$ yields $q^*\mathcal{P}_a \otimes p^*\mathcal{P}_a$, respectively $\widehat{m}^*\mathcal{P}_a$, where $p, q : \widehat{A} \times \widehat{A} \longrightarrow \widehat{A}$ are the two projections. Since $\mathcal{P}$ is also the Poincaré bundle for $\widehat{A}$ with A viewed as the dual of $\widehat{A}$, the fibre $\mathcal{P}_a$ is in $\mathrm{Pic}^0(\widehat{A})$. Thus, $q^*\mathcal{P}_a \otimes p^*\mathcal{P}_a \simeq \widehat{m}^*\mathcal{P}_a$ (cf. Example 9.6). Restricting both line bundles to $\{\hat{e}\} \times A \times \{\hat{e}\}$ yields in both cases the trivial line bundle. Thus, $\pi_{12}^*\mathcal{P} \otimes \pi_{23}^*\mathcal{P} \simeq \widehat{m}_{13}^*\mathcal{P}$. □

More algebraically, the pair $(\mathcal{P}, \widehat{A})$ represents the *Picard functor* Pic^0_A: To any variety S one associates the set

$$\mathrm{Pic}^0_A(S) := \{M \in \mathrm{Pic}(S \times A) \mid M_s \in \mathrm{Pic}^0(A) \text{ for every closed } s \in S\}/\sim .$$

Here, $M \sim M'$ if there exists a line bundle L on S such that $M \otimes q^*L \simeq M'$. The functor is contravariant by pulling-back a line bundle M via a given morphism $f : T \longrightarrow S$, i.e. $(f \times \mathrm{id}_A)^* : \mathrm{Pic}^0_A(S) \longrightarrow \mathrm{Pic}^0_A(T)$.

Theorem 9.14 *The dual variety $\widehat{A}$ is a smooth projective variety that represents the Picard functor Pic^0_A, i.e. there exists a natural isomorphism $\mathrm{Pic}^0_A \simeq \mathrm{Hom}(\ \ , \widehat{A})$. The Poincaré bundle $\mathcal{P} \in \mathrm{Pic}^0_A(\widehat{A})$ corresponds to $\mathrm{id}_{\widehat{A}}$. See [85, Ch.II.8].*

In other words, for any $M \in \mathrm{Pic}^0_A(S)$ there exists a unique morphism

$$f_M : S \longrightarrow \widehat{A}$$

with $M \sim (f_M \times \mathrm{id}_A)^*\mathcal{P}$. Moreover, if the restriction of M to $S \times \{e\}$ is trivial, then $M \simeq (f_M \times \mathrm{id}_A)^*\mathcal{P}$. Note that in particular, $f_M(s) \in \widehat{A}$ corresponds to the line bundle on A given by $M|_{\{s\} \times A}$.

With this general result at hand, one describes $\varphi_L : A \longrightarrow \widehat{A}$, $a \longmapsto t_a^*L \otimes L^*$ (see (9.4)) more functorially as follows.

Consider the line bundle $\mathcal{Q} := m^*L \otimes (L^* \boxtimes L^*)$ on $A \times A$. Restricted to $\{a\} \times A$ it yields the line bundle $t_a^*L \otimes L^* \in \mathrm{Pic}^0(A)$. Hence there exists a morphism $\varphi_L := f_{\mathcal{Q}} : A \longrightarrow \widehat{A}$ such that $(\varphi_L \times \mathrm{id}_A)^*\mathcal{P} \sim \mathcal{Q}$. In particular, $\varphi_L(a) \in \widehat{A}$ corresponds to the line bundle $t_a^*L \otimes L^*$. In fact, since the restriction of $m^*L \otimes (L^* \boxtimes L^*)$ to $A \times \{e\}$ is trivial, one has

$$(\varphi_L \times \mathrm{id}_A)^*\mathcal{P} \simeq m^*L \otimes (L^* \boxtimes L^*). \tag{9.8}$$

Exercise 9.15 Show that $\varphi_L(-a) = \varphi_{L^*}(a)$ or, in other words, $\varphi_L \circ \iota = \varphi_{L^*}$. Also note that $\varphi_L(-a) = \varphi_L(a)^*$, which can also be written as $\varphi_L \circ \iota = \hat{\iota}^* \circ \varphi_L$ once $\mathrm{Pic}^0(A)$ has been identified with the dual abelian variety $\widehat{A}$.

Let us next study the homomorphism φ_L for L ample.

Lemma 9.16 *If L is an ample line bundle on A, then $\varphi_L : A \longrightarrow \widehat{A}$ is finite, i.e. an isogeny.*

Proof Suppose φ_L is not finite. Then there exists a curve $C \subset A$ contracted by φ_L to a point in $\widehat{A}$. In other words, for any closed point $a \in C \subset A$ one has $t_a^*L \simeq L$.

Consider the projection $p : A \times C \longrightarrow C$. For any $a \in C$ one has $m^*L|_{A\times\{a\}} \simeq t_a^*L \simeq L$. By the see-saw principle this shows that $m^*L|_{A\times C} \simeq q^*L \otimes p^*N$ for some line bundle N on C. On the other hand, $m^*L|_{\{e\}\times C} \simeq L|_C$ and, therefore, $N \simeq L|_C$.

Thus, $m^*L|_{A\times C} \simeq q^*L \otimes p^*(L|_C)$, which is ample. This contradicts the fact that the restriction of m^*L to $m^{-1}(e)$, which is the curve $\{(-a, a) \mid a \in C\}$, is trivial. □

A *polarized abelian variety* is a pair (A, L) consisting of an abelian variety A and an ample line bundle L. Often, the pair (A, φ_L) is called a polarized abelian variety. Note that L and $L \otimes M$ with $M \in \mathrm{Pic}^0(A)$ define the same $\varphi_L : A \longrightarrow \widehat{A}$.

The degree of the isogeny φ_L can be explicitly computed as follows.

Lemma 9.17 *Let (A, L) be a polarized abelian variety. Then* $\deg(\varphi_L) = \chi(L)^2$.

Proof Firstly, $\chi(\widehat{A} \times A, \mathcal{P}) = (1/(2g)!) \int c_1(\mathcal{P})^{2g} = (-1)^g$ (by the Hirzebruch–Riemann–Roch formula) and (9.7).

Secondly, $\chi(A \times A, (\varphi_L \times \mathrm{id})^*\mathcal{P}) = \deg(\varphi_L) \cdot \chi(\widehat{A} \times A, \mathcal{P})$. Together with $(\varphi_L \times \mathrm{id})^*\mathcal{P} \simeq m^*L \otimes (L^* \boxtimes L^*)$ (see (9.8)), this yields

$$(-1)^g \cdot \deg(\varphi_L) = \chi(A \times A, m^*L \otimes (L^* \boxtimes L^*)).$$

Now use that $p_*(m^*L \otimes q^*L^*)$ is concentrated in the finitely many points of the kernel of φ_L (see Lemma 9.9) to conclude that

$$\begin{aligned}\chi(A \times A, m^*L \otimes (L^* \boxtimes L^*)) &= \chi(A, p_*(m^*L \otimes q^*L^*) \otimes L^*)\\ &= \chi(A, p_*(m^*L \otimes q^*L^*))\\ &= \chi(A \times A, m^*L \otimes q^*L^*).\end{aligned}$$

Altogether, we obtain

$$\begin{aligned}(-1)^g \cdot \deg(\varphi_L) &= \chi(A \times A, m^*L \otimes q^*L^*)\\ &= \chi(A \times A, (m \times \mathrm{id})^*(L \boxtimes L^*))\\ &= \chi(A \times A, L \boxtimes L^*) = \chi(A, L) \cdot \chi(A, L^*)\\ &= (-1)^g \chi(L)^2.\end{aligned}$$

For the last equality we use again the Hirzebruch–Riemann–Roch formula. □

In fact, due to the vanishing of the higher cohomology groups $H^i(A, L)$, $i > 0$, for any ample line bundle L (e.g. by Kodaira vanishing), one also has $\deg(\varphi_L) = h^0(A, L)^2$.

Definition 9.18 *A* principally polarized abelian variety *is a polarized abelian variety* (A, L) *with* $\deg(\varphi_L) = 1$, *i.e.* $\varphi_L : A \xrightarrow{\sim} \widehat{A}$, *or, equivalently,* $\chi(L) = 1$.

9.2 The Poincaré bundle as a Fourier–Mukai kernel

Historically, the starting point of the theory of Fourier–Mukai transforms is the following result due to Mukai.

Proposition 9.19 **(Mukai)** *If* $\mathcal{P}$ *is the Poincaré bundle on* $A \times \widehat{A}$, *then*

$$\Phi_{\mathcal{P}} : \mathrm{D}^{\mathrm{b}}(\widehat{A}) \longrightarrow \mathrm{D}^{\mathrm{b}}(A)$$

is an equivalence.

Moreover, the composition

$$\mathrm{D}^{\mathrm{b}}(\widehat{A}) \xrightarrow{\Phi_{\mathcal{P}}} \mathrm{D}^{\mathrm{b}}(A) \xrightarrow{\Phi_{\mathcal{P}}} \mathrm{D}^{\mathrm{b}}(\widehat{A})$$

is isomorphic to $\hat{\iota}^* \circ [-g]$, *where* $g = \dim(A)$. *See [79].*

Proof Let us apply Proposition 7.1. Choose two closed points $\alpha, \beta \in \widehat{A}$. Then,

$$\Phi_{\mathcal{P}}(k(\alpha)) \simeq \mathcal{P}_\alpha \quad \text{and} \quad \Phi_{\mathcal{P}}(k(\beta)) \simeq \mathcal{P}_\beta.$$

are line bundles on A. Clearly, $\mathrm{Hom}(\mathcal{P}_\alpha, \mathcal{P}_\beta[i]) = H^i(A, \mathcal{P}_\alpha^* \otimes \mathcal{P}_\beta)$. Thus,

$$\mathrm{Hom}(\mathcal{P}_\alpha, \mathcal{P}_\beta[i]) = 0$$

for $i \notin [0, g]$. Moreover, for $i = 0$ and $\alpha = \beta$ this equals $H^0(A, \mathcal{O}_A) = k$.

Thus, it suffices to verify that $\mathrm{Hom}(\mathcal{P}_\alpha, \mathcal{P}_\beta[i]) = 0$ for $\alpha \neq \beta$ and all i. But in this case, $\mathcal{P}_\alpha, \mathcal{P}_\beta \in \mathrm{Pic}^0(A)$ are non-isomorphic and hence $H^i(A, \mathcal{P}_\alpha^* \otimes \mathcal{P}_\beta) = 0$ (see Lemma 9.9). Hence, $\Phi_{\mathcal{P}}$ is fully faithful and, since the canonical bundles of A and $\widehat{A}$ are trivial, indeed an equivalence.

Let us now study the composition, which is isomorphic to the Fourier–Mukai transform with kernel $\mathcal{R} := \pi_{13*}(\pi_{12}^*\mathcal{P} \otimes \pi_{23}^*\mathcal{P})$ (see Proposition 5.10), where we use the following diagram

$$\begin{array}{ccccc} \widehat{A} \times A & \xleftarrow{\pi_{12}} & \widehat{A} \times A \times \widehat{A} & \xrightarrow{\pi_{23}} & A \times \widehat{A} \\ & & \Big\downarrow{\scriptstyle \pi_{13}} & & \\ & & \widehat{A} \times \widehat{A}. & & \end{array}$$

In order to determine the support of $\mathcal{R}$, let us study the cohomology of $\pi_{12}^*\mathcal{P}\otimes\pi_{23}^*\mathcal{P}$ on the fibres of π_{13}. For a closed point $(\alpha,\beta)\in\widehat{A}\times\widehat{A}$ the restriction of $\pi_{12}^*\mathcal{P}\otimes\pi_{23}^*\mathcal{P}$ to $\pi_{13}^{-1}(\alpha,\beta)$ is the line bundle $\mathcal{P}_\alpha\otimes\mathcal{P}_\beta\in\mathrm{Pic}^0(A)$. By Lemma 9.9 we know that a non-trivial line bundle $L\in\mathrm{Pic}^0(A)$ has no cohomology whatsoever. Thus, $H^*(A,\mathcal{P}_\alpha\otimes\mathcal{P}_\beta)\neq 0$ if and only if $\mathcal{P}_\alpha\otimes\mathcal{P}_\beta\simeq\mathcal{O}$, i.e. if and only if $\beta=-\alpha$. In particular, $\mathrm{supp}(\mathcal{R})$ is contained in the graph $\Gamma_{\hat{\iota}}$ of $\hat{\iota}$.

We will show that $\mathcal{R}$ is isomorphic to the trivial line bundle on $\Gamma_{\hat{\iota}}$ shifted by $[-g]$. Recall that $\pi_{12}^*\mathcal{P}\otimes\pi_{23}^*\mathcal{P}$ is isomorphic to $\widehat{m}_{13}^*\mathcal{P}$ (see Lemma 9.13).

This can be combined with the fibre product diagram

$$\begin{array}{ccc}
\widehat{A}\times A\times\widehat{A} & \xrightarrow{\ \widehat{m}_{13}\ } & \widehat{A}\times A \\
{\scriptstyle\pi_{13}}\downarrow & & \downarrow{\scriptstyle q} \\
\widehat{A}\times\widehat{A} & \xrightarrow[\ \widehat{m}\ ]{} & \widehat{A}.
\end{array}$$

Since $\widehat{m}$ is flat, base change yields

$$\mathcal{R}=\pi_{13_*}(\pi_{12}^*\mathcal{P}\otimes\pi_{23}^*\mathcal{P})\simeq\pi_{13*}\widehat{m}_{13}^*\mathcal{P}\simeq\widehat{m}^*q_*\mathcal{P}.$$

For the final step one shows that $q_*\mathcal{P}\simeq k(\hat{e})[-g]$. This immediately then yields $\mathcal{R}\simeq\widehat{m}^*k(\hat{e})[-g]\simeq\mathcal{O}_{\Gamma_{\hat{\iota}}}[-g]$, which is what we had set out to prove.

To prove $q_*\mathcal{P}\simeq k(\hat{e})[-g]$, we shall use that $\Phi_{\mathcal{P}}:\mathrm{D^b}(A)\longrightarrow\mathrm{D^b}(\widehat{A})$ is fully faithful and that the support of $R^iq_*\mathcal{P}$ is contained in $\{\hat{e}\}$ (cf. the argument above that showed that $\mathcal{R}$ is concentrated in $\Gamma_{\hat{\iota}}$). Thus it suffices to show that $\mathrm{Hom}(\mathcal{O}_{\widehat{A}},R^iq_*\mathcal{P})=0$ for $i\neq g$ and $\mathrm{Hom}(\mathcal{O}_{\widehat{A}},R^gq_*\mathcal{P})=k$. To this end, we consider the spectral sequence

$$E_2^{r,s}=\mathrm{Ext}^r(\mathcal{O}_{\widehat{A}},R^sq_*\mathcal{P})\Rightarrow\mathrm{Ext}^{r+s}(\mathcal{O}_{\widehat{A}},q_*\mathcal{P}).$$

As every $R^sq_*\mathcal{P}$ has support contained in $\{\hat{e}\}$, the $E_2^{r,s}$-terms are all trivial except for $r=0$. In particular, $E_2^{r,s}=E_\infty^{r,s}$ for all r,s.

For the limit one computes

$$\begin{aligned}
\mathrm{Ext}^{r+s}(\mathcal{O}_{\widehat{A}},q_*\mathcal{P}) &\simeq\mathrm{Hom}(\mathcal{O}_{\widehat{A}},q_*\mathcal{P}[r+s])\\
&\simeq\mathrm{Hom}(\Phi_{\mathcal{P}}(k(e)),\Phi_{\mathcal{P}}(\mathcal{O}_A)[r+s])\\
&\simeq\mathrm{Hom}(k(e),\mathcal{O}_A[r+s])\simeq H^{g-r-s}(A,k(e))^*.
\end{aligned}$$

Hence, $\mathrm{Hom}(\mathcal{O}_{\widehat{A}},R^sq_*\mathcal{P})=H^{g-s}(A,k(e))$, which implies the result. □

Remark 9.20 The original argument of Mukai did not, of course, use Proposition 7.1. Clearly, once the composition has been described as $\hat{\iota}^*\circ[-g]$, one also finds that $\Phi_{\mathcal{P}}$ is an equivalence (use the symmetry of the situation with respect to A and $\widehat{A}$).

Thus, in the above proof, the only thing that has to be changed in order to avoid using Proposition 7.1 is the argument proving $q_*\mathcal{P} \simeq k(\hat{e})[-g]$. In the proof given above, we have used that $\Phi_{\mathcal{P}}$ is fully faithful, so Mukai in his original approach had to circumvent this. The techniques employed by Mukai (the full machinery of base change is used) are important in other situations too, so the reader might want to go back to [79] to see how it works without invoking Proposition 7.1.

Exercise 9.21 Use Corollary 5.23 and the techniques of the last proof to conclude that for any $a \in A$ and $\alpha \in \widehat{A}$ one has

$$\text{i)} \quad (\mathcal{P}_\alpha^* \otimes (\ \ )) \circ \Phi_{\mathcal{P}} \simeq \Phi_{\mathcal{P}} \circ t_\alpha^* : \mathrm{D^b}(\widehat{A}) \xrightarrow{\sim} \mathrm{D^b}(A)$$

$$\text{ii)} \quad \Phi_{\mathcal{P}} \circ (\mathcal{P}_\alpha \otimes (\ \ )) \simeq t_\alpha^* \circ \Phi_{\mathcal{P}} : \mathrm{D^b}(A) \xrightarrow{\sim} \mathrm{D^b}(\widehat{A}).$$

See [79].

Remark 9.22 i) Note that if $\widehat{A} \not\simeq A$, which happens quite frequently, the proposition yields derived equivalent varieties which are not isomorphic and not even birational.

ii) For elliptic curves the fact that the derived categories are equivalent is of course trivial, as the dual variety of an elliptic curve E is isomorphic to E.

However, the functor $\Phi_{\mathcal{P}} : \mathrm{D^b}(E) \longrightarrow \mathrm{D^b}(E)$ is nevertheless of considerable interest as it relates torsion sheaves, e.g. $k(\alpha)$, to vector bundles (the line bundles $\mathcal{P}_\alpha$). This can be used to give a new interpretation of certain results on vector bundles on elliptic curves. See [96, Ch.14].

The cohomological Fourier–Mukai transform

$$\Phi_{\mathcal{P}}^H : H^*(A, \mathbb{Q}) \longrightarrow H^*(\widehat{A}, \mathbb{Q})$$

can be compared with the natural isomorphism given by Poincaré duality.

More precisely, since $\Phi_{\mathcal{P}}^H : H^n(A, \mathbb{Q}) \longrightarrow H^*(\widehat{A}, \mathbb{Q})$ is given by $\mathrm{ch}(\mathcal{P}) = \exp(\mathrm{c}_1(\mathcal{P})) \in H^*(A \times \widehat{A}, \mathbb{Q})$ and $\mathrm{c}_1(\mathcal{P})^k \in H^k(A, \mathbb{Z}) \otimes H^k(\widehat{A}, \mathbb{Z})$, one finds that $\Phi_{\mathcal{P}}^H(H^n(A, \mathbb{Q})) \subset H^{2g-n}(\widehat{A}, \mathbb{Q})$.

On the other hand, integral Poincaré duality yields canonical isomorphisms

$$\mathrm{PD}_n : H^n(A, \mathbb{Z}) \xrightarrow{\sim} H^{2g-n}(A, \mathbb{Z})^* \xrightarrow{\sim} H^{2g-n}(\widehat{A}, \mathbb{Z}).$$

Lemma 9.23 *Poincaré duality and the Fourier–Mukai transform with kernel* $\mathcal{P}$ *compare via*

$$\Phi_{\mathcal{P}}^{H} = (-1)^{\frac{n(n+1)}{2}+g} \cdot \mathrm{PD}_n : H^n(A,\mathbb{Q}) \xrightarrow{\sim} H^{2g-n}(\widehat{A},\mathbb{Q}) = H^{2g-n}(A,\mathbb{Q})^*.$$

Proof The easiest way to see this is by introducing a basis $\{e_i\}$ of $H^1(A,\mathbb{Z})$ and to express all homomorphisms involved in this basis.

Let $J = (j_1 < \ldots < j_n)$ and $I = (i_1 < \ldots < i_{2g-n})$ with $I \cup J = \{1,\ldots,2g\}$. We shall write e_J instead of $e_{j_1} \wedge \ldots \wedge e_{j_n}$, etc. Then $\mathrm{PD}_n(e_J) = \varepsilon \cdot e_I^*$, where the sign ε can be determined by $\varepsilon = \varepsilon \cdot e_I^*(e_I) = \int e_J \wedge e_I$.

In order to express $\Phi_{\mathcal{P}}^{H}(e_J)$, one uses formula (9.6)

$$\mathrm{c}_1(\mathcal{P})^n = \sum_{\ell_1<\ldots<\ell_n} (-1)^{n(n-1)/2} \cdot n! \cdot (e_{\ell_1} \wedge \ldots \wedge e_{\ell_n}) \wedge (e_{\ell_1}^* \wedge \ldots \wedge e_{\ell_n}^*).$$

Hence,

$$\begin{aligned} \Phi_{\mathcal{P}}^{H}(e_J) &= \Phi^{H}_{\mathrm{c}_1(\mathcal{P})^{2g-n}/(2g-n)!}(e_J) \\ &= (-1)^{(2g-n)(2g-n-1)/2}\left(\int (e_J \wedge e_I)\right) e_I^*. \end{aligned}$$

The remaining sign verification is left to the reader. □

Corollary 9.24 *The cohomological Fourier–Mukai transform associated to the Poincaré bundle defines an isomorphism of integral(!) cohomology*

$$\Phi_{\mathcal{P}}^{H} : H^*(A,\mathbb{Z}) \xrightarrow{\sim} H^*(\widehat{A},\mathbb{Z}).$$

Moreover, the square

$$H^n(\widehat{A},\mathbb{Z}) \xrightarrow{\Phi_{\mathcal{P}}^{H}} H^{2g-n}(A,\mathbb{Z}) \xrightarrow{\Phi_{\mathcal{P}}^{H}} H^n(\widehat{A},\mathbb{Z})$$

is given by multiplication with $(-1)^{n+g}$.

Proof The first assertion follows from the above comparison with Poincaré duality, which is defined over the integers.

The second assertion could either be seen as a corollary to Proposition 9.19, as $\iota^* \circ [-g]$ acts as $(-1)^{n+g}$ on $H^n(A,\mathbb{Z})$, or directly proved by the sign check $(-1)^{n(n+1)/2+g+(2g-n)(2g-n+1)/2+g} = (-1)^{n+g}$. □

9.3 Sl_2-action

All the results of this sections are again due to Mukai.

It has been mentioned that a general Fourier–Mukai transform is not compatible with tensor product and it is easy to check that the Fourier–Mukai transform with kernel the Poincaré bundle on $\widehat{A} \times A$ is no exception. However, there is another multiplicative structure on the derived category of an abelian variety and the Fourier–Mukai transform $\Phi_{\mathcal{P}}$ relates it to the tensor product.

Definition 9.25 *Let A be an abelian variety. Then one defines the* convolution *as the bifunctor*

$$*: \ \mathrm{D^b}(A) \times \mathrm{D^b}(A) \longrightarrow \mathrm{D^b}(A) \quad \text{as} \ \mathcal{F}^\bullet * \mathcal{E}^\bullet := m_*(\mathcal{F}^\bullet \boxtimes \mathcal{E}^\bullet).$$

Note that the convolution is the composition of the right derived functor Rm_* and the bifunctor $\boxtimes$, the latter of which descends from the homotopy categories without deriving.

Exercise 9.26 Let $f: B \longrightarrow A$ be a homomorphism of abelian varieties. Show that $f^*\mathcal{F}^\bullet * f^*\mathcal{E}^\bullet \simeq f^*(\mathcal{F}^\bullet * \mathcal{E}^\bullet)$.

Lemma 9.27 *Let $\Phi_{\mathcal{P}}: \mathrm{D^b}(\widehat{A}) \longrightarrow \mathrm{D^b}(A)$ be the Fourier–Mukai functor with kernel $\mathcal{P}$. Then there exist functorial isomorphisms*

$$\Phi_{\mathcal{P}}(\mathcal{F}^\bullet * \mathcal{E}^\bullet) \simeq \Phi_{\mathcal{P}}(\mathcal{F}^\bullet) \otimes \Phi_{\mathcal{P}}(\mathcal{E}^\bullet)$$

and

$$\Phi_{\mathcal{P}}(\mathcal{F}^\bullet \otimes \mathcal{E}^\bullet) \simeq \Phi_{\mathcal{P}}(\mathcal{F}^\bullet) * \Phi_{\mathcal{P}}(\mathcal{E}^\bullet)[g].$$

Proof We use the following commutative diagram:

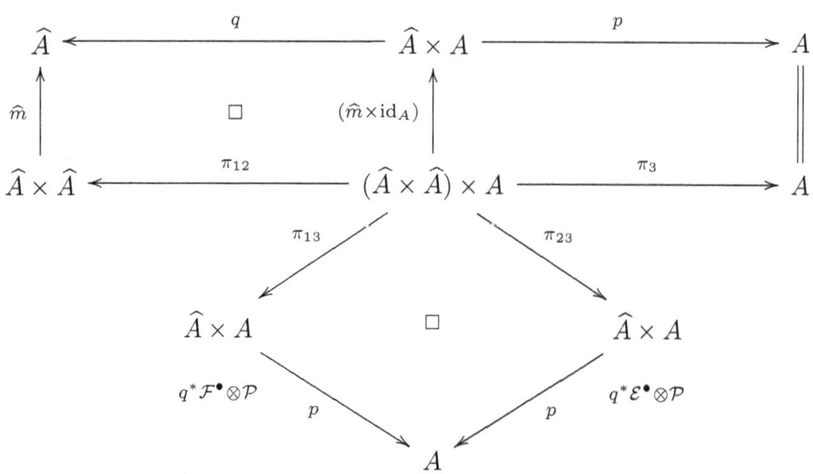

Then one works through the following series of functorial isomorphisms

$$\begin{aligned}
\Phi_{\mathcal{P}}(\mathcal{F}^\bullet * \mathcal{E}^\bullet) &= p_*\left(q^*\widehat{m}_*(\mathcal{F}^\bullet \boxtimes \mathcal{E}^\bullet)\otimes\mathcal{P}\right)\\
&\simeq p_*((\widehat{m}\times \mathrm{id}_A)_*\pi_{12}^*(\mathcal{F}^\bullet \boxtimes \mathcal{E}^\bullet)\otimes\mathcal{P}) \quad \text{(use flat base change)}\\
&\simeq p_*(\widehat{m}\times \mathrm{id}_A)_*\left(\pi_{12}^*(\mathcal{F}^\bullet \boxtimes \mathcal{E}^\bullet)\otimes(\widehat{m}\times \mathrm{id}_A)^*(\mathcal{P})\right)\\
&\simeq p_*(\widehat{m}\times \mathrm{id}_A)_*(\pi_{13}^*(q^*\mathcal{F}^\bullet\otimes\mathcal{P})\otimes\pi_{23}^*(q^*\mathcal{E}^\bullet\otimes\mathcal{P}))\\
&\qquad\qquad \text{(use } (\widehat{m}\times\mathrm{id}_A)^*\mathcal{P}\simeq\pi_{13}^*\mathcal{P}\otimes\pi_{23}^*\mathcal{P})\\
&\simeq p_*\pi_{13_*}\left(\pi_{13}^*(q^*\mathcal{F}^\bullet\otimes\mathcal{P})\otimes\pi_{23}^*(q^*\mathcal{E}^\bullet\otimes\mathcal{P})\right)\\
&\qquad\qquad \text{(use } p\circ(\widehat{m},\mathrm{id}_A)=\pi_3=p\circ\pi_{13})\\
&\simeq p_*(q^*\mathcal{F}^\bullet\otimes\mathcal{P}\otimes\pi_{13_*}\pi_{23}^*(q^*\mathcal{E}^\bullet\otimes\mathcal{P}))\\
&\simeq p_*(q^*\mathcal{F}^\bullet\otimes\mathcal{P}\otimes p^*p_*(q^*\mathcal{E}^\bullet\otimes\mathcal{P})) \quad \text{(use flat base change)}\\
&\simeq p_*(q^*\mathcal{F}^\bullet\otimes\mathcal{P})\otimes p_*(q^*\mathcal{E}^\bullet\otimes\mathcal{P}).
\end{aligned}$$

This proves the first assertion. The second one is deduced from it by applying $\Phi_{\mathcal{P}} : \mathrm{D^b}(A) \longrightarrow \mathrm{D^b}(\widehat{A})$ to both sides.

Indeed, if we set $\mathcal{G}^\bullet := \Phi_{\mathcal{P}}(\mathcal{F}^\bullet)$ and $\mathcal{H}^\bullet := \Phi_{\mathcal{P}}(\mathcal{E}^\bullet)$ and apply Proposition 9.19, then

$$\begin{aligned}
\Phi_{\mathcal{P}}(\mathcal{G}^\bullet\otimes\mathcal{H}^\bullet) &\simeq \Phi_{\mathcal{P}}(\Phi_{\mathcal{P}}(\mathcal{F}^\bullet)\otimes\Phi_{\mathcal{P}}(\mathcal{E}^\bullet))\\
&\simeq \Phi_{\mathcal{P}}(\Phi_{\mathcal{P}}(\mathcal{F}^\bullet * \mathcal{E}^\bullet))\\
&\simeq \iota^*(\mathcal{F}^\bullet * \mathcal{E}^\bullet)[-g].
\end{aligned}$$

Since $\Phi_{\mathcal{P}}(\mathcal{G}^\bullet)\simeq\iota^*\mathcal{F}^\bullet[-g]$ by Proposition 9.19 and similarly $\Phi_{\mathcal{P}}(\mathcal{H}^\bullet)\simeq\iota^*\mathcal{E}^\bullet[-g]$, one obtains

$$\Phi_{\mathcal{P}}(\mathcal{G}^\bullet\otimes\mathcal{H}^\bullet) = \iota^*\left(\iota^*\Phi_{\mathcal{P}}(\mathcal{G}^\bullet)[g] * \iota^*\Phi_{\mathcal{P}}(\mathcal{H}^\bullet)[g]\right)[-g] \simeq \left(\Phi_{\mathcal{P}}(\mathcal{G}^\bullet) * \Phi_{\mathcal{P}}(\mathcal{H}^\bullet)\right)[g],$$

where we use $m_*\circ(\iota\times\iota)^* = \iota^*\circ m_*$. As all $\mathcal{G}^\bullet$ and $\mathcal{H}^\bullet$ in $\mathrm{D^b}(A)$ are isomorphic to some object of the form $\Phi_{\mathcal{P}}(\mathcal{F}^\bullet)$, respectively $\Phi_{\mathcal{P}}(\mathcal{E}^\bullet)$, the second assertion follows. (We actually prove the assertion for $\Phi_{\mathcal{P}} : \mathrm{D^b}(A)\longrightarrow\mathrm{D^b}(\widehat{A})$, but the situation is symmetric.) □

Next let M be a non-degenerate line bundle on A, i.e. a line bundle such that $\varphi_M : A \longrightarrow \widehat{A}$ is an isogeny. Due to Lemma 9.16 this is the case if M is ample. The convolution with M can be expressed in terms of the Fourier–Mukai transform $\Phi_{\mathcal{P}} : \mathrm{D^b}(A)\longrightarrow\mathrm{D^b}(\widehat{A})$ as follows.

Lemma 9.28 *There exists a functorial isomorphism*

$$\mathcal{F}^\bullet * M \simeq M\otimes\varphi_M^*\Phi_{\mathcal{P}}(\iota^*\mathcal{F}^\bullet\otimes M).$$

Proof To prove the assertion, one introduces the morphisms $\xi : A\times A\longrightarrow A\times A$, $(a,b)\longmapsto(a,a+b)$ and $d : A\times A\longrightarrow A$, $(a,b)\longmapsto b-a$. They satisfy

$$d\circ\xi=\pi_2,\ \ \pi_1\circ\xi=\pi_1,\ \text{ and } \pi_2\circ\xi=m.$$

Using this (and the obvious $\xi_*\xi^* = \mathrm{id}$) one computes

$$\begin{aligned}\mathcal{F}^\bullet * M &\simeq m_*(\pi_1^*\mathcal{F}^\bullet \otimes \pi_2^* M) \sim m_*(\xi^*\pi_1^*\mathcal{F}^\bullet \otimes \xi^* d^* M)\\ &\simeq \pi_{2*}\xi_*\xi^*(\pi_1^*\mathcal{F}^\bullet \otimes d^*M) \simeq \pi_{2*}(\pi_1^*\mathcal{F}^\bullet \otimes d^* M).\end{aligned}$$

On the other hand, $m^*M \simeq (M \boxtimes M) \otimes (1 \times \varphi_M)^*\mathcal{P}$ by formula (9.8) and hence $d^*M \simeq (\iota^*M \boxtimes M) \otimes (\iota \times \varphi_M)^*\mathcal{P}$. Thus,

$$\begin{aligned}\mathcal{F}^\bullet * M &\simeq \pi_{2*}(\pi_1^*\mathcal{F}^\bullet \otimes (\iota^* M \boxtimes M) \otimes (\iota \times \varphi_M)^*\mathcal{P})\\ &\simeq \pi_{2*}(\pi_1^*(\iota^*M \otimes \mathcal{F}^\bullet) \otimes \pi_2^* M \otimes (\iota \times \varphi_M)^*\mathcal{P})\\ &\simeq \pi_{2*}((\iota \times \varphi_M)^*(\pi_1^*(M \otimes \iota^*\mathcal{F}^\bullet) \otimes \mathcal{P})) \otimes M\\ &\simeq \varphi_M^*\pi_{2*}(\pi_1^*(M \otimes \iota^*\mathcal{F}^\bullet) \otimes \mathcal{P}) \otimes M.\end{aligned}$$

For the last isomorphism we use $\pi_{2*} \circ (\iota \times \varphi_M)^* = \pi_{2*} \circ (\iota \times \mathrm{id})^* \circ (\mathrm{id} \times \varphi_M)^*$, $\pi_{2*} = \pi_{2*} \circ (\iota \times \mathrm{id})_*$, and $\pi_{2*} \circ (\mathrm{id} \times \varphi_M)^* = \varphi_M^* \circ \pi_{2*}$ (which follows from flat base change). □

Let (A, L) be a principally polarized abelian variety. Identifying A with $\widehat{A}$ via the induced isomorphism $\varphi_L : A \xrightarrow{\sim} \widehat{A}$ one may consider the Fourier–Mukai transform $\Phi_{\mathcal{P}}$ as an autoequivalence $\Phi : \mathrm{D^b}(A) \longrightarrow \mathrm{D^b}(A)$. To be precise, we let Φ be the Fourier–Mukai transform with kernel $(\mathrm{id} \times \varphi_L)^*\mathcal{P}$ or, equivalently

$$\Phi = \varphi_L^* \circ \Phi_{\mathcal{P}} : \mathrm{D^b}(A) \xrightarrow{\ \sim\ } \mathrm{D^b}(A).$$

Lemma 9.29 *Under these assumptions $\Phi(L) \simeq L^*$.*

Proof By the definition of φ (see formula (9.8)) one has $(\mathrm{id} \times \varphi_L)^*\mathcal{P} \simeq m^*L \otimes (L^* \boxtimes L^*)$. Hence, $\Phi(L) \simeq p_*(q^*L \otimes m^*L \otimes (L^* \boxtimes L^*)) \simeq p_*(m^*L) \otimes L^*$. Thus, it suffices to show that $p_*m^*L \simeq \mathcal{O}$.

Since (A, L) is a principally polarized abelian variety, one has a unique (up to scaling) non-trivial section $s : \mathcal{O} \longrightarrow L$ (see Definition 9.18). The pull-back yields a section $m^*s : \mathcal{O}_{A\times A} \longrightarrow m^*L$. Its restriction to any fibre $p^{-1}(a) \simeq A$ is a non-trivial section $m^*s|_{A\times\{a\}} : \mathcal{O}_A \longrightarrow t_a^*L$.

Since t_a^*L is again a principal polarization, the section is unique and, therefore, induces a bijection $H^0(p^{-1}(a), \mathcal{O}) \simeq H^0(p^{-1}(a), m^*L|_{p^{-1}(a)})$. Hence, $\mathcal{O} \simeq p_*\mathcal{O} \simeq R^0p_*m^*L$. This is enough, as the higher cohomology groups of t_a^*L are all trivial and hence $R^ip_*m^*L = 0$ for $i > 0$. □

Proposition 9.30 **(Mukai)** *Let (A, L) be a principally polarized abelian variety. If $\Phi : \mathrm{D^b}(A) \longrightarrow \mathrm{D^b}(A)$ denotes the Fourier–Mukai functor with kernel $(\mathrm{id} \times \varphi_L)^*\mathcal{P}$, then*

$$\Phi^4 \simeq [-2g] \quad \textit{and} \quad (L \otimes (\ \ ) \circ \Phi)^3 \simeq [-g].$$

Proof We start out with the following identity

$$\Phi^2 \simeq \iota^*[-g] \tag{9.9}$$

which follows from Proposition 9.19 and the commutativity of the diagram

$$\begin{array}{ccccc} & & \mathrm{D^b}(A) & & \\ & \overset{\varphi_L^*}{\nearrow} & & \overset{\Phi_{\mathcal{P}}}{\searrow} & \\ \mathrm{D^b}(\widehat{A}) & & \circlearrowleft & & \mathrm{D^b}(\widehat{A}) \\ & \underset{\Phi_{\mathcal{P}}}{\searrow} & & \underset{\varphi_L^{*\,-1}}{\nearrow} & \\ & & \mathrm{D^b}(A) & & \end{array}$$

The latter can be seen by writing down the kernels given by $(\varphi_L \times \mathrm{id})_*\mathcal{P}$, respectively $(\mathrm{id} \times \varphi_L)_*\mathcal{P}$, where $\mathcal{P}$ is considered as a line bundle on $A \times \widehat{A}$ in the first and as a line bundle on $\widehat{A} \times A$ in the second case. (Remember that the Poincaré bundle is universal in both directions.) The first assertion now follows immediately.

Let $\mathcal{F}^\bullet \in \mathrm{D^b}(A)$. Then, invoking Lemmas 9.27, 9.29, Exercise 9.26, and Proposition 9.19 one obtains

$$\begin{aligned} &\Phi(L \otimes \Phi(\mathcal{F}^\bullet)) \\ &\quad \simeq \varphi_L^* \Phi_{\mathcal{P}}(L \otimes \varphi_L^* \Phi_{\mathcal{P}}(\mathcal{F}^\bullet)) \\ &\quad \overset{9.27}{\simeq} \varphi_L^*(\Phi_{\mathcal{P}}(L) * \Phi_{\mathcal{P}} \varphi_L^* \Phi_{\mathcal{P}}(\mathcal{F}^\bullet))[g] \\ &\quad \overset{9.26}{\simeq} \varphi_L^* \Phi_{\mathcal{P}}(L) * \varphi_L^* \Phi_{\mathcal{P}} \varphi_L^* \Phi_{\mathcal{P}}(\mathcal{F}^\bullet)[g] \simeq \Phi(L) * \Phi^2(\mathcal{F}^\bullet)[g] \\ &\quad \overset{9.29}{\simeq} L^* * \Phi^2(\mathcal{F}^\bullet)[g] \overset{(9.9)}{\simeq} L^* * \iota^* \mathcal{F}^\bullet[-g][g] \simeq L^* * \iota^* \mathcal{F}^\bullet \\ &\quad \overset{9.28}{\simeq} L^* \otimes \varphi_{L^*}^* \Phi_{\mathcal{P}}(\mathcal{F}^\bullet \otimes L^*). \end{aligned}$$

Then applying Φ once more yields

$$\begin{aligned} &\Phi(L \otimes \Phi(L \otimes \Phi(\mathcal{F}^\bullet))) \\ &\quad \simeq \Phi(\varphi_{L^*}^* \Phi_{\mathcal{P}}(\mathcal{F}^\bullet \otimes L^*)) \\ &\quad \overset{9.15}{\simeq} \Phi(\iota^* \varphi_L^* \Phi_{\mathcal{P}}(\mathcal{F}^\bullet \otimes L^*)) \simeq \Phi(\iota^* \Phi(\mathcal{F}^\bullet \otimes L^*)) \\ &\quad \simeq \varphi_L^* \Phi_{\mathcal{P}} \iota^* \Phi(\mathcal{F}^\bullet \otimes L^*) \overset{9.11}{\simeq} \varphi_L^* \hat{\iota}^* \Phi_{\mathcal{P}} \Phi(\mathcal{F}^\bullet \otimes L^*) \\ &\quad \overset{9.15}{\simeq} \iota^* \varphi_L^* \Phi_{\mathcal{P}} \Phi(\mathcal{F}^\bullet \otimes L^*) \simeq \iota^* \Phi^2(\mathcal{F}^\bullet \otimes L^*) \\ &\quad \overset{(9.9)}{\simeq} \iota^* \iota^* (\mathcal{F}^\bullet \otimes L^*)[-g] \simeq \mathcal{F}^\bullet \otimes L^*[-g]. \end{aligned}$$

□

Remark 9.31 One interpretation of the above results is that modulo shifts, the elements of the group $\mathrm{Sl}_2(\mathbb{Z})$ act naturally as autoequivalences on the derived category $\mathrm{D^b}(A)$ of a principally polarized abelian variety. Indeed, $\mathrm{Sl}_2(\mathbb{Z})$ is generated by the two matrices

$$S = \begin{pmatrix} 0 & -1 \\ 1 & 0 \end{pmatrix} \quad \text{and} \quad T = \begin{pmatrix} 1 & 0 \\ -1 & 1 \end{pmatrix}$$

with the generating relations $S^4 = 1$ and $(T \circ S)^3 = 1$. In this picture

$$S \longleftrightarrow \Phi$$

$$T \longleftrightarrow L \otimes (\ \ ).$$

In Mukai's original paper Φ and $L \otimes (\ \ )$ are rather set into correspondence with the matrices $\begin{pmatrix} 0 & 1 \\ -1 & 0 \end{pmatrix}$, respectively $\begin{pmatrix} 1 & 1 \\ 0 & 1 \end{pmatrix}$. The one given above turns out to fit better in the context of the discussion of Section 9.5, where we shall describe the full group of autoequivalences of an arbitrary abelian variety.

Exercise 9.32 Give a direct proof of the induced cohomological identities $((L \otimes (\ \ ))^H \circ \Phi^H)^3 = (-1)^g$ and $(\Phi^H)^4 = \mathrm{id}$, e.g. for $g = 1$. The more adventurous reader, not afraid of signs (e.g. in Lemma 9.23), may attack the general case.

9.4 Derived equivalences of abelian varieties

The aim of this section and the next one is to give a geometric interpretation of any derived equivalence $\Phi : \mathrm{D^b}(A) \xrightarrow{\sim} \mathrm{D^b}(B)$ of two abelian varieties A and B. This will be done by associating to Φ a derived equivalence of $A \times \widehat{A}$ and $B \times \widehat{B}$, which, as it turns out, is in fact given by an isomorphism $A \times \widehat{A} \simeq B \times \widehat{B}$. The same approach also allows a full description of the group of autoequivalences. The construction is quite involved and somewhat miraculous, but it leads to a complete understanding of the situation. As will become clear immediately, almost nothing of what will be said here has a chance to generalize to other types of varieties. Essentially all techniques used in this section are due to Orlov and Polishchuk.

Let us begin with a closer examination of two very special equivalences. Firstly, consider the automorphism

$$\mu : A \times A \longrightarrow A \times A, \qquad (a_1, a_2) \longmapsto (a_1 + a_2, a_2)$$

defined for any abelian variety A. Secondly, we use the Poincaré bundle $\mathcal{P}$ on $\widehat{A} \times A$ to define the equivalences $\Phi_{\mathcal{P}} : \mathrm{D^b}(\widehat{A}) \xrightarrow{\sim} \mathrm{D^b}(A)$ and $\mathrm{id} \times \Phi_{\mathcal{P}} : \mathrm{D^b}(A \times \widehat{A}) \xrightarrow{\sim} \mathrm{D^b}(A \times A)$.

Examples 9.33 Let us try to understand the composition

$$\mu_* \circ (\mathrm{id} \times \Phi_{\mathcal{P}}) : \mathrm{D}^{\mathrm{b}}(A \times \widehat{A}) \longrightarrow \mathrm{D}^{\mathrm{b}}(A \times A)$$

by computing the image of $k(a) \boxtimes k(\alpha)$, where $a \in A$ and $\alpha \in \widehat{A}$ corresponds to the line bundle $L := \mathcal{P}_\alpha$ on A.

By definition, $(\mathrm{id} \times \Phi_{\mathcal{P}})(k(a) \boxtimes k(\alpha)) \simeq k(a) \boxtimes L$ and $\mu_*(k(a) \boxtimes L) = \mu_*(i_{a*}L)$, where $i_a : A \longrightarrow A \times A$, $a' \longmapsto (a, a')$. Since the image of the composition $\mu \circ i_a$, which is a closed embedding, is just the graph $\Gamma_{-a} := \Gamma_{t_{-a}}$ of t_{-a}, one obtains $\mu_*(k(a) \boxtimes L) \simeq (\mathcal{O} \boxtimes L) \otimes \mathcal{O}_{\Gamma_{-a}}$.

Therefore,

$$\mu_*(\mathrm{id} \times \Phi_{\mathcal{P}})(k(a) \boxtimes k(\alpha)) \simeq (\mathcal{O} \boxtimes L) \otimes \mathcal{O}_{\Gamma_{-a}}.$$

This is an object in $\mathrm{D}^{\mathrm{b}}(A \times A)$ and thus gives rise to a Fourier–Mukai transform $\mathrm{D}^{\mathrm{b}}(A) \longrightarrow \mathrm{D}^{\mathrm{b}}(A)$. Clearly, this is nothing but the composition $(L \otimes (\ )) \circ t_{-a*}$.

Let $\Phi_{\mathcal{E}} : \mathrm{D}^{\mathrm{b}}(A) \xrightarrow{\sim} \mathrm{D}^{\mathrm{b}}(B)$ be a derived equivalence between two abelian varieties given as a Fourier–Mukai transform with kernel $\mathcal{E}$ (which is uniquely determined!). By general results (see Proposition 4.1) we know that A and B are of the same dimension, say g. Also recall that $\mathcal{E}_{\mathrm{R}} = \mathcal{E}^\vee[g]$ (as the canonical bundle of an abelian variety is trivial) and that the induced Fourier–Mukai transform $\Phi_{\mathcal{E}_{\mathrm{R}}} : \mathrm{D}^{\mathrm{b}}(B) \xrightarrow{\sim} \mathrm{D}^{\mathrm{b}}(A)$ is quasi-inverse to $\Phi_{\mathcal{E}}$.

We will, however, be more interested in the induced Fourier–Mukai transform in the opposite direction $\Phi_{\mathcal{E}_{\mathrm{R}}} : \mathrm{D}^{\mathrm{b}}(A) \xrightarrow{\sim} \mathrm{D}^{\mathrm{b}}(B)$, which is also an equivalence (see the proof of Proposition 6.1 or Remark 7.7), and in the product equivalence $\Phi_{\mathcal{E}} \times \Phi_{\mathcal{E}_{\mathrm{R}}} : \mathrm{D}^{\mathrm{b}}(A \times A) \xrightarrow{\sim} \mathrm{D}^{\mathrm{b}}(B \times B)$ (cf. Exercise 5.20). Note that

$$\mathrm{D}^{\mathrm{b}}(A) \xrightarrow{\ \Phi_{\mathcal{E}_{\mathrm{R}}}\ } \mathrm{D}^{\mathrm{b}}(B) \xrightarrow{\ \Phi_{\mathcal{E}}\ } \mathrm{D}^{\mathrm{b}}(A)$$

is isomorphic to the identity.

Definition 9.34 *To the equivalence $\Phi_{\mathcal{E}}$ one associates the equivalence*

$$F_{\mathcal{E}} : \mathrm{D}^{\mathrm{b}}(A \times \widehat{A}) \xrightarrow{\ \sim\ } \mathrm{D}^{\mathrm{b}}(B \times \widehat{B})$$

given as the composition

$$\begin{array}{ccc} \mathrm{D^b}(A \times \widehat{A}) & \xrightarrow{\quad F_{\mathcal{E}} \quad} & \mathrm{D^b}(B \times \widehat{B}) \\ {\scriptstyle \mathrm{id}\times\Phi_{\mathcal{P}_A}} \downarrow & & \uparrow {\scriptstyle (\mathrm{id}\times\Phi_{\mathcal{P}_B})^{-1}} \\ \mathrm{D^b}(A \times A) & & \mathrm{D^b}(B \times B) \\ {\scriptstyle \mu_{A*}} \downarrow & & \uparrow {\scriptstyle \mu_B^*} \\ \mathrm{D^b}(A \times A) & \xrightarrow[\Phi_{\mathcal{E}}\times\Phi_{\mathcal{E}_{\mathrm{R}}}]{} & \mathrm{D^b}(B \times B). \end{array}$$

Remark 9.35 The key idea of everything that follows is that by passing from $\Phi_{\mathcal{E}} \times \Phi_{\mathcal{E}_{\mathrm{R}}} : \mathrm{D^b}(A \times A) \xrightarrow{\sim} \mathrm{D^b}(B \times B)$ to $F_{\mathcal{E}} : \mathrm{D^b}(A \times \widehat{A}) \xrightarrow{\sim} \mathrm{D^b}(B \times \widehat{B})$ the situation becomes, for some mysterious reason, more geometric.

Lemma 9.36 *The construction $\Phi_{\mathcal{E}} \longmapsto F_{\mathcal{E}}$ is compatible with composition, i.e. if $\Phi_{\mathcal{G}} : \mathrm{D^b}(A) \xrightarrow{\sim} \mathrm{D^b}(C)$ is the composition of*

$$\Phi_{\mathcal{E}} : \mathrm{D^b}(A) \xrightarrow{\ \sim\ } \mathrm{D^b}(B) \quad \textit{and} \quad \Phi_{\mathcal{F}} : \mathrm{D^b}(B) \xrightarrow{\ \sim\ } \mathrm{D^b}(C)$$

then $F_{\mathcal{G}} \simeq F_{\mathcal{F}} \circ F_{\mathcal{E}}$.

Proof By the very construction of F, the assertion follows immediately from $(\Phi_{\mathcal{F}} \times \Phi_{\mathcal{F}_{\mathrm{R}}}) \circ (\Phi_{\mathcal{E}} \times \Phi_{\mathcal{E}_{\mathrm{R}}}) = \Phi_{\mathcal{G}} \times \Phi_{\mathcal{G}_{\mathrm{R}}}$. The latter is a consequence of the assumption $\Phi_{\mathcal{G}} = \Phi_{\mathcal{F}} \circ \Phi_{\mathcal{E}}$ and Remark 5.11. □

As $F_{\mathcal{E}} = \mathrm{id}$ for $\Phi_{\mathcal{E}} = \mathrm{id}$, the lemma yields in particular the following

Corollary 9.37 *The map*

$$\mathrm{Aut}(\mathrm{D^b}(A)) \longrightarrow \mathrm{Aut}(\mathrm{D^b}(A \times \widehat{A})), \qquad \Phi_{\mathcal{E}} \longmapsto F_{\mathcal{E}}$$

is a group homomorphism. □

Examples 9.38 In the following we compute $F_{\mathcal{E}}$ explicitly in a few important cases.

i) Suppose

$$\Phi_{\mathcal{E}} = M \otimes (\ \ ) : \mathrm{D^b}(A) \xrightarrow{\ \sim\ } \mathrm{D^b}(A).$$

for some line bundle M on A. Thus, $\mathcal{E} \simeq \Delta_* M$. Clearly, the inverse functor is $M^* \otimes (\ )$ and hence $\mathcal{E}_{\mathrm{R}} \simeq \Delta_*(M^*)$. Thus, $\Phi_{\mathcal{E}} \times \Phi_{\mathcal{E}_{\mathrm{R}}} : \mathrm{D^b}(A \times A) \xrightarrow{\sim} \mathrm{D^b}(A \times A)$ is just $(M \boxtimes M^*) \otimes (\ )$.

In order to describe $F := F_{\mathcal{E}}$ in this case, we shall study $F(k(a) \boxtimes k(\alpha))$ for all closed points $(a, \alpha) \in A \times \widehat{A}$. It has been shown (see Example 9.33) that $\mu_*(\mathrm{id} \times \Phi_{\mathcal{P}})(k(a) \boxtimes k(\alpha)) \simeq (\mathcal{O} \boxtimes L) \otimes \mathcal{O}_{\Gamma_{-a}}$ (which is the Fourier–Mukai kernel for $(L \otimes (\ )) \circ t_{-a*}$). Here, $L = \mathcal{P}_\alpha$ as before. Hence,

$$(\Phi_{\mathcal{E}} \times \Phi_{\mathcal{E}_{\mathrm{R}}})(\mu_*(\mathrm{id} \times \Phi_{\mathcal{P}})(k(a) \boxtimes k(\alpha))) \simeq (M \boxtimes (M^* \otimes L)) \otimes \mathcal{O}_{\Gamma_{-a}}.$$

Indeed, by Exercise 5.13 $(\Phi_{\mathcal{E}} \times \Phi_{\mathcal{E}_{\mathrm{R}}})((\mathcal{O} \boxtimes L) \otimes \mathcal{O}_{\Gamma_{-a}})$ is the kernel of the equivalence $\mathcal{F}^\bullet \longmapsto (M^* \otimes L) \otimes t_{-a*}(M \otimes \mathcal{F}^\bullet)$.

If $M \in \mathrm{Pic}^0(A)$, then $t_{a*}M \simeq M$ and hence $(M \boxtimes M^*) \otimes \mathcal{O}_{\Gamma_{-a}} \simeq \mathcal{O}_{\Gamma_{-a}}$. Thus in this case one immediately finds $F(k(a) \boxtimes k(\alpha)) = k(a) \boxtimes k(\alpha)$ for all closed points $(a, \alpha) \in A \times \widehat{A}$.

Now, by Corollary 5.23 we conclude that under the hypothesis that $M \in \mathrm{Pic}^0(A)$ the equivalence F is isomorphic to $N \otimes (\ )$ for some line bundle N on $A \times \widehat{A}$. Thus, all that is needed to get a complete description of F is to determine this line bundle N, which is $N \simeq F(\mathcal{O}_{A \times \widehat{A}})$.

Using $p_*\mathcal{P} \simeq k(e)[-g]$ (see the proof of Proposition 9.19), one computes first

$$(\mathrm{id} \times \Phi_{\mathcal{P}})(\mathcal{O}) \simeq \mathcal{O} \boxtimes \Phi_{\mathcal{P}}(\mathcal{O}) \simeq \mathcal{O} \boxtimes k(e)[-g].$$

Next, $\mu_*(\mathcal{O} \boxtimes k(e)[-g]) \simeq \mathcal{O} \boxtimes k(e)[-g]$ and

$$(\Phi_{\mathcal{E}} \times \Phi_{\mathcal{E}_{\mathrm{R}}})(\mathcal{O} \boxtimes k(e)[-g]) \simeq M \boxtimes k(e)[-g].$$

Hence,

$$F(\mathcal{O}) \simeq (\mathrm{id} \times \Phi_{\mathcal{P}})^{-1}(\mu^*(\mathcal{O} \boxtimes k(e)[-g])) \simeq M \boxtimes \mathcal{O}.$$

Therefore, if $M \in \mathrm{Pic}^0(A)$, then the equivalence $F : \mathrm{D^b}(A \times \widehat{A}) \xrightarrow{\sim} \mathrm{D^b}(A \times \widehat{A})$ associated to $M \otimes (\ ) : \mathrm{D^b}(A) \xrightarrow{\sim} \mathrm{D^b}(A)$ is isomorphic to $(M \boxtimes \mathcal{O}) \otimes (\ )$.

We also mention here that for $M \notin \mathrm{Pic}^0(A)$ the induced equivalence is still of the form $F = (N \otimes (\ \ )) \circ f_*$ only f is not the identity. See Example 9.40.

ii) In the second example we consider the case of the autoequivalence

$$\Phi_{\mathcal{E}} = t_{a_0 *} : \mathrm{D^b}(A) \xrightarrow{\ \sim\ } \mathrm{D^b}(A)$$

for some point $a_0 \in A$. In particular, its kernel is $\mathcal{E} \simeq \mathcal{O}_{\Gamma_{a_0}}$ and also $\mathcal{E}_{\mathrm{R}} \simeq \mathcal{O}_{\Gamma_{a_0}}$. We follow the same strategy as above and try to compute the image $F(k(a) \boxtimes k(\alpha))$ for any closed point $(a, \alpha) \in A \times \widehat{A}$.

Using again Example 9.33 and $t_{a_0*}L \simeq L$ for $L = \mathcal{P}_\alpha \in \mathrm{Pic}^0(A)$, one finds

$$\begin{aligned}(\Phi_{\mathcal{E}} \times \Phi_{\mathcal{E}_{\mathrm{R}}})(\mu_*(\mathrm{id} \times \Phi_{\mathcal{P}})(k(a) \boxtimes k(\alpha))) &\simeq (t_{a_0*} \times t_{a_0*})((\mathcal{O} \boxtimes L) \otimes \mathcal{O}_{\Gamma_{-a}}) \\ &\simeq (\mathcal{O} \boxtimes L) \otimes \mathcal{O}_{\Gamma_{-a}}.\end{aligned}$$

So, once more F sends $k(a) \boxtimes k(\alpha)$ to itself and, therefore $F \simeq N \otimes (\ )$ for some line bundle N on $A \times \widehat{A}$. To compute N, one first shows

$$(t_{a_0*} \times t_{a_0*})(\mu_*(\mathrm{id} \times \Phi_{\mathcal{P}})(\mathcal{O})) \simeq (t_{a_0*} \times t_{a_0*})(\mathcal{O} \boxtimes k(e)[-g]) \\ \simeq \mathcal{O} \boxtimes k(a_0)[-g].$$

Since $\mu^*(\mathcal{O} \boxtimes k(a_0)[-g]) \simeq \mathcal{O} \boxtimes k(a_0)[-g]$ and $\Phi_{\mathcal{P}}(L_0^*) \simeq k(a_0)[-g]$, where L_0 is the line bundle on $\widehat{A}$ corresponding to $a_0 \in A$ (use the same arguments as in the proof of Proposition 9.19), one finds $F(\mathcal{O}) \simeq \mathcal{O} \boxtimes L_{a_0}^*$.

Therefore, the equivalence $F : \mathrm{D^b}(A \times \widehat{A}) \xrightarrow{\sim} \mathrm{D^b}(A \times \widehat{A})$ induced by $t_{a_0*} : \mathrm{D^b}(A) \xrightarrow{\sim} \mathrm{D^b}(A)$ is given by the tensor product with $\mathcal{O} \boxtimes L_0^*$, where $L_0 = \mathcal{P}|_{\{a_0\} \times \widehat{A}} \in \mathrm{Pic}^0(\widehat{A})$.

iii) Corollary 9.37 allows to combine i) and ii) as follows. Let $(a, \alpha) \in A \times \widehat{A}$ and

$$\Phi_{(a,\alpha)} := (L \otimes (\ )) \circ t_{a*} : \mathrm{D^b}(A) \xrightarrow{\sim} \mathrm{D^b}(A),$$

where $L = \mathcal{P}_\alpha$. The induced equivalence F is given by

$$F_{(a,\alpha)} \simeq L \boxtimes L_0^* \otimes (\ \ ) : \mathrm{D^b}(A \times \widehat{A}) \xrightarrow{\sim} \mathrm{D^b}(A \times \widehat{A}),$$

where $L_0 = \mathcal{P}|_{\{a\} \times \widehat{A}}$. In particular, its kernel is given by $\Delta_*(L \boxtimes L_0^*)$ where Δ is the diagonal embedding of $A \times \widehat{A}$.

iv) Consider a simple shift functor $\mathrm{D^b}(A) \xrightarrow{\sim} \mathrm{D^b}(A)$, $\mathcal{F}^\bullet \longmapsto \mathcal{F}^\bullet[n]$. Then the induced $F : \mathrm{D^b}(A \times A) \xrightarrow{\sim} \mathrm{D^b}(A \times A)$ is isomorphic to the identity. This follows from $(\mathcal{O}_\Delta[n])_\mathrm{R} \simeq \mathcal{O}_\Delta[-n]$.

v) For completeness sake we also consider the case of the equivalence

$$\Phi_{\mathcal{P}} : \mathrm{D^b}(A) \xrightarrow{\sim} \mathrm{D^b}(\widehat{A})$$

given by the Poincaré bundle. It is however much easier to view this as a consequence of the general results that will be proved below, so this will become Exercise 9.52. At any rate, the result is that F is given as $(\mathcal{P} \otimes (\ \ )) \circ f_{\mathcal{P}*}$, with $f_{\mathcal{P}} : A \times \widehat{A} \xrightarrow{\sim} \widehat{A} \times A$, $(a, \alpha) \longmapsto (-\alpha, a)$. (Note for the record that $\mathcal{P} \notin \mathrm{Pic}^0(\widehat{A} \times A)$.)

In all examples, we observed that $F_{\mathcal{E}}$ sends closed points to closed points and is therefore given, up to a twist by a line bundle, by an automorphism (see

Corollary 5.23). In fact, the automorphism in the examples i)–iv) was always the identity. More generally one has

Proposition 9.39 (Orlov) *Let* $\Phi_{\mathcal{E}} : \mathrm{D^b}(A) \xrightarrow{\sim} \mathrm{D^b}(B)$ *be an equivalence. Then the associated equivalence* $F_{\mathcal{E}} : \mathrm{D^b}(A \times \widehat{A}) \xrightarrow{\sim} \mathrm{D^b}(B \times \widehat{B})$ *is of the form*

$$F_{\mathcal{E}} \simeq (N_{\mathcal{E}} \otimes (\ )) \circ f_{\mathcal{E}*}$$

with $N_{\mathcal{E}} \in \mathrm{Pic}(B \times \widehat{B})$ *and* $f_{\mathcal{E}} : A \times \widehat{A} \xrightarrow{\sim} B \times \widehat{B}$ *an isomorphism of abelian varieties. See [93].*

Proof The proof is split into three steps.

i) In the first step we show $F_{\mathcal{E}}(k(e) \boxtimes k(\hat{e})) \simeq k(e) \boxtimes k(\hat{e})$. By Example 9.33 we have $\mu_*(\mathrm{id} \times \Phi_{\mathcal{P}})(k(e) \boxtimes k(\hat{e})) \simeq \mathcal{O}_{\Delta_A}$. Let $\mathcal{G} := (\Phi_{\mathcal{E}} \times \Phi_{\mathcal{E}_{\mathrm{R}}})(\mathcal{O}_{\Delta_A})$. Then $\Phi_{\mathcal{G}} : \mathrm{D^b}(B) \longrightarrow \mathrm{D^b}(B)$ is isomorphic to the composition

$$\mathrm{D^b}(B) \xrightarrow{\Phi_{\mathcal{E}}^{-1}} \mathrm{D^b}(A) \xrightarrow{\Phi_{\mathcal{O}_{\Delta_A}} = \mathrm{id}} \mathrm{D^b}(A) \xrightarrow{\Phi_{\mathcal{E}_{\mathrm{R}}}} \mathrm{D^b}(B)$$

(see Exercise 5.13). Hence, $\Phi_{\mathcal{G}} = \mathrm{id}$ and, therefore, $\mathcal{G} = \mathcal{O}_{\Delta_B}$. This is enough to conclude $F_{\mathcal{E}}(k(e) \boxtimes k(\hat{e})) \simeq k(e) \boxtimes k(\hat{e})$.

ii) Here we just recall a very general fact (see Corollary 6.14). If an equivalence $\Phi : \mathrm{D^b}(X) \xrightarrow{\sim} \mathrm{D^b}(Y)$ sends a closed point $x_0 \in X$ to a closed point $y_0 \in Y$, i.e. $\Phi(k(x_0)) \simeq k(y_0)$, then there exists an open neighbourhood $x_0 \in U \subset X$ such that for any closed point $x \in U$ there exists a closed point $y \in Y$ with $\Phi(k(x)) \simeq k(y)$.

iii) In this final step we show that F sends closed points to closed points. For this, we use the existence of the Zariski open subset $(e, \hat{e}) \in U \subset A \times \widehat{A}$ as in ii). Note that any other point $(a, \alpha) \in A \times \widehat{A}$ can be written as $(a, \alpha) = (a_1, \alpha_1) + (a_2, \alpha_2)$ with $(a_i, \alpha_i) \in U$. (This is a general fact for open non-empty subsets U of an abelian variety C. If $x \in C \setminus U$, then the image of the open immersion $t_x : \iota(U) \longrightarrow C$ meets U. Hence $x - y_1 = y_2$ for certain $y_1, y_2 \in U$.) By definition of $U \subset A \times \widehat{A}$, there exist points $(b_i, \beta_i) \in B \times \widehat{B}$, $i = 1, 2$, with $F_{\mathcal{E}}(k(a_i) \boxtimes k(\alpha_i)) = k(b_i) \boxtimes k(\beta_i)$. We denote the line bundles corresponding to β_i, $i = 1, 2$, by M_i.

Then define

$$\begin{aligned} \mathcal{G} :&= (\Phi_{\mathcal{E}} \times \Phi_{\mathcal{E}_{\mathrm{R}}})(\mu_{A*}(\mathrm{id} \times \Phi_{\mathcal{P}_A})(k(a) \boxtimes k(\alpha))) \\ &\simeq (\Phi_{\mathcal{E}} \times \Phi_{\mathcal{E}_{\mathrm{R}}})((\mathcal{O} \boxtimes L) \otimes \mathcal{O}_{\Gamma_{-a}}). \end{aligned}$$

As before $L := \mathcal{P}_\alpha$ which can now be written as $L \simeq L_1 \otimes L_2$ with $L_i = \mathcal{P}|_{A \times \{\alpha_i\}}$. Similarly, we let $M_i \in \mathrm{Pic}(B)$ correspond to $\beta_i \in B$.

The induced Fourier–Mukai transform $\Phi_{\mathcal{G}} : \mathrm{D^b}(B) \longrightarrow \mathrm{D^b}(B)$ is isomorphic to the composition

$$\mathrm{D^b}(B) \xrightarrow{\Phi_{\mathcal{E}}} \mathrm{D^b}(A) \xrightarrow{(L \otimes (\ )) \circ t_{-a*}} \mathrm{D^b}(A) \xrightarrow{\Phi_{\mathcal{E}_{\mathrm{R}}}} \mathrm{D^b}(B)$$

(see Exercise 5.13). Hence,

$$\begin{aligned}\Phi_{\mathcal{G}} &= \Phi_{\mathcal{E}_{\mathrm{R}}} \circ (L_1 \otimes (\ )) \circ t_{-a_1*} \circ (L_2 \otimes (\ )) \circ t_{-a_2*} \circ \Phi_{\mathcal{E}} \\ &= \Phi_{\mathcal{E}_{\mathrm{R}}} \circ (L_1 \otimes (\ )) \circ t_{-a_1*} \circ \Phi_{\mathcal{E}} \circ \Phi_{\mathcal{E}_{\mathrm{R}}} \circ (L_2 \otimes (\ )) \circ t_{-a_2*} \circ \Phi_{\mathcal{E}} \\ &= (M_1 \otimes (\ )) \circ t_{-b_1*} \circ (M_2 \otimes (\ )) \circ t_{-b_2*} \\ &= ((M_1 \otimes M_2) \otimes (\ )) \circ t_{-b_1-b_2*}\end{aligned}$$

(Don't get confused with the directions of the Fourier–Mukai functor. Write down the above diagram to make sure, e.g. $\Phi_{\mathcal{E}}$ means $\Phi_{\mathcal{E}} : \mathrm{D^b}(B) \longrightarrow \mathrm{D^b}(A)$.)

This is enough to conclude

$$F_{\mathcal{E}}(k(a) \boxtimes k(\alpha)) = k(b_1 + b_2) \boxtimes k(\beta_1 + \beta_2).$$

Therefore, $F_{\mathcal{E}}$ is up to twist by a line bundle $N_{\mathcal{E}}$ on $B \times \widehat{B}$ induced by an isomorphism $f_{\mathcal{E}} : A \times \widehat{A} \xrightarrow{\sim} B \times \widehat{B}$. In fact, the above calculation indeed shows that $f_{\mathcal{E}}$ respects the group structure. □

Examples 9.40 Coming back to Example 9.38, i) one finds that for $M \notin \mathrm{Pic}^0(A)$ the induced $F_{\Delta_*M} = (f_{\Delta_*M}, N_{\Delta_*M})$ satisfies $f_{\Delta_*M} \neq \mathrm{id}$.

Exercise 9.41 Consider the composition $\Phi_{\mathcal{G}} = \Phi_{\mathcal{E}} \circ \Phi_{\mathcal{F}}$ of two equivalences

$$\Phi_{\mathcal{F}} : \mathrm{D^b}(A) \xrightarrow{\sim} \mathrm{D^b}(B) \quad \text{and} \quad \Phi_{\mathcal{E}} : \mathrm{D^b}(B) \xrightarrow{\sim} \mathrm{D^b}(C).$$

Show that for the induced $(N_{\mathcal{F}}, f_{\mathcal{E}})$, $(N_{\mathcal{E}}, f_{\mathcal{E}})$, and $(N_{\mathcal{G}}, f_{\mathcal{G}})$ one has

$$f_{\mathcal{G}} = f_{\mathcal{E}} \circ f_{\mathcal{F}} \ \text{ and } \ N_{\mathcal{G}} \simeq N_{\mathcal{E}} \otimes f_{\mathcal{E}*}N_{\mathcal{F}}.$$

Here are a few immediate consequences of the proposition. The first one roughly says that the number of Fourier–Mukai partners of an abelian variety is finite.

Corollary 9.42 *To any abelian variety A there exist, up to isomorphisms, only a finite number of derived equivalent abelian varieties B.*

Proof If B is an abelian variety with $\mathrm{D^b}(A) \simeq \mathrm{D^b}(B)$, then $A \times \widehat{A} \simeq B \times \widehat{B}$. In particular, any such abelian variety B is a direct factor of $A \times \widehat{A}$. A standard argument shows that any abelian variety has, up to the action of automorphisms, only a finite number of direct factors (see [78, V, 18.7]). □

The second one is a generalization of Corollary 9.24.

Corollary 9.43 *Let* $\Phi_{\mathcal{E}} : \mathrm{D^b}(A) \xrightarrow{\sim} \mathrm{D^b}(B)$ *be a derived equivalence of abelian varieties. Then the induced cohomological Fourier–Mukai transform defines an isomorphism of the integral(!) cohomology*

$$\Phi_{\mathcal{E}}^H : H^*(A, \mathbb{Z}) \xrightarrow{\sim} H^*(B, \mathbb{Z}).$$

Proof Imitating what has been said about the powers $\mathrm{c}_1(\mathcal{P})^n$, one first observes that the Chern character $\mathrm{ch}(L)$ of a line bundle L on an abelian variety is always integral.

Thus, the induced equivalence $F_{\mathcal{E}} : \mathrm{D^b}(A \times \widehat{A}) \xrightarrow{\sim} \mathrm{D^b}(B \times \widehat{B})$, which is a composition of $f_{\mathcal{E}*}$ and the tensor product with $N_{\mathcal{E}}$, yields an integral isomorphism

$$F_{\mathcal{E}}^H : H^*(A \times \widehat{A}, \mathbb{Z}) \xrightarrow{\sim} H^*(B \times \widehat{B}, \mathbb{Z}).$$

Using Corollary 9.24 and the diagram in Definition 9.34 that defines $F_{\mathcal{E}}$, this implies that also

$$\Phi_{\mathcal{E} \boxtimes \mathcal{E}_{\mathrm{R}}}^H = \Phi_{\mathcal{E}}^H \otimes \Phi_{\mathcal{E}_{\mathrm{R}}}^H : H^*(A \times A, \mathbb{Z}) \xrightarrow{\sim} H^*(B \times B, \mathbb{Z}),$$

which clearly suffices to conclude. □

The construction of the morphism $f_{\mathcal{E}}$ associated to any equivalence $\Phi_{\mathcal{E}}$ seems rather mysterious. Here is another view of it, which might help to understand it. We shall use the notation

$$\Phi_{(a,\alpha)} := (L \otimes (\ \ )) \circ t_{a*} : \mathrm{D^b}(A) \xrightarrow{\sim} \mathrm{D^b}(A)$$

for a closed point $(a, \alpha) \in A \times \widehat{A}$ and similarly

$$\Phi_{(b,\beta)} : \mathrm{D^b}(B) \xrightarrow{\sim} \mathrm{D^b}(B)$$

for a closed point $(b, \beta) \in B \times \widehat{B}$. The induced equivalences $F_{(a,\alpha)}$ have been computed in Examples 9.38.

Corollary 9.44 *Suppose* $\Phi_{\mathcal{E}} : \mathrm{D^b}(A) \xrightarrow{\sim} \mathrm{D^b}(B)$ *is an equivalence and the induced isomorphism is* $f_{\mathcal{E}} : A \times \widehat{A} \xrightarrow{\sim} B \times \widehat{B}$. *Then* $f_{\mathcal{E}}(a, \alpha) = (b, \beta)$ *if and only if*

$$\Phi_{(b,\beta)} \circ \Phi_{\mathcal{E}} \simeq \Phi_{\mathcal{E}} \circ \Phi_{(a,\alpha)} : \mathrm{D^b}(A) \xrightarrow{\sim} \mathrm{D^b}(B)$$

Proof Due to Example 9.33, $\mu_* \circ (\mathrm{id} \times \Phi_{\mathcal{P}})$ sends a closed point (a, α) to $(\mathcal{O} \boxtimes L) \otimes \mathcal{O}_{\Gamma_{-a}}$. A similar calculation applies to (b, β).

Thus, $F_{\mathcal{E}}(k(a, \alpha)) \simeq k(b, \beta)$ (or, equivalently $f_{\mathcal{E}}(a, \alpha) = (b, \beta)$) if and only if

$$(\Phi_{\mathcal{E}} \times \Phi_{\mathcal{E}_{\mathrm{R}}})((\mathcal{O} \boxtimes L) \otimes \mathcal{O}_{\Gamma_{-a}}) \simeq (\mathcal{O} \boxtimes M) \otimes \mathcal{O}_{\Gamma_{-b}}, \qquad (9.10)$$

where $M = \mathcal{P}_\beta$. In order to apply Exercise 5.13, we swap the two factors in (9.10) and rewrite it as

$$(\Phi_{\mathcal{E}_{\mathrm{R}}} \times \Phi_{\mathcal{E}})((L \boxtimes \mathcal{O}) \otimes \mathcal{O}_{\Gamma_a}) \simeq (M \boxtimes \mathcal{O}) \otimes \mathcal{O}_{\Gamma_b}. \qquad (9.11)$$

Since $\Phi_{(a,\alpha)}$ is the Fourier–Mukai transform with kernel $(L \boxtimes \mathcal{O}) \otimes \mathcal{O}_{\Gamma_a}$, Exercise 5.13 now says that (9.11) is equivalent to the commutativity of

$$\begin{array}{ccc} \mathrm{D}^{\mathrm{b}}(A) & \xleftarrow{\Phi_{\mathcal{E}_{\mathrm{R}}}} & \mathrm{D}^{\mathrm{b}}(B) \\ {\scriptstyle \Phi_{(a,\alpha)}}\downarrow & & \downarrow{\scriptstyle \Phi_{(b,\beta)}} \\ \mathrm{D}^{\mathrm{b}}(A) & \xrightarrow[\Phi_{\mathcal{E}}]{} & \mathrm{D}^{\mathrm{b}}(B). \end{array}$$

With $\Phi_{\mathcal{E}_{\mathrm{R}}} \simeq \Phi_{\mathcal{E}}^{-1}$ this is just saying $\Phi_{(b,\beta)} \circ \Phi_{\mathcal{E}} \simeq \Phi_{\mathcal{E}} \circ \Phi_{(a,\alpha)}$. □

The following digression shows that behind Proposition 9.39 there is a general principle. In some sense, to be made precise below, Proposition 9.39 holds for arbitrary projective varieties. The following remarks sketch the principal ideas of an unpublished result of R. Rouquier.

Suppose

$$F : \mathrm{D}^{\mathrm{b}}(X) \xrightarrow{\sim} \mathrm{D}^{\mathrm{b}}(Y)$$

is an equivalence. Clearly, F induces an isomorphism between the groups of autoequivalences

$$\mathrm{Aut}(\mathrm{D}^{\mathrm{b}}(X)) \xrightarrow{\sim} \mathrm{Aut}(\mathrm{D}^{\mathrm{b}}(Y)), \qquad \Phi \longmapsto F^*\Phi := F \circ \Phi \circ F^{-1}.$$

Equivalently, this is given by the following diagram

$$\begin{array}{ccc} \mathrm{D}^{\mathrm{b}}(X) & \xleftarrow{F^{-1}} & \mathrm{D}^{\mathrm{b}}(Y) \\ {\scriptstyle \Phi}\downarrow & & \downarrow{\scriptstyle F^*\Phi} \\ \mathrm{D}^{\mathrm{b}}(X) & \xrightarrow[F]{} & \mathrm{D}^{\mathrm{b}}(Y). \end{array}$$

Thinking of F as a Fourier–Mukai transform $\Phi_{\mathcal{E}}$, the map $\Phi_{\mathcal{R}} \longmapsto F^*\Phi_{\mathcal{R}} = F \circ \Phi_{\mathcal{R}} \circ F^{-1}$ is on the level of Fourier–Mukai kernels described by

$$\mathcal{R} \longmapsto F^*\mathcal{R} := (\Phi_{\mathcal{E}_{\mathrm{R}}} \times \Phi_{\mathcal{E}})(\mathcal{R}).$$

Here, $(\Phi_{\mathcal{E}_{\mathrm{R}}} \times \Phi_{\mathcal{E}}) : \mathrm{D^b}(X \times X) \longrightarrow \mathrm{D^b}(Y \times Y)$ is the Fourier–Mukai transform with kernel $\mathcal{E}_{\mathrm{R}} \boxtimes \mathcal{E} \in \mathrm{D^b}((X \times Y) \times (X \times Y))$. See Exercise 5.13.

We will be interested in the 'neighbourhood' of the identity id : $\mathrm{D^b}(X) \xrightarrow{\sim} \mathrm{D^b}(X)$. To this end we consider the semi-direct product $\mathrm{Aut}(X) \ltimes \mathrm{Pic}(X)$ as a subgroup of $\mathrm{Aut}(\mathrm{D^b}(X))$ by associating to $(\varphi, L) \in \mathrm{Aut}(X) \ltimes \mathrm{Pic}(X)$ the equivalence $\Phi_{(\varphi,L)} := (L \otimes (\ \ )) \circ \varphi_*$, the kernel of which is of the form $(\mathrm{id} \times \varphi)_* L \in \mathrm{D^b}(X \times X)$.

Clearly, $F^*\Phi_{(\mathrm{id}_X,\mathcal{O})} \simeq \Phi_{(\mathrm{id}_Y,\mathcal{O}_Y)}$. In other words, $F^*\mathcal{O}_\Delta \simeq \mathcal{O}_\Delta$. In particular, the image of the kernel defining $\mathrm{id} = \Phi_{(\mathrm{id}_X,\mathcal{O})}$ is isomorphic to a line bundle concentrated on the graph of an automorphism of Y. This then will be true for small deformations of $(\mathrm{id}_X, \mathcal{O})$, i.e. for any (φ, L) contained in a small neighbourhood of $\mathrm{Aut}(X) \ltimes \mathrm{Pic}(X)$, which is an algebraic group, the image $F^*\Phi_{(\varphi,L)}$ will be again of the form $\Phi_{(\psi,M)}$ with $(\psi, M) \in \mathrm{Aut}(Y) \ltimes \mathrm{Pic}(Y)$.

As any open neighbourhood of the identity $\mathrm{id}_X \in \mathrm{Aut}(X) \ltimes \mathrm{Pic}(X)$ generates the connected component $\mathrm{Aut}^0(X) \ltimes \mathrm{Pic}^0(X)$, the map $\Phi \longmapsto F^*\Phi$ induces a map

$$\mathrm{Aut}^0(X) \ltimes \mathrm{Pic}^0(X) \longrightarrow \mathrm{Aut}^0(Y) \ltimes \mathrm{Pic}^0(Y).$$

Using the same argument for the inverse F^{-1}, one shows that it is in fact an isomorphism.

These are the main ideas to prove the following result, whose complete proof needs to address a few more technical details.

Proposition 9.45 (Rouquier) *Any equivalence* $F : \mathrm{D^b}(X) \xrightarrow{\sim} \mathrm{D^b}(Y)$ *induces an isomorphism of algebraic groups*

$$F^* : \mathrm{Aut}^0(X) \ltimes \mathrm{Pic}^0(X) \xrightarrow{\sim} \mathrm{Aut}^0(Y) \ltimes \mathrm{Pic}^0(Y). \tag{9.12}$$

For abelian varieties A and B as considered earlier, this is exactly what is expressed by Proposition 9.39. Indeed, one has $A \simeq \mathrm{Aut}^0(A)$ via $a \longmapsto t_{a*}$ and $\widehat{A} \simeq \mathrm{Pic}^0(A)$. As Corollary 9.44 shows, the isomorphism $f_{\mathcal{E}} : A \times \widehat{A} \xrightarrow{\sim} B \times \widehat{B}$ induced by any equivalence $F := \Phi_{\mathcal{E}} : \mathrm{D^b}(A) \xrightarrow{\sim} \mathrm{D^b}(B)$ is nothing but F^*, i.e.

$$f_{\mathcal{E}} \simeq F^* : A \times \widehat{A} = \mathrm{Aut}^0(A) \ltimes \mathrm{Pic}^0(A) \xrightarrow{\sim} \mathrm{Aut}^0(B) \ltimes \mathrm{Pic}^0(B) = B \times \widehat{B}.$$

One last remark on the general case, the isomorphism (9.12) can be seen as the geometric realization of the isomorphism of the first Hochschild cohomology (see the discussion on p. 140)

$$HH^1(X) = H^0(X, \mathcal{T}_X) \oplus H^1(X, \mathcal{O}_X) \simeq H^0(Y, \mathcal{T}_Y) \oplus H^1(Y, \mathcal{O}_Y) = HH^1(Y)$$

induced by the equivalence. Indeed, the tangent spaces of $\mathrm{Aut}(X)$ and $\mathrm{Pic}(X)$ are nothing but $H^0(X, \mathcal{T}_X)$, respectively $H^1(X, \mathcal{O}_X)$.

Let us now pursue our discussion for abelian varieties. As we will show next, this interpretation of $f_{\mathcal{E}}$ provided by Corollary 9.44 will lead to a characterization of all isomorphisms $f_{\mathcal{E}}$ that might occur.

Any isomorphism $f : A \times \widehat{A} \xrightarrow{\sim} B \times \widehat{B}$ can be written as $f = \begin{pmatrix} f_1 & f_2 \\ f_3 & f_4 \end{pmatrix}$ and one associates to it the isomorphism $\tilde{f} : B \times \widehat{B} \xrightarrow{\sim} A \times \widehat{A}$

$$\tilde{f} =: \begin{pmatrix} \widehat{f_4} & -\widehat{f_2} \\ -\widehat{f_3} & \widehat{f_1} \end{pmatrix}.$$

(Note that we tacitly use the isomorphisms $A \simeq \widehat{\widehat{A}}$ and $B \simeq \widehat{\widehat{B}}$.)

The following subgroup of isomorphisms $A \times \widehat{A} \xrightarrow{\sim} B \times \widehat{B}$ was independently studied by Mukai and Polishchuk.

Definition 9.46 *By $U(A \times \widehat{A}, B \times \widehat{B})$ one denotes the subgroup of isomorphisms $f : A \times \widehat{A} \xrightarrow{\sim} B \times \widehat{B}$ with $\tilde{f} = f^{-1}$.*

Corollary 9.47 *The isomorphism $f_{\mathcal{E}} : A \times \widehat{A} \xrightarrow{\sim} B \times \widehat{B}$ associated to an equivalence $\Phi_{\mathcal{E}} : \mathrm{D^b}(A) \xrightarrow{\sim} \mathrm{D^b}(B)$ is contained in $U(A \times \widehat{A}, B \times \widehat{B})$.*

Proof If we denote as before the autoequivalence of $\mathrm{D^b}(A \times \widehat{A})$ induced by $\Phi_{(a,\alpha)}$ by $F_{(a,\alpha)}$, then Corollary 9.37 and Corollary 9.44 imply

$$F_{\mathcal{E}} \circ F_{(a,\alpha)} = F_{f_{\mathcal{E}}(a,\alpha)} \circ F_{\mathcal{E}}. \tag{9.13}$$

The closed point $a \in A$ corresponds to a line bundle L_0 on $\widehat{A}$ and $\alpha \in \widehat{A}$ to a line bundle L on A. Similarly, $(b, \beta) = f_{\mathcal{E}}(a, \alpha)$ gives rise to $(M, M_0) \in \mathrm{Pic}(B \times \widehat{B})$. Thus, (9.13) reads (see Examples 9.38, iii))

$$(N_{\mathcal{E}} \otimes (\ \ )) \circ f_{\mathcal{E}*} \circ (L \boxtimes L_0^* \otimes (\ \ )) \simeq (M \boxtimes M_0^* \otimes (\ \ )) \circ (N_{\mathcal{E}} \otimes (\ \ )) \circ f_{\mathcal{E}*}$$

or, equivalently, $f_{\mathcal{E}*}(L \boxtimes L_0^*) \simeq M \boxtimes M_0^*$. The latter translates to $\widehat{f_{\mathcal{E}}}(\beta, -b) = (\alpha, -a)$ or further to $\tilde{f}_{\mathcal{E}}(b, \beta) = (a, \alpha)$. As $(b, \beta) = f_{\mathcal{E}}(a, \alpha)$, this proves $\tilde{f}_{\mathcal{E}} = f_{\mathcal{E}}^{-1}$ on all closed points which is enough. □

The following result is originally due to Polishchuk. An alternative proof was given by Orlov.

Proposition 9.48 (Orlov, Polishchuk) *Consider two abelian varieties A and B. Any $f \in U(A \times \widehat{A}, B \times \widehat{B})$ is of the form $f = f_{\mathcal{E}}$ for some equivalence $\Phi_{\mathcal{E}} : \mathrm{D^b}(A) \xrightarrow{\sim} \mathrm{D^b}(B)$. See [93, 96].*

We refrain from giving any indications of the proof. However, looking at Lemma 9.51 below one gets the impression that once f and an appropriate line bundle N are given, a potential kernel $\mathcal{E}$ can be constructed easily, well, at least the modification $\mathcal{E}^\vee(e_A, e_B) \otimes \mathcal{E}$.

Orlov's proof makes extensive use of semi-homogenous vector bundles on abelian varieties. Polishchuk's approach is explained in [96, Ch.15]. Both would lead us too far astray.

The next corollary summarizes the discussion of this section and gives a complete answer to the question of when two abelian varieties are derived equivalent.

Corollary 9.49 *Two abelian varieties A and B define equivalent derived categories $\mathrm{D^b}(A)$ and $\mathrm{D^b}(B)$ if and only if there exists an isomorphism $f : A \times \widehat{A} \xrightarrow{\sim} B \times \widehat{B}$ with $\tilde{f} = f^{-1}$:*

$$\mathrm{D^b}(A) \simeq \mathrm{D^b}(B) \quad \Longleftrightarrow \quad U(A \times \widehat{A}, B \times \widehat{B}) \neq \varnothing.$$

The corollary can be rephrased in terms of Hodge structures of weight one. An abelian variety A is determined by its weight-one Hodge structure on $H^1(A, \mathbb{Z})$. This applies also to $A \times \widehat{A}$, which corresponds to the induced weight-one Hodge structure on $H^1(A \times \widehat{A}, \mathbb{Z}) = H^1(A, \mathbb{Z}) \oplus H^1(A, \mathbb{Z})^*$.

Moreover, the lattice $H^1(A, \mathbb{Z}) \oplus H^1(A, \mathbb{Z})^*$ comes with a natural quadratic form given by the dual pairing:

$$q_A((a, \alpha)) := 2\alpha(a).$$

Corollary 9.50 *Two abelian varieties A and B are derived equivalent if and only if there exists a Hodge isometry*

$$f : H^1(A \times \widehat{A}, \mathbb{Z}) \xrightarrow{\ \sim\ } H^1(B \times \widehat{B}, \mathbb{Z}).$$

See [41].

Proof It clearly suffices to show that the condition f being an isometry, i.e. $q_B(f(a, \alpha)) = q_A((a, \alpha))$, is equivalent to the condition $\tilde{f} = f^{-1}$.

After choosing a basis for $H^1(A, \mathbb{Z})$ and $H^1(B, \mathbb{Z})$, the quadratic forms q_A and q_B correspond to the matrix $\begin{pmatrix} 0 & 1 \\ 1 & 0 \end{pmatrix}$. Thus, $f = \begin{pmatrix} f_1 & f_2 \\ f_3 & f_4 \end{pmatrix}$ is an isometry if and only if

$$\begin{pmatrix} f_1^t & f_3^t \\ f_1^t & f_4^t \end{pmatrix} \begin{pmatrix} 0 & 1 \\ 1 & 0 \end{pmatrix} \begin{pmatrix} f_1 & f_2 \\ f_3 & f_4 \end{pmatrix} = \begin{pmatrix} 0 & 1 \\ 1 & 0 \end{pmatrix}.$$

The assertion follows immediately from the relations

$$f_1^t = \widehat{f}_1, \ f_4^t = \widehat{f}_4, \ f_2^t = -\widehat{f}_2, \quad \text{and} \quad f_3^t = -\widehat{f}_3.$$

Note that $\widehat{f}_2$ is actually defined as $\widehat{f}_2 : \widehat{B} \longrightarrow \widehat{\widehat{A}} \xrightarrow{\sim} A$, where the isomorphism is given by the Poincaré bundle. Then use Remark 9.12. □

In the remainder we will discuss yet another result of Orlov saying that any derived equivalence between abelian varieties is a Fourier–Mukai transform with a shifted sheaf as a Fourier–Mukai kernel. This part is independent of the rest of this section, but it will come in at a crucial step in the next one where we build upon the above discussion to determine the group of autoequivalences of the derived category of an abelian variety.

Let us start with the following technical result. We let $\Phi_{\mathcal{E}} : \mathrm{D^b}(A) \xrightarrow{\sim} \mathrm{D^b}(B)$ be an equivalence and $F_{\mathcal{E}} : \mathrm{D^b}(A \times \widehat{A}) \xrightarrow{\sim} \mathrm{D^b}(B \times \widehat{B})$ be the induced equivalence as introduced above. Then $F_{\mathcal{E}}$ can be described as a Fourier–Mukai transform $\Phi_{\mathcal{I}(\mathcal{E})}$, where $\mathcal{I}(\mathcal{E})$ is a line bundle $N_{\mathcal{E}}$ on the graph of a certain automorphism $f_{\mathcal{E}} : A \times \widehat{A} \xrightarrow{\sim} B \times \widehat{B}$.

Lemma 9.51 *If $\pi : A \times \widehat{A} \times B \times \widehat{B} \longrightarrow A \times B$ is the natural projection, then*

$$K_{\mathcal{E}} := \pi_* \mathcal{I}(\mathcal{E}) \simeq \mathcal{E}^{\vee}(e_A, e_B) \otimes \mathcal{E}.$$

Here, $\mathcal{E}^{\vee}(e_A, e_B) = (e_A, e_B)^* \mathcal{E}^{\vee}$ is the fibre of the complex $\mathcal{E}^{\vee}$ in the origin.

Proof As $F_{\mathcal{E}}$ has been introduced as a composition of a number of Fourier–Mukai transforms, its kernel $\mathcal{I}(\mathcal{E})$ can be described by the methods of Section 5.1 as a direct image of the tensor products $\mathcal{R}$ of the various kernels. Instead of writing this out, let us try to explain this by the following picture.

We shall use $\Gamma_{\mu_A} = \{(a_1, a_2, a_1 + a_2, a_2)\}$ and $\Gamma_{\mu_B} = \{(b_1 + b_2, b_2, b_1, b_2)\}$.

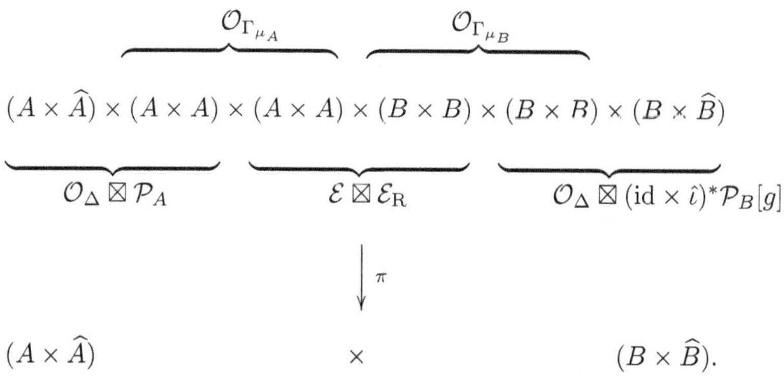

The projection π can be decomposed as follows

$$\begin{array}{c}
(A\times\widehat{A})\times(A\times A)\times(A\times A)\times(B\times B)\times(B\times B)\times(B\times\widehat{B})\\
\Big\downarrow \text{i)}\\
A\times(A\times A)\times(A\times A)\times(B\times B)\times(B\times B)\times B\\
\Big\downarrow \text{ii)}\\
A\qquad\times\qquad(A\times A)\times(B\times B)\qquad\times\qquad B\\
\Big\downarrow \text{iii)}\\
A\qquad\qquad\qquad\qquad\times\qquad\qquad\qquad\qquad B.
\end{array}$$

The direct image of $\mathcal{R}$ under i) yields

$$\underbrace{(A\quad)\times(A\times \rlap{$\overbrace{}^{\mathcal{O}_{\Gamma_{\mu_A}}}$}A)}_{\mathcal{O}_\Delta\boxtimes k(e_A)[-g]}\times\underbrace{(A\times A)\times(B\times B)}_{\mathcal{E}\boxtimes\mathcal{E}_{\mathrm{R}}}\times\underbrace{\llap{$\overbrace{}^{\mathcal{O}_{\Gamma_{\mu_B}}}$}(B\times B)\times(B\quad)}_{\mathcal{O}_\Delta\boxtimes k(e_B)[-g][g]}$$

This follows from $p_*\mathcal{P} = k(e)[-g]$ which has been shown in the proof of Proposition 9.19.

Next use that the tensor product $(\mathcal{O}_\Delta\boxtimes k(e_A)[-g])\otimes\mathcal{O}_{\Gamma_{\mu_A}}$ is isomorphic to the structure sheaf of the subvariety $\{(a,a,e,a,e)\mid a\in A\}$ shifted by $[-g]$ and similarly for the corresponding tensor product on the B-side.

Thus, the direct image under ii) yields

$$\overbrace{A\times(A\times A)}^{\mathcal{O}_{\{(a,a,e)\}}[-g]}\times\overbrace{(B\times B)\times B}^{\mathcal{O}_{\{(b,e,b)\}}}$$
$$\text{with } \underbrace{(A\times A)\times(B\times B)}_{\mathcal{E}\boxtimes\mathcal{E}_{\mathrm{R}}}.$$

The direct image under the last projection iii) turns this into the desired $\mathcal{E}\otimes\mathcal{E}_{\mathrm{R}}(e_A,e_B)[-g]\simeq\mathcal{E}\otimes\mathcal{E}^\vee(e_A,e_B)$. □

Exercise 9.52 Prove the description of $f_{\mathcal{P}}$ and $N_{\mathcal{P}}$ given in Example 9.38, v). See also Exercise 9.21.

Proposition 9.53 (Orlov) *Let* $\Phi_{\mathcal{E}}:\mathrm{D^b}(A)\xrightarrow{\sim}\mathrm{D^b}(B)$ *be an equivalence between the derived categories of two abelian varieties. Then up to a shift* $\mathcal{E}$ *is isomorphic to a sheaf. See [93].*

Proof After shifting $\mathcal{E}$, we may assume that $\mathcal{H}^0(\mathcal{E}) \neq 0$ and $\mathcal{H}^i(\mathcal{E}) = 0$ for $i > 0$. We have to show that in this case $\mathcal{E}$ is isomorphic to a sheaf. Let j be minimal with $\mathcal{H}^j(\mathcal{E}) \neq 0$. Then there exists a non-trivial homomorphism $\mathcal{H}^j(\mathcal{E})[-j] \longrightarrow \mathcal{E}$ and, therefore, also a non-trivial $\mathcal{E}^\vee \longrightarrow \mathcal{H}^j(\mathcal{E})^\vee[j]$. In order to use this information, we have to study the dual $\mathcal{E}^\vee$.

Since $\mathcal{E}$ is a sheaf if and only if $t_{(a,b)*}\mathcal{E}$ is one, we may assume that $(e_A, e_B) \in \mathrm{supp}(\mathcal{H}^k(\mathcal{E}^\vee))$, which will come in handy later.

Suppose k is maximal with $\mathcal{H}^k(\mathcal{E}^\vee) \neq 0$. If $\mathcal{H}^j(\mathcal{E})^\vee$ is concentrated in degree $\geq \ell$, then $k \geq \ell - j$. This can be seen by using the spectral sequence

$$E_2^{p,q} = \mathcal{E}xt^p(\mathcal{H}^{-q}(\mathcal{E}), \mathcal{O}) = \mathcal{H}^p(\mathcal{H}^{-q}(\mathcal{E})^\vee) \Rightarrow \mathcal{E}xt^{p+q}(\mathcal{E}, \mathcal{O}) = \mathcal{H}^{p+q}(\mathcal{E}^\vee).$$

By the previous lemma $K_\mathcal{E} := \pi_*\mathcal{I}(\mathcal{E}) \simeq \mathcal{E}^\vee(e_A, e_B) \otimes \mathcal{E}$ with $\mathcal{I}(\mathcal{E})$ a line bundle on the graph of $f_\mathcal{E}$. Denote the codimension of $\pi(\Gamma_{f_\mathcal{E}}) \subset A \times B$ by d. Hence, $\mathcal{E}$ is concentrated in codimension $\geq d$. This implies that $\mathcal{H}^j(\mathcal{E})^\vee$ is concentrated in degree $\geq d$ (see the explanation on p. 78) and hence $k \geq d - j$.

The complex $\mathcal{E}$ is concentrated in $[j, 0]$ with $\mathcal{H}^0(\mathcal{E}) \neq 0$ and $\mathcal{E}^\vee(e_A, e_B)$ is concentrated in degree $\leq k$ with non-trivial cohomology in degree k. Thus, $K_\mathcal{E} = \mathcal{E}^\vee(e_A, e_B) \otimes \mathcal{E}$ has non-trivial cohomology in degree k.

On the other hand, since the graph of $f_\mathcal{E}$ is of dimension $2g$, the fibres of $\Gamma_{f_\mathcal{E}} \longrightarrow A \times B$ (which as a homomorphism of abelian varieties is smooth) have all dimension d. In particular, $K_\mathcal{E}$ as the direct image of a line bundle on $\Gamma_{f_\mathcal{E}}$ is concentrated in degree $\leq d$. Therefore, $k \leq d$ and hence $d \geq k \geq d - j$. This yields $j \geq 0$ and thus $j = 0$. □

Remark 9.54 The case of abelian varieties is very special. Already for K3 surfaces, a Fourier–Mukai kernel is, in general, not simply a sheaf. Examples for (auto)equivalences with genuine complexes as Fourier–Mukai kernels have been encountered already in Chapter 8.

9.5 Autoequivalences of abelian varieties

Orlov pushed the techniques further to give a complete description of the group of all autoequivalences of $\mathrm{D}^\mathrm{b}(A)$ for any abelian variety A.

The results of the previous section applied to the case $A = B$ show that the map $\mathrm{Aut}(\mathrm{D}^\mathrm{b}(A)) \longrightarrow \mathrm{Aut}(\mathrm{D}^\mathrm{b}(A \times \widehat{A}))$, $\Phi_\mathcal{E} \longmapsto F_\mathcal{E}$, factorizes via $\mathrm{Pic}(A \times \widehat{A}) \rtimes \mathrm{Aut}(A \times \widehat{A})$. This is the map $\Phi_\mathcal{E} \longmapsto (N_\mathcal{E}, f_\mathcal{E})$. Indeed, the composition $\Phi_\mathcal{E} \circ \Phi_\mathcal{F}$ is mapped to $(N_\mathcal{E} \otimes f_{\mathcal{E}*}N_\mathcal{F}, f_\mathcal{E} \circ f_\mathcal{F})$ (see Exercise 9.41). In particular, the further projection to $f_\mathcal{E}$ is indeed a group homomorphism.

Proposition 9.55 **(Orlov)** *The kernel of the natural map*

$$\mathrm{Aut}(\mathrm{D}^\mathrm{b}(A)) \longrightarrow \mathrm{Aut}(A \times \widehat{A}) \ , \quad \Phi_\mathcal{E} \longmapsto f_\mathcal{E}$$

is isomorphic to the group $\mathbb{Z} \oplus (A \times \widehat{A})$ *generated by shifts* $[n]$, *translations* t_{a*}, *and tensor products* $L \otimes (\ )$ *with* $L \in \mathrm{Pic}^0(A)$. *See [93].*

Proof In Examples 9.38 we have seen that shifts, translations and tensor products with $L \in \mathrm{Pic}^0(A)$ all induce the identity $f_{\mathcal{E}}$. So, all are contained in the kernel.

Suppose now that $f_{\mathcal{E}}$ is the identity. In other words, $F_{\mathcal{E}}$ is the tensor product with a line bundle $N_{\mathcal{E}}$ on $A \times \widehat{A}$, which will be considered as a line bundle on the diagonal $\Delta_{A\times\widehat{A}} \subset A \times \widehat{A} \times A \times \widehat{A}$. Therefore, $K_{\mathcal{E}}$, which is the direct image of this line bundle on the diagonal, has support in Δ_A. By Lemma 9.51 the same holds for $\mathcal{E}^\vee(e,e) \otimes \mathcal{E}$.

Suppose that $\mathcal{E}^\vee(e,e) \neq 0$. Then $\mathcal{E}$ is supported on Δ_A. In other words, $\Phi_{\mathcal{E}} = E \otimes (\ )$ for some shifted coherent sheaf E on A (cf. Proposition 9.53). A closer inspection, e.g. testing $\Phi_{\mathcal{E}}$ on points, shows that E must be a shifted line bundle $M[n]$.

As was explained in Example 9.40, $M \otimes (\ )$ yields the identity $f_{\mathcal{E}} = \mathrm{id}$ if and only if $M \in \mathrm{Pic}^0(A)$.

In order to ensure that $\mathcal{E}^\vee(e,e) \neq 0$, we just translate. Indeed, if $\mathcal{E}^\vee(a,b) \neq 0$ and $\Phi_{\mathcal{F}^\vee} := t_{-a*} \circ \Phi_{\mathcal{E}^\vee} \circ t_{-b*}$, then $\mathcal{F}^\vee(e,e) \neq 0$. □

Remark 9.56 Previously, we have studied the homomorphism

$$\Phi_{\mathcal{E}} \longmapsto F_{\mathcal{E}} = (N_{\mathcal{E}}, f_{\mathcal{E}}) \in \mathrm{Pic}(A \times \widehat{A}) \rtimes \mathrm{Aut}(A \times \widehat{A}).$$

In general, $N_{\mathcal{E}}$ is not of degree zero. The proposition however shows that

$$\{\bar{F}_{\mathcal{E}}\} \longrightarrow \mathrm{Aut}(A \times \widehat{A}),$$

is injective, where $\bar{F}_{\mathcal{E}}$ is the image of $F_{\mathcal{E}}$ in $H^2(A \times \widehat{A}) \rtimes \mathrm{Aut}(A \times \widehat{A})$ under the projection induced by $\mathrm{c}_1 : \mathrm{Pic}(A \times \widehat{A}) \twoheadrightarrow \mathrm{Pic}(A \times \widehat{A})/\mathrm{Pic}^0(A \times \widehat{A}) \subset H^2(A \times \widehat{A})$.

Together with Proposition 9.48 and using the short-hand $U(A \times \widehat{A})$ for $U(A \times \widehat{A}, A \times \widehat{A})$ one obtains a complete description of the group of autoequivalences of an abelian variety.

Corollary 9.57 *The map* $\Phi_{\mathcal{E}} \mapsto f_{\mathcal{E}}$ *induces a short exact sequence*

$$0 \longrightarrow \mathbb{Z} \oplus (A \times \widehat{A}) \longrightarrow \mathrm{Aut}(\mathrm{D^b}(A)) \longrightarrow U(A \times \widehat{A}) \longrightarrow 1.$$

□

The proposition also shows that the subgroup $\mathbb{Z} \oplus (A \times \widehat{A}) \subset \mathrm{Aut}(\mathrm{D^b}(A))$ is actually normal. In particular, if we denote as before the equivalence associated to a closed point $(a, \alpha) \in A \times \widehat{A}$ by $\Phi_{(a,\alpha)}$, then for any other equivalence $\Phi_{\mathcal{E}} \in \mathrm{Aut}(\mathrm{D^b}(A))$ the normalizer $\Phi_{\mathcal{E}} \circ \Phi_{(a,\alpha)} \circ \Phi_{\mathcal{E}}^{-1}$ is again of the form $\Phi_{(b,\beta)}$ up to shift. This leads to the following alternative description of the automorphism $f_{\mathcal{E}}$, which is just a repetition of Corollary 9.44 in the case of autoequivalences.

Corollary 9.58 *Up to shift one has*

$$\Phi_{\mathcal{E}} \circ \Phi_{(a,\alpha)} \circ \Phi_{\mathcal{E}}^{-1} \simeq \Phi_{f_{\mathcal{E}}(a,\alpha)}$$

for any closed point $(a,\alpha) \in A \times \widehat{A}$.

Proof Indeed, Corollary 9.37 implies

$$F_{\mathcal{E}} \circ F_{(a,\alpha)} = F_{(b,\beta)} \circ F_{\mathcal{E}} \tag{9.14}$$

for some (b,β). The closed point $a \in A$ corresponds to a line bundle L_0 on $\widehat{A}$ and $\alpha \in \widehat{A}$ to a line bundle L on A. Similarly, (b,β) gives rise to $(M, M_0) \in \mathrm{Pic}(B \times \widehat{B})$. Thus, (9.14) reads (see Examples 9.38, iii))

$$(N_{\mathcal{E}} \otimes (\ \ )) \circ f_{\mathcal{E}*} \circ (L \boxtimes L_0^* \otimes (\ \ )) \simeq (M \boxtimes M_0^* \otimes (\ \ )) \circ (N_{\mathcal{E}} \otimes (\ \ )) \circ f_{\mathcal{E}*}$$

or, equivalently, $f_{\mathcal{E}*}(L \boxtimes L_0^*) \simeq M \boxtimes M_0^*$. The latter translates to $\widehat{f}(\beta,-b) = (\alpha,-a)$ or further to $\tilde{f}(b,\beta) = (a,\alpha)$. Using $\tilde{f} = f^{-1}$ yields the assertion $f(a,\alpha) = (b,\beta)$. □

Let us try to clarify the relation between Mukai's $\mathrm{Sl}_2(\mathbb{Z})$-action and the above description of $\mathrm{Aut}(\mathrm{D^b}(A))$ in the case of a principally polarized abelian variety (A, φ_L).

We invoke Example 9.38, v) to see that for $\Phi = \varphi_L^* \circ \Phi_{\mathcal{P}} \in \mathrm{Aut}(\mathrm{D^b}(A))$ one has

$$\Phi \longmapsto f = \begin{pmatrix} 0 & -1 \\ 1 & 0 \end{pmatrix} : A \times A \longrightarrow A \times A.$$

The other generator of the $\mathrm{Sl}_2(\mathbb{Z})$-action in Section 9.3 is given by the autoequivalence $L \otimes (\ \ )$. Using Corollary 9.58 one easily computes that for this autoequivalence one has

$$L \otimes (\ \ ) \longmapsto f = \begin{pmatrix} 1 & 0 \\ -1 & 1 \end{pmatrix} : A \times A \longrightarrow A \times A.$$

Exercise 9.59 Prove this.

This description fits nicely with the one given in Remark 9.31. In order to incorporate the shift functor one introduces a $\mathbb{Z}$-cover of Sl_2.

Definition 9.60 *Denote by* $\widetilde{\mathrm{Sl}}_2(\mathbb{Z})$ *the group that is generated by three elements* A_1, A_2, *and* t *satisfying the relations*

$$(A_1 \cdot A_2)^3 = t^g, \quad A_2^4 = t^{2g}, \quad \textit{and} \quad A_i \cdot t = t \cdot A_i.$$

(For a discussion of the group $\widetilde{\mathrm{Sl}}_2$, especially from the symplectic point of view, in the case $g = 1$ see [106].)

Clearly, setting $t = 1$, $A_1 = \begin{pmatrix} 1 & 0 \\ -1 & 1 \end{pmatrix}$, and $A_2 = \begin{pmatrix} 0 & -1 \\ 1 & 0 \end{pmatrix}$ defines a surjection $\widetilde{\mathrm{Sl}}_2(\mathbb{Z}) \twoheadrightarrow \mathrm{Sl}_2(\mathbb{Z})$. On the other hand, Mukai's results of Section 9.3 just show that sending $t \longmapsto [-1]$, $A_1 \longmapsto L \otimes (\ \ )$, and $A_2 \longmapsto \Phi$ defines a group homomorphism $\widetilde{\mathrm{Sl}}_2(\mathbb{Z}) \longrightarrow \mathrm{Aut}(\mathrm{D^b}(A))$.

Eventually one obtains a diagram

$$\begin{array}{ccccccccc} 0 & \longrightarrow & \mathbb{Z} \oplus (A \times \widehat{A}) & \longrightarrow & (A \times \widehat{A}) \rtimes \widetilde{\mathrm{Sl}}_2(\mathbb{Z}) & \longrightarrow & \mathrm{Sl}_2(\mathbb{Z}) & \longrightarrow & 1 \\ & & \| & & \downarrow & & \downarrow & & \\ 0 & \longrightarrow & \mathbb{Z} \oplus (A \times \widehat{A}) & \longrightarrow & \mathrm{Aut}(\mathrm{D^b}(A)) & \longrightarrow & U(A \times \widehat{A}) & \longrightarrow & 1. \end{array}$$

Here, the inclusion $\mathrm{Sl}_2(\mathbb{Z}) \subset U(A \times \widehat{A}) \subset \mathrm{Aut}(A \times A)$ is the natural one, which also explains the semi-direct product $(A \times \widehat{A}) \rtimes \widetilde{\mathrm{Sl}}_2(\mathbb{Z})$ by adding that t acts as the identity on $A \times \widehat{A}$.

Note that for a generic principally polarized abelian variety the inclusion $\mathrm{Sl}_2(\mathbb{Z}) \subset U(A \times \widehat{A}) \subset \mathrm{Aut}(A \times A)$ is an equality.

We conclude this section with a few comments on the relation between the two representations of the group of autoequivalences encountered so far. In this chapter, we have studied in length

$$\gamma : \mathrm{Aut}(\mathrm{D^b}(X)) \longrightarrow U(A \times \widehat{A}), \qquad \Phi_{\mathcal{E}} \longmapsto f_{\mathcal{E}},$$

whereas in the general context we were looking at

$$\rho : \mathrm{Aut}(\mathrm{D^b}(A)) \longrightarrow \mathrm{Gl}(H^*(A, \mathbb{Z})), \qquad \Phi_{\mathcal{E}} \longmapsto \Phi^H_{\mathcal{E}}.$$

Although $U(A \times \widehat{A})$ can be interpreted as a subgroup of

$$\mathrm{Gl}\left(H^1(A, \mathbb{Z}) \oplus H^1(\widehat{A}, \mathbb{Z})\right) \simeq \mathrm{Gl}\left(H^1(A, \mathbb{Z}) \oplus H^{2g-1}(A, \mathbb{Z})\right)$$

these two representations have quite a different flavour. A detailed discussion, involving Spin-representation, can be found in [41]. We just mention the following

Corollary 9.61 *There exists a homomorphism* $\lambda : \mathrm{Im}(\rho) \longrightarrow U(A \times \widehat{A})$ *giving rise to the commutative diagram*

$$\begin{array}{ccccc} & & U(A \times \widehat{A}) & & \\ & \overset{\gamma}{\nearrow} & \uparrow \lambda & & \\ \mathrm{Aut}(\mathrm{D^b}(A)) & & & & \\ & \underset{\rho}{\searrow} & & & \\ & & \mathrm{Im}(\rho) & \hookrightarrow & \mathrm{Gl}(H^*(A, \mathbb{Z})). \end{array}$$

Its kernel has order two and is spanned by the image of the shift functor.

Proof In order to prove the existence of λ one simply shows that $\Phi_{\mathcal{E}}^H = \mathrm{id}$ implies $f_{\mathcal{E}} = \mathrm{id}$. This follows from the cohomological version of the diagram in Definition 9.34. Indeed, if $\Phi_{\mathcal{E}}^H = \mathrm{id}$, then also $\Phi_{\mathcal{E}_{\mathrm{R}}}^H = \mathrm{id}$ and thus $F_{\mathcal{E}}^H = \mathrm{id}$. The latter suffices to conclude $f_{\mathcal{E}} = \mathrm{id}$.

By Proposition 9.55 any autoequivalence $\Phi_{\mathcal{E}}$ in the kernel of γ is contained in $\mathbb{Z} \oplus (A \times \widehat{A})$, i.e. up to shift it is of the form $\Phi_{(a,\alpha)}$. It is straightforward to show that the $\Phi_{(a,\alpha)}$ act trivially on cohomology. As the shift functor $\mathcal{F}^\bullet \longmapsto \mathcal{F}^\bullet[1]$ acts by a global sign, this proves the description of the kernel of λ. □

Another way to view this result is in terms of the following diagram:

$$\begin{array}{ccccccccc} & & \mathbb{Z}/2\mathbb{Z} & & & & & & \\ & & \uparrow & & & & & & \\ 0 & \longrightarrow & \mathbb{Z} \oplus (A \times \widehat{A}) & \longrightarrow & \mathrm{Aut}(\mathrm{D^b}(A)) & \xrightarrow{\gamma} & U(A \times \widehat{A}) & \longrightarrow & 1 \\ & & \uparrow & & \| & & \uparrow \lambda & & \\ 0 & \longrightarrow & 2\mathbb{Z} \oplus (A \times \widehat{A}) & \longrightarrow & \mathrm{Aut}(\mathrm{D^b}(A)) & \xrightarrow{\rho} & \mathrm{Im}(\rho) & \longrightarrow & 1 \\ & & & & & & \uparrow & & \\ & & & & & & \mathbb{Z}/2\mathbb{Z}. & & \end{array}$$

10

K3 SURFACES

K3 surfaces play a central rôle in the classification of algebraic surfaces. They are positioned halfway between ruled surfaces and surfaces of general type. Similarly to abelian surfaces, they have a rich internal geometry and a highly interesting moduli theory.

One of the landmarks in the theory of K3 surfaces is the global Torelli theorem (conjectured by Andreotti and Weil), asserting that two K3 surfaces are isomorphic if and only if they have isomorphic periods.

More recent work of Mukai and Orlov provide a derived version of the global Torelli theorem. This chapter presents a rather detailed account of this beautiful result.

Their techniques also allow us to give an almost complete description of the cohomological action of the group of autoequivalences of the derived category of a K3 surface. Surprisingly, cohomologically trivial autoequivalences elude us for the time being. In this respect, the theory of derived categories of K3 surfaces deviates from the otherwise very similar theory for abelian surfaces.

In Section 10.1 we provide basic definitions and fundamental facts from K3 surface theory. Section 10.2 proves the derived global Torelli theorem. As moduli spaces of stable sheaves on K3 surfaces are crucial for the argument, a brief outline of their theory is presented in Section 10.3.

10.1 Recap: K3 surfaces

We briefly recall the fundamental facts on K3 surfaces needed in the sequel. A detailed account of the theory can be found in [1, 5].

Definition 10.1 *A* K3 surface *is a compact complex surface X with trivial canonical bundle, i.e. $\omega_X \simeq \mathcal{O}_X$, and $H^1(X, \mathcal{O}_X) = 0$.*

Corollary 10.2 *Let X be a K3 surface. Any smooth projective variety Y which is D-equivalent to X is a K3 surface.*

Proof Due to Proposition 4.1 one knows that Y is also a surface with trivial canonical bundle. Proposition 5.39 for $i = -1$ yields $(h^{0,1} + h^{1,2})(X) = (h^{0,1} + h^{1,2})(Y)$. Hodge theory and Serre duality then show $h^{0,1} = h^{1,2}$. Hence, $h^{1,0}(X) = h^{1,0}(Y)$. Thus, Y is a K3 surface. □

• Any K3 surface is Kähler, but most of them are not algebraic. The algebraic ones are nevertheless dense in the moduli space of all K3 surfaces. We are ultimately interested in derived categories of K3 surfaces, which are sensitive invariants only when the variety is algebraic. So, the algebraically inclined reader might assume all K3 surfaces in this section to be algebraic, although for the moment this additional assumption is superfluous.

Examples 10.3 Here are the two most famous ones:

i) A smooth quartic hypersurface $X \subset \mathbb{P}^3$ is a K3 surface, e.g. the *Fermat surface* defined by $x_0^4 + x_1^4 + x_2^4 + x_3^4 = 0$.

Indeed, the canonical bundle can be computed by the adjunction formula $\omega_X \simeq (\omega_{\mathbb{P}^3} \otimes \mathcal{O}(4))|_X \simeq \mathcal{O}_X$ and $H^1(X, \mathcal{O}_X) = 0$ follows from the structure sheaf sequence for X and the vanishing $H^1(\mathbb{P}^3, \mathcal{O}_{\mathbb{P}^3}) = H^2(\mathbb{P}^3, \mathcal{O}(-4)) = 0$.

ii) Let A be an abelian surface. Then the minimal resolution $X \longrightarrow A/\iota$ of the quotient A/ι is a K3 surface, the *Kummer surface* of A. Here, $\iota : A \longrightarrow A$ is the involution $a \longmapsto -a$.

• By definition, $\chi(X, \mathcal{O}_X) = 2$ and the Noether formula thus yields

$$2 = \chi(X, \mathcal{O}_X) = \frac{c_1^2(X) + c_2(X)}{12} = \frac{c_2(X)}{12}.$$

Interpreting $c_2(X)$ as the topological Euler number $e(X) = \sum(-1)^i b_i(X)$ yields

$$e(X) = 24.$$

Hodge decomposition $H^1(X, \mathbb{C}) = H^{1,0}(X) \oplus H^{0,1}(X)$ and the identification $H^{0,1}(X) \simeq H^1(X, \mathcal{O}_X)$ shows $b_1(X) = 0$ and hence, by Poincaré duality, also $b_3(X) = 0$. Put together one has

$$b_0(X) = b_4(X) = 1, \quad b_1(X) = b_3(X) = 0, \text{ and } b_2(X) = 22.$$

• The hypothesis that $c_1(X) = 0$ can be used to show that the intersection pairing

$$(\ ,\ ) : H^2(X, \mathbb{Z}) \times H^2(X, \mathbb{Z}) \longrightarrow \mathbb{Z}$$

is even, i.e. for all $\alpha \in H^2(X, \mathbb{Z})$ one has $\alpha^2 = (\alpha, \alpha) \in 2\mathbb{Z}$. For algebraic classes $\alpha = c_1(L)$ this can be seen as a consequence of the Riemann–Roch formula $\chi(L) = c_1^2(L)/2 + 2$. For topologists the evenness of the intersection form follows from the vanishing of the second Stiefel–Whitney class.

Thus, $(H^2(X, \mathbb{Z}), (\ ,\ ))$ is an even unimodular lattice of rank 22. (We tacitly use that $H^2(X, \mathbb{Z})$ is torsion free, which follows from the fact that any K3 surface is simply connected. The latter statement is not completely trivial.)

In order to use classification theory of unimodular lattices, one also has to determine the index of the intersection pairing, which can readily be computed by applying the Hodge index theorem. One finds that the intersection pairing has three positive eigenvalues. Hence,

$$(H^2(X, \mathbb{Z}), (\ ,\ )) \simeq 2(-E_8) \oplus 3U,$$

where E_8 is the unique positive definite even unimodular lattice of rank eight and U is the hyperbolic plane, i.e. the free group of rank two generated by two isotropic vectors e_1, e_2 with $(e_1, e_2) = 1$. (For the classification of unimodular lattices see, e.g. [107].)

Note that the exponential sequence yields an injection $\mathrm{Pic}(X) \subset H^2(X, \mathbb{Z})$. This inclusion is strict. The Picard number $\rho(X)$, i.e. the rank of $\mathrm{Pic}(X)$, can in fact vary between 0 and 20. For algebraic K3 surfaces it is, of course, at least one.

• The most interesting structure associated to a K3 surface is its weight-two Hodge structure on $H^2(X, \mathbb{Z})$ given by the decomposition

$$H^2(X, \mathbb{C}) = H^{2,0}(X) \oplus H^{1,1}(X) \oplus H^{0,2}(X).$$

Using $H^{2,0}(X) \simeq H^0(X, \Omega_X^2) = H^0(X, \omega_X) \simeq H^0(X, \mathcal{O}_X) = \mathbb{C}$, one finds that $h^{2,0}(X) = h^{0,2}(X) = 1$. (In fact, this proves the above upper bound for the Picard number, as $\mathrm{Pic}(X) \subset H^{1,1}(X)$.)

Since the Hodge decomposition is orthogonal with respect to the intersection pairing, it is in fact completely determined by the complex line $H^{2,0}(X) \subset H^2(X, \mathbb{C})$.

Let us next state the most important single theorem for K3 surfaces.

Theorem 10.4 **(Global Torelli)** *Two K3 surfaces X and Y are isomorphic if and only if there exists a Hodge isometry $\varphi : H^2(X, \mathbb{Z}) \xrightarrow{\sim} H^2(Y, \mathbb{Z})$.*

If φ maps at least one Kähler class on X to a Kähler class on Y, then there exists a unique isomorphism $f : X \xrightarrow{\sim} Y$ with $f_* = \varphi$.

A *Hodge isometry* is a group isomorphism that respects the intersection product and maps $H^{2,0}(X)$ to $H^{2,0}(Y)$. The *period* of a K3 surface X is by definition the natural weight-two Hodge structure on the lattice $H^2(X, \mathbb{Z})$. Thus, the global Torelli theorem asserts that two K3 surfaces are isomorphic if and only if their periods are isomorphic.

The second assertion allows us to describe the automorphism group $\mathrm{Aut}(X)$ as the group of Hodge isometries $H^2(X, \mathbb{Z}) \simeq H^2(X, \mathbb{Z})$ respecting Kähler classes.

• The above considerations show the importance of a detailed understanding of the *Kähler cone* $\mathcal{K}_X \subset H^{1,1}(X, \mathbb{R})$ of all Kähler classes. The proof of the following fact is intimately related to the proof of the global Torelli theorem.

Theorem 10.5 *The Kähler cone $\mathcal{K}_X$ of a K3 surface X is a connected component of the open cone of all classes $\alpha \in H^{1,1}(X, \mathbb{R})$ with $(\alpha, \alpha) > 0$ and $\int_C \alpha > 0$ for all smooth rational curves $\mathbb{P}^1 \simeq C \subset X$.*

Since a line bundle L is ample if and only if $c_1(L)$ is a Kähler class, the theorem can also be read as a description of the ample cone.

If $C \subset X$ is a smooth rational curve, then $[C] \in H^2(X, \mathbb{Z})$ is a (-2)-*class*, i.e.

$$([C], [C]) = -2.$$

Every (-2)-class $\delta \in H^2(X,\mathbb{Z})$ defines a reflection

$$s_\delta : H^2(X,\mathbb{Z}) \longrightarrow H^2(X,\mathbb{Z}) \ , \quad \alpha \longmapsto \alpha + (\alpha,\delta)\delta.$$

If δ is of type $(1,1)$, e.g. if $\delta = [C]$, the induced reflection is a Hodge isometry. It can be shown that for any class $\alpha \in H^{1,1}(X)$ with $(\alpha,\alpha) > 0$, there exists a sequence of (-2)-curves $C_1,\ldots,C_n \subset X$ such that

$$\pm(s_{[C_1]} \circ \ldots \circ s_{[C_n]})(\alpha) \in \mathcal{K}_X.$$

These arguments can also be used to show that any given Hodge isometry $H^2(X,\mathbb{Z}) \simeq H^2(Y,\mathbb{Z})$ becomes, after composing with a suitable number of reflections of the type $s_{[C]}$, a Hodge isometry that sends a Kähler class to a Kähler class (up to sign).

• We shall also need the *Mukai pairing* on $H^*(X,\mathbb{Z})$ introduced in Section 5.2. To view it as an extension of the intersection pairing on $H^2(X,\mathbb{Z})$ we will however change the original definition by a sign. Thus, we shall use the following convention

$$\langle(\alpha_0,\alpha_1,\alpha_2),(\beta_0,\beta_1,\beta_2)\rangle := \alpha_1.\beta_1 - \alpha_0.\beta_2 - \alpha_2.\beta_0 \in \mathbb{Z}. \tag{10.1}$$

Here, $\alpha_i, \beta_i \in H^{2i}(X,\mathbb{Z})$.

With this intersection pairing, $H^*(X,\mathbb{Z})$ becomes a unimodular even lattice abstractly isomorphic to $2(-E_8) \oplus 4U$.

Mukai also introduced a weight-two Hodge structure on $H^*(X,\mathbb{Z})$ by declaring $H^0(X,\mathbb{C}) \oplus H^4(X,\mathbb{C})$ to be of type $(1,1)$ and keeping the standard Hodge structure on $H^2(X,\mathbb{C})$. In the sequel, we shall write $\widetilde{H}(X,\mathbb{Z})$ for $H^*(X,\mathbb{Z})$ endowed with the Mukai pairing and this weight-two Hodge structure. So, with this notation we have

$$\widetilde{H}^{1,1}(X) = (H^0 \oplus H^4)(X) \oplus H^{1,1}(X) \ \text{ and } \ \widetilde{H}^{2,0}(X) = H^{2,0}(X).$$

• The *Mukai vector* (see Definition 5.28) of a sheaf $\mathcal{E}$ (or a complex of sheaves) is by definition the class

$$v(\mathcal{E}) = \mathrm{ch}(\mathcal{E}).\sqrt{\mathrm{td}(X)} = (v_0(\mathcal{E}), v_1(\mathcal{E}), v_2(\mathcal{E})) \in \widetilde{H}(X,\mathbb{Z}),$$

which is, with respect to Mukai's weight-two Hodge structure, of type $(1,1)$. In other words, $v(\mathcal{E}) \in \widetilde{H}^{1,1}(X)$.

Let us make this a bit more explicit. Firstly, since $\mathrm{c}_1(X) = 0$ and $2 = \chi(X,\mathcal{O}_X) = \mathrm{td}_2(X)$, one has $\mathrm{td}(X) = (1,0,2)$ and, therefore, $\sqrt{\mathrm{td}(X)} = (1,0,1)$. This then yields

$$v(\mathcal{E}) = \left(\mathrm{rk}(\mathcal{E}), \mathrm{c}_1(\mathcal{E}), \mathrm{rk}(\mathcal{E}) + \mathrm{c}_1^2(\mathcal{E})/2 - \mathrm{c}_2(\mathcal{E})\right).$$

Note that due to the fact that the intersection pairing on $H^2(X,\mathbb{Z})$ is even, the Mukai vector is indeed an integral cohomology class.

The Mukai vector $v(\mathcal{E}^\bullet) \in H^*(X \times Y, \mathbb{Q})$ for an object $\mathcal{E}^\bullet \in \mathrm{D^b}(X \times Y)$ on the product of two K3 surfaces X and Y will also be important, as the cohomological Fourier–Mukai transform is defined in terms of it. As will be shown in Lemma 10.6, we again have $v(\mathcal{E}^\bullet) \in H^*(X \times Y, \mathbb{Z})$.

10.2 Derived equivalence of K3 surfaces

The aim of this section is to derive a cohomological criterion that decides when two K3 surfaces have equivalent derived categories. The result will be a natural generalization of the global Torelli theorem, which asserts that two K3 surfaces are isomorphic if and only if there exists a Hodge isometry $H^2(X,\mathbb{Z}) \simeq H^2(Y,\mathbb{Z})$ (see Theorem 10.4).

The following is a technical lemma, which will permit us to work on the level of integral cohomology.

Lemma 10.6 **(Mukai)** *Let X and Y be two K3 surfaces. Then the Mukai vector of any object $\mathcal{E}^\bullet \in \mathrm{D^b}(X \times Y)$ is an integral cohomology class $v(\mathcal{E}^\bullet) \in H^*(X \times Y, \mathbb{Z})$. See [80].*

Proof We actually show that $\mathrm{ch}(\mathcal{E}^\bullet)$ is integral. Since the square root of the Todd genus can be computed by

$$\sqrt{\mathrm{td}(X \times Y)} = q^*\sqrt{\mathrm{td}(X)}.p^*\sqrt{\mathrm{td}(Y)} = q^*(1,0,1).p^*(1,0,1),$$

this is certainly enough.

Let us write

$$\mathrm{ch}(\mathcal{E}^\bullet) = \left(\mathrm{rk}(\mathcal{E}^\bullet), \mathrm{c}_1(\mathcal{E}^\bullet), (1/2)(\mathrm{c}_1^2(\mathcal{E}^\bullet) - 2\mathrm{c}_2(\mathcal{E}^\bullet)), \mathrm{ch}_3(\mathcal{E}^\bullet), \mathrm{ch}_4(\mathcal{E}^\bullet)\right).$$

Clearly, $\mathrm{rk}(\mathcal{E}^\bullet)$ and $\mathrm{c}_1(\mathcal{E}^\bullet)$ are integral. Moreover, $\mathrm{c}_1(\mathcal{E}^\bullet) \in H^2(X \times Y, \mathbb{Z}) = H^2(X,\mathbb{Z}) \oplus H^2(Y,\mathbb{Z})$ (Künneth decomposition) can be written as $q^*\alpha + p^*\beta$ with $\alpha \in H^2(X,\mathbb{Z})$ and $\beta \in H^2(Y,\mathbb{Z})$. Hence, $\mathrm{c}_1^2(\mathcal{E}^\bullet) = q^*\alpha^2 + 2q^*\alpha.p^*\beta + p^*\beta^2$. Since the intersection pairing on X and Y is even, this shows that $\mathrm{c}_1^2(\mathcal{E}^\bullet)$ is divisible by two.

It remains to show that ch_3 and ch_4 are integral. The main idea is to use the Grothendieck–Riemann–Roch formula

$$\mathrm{ch}(p_*\mathcal{E}^\bullet) = p_*\left(\mathrm{ch}(\mathcal{E}^\bullet).q^*\mathrm{td}(X)\right)$$

and the integrality of the Chern characters on the K3 surfaces X and Y.

Write $\mathrm{ch}(\mathcal{E}^\bullet)$ according to the Künneth decomposition as

$$\mathrm{ch}(\mathcal{E}^\bullet) = \sum_{r,s \le 4} e^{r,s}$$

with $e^{r,s} \in H^r(X) \otimes H^s(Y)$. By what we have just seen, the component $e^{r,s}$ is integral for $r + s \le 4$. Using $\mathrm{td}(X) = (1,0,2)$ we obtain $\mathrm{c}_1(p_*(\mathcal{E}^\bullet)) = \int_X e^{4,2} + 2e^{0,2}$, from which we deduce that $e^{4,2}$ is integral.

Using the Grothendieck–Riemann–Roch formula again, this time with respect to the first projection q, one similarly proves that $e^{2,4}$ is integral and, therefore, that $\mathrm{ch}_3(\mathcal{E}^\bullet) \in H^6(X \times Y, \mathbb{Z})$. Similarly, one uses $\mathrm{ch}_2(p_*\mathcal{E}^\bullet) = \int_X e^{4,4} + 2e^{0,4}$ to deduce that $\mathrm{ch}_4(\mathcal{E}^\bullet) = e^{4,4}$ is integral. □

Corollary 10.7 **(Mukai)** *If $\Phi_{\mathcal{E}^\bullet} : \mathrm{D}^{\mathrm{b}}(X) \xrightarrow{\sim} \mathrm{D}^{\mathrm{b}}(Y)$ is an equivalence between the derived categories of two K3 surfaces, then the induced map on the cohomology defines a Hodge isometry*

$$\Phi^H_{\mathcal{E}^\bullet} : \widetilde{H}(X, \mathbb{Z}) \xrightarrow{\sim} \widetilde{H}(Y, \mathbb{Z}).$$

Proof The cohomological Fourier–Mukai transform is

$$\Phi^H_{\mathcal{E}^\bullet} : H^*(X, \mathbb{Q}) \xrightarrow{\sim} H^*(Y, \mathbb{Q}) \ , \quad \alpha \longmapsto p_*(q^*\alpha . v(\mathcal{E}^\bullet))$$

(see Remark 5.30). The above lemma now shows that it maps an integral class to an integral class.

Applying the same argument to the inverse functor, which is also given as a Fourier–Mukai transform, proves that indeed $\Phi^H_{\mathcal{E}^\bullet} : H^*(X, \mathbb{Z}) \xrightarrow{\sim} H^*(Y, \mathbb{Z})$.

The compatibility with the Mukai pairing was shown in a broader context in Proposition 5.44.

In order to conclude that $\Phi^H_{\mathcal{E}^\bullet}$ is a Hodge isometry, one has to show that it(s $\mathbb{C}$-linear extension) maps $H^{2,0}(X)$ to $H^{2,0}(Y)$. But this follows again from the general fact proved in Proposition 5.39. □

Remarks 10.8 i) Of course, in general $\Phi^H_{\mathcal{E}^\bullet}$ will not preserve the cohomological degree. In particular, one should not expect $\Phi^H_{\mathcal{E}^\bullet}$ to define a Hodge isometry $H^2(X, \mathbb{Z}) \simeq H^2(Y, \mathbb{Z})$. (The latter would imply that X and Y are isomorphic by Theorem 10.4, but there are definitely cases where this is not true.)

ii) The corollary is the rather trivial analogue of the corresponding, much deeper fact for abelian varieties. Recall that if $\mathrm{D}^{\mathrm{b}}(A) \simeq \mathrm{D}^{\mathrm{b}}(B)$ then $A \times \widehat{A} \simeq B \times \widehat{B}$ or, in other words, there exists an isomorphism of Hodge structures $H^1(A, \mathbb{Z}) \oplus H^{2g-1}(A, \mathbb{Z}) \simeq H^1(B, \mathbb{Z}) \oplus H^{2g-1}(B, \mathbb{Z})$ (see Proposition 9.39).

Before trying to find a criterion that decides exactly when two K3 surfaces are derived equivalent, let us discuss a few known (auto)equivalences.

Examples 10.9 i) Let $L \in \mathrm{Pic}(X)$. Then

$$L \otimes (\ ) : \mathrm{D}^{\mathrm{b}}(X) \longrightarrow \mathrm{D}^{\mathrm{b}}(X)$$

is an autoequivalence with kernel $\iota_* L$, where $\iota : X \longrightarrow X \times X$ is the diagonal embedding. The induced action on the cohomology is given by multiplication with the Chern character

$$\mathrm{ch}(L) = \exp(\mathrm{c}_1(L)) : \widetilde{H}(X,\mathbb{Z}) \xrightarrow{\sim} \widetilde{H}(X,\mathbb{Z}).$$

ii) By definition, the structure sheaf $\mathcal{O}_X$ on a K3 surface X is a spherical object (see Examples 8.10). It thus induces an equivalence

$$T_{\mathcal{O}_X} : \mathrm{D^b}(X) \xrightarrow{\sim} \mathrm{D^b}(X).$$

Recall that $T_{\mathcal{O}_X}$ is defined as the Fourier–Mukai transform whose kernel is the cone of the natural map $\mathcal{O}_{X\times X} \longrightarrow \mathcal{O}_\Delta$ (which is just the shifted ideal sheaf of the diagonal, but this is not helpful at this point).

The description of the kernel as a cone allows one to compute the cohomological Fourier–Mukai $T^H_{\mathcal{O}_X}$ easily as $\mathrm{id} - \Phi^H_{\mathcal{O}_{X\times X}}$. The Mukai vector of $\mathcal{O}_{X\times X}$ is $v = q^*(1,0,1).p^*(1,0,1)$ and, hence, $\Phi^H_{\mathcal{O}_{X\times X}}(\alpha) = -\langle\alpha,(1,0,1)\rangle(1,0,1)$. Thus,

$$T^H_{\mathcal{O}_X}(\alpha) = \alpha + \langle\alpha,(1,0,1)\rangle(1,0,1),$$

which is nothing but the reflection $s_{(1,0,1)}$ with respect to the (-2) vector $(1,0,1)$ (see Lemma 8.12 for the general assertion).

More precisely, $T^H_{\mathcal{O}_X}$ is the identity on $H^2(X,\mathbb{Z})$ and interchanges the generator of $H^0(X,\mathbb{Z})$ and $H^4(X,\mathbb{Z})$ up to sign, i.e. $e_0 \longmapsto -e_4$ and $e_4 \longmapsto -e_0$.

iii) If X contains a smooth rational curve $C \subset X$, then all sheaves of the form $\mathcal{O}_C(n)$ are spherical objects in $\mathrm{D^b}(X)$. We are most interested in $\mathcal{O}_C(-1)$, as the equivalence defined by it induces a particularly nice action on cohomology.

Indeed, the Mukai vector $v(\mathcal{O}_C(-1))$ is easily computed as $(0,[C],0)$. Hence, $T^H_{\mathcal{O}_C(-1)}$ sends a class α to $\alpha + \langle\alpha,(0,[C],0)\rangle(0,[C],0)$. In other words,

$$T^H_{\mathcal{O}_C(-1)} = s_{[C]}.$$

As was recalled in Section 10.1, these reflections are of utmost importance in the study of K3 surfaces.

The next result is the analogue of Corollary 9.50.

Proposition 10.10 **(Mukai, Orlov)** *Two K3 surfaces X and Y are derived equivalent if and only if there exists a Hodge isometry $\widetilde{H}(X,\mathbb{Z}) \simeq \widetilde{H}(Y,\mathbb{Z})$. See [80, 92].*

Proof Corollary 10.7 shows the 'only if' direction. Let us prove the 'if' which is due to Orlov.

i) Suppose there exists a Hodge isometry

$$\varphi : \widetilde{H}(X,\mathbb{Z}) \xrightarrow{\sim} \widetilde{H}(Y,\mathbb{Z})$$

with $\varphi(0,0,1) = \pm(0,0,1)$. Then φ induces a Hodge isometry $H^2(X,\mathbb{Z}) \simeq H^2(Y,\mathbb{Z})$ and, by the global Torelli theorem 10.4, this yields $X \simeq Y$. In particular, $\mathrm{D^b}(X) \simeq \mathrm{D^b}(Y)$.

ii) Next, one assumes that $v := (r,\ell,s) := \varphi(0,0,1)$ satisfies $r \neq 0$. Changing φ by a sign, we may assume $r > 0$. Note that $v' := \varphi(-1,0,0)$ satisfies $\langle v', v\rangle = 1$. Since $\langle v, v\rangle = 0$, the following general fact from the theory of moduli spaces of stable sheaves on K3 surfaces applies:

• If Y is a K3 surface and $v, v' \in \widetilde{H}^{1,1}(Y,\mathbb{Z})$ with $\langle v, v\rangle = 0$ and $\langle v, v'\rangle = 1$, then there exists another K3 surface M and a sheaf $\mathcal{E}$ on $Y \times M$ satisfying the hypothesis of the equivalence criterion Proposition 7.1 such that for all $m \in M$ the Mukai vector of $\mathcal{E}_m$ equals v. (A few indications of how this is obtained can be found in Section 10.3, e.g. Proposition 10.24. Note that in particular, the associated Fourier–Mukai transform induces an equivalence $\Phi_{\mathcal{E}} : \mathrm{D^b}(Y) \xrightarrow{\sim} \mathrm{D^b}(M)$.)

In our case, we consider the composition

$$\psi : \widetilde{H}(X,\mathbb{Z}) \xrightarrow{\ \varphi\ } \widetilde{H}(Y,\mathbb{Z}) \xrightarrow{\ \Phi^H_{\mathcal{E}}\ } \widetilde{H}(M,\mathbb{Z}),$$

which satisfies $\psi(0,0,1) = (0,0,1)$. According to i) this proves $X \simeq M$. Any such isomorphism composed with the inverse of the equivalence $\Phi_{\mathcal{E}} : \mathrm{D^b}(Y) \xrightarrow{\sim} \mathrm{D^b}(M)$ yields an equivalence $\mathrm{D^b}(X) \simeq \mathrm{D^b}(Y)$.

iii) Finally, we treat the case that $v := \varphi(0,0,1)$ is of the form $(0,\ell,s)$ with $\ell \neq 0$. Applying the Hodge isometry $\exp(\mathrm{c}_1(L))$ for some $L \in \mathrm{Pic}(Y)$ yields the vector $\exp(\mathrm{c}_1(L))(0,\ell,s) = (0,\ell,s + (\mathrm{c}_1(L),\ell))$. As $\ell \neq 0$, one can always choose L such that $s + (\mathrm{c}_1(L),\ell) \neq 0$. Hence $(r',\ell',s') := T^H_{\mathcal{O}_Y}(v)$ satisfies $r' \neq 0$. Then continue as in ii). □

Remark 10.11 The description of Fourier–Mukai partners of a given K3 surface in terms of Hodge isometries combined with some lattice theory allows us to show that a K3 surface admits only finitely many Fourier–Mukai partners (which are all K3 surfaces).

It is more complicated to actually count the number of Fourier–Mukai partners. This has been successfully done in [48, 49, 108]. In particular, it has been shown that for any positive integer N there exists a K3 surface X with at least N Fourier–Mukai partners (up to isomorphism).

The following result is obtained by reinspection of the above proof and keeping track of how the original Hodge isometry is modified in the course of the argument. It has been independently observed by Hosono et al. and Ploog (see [47, 95]). Why $\pm\mathrm{id}_{H^2}$ should appear in the statement was first explained by Szendrői in [110] (see also [54]).

Corollary 10.12 *Let φ be a Hodge isometry of the Mukai lattice $\widetilde{H}(X,\mathbb{Z})$ of a K3 surface X, i.e. $\varphi \in \mathrm{Aut}(\widetilde{H}(X,\mathbb{Z}))$. Then there exists an autoequivalence*

$\Phi_{\mathcal{E}^\bullet} : \mathrm{D^b}(X) \xrightarrow{\sim} \mathrm{D^b}(X)$ *with*

$$\Phi^H_{\mathcal{E}^\bullet} = \varphi \circ (\pm \mathrm{id}_{H^2}).$$

Proof Consider more generally a Hodge isometry $\varphi : \widetilde{H}(X,\mathbb{Z}) \xrightarrow{\sim} \widetilde{H}(Y,\mathbb{Z})$. We shall prove that $\varphi \circ (\pm \mathrm{id}_{H^2})$ is induced by an equivalence.

In order to apply the global Torelli theorem, we will pass from the given φ to a new Hodge isometry $\varphi' := \Phi^H \circ \varphi$, where Φ^H is induced by an equivalence $\Phi : \mathrm{D^b}(Y) \xrightarrow{\sim} \mathrm{D^b}(M)$ with M a certain K3 surface. Clearly, $\varphi \circ (\pm \mathrm{id}_{H^2})$ is induced by an equivalence if and only if this is true for $\varphi' \circ (\pm \mathrm{id}_{H^2})$. Which equivalence Φ is appropriate depends on φ.

There are in fact three cases, which correspond to the cases i)–iii) in the proof of the proposition. Again, everything will eventually be reduced to i). Here are the details:

i) Suppose $\varphi(0,0,1) = \pm(0,0,1)$. Then $(r,\ell,s) := \pm\varphi(1,0,0)$ satisfies $r = 1$ and $s = \ell^2/2$. In other words, $\varphi(1,0,0) = \pm\exp(\mathrm{c}_1(L))$ for a certain line bundle L on Y. Then set

$$\Phi := \begin{cases} L^* \otimes (\ ) & \text{if } \varphi(0,0,1) = (0,0,1) \\ L^*[1] \otimes (\ ) & \text{if } \varphi(0,0,1) = -(0,0,1). \end{cases}$$

The composition $\varphi' := \Phi^H \circ \varphi$ is not only a Hodge isometry mapping $(0,0,1)$ to $(0,0,1)$, but it in fact respects the degree decomposition, i.e.

$$\varphi' = \mathrm{id}_{H^0} \oplus \varphi'_2 \oplus \mathrm{id}_{H^4}.$$

(By abuse of notation we identify $H^0(X) = H^0(Y)$ and similarly for H^4.)

After composing with a suitable number of reflections $s_{[C_i]}$, $i = 1, \ldots, n$, where $C_i \subset Y$ are smooth rational curves, we find an isomorphism $f : X \xrightarrow{\sim} Y$ realizing $\varphi'_2 : H^2(X,\mathbb{Z}) \simeq H^2(Y,\mathbb{Z})$ up to sign.

Therefore,

$$s_{[C_1]} \circ \ldots \circ s_{[C_n]} \circ \varphi' = f_* \circ (\pm \mathrm{id}_{H^2}).$$

Thus, φ is the Hodge isometry $(s_{[C_n]} \circ \ldots \circ s_{[C_1]} \circ \Phi^H)^{-1} \circ f_* \circ (\pm \mathrm{id}_{H^2})$ which is induced by an equivalence. This proves the assertion in this case.

ii) If $v := (r,\ell,s) := \varphi(0,0,1)$ satisfies $r \neq 0$, then define an equivalence $\Phi : \mathrm{D^b}(Y) \xrightarrow{\sim} \mathrm{D^b}(M)$ by

$$\Phi := \begin{cases} \Phi_{\mathcal{E}} & \text{if } r > 0 \\ \Phi_{\mathcal{E}[1]} & \text{if } r < 0, \end{cases}$$

where $\mathcal{E}$ is the universal family of stable sheaves over $Y \times M$ with M the moduli space of stable sheaves with Mukai vector v, respectively $-v$ (see the proof of the proposition). Then $\varphi' := \Phi^H \circ \varphi$ satisfies the condition in i) and one proceeds as in this case.

iii) Finally, consider a Hodge isometry φ with $\varphi(0,0,1) = (0,\ell,s)$ with $\ell \neq 0$. Then let Φ be the following composition of derived equivalences

$$\Phi : \mathrm{D^b}(Y) \xrightarrow{(L\otimes\ )} \mathrm{D^b}(Y) \xrightarrow{T_{\mathcal{O}_Y}} \mathrm{D^b}(Y).$$

Then $\varphi' := \Phi^H \circ \varphi$ satisfies the condition of ii) and we proceed as before. □

Going a bit deeper into the construction of the moduli spaces allows us to find a sufficient condition for a given Hodge isometry to lift to an equivalence. For a change, we state this improvement of Corollary 10.12 for Hodge isometries between different K3 surfaces.

Corollary 10.13 *Suppose $\varphi : \widetilde{H}(X,\mathbb{Z}) \xrightarrow{\sim} \widetilde{H}(Y,\mathbb{Z})$ is a Hodge isometry. If φ respects the natural orientation of the positive directions, then there exists an equivalence $\Phi : \mathrm{D^b}(X) \xrightarrow{\sim} \mathrm{D^b}(Y)$ with*

$$\Phi^H = \varphi.$$

Before going into the proof, we have to explain what is meant by preserving the natural orientation of the positive directions.

The Mukai lattice $\widetilde{H}(X,\mathbb{Z})$ has signature $(4,20)$. Consider two four-dimensional subspaces $V_1, V_2 \subset \widetilde{H}(X,\mathbb{R})$ for which the restriction of the Mukai pairing $\langle\ ,\ \rangle|_{V_i}$, $i = 1,2$, is positive definite. We say that chosen orientations on V_1 and V_2 are equivalent if they coincide under the orthogonal projection

$$V_1 \hookrightarrow \widetilde{H}(X,\mathbb{R}) \longrightarrow V_2.$$

which is necessarily bijective.

By definition, an orientation of the four positive directions of $\widetilde{H}(X,\mathbb{Z})$ is an equivalence class of such orientations, i.e. given by the choice of a positive four-space with an orientation.

It turns out that a *natural orientation* of the four positive directions in $\widetilde{H}(X,\mathbb{Z})$ exists, depending only on the complex structure given by the K3 surface X. Indeed, for any ample (or Kähler) class $\alpha \in H^{1,1}(X)$ one considers the four classes $\mathrm{Re}(\sigma), \mathrm{Im}(\sigma), \mathrm{Re}(\exp(i\alpha))$, and $\mathrm{Im}(\exp(i\alpha))$. Here, σ spans $H^{2,0}(X)$ and $\exp(i\alpha) = 1 + i\alpha - \alpha^2/2$, which is a cohomology class of mixed degree. It is not difficult to verify that these four classes are pairwise orthogonal and of positive square. Hencc, they span a positive four-space and the choice of the basis induces an orientation.

Since the set of Kähler classes $\alpha \in H^{1,1}(X)$ is connected and changing σ by a complex scalar does not change the orientation of the plane $\mathrm{Re}(\sigma), \mathrm{Im}(\sigma)$, the orientation of the four positive directions defined in this way does not depend either on α or on σ.

It should now be clear what is meant by φ preserving the natural orientation of the four positive directions.

We also note that we can similarly define the natural orientation of the two positive directions in $\widetilde{H}^{1,1}$ by $1-\alpha^2/2, \alpha$. Since any Hodge isometry respects the Hodge decomposition, i.e. maps $\mathrm{Re}(\sigma_X)$ and $\mathrm{Im}(\sigma_X)$ to the analogous classes on Y, it respects the natural orientation of the four positive directions if and only if it respects the orientation of the two positive directions in $\widetilde{H}^{1,1}(X,\mathbb{Z})$.

Proof The argument is taken from [54].

Going back to the proof of the previous corollary we find that in the course of the arguments the original Hodge isometry had to be changed several times by Hodge isometries that either occur in the list of Examples 10.9 or are induced by a fine moduli space. The Hodge isometry that was obtained eventually respected the degree and mapped a Kähler class to a Kähler class up to a sign. Then the global Torelli could be applied. Thus it remains to verify the following assertions:

i) A Hodge isometry $\widetilde{H}(X,\mathbb{Z}) \xrightarrow{\sim} \widetilde{H}(Y,\mathbb{Z})$ respecting the degree maps the positive cone to the positive cone if and only if it respects the natural orientation of the positive directions.
ii) Any Hodge isometry provided by one of the Examples 10.9 or by a fine moduli space as in Proposition 10.25 respects the natural orientation of the positive directions.

The first of these two conditions is easily verified, for $\exp(i\alpha)$ under a degree preserving Hodge isometry f is mapped to $\exp(if(\alpha))$ and, in particular, its real part remains unchanged.

The explicit description of the Hodge isometries in Examples 10.9 allows us to verify directly that they preserve the natural orientation. We leave this as Exercise 10.14 to the reader. Thus, ii) needs to be checked only for Hodge isometries induced by universal families of stable sheaves.

So, let us consider the case of a universal family $\mathcal{E}$ of stable sheaves living on the product $X \times M$ and the induced Fourier–Mukai transform

$$\Phi := \Phi_{\mathcal{E}} : \mathrm{D^b}(X) \xrightarrow{\sim} \mathrm{D^b}(M).$$

We shall use the description of a certain ample line bundle on the moduli space that comes automatically with the construction (following Simpson's method). We have to refer to [53] for its proof (see however the few scattered remarks in the next section) but the outcome is:

For any fixed $n \gg 0$ the line bundle

$$\det(\Phi(\mathcal{O}(k)))^{\otimes\chi(0)} \otimes \det(\Phi(\mathcal{O}))^{\otimes-\chi(k)}$$

is ample. Here, $\chi(k)$ is the Hilbert polynomial $\chi(E(k))$ of a stable sheaf E with Mukai vector v. See the explanations in Remark 10.21. To simplify, we assumed that the twist m considered there is superfluous, i.e. $m = 0$. This is allowed, as passing from $\mathcal{E}$ to $\mathcal{E} \otimes q^*\mathcal{O}(m)$ corresponds on the level of Fourier–Mukai functors to composing Φ with $\mathcal{O}(m) \otimes (\ \ )$, which does not change anything for the question of orientation.

If one denotes

$$u := \chi(0)\exp(kh) - \chi(k)(1,0,0),$$

then the degree zero part of $\Phi^H(u)$ is trivial and the degree two part $\Phi^H(u)_2 \in H^2(M,\mathbb{Z})$ is the first Chern class of the ample line bundle. This is the extra positivity needed to ensure that Φ respects the orientation.

Indeed, if we write $\Phi^H(\exp(ih)) = \lambda\exp(b+ai)$ for some $\lambda \in \mathbb{C}$ and $a, b \in H^2(M,\mathbb{Q})$, then automatically $a \in \pm\mathcal{C}_M$ and Φ respects the orientation if and only if $a \in \mathcal{C}_M$. The latter is then proved by writing $a = \Phi(u')_2$ (up to scaling) for a certain explicit u' with $\Phi^H(u')_0 = 0$ and computing explicitly

$$\langle a, \Phi^H(u)_2\rangle = \langle \Phi^H(u'), \Phi^H(u)\rangle = \langle u', u\rangle > 0.$$

For details see the original argument in [54]. □

Exercise 10.14 Prove that the Hodge isometries induced by the equivalences described in Examples 10.9, i)–iii) are orientation preserving.

Remark 10.15 There are two interesting problems concerning derived equivalences of K3 surfaces that are still open for the time being.

i) Let $\Phi : \mathrm{D^b}(X) \xrightarrow{\sim} \mathrm{D^b}(Y)$ be an arbitrary derived equivalence between two K3 surfaces. Does

$$\Phi^H : \widetilde{H}(X,\mathbb{Z}) \xrightarrow{\ \sim\ } \widetilde{H}(Y,\mathbb{Z})$$

preserve the natural orientation of the positive directions?

For an affirmative answer, it would suffice to show that $\mathrm{id}_{H^0\oplus H^4} \oplus (-\mathrm{id}_{H^2})$ is not induced by any equivalence (cf. Corollary 10.12).

Note that the almost surjectivity of $\mathrm{Aut}(\mathrm{D^b}(X)) \longrightarrow \mathrm{Aut}(\widetilde{H}(X,\mathbb{Z}))$ is analogous to the surjectivity of $\mathrm{Aut}(\mathrm{D^b}(A)) \longrightarrow U(A \times \widehat{A})$ for any abelian variety A.

ii) What can be said about the subgroup of autoequivalences of $\mathrm{D^b}(X)$ acting trivially on cohomology, i.e.

$$\mathrm{Ker}\left(\mathrm{Aut}(\mathrm{D^b}(X)) \longrightarrow \mathrm{Aut}(\widetilde{H}(X,\mathbb{Z}))\right) = ?$$

We know that this subgroup is not trivial, for obvious reasons, because $\mathcal{E}^\bullet \longmapsto \mathcal{E}^\bullet[2]$ is contained in it, but also due to the existence of non-trivial elements induced by spherical twists, e.g. $T^2_{\mathcal{O}}$.

Recently, Bridgeland [21] proposed a completely new approach to this problem. He conjectures that this subgroup can be described as the fundamental group of a certain open subset in $\widetilde{H}^{1,1}(X,\mathbb{Q}) \otimes \mathbb{C}$.

10.3 Recap: Moduli spaces of sheaves

This is a very rough sketch of some aspects of the theory of moduli spaces of semi-stable sheaves on projective varieties. For more details we refer to [53]. In fact, the discussion will be tailor-made for the application to K3 surfaces discussed in the previous section.

In the following we let X be a K3 surface with an ample line bundle $\mathcal{O}(1)$ whose first Chern class is denoted $h = \mathrm{c}_1(\mathcal{O}(1))$.

Definition 10.16 *Let E be a coherent sheaf of positive rank on X. Then E is* semi-stable *if for any subsheaf $F \subset E$ one has*

$$\chi(F(n)) \leq \chi(E(n)) \cdot (\mathrm{rk}(F)/\mathrm{rk}(E)) \text{ for } n \gg 0.$$

We say that E is stable *if the strict inequality holds whenever F is a proper non-trivial subsheaf.*

Exercise 10.17 Show that any stable sheaf is simple, i.e. $\mathrm{Hom}(E, E) = \mathbb{C}$.

Moduli functor Ideally, one would like to paramatrize all sheaves E on X with a given Mukai vector (r, ℓ, s). This can be done in a reasonable way only if we restrict to the smaller class of semi-stable sheaves. More precisely, one considers the *moduli functor*:

$$\mathcal{M}_v(h) : \mathrm{Sch}_k \longrightarrow \mathrm{Set}$$

that associates to any scheme of finite type T over $k = \mathbb{C}$ the set of equivalence classes of all T-flat coherent sheaves $\mathcal{E}$ on $X \times T$ such that for any closed point $t \in T$ the sheaf $\mathcal{E}_t$ on X is semi-stable (with respect to h) with $v(\mathcal{E}_t) = v$.

Two such sheaves $\mathcal{E}$ and $\mathcal{E}'$ are called equivalent if there exists a line bundle L on T such that $\mathcal{E} \simeq \mathcal{E}' \otimes p^*L$, where $p : X \times T \longrightarrow T$ is the second projection.

Fine moduli space If $\mathcal{M}_v(h)$ is representable by a scheme $M_v(h)$, then $M_v(h)$ is called a *fine moduli space.* Recall that $M_v(h)$ represents $\mathcal{M}_v(h)$ if there exists a functor isomorphism

$$\mathcal{M}_v(h) \simeq \mathrm{Mor}(\ \ , M_v(h)).$$

The identity $\mathrm{id} \in \mathrm{Mor}(M_v(h), M_v(h))$ induces a *universal family* $\mathcal{E}$ on $X \times M_v(h)$, which is unique up to a twist by a line bundle on $M_v(h)$.

Also note that any family $\mathcal{F}$ on $X \times T$ as above induces a classifying morphism $\varphi_{\mathcal{F}} : T \longrightarrow M_v(h)$ such that $(\varphi \times \mathrm{id}_X)^*\mathcal{E}$ and $\mathcal{F}$ are equivalent. The isomorphism applied to $\mathrm{Spec}(\mathbb{C})$ shows that the closed points of a fine moduli space $M_v(h)$ parametrize all semi-stable sheaves of Mukai vector v.

Coarse moduli space However, often a fine moduli space does not exist for various reasons that will be explained shortly. A weaker notion is introduced as follows.

A *coarse moduli space* is a scheme $M_v(h)$ together with a functor morphism

$$\mathcal{M}_v(h) \longrightarrow \mathrm{Mor}(\ , M_v(h))$$

such that the induced map

$$\mathcal{M}_v(h)(\mathrm{Spec}(k)) \longrightarrow M_v(h)(k)$$

defines a bijection of S-equivalence classes of semi-stable sheaves and closed points of $M_h(v)$. We omit the definition of S-equivalence, although the basic idea behind it can be observed in the example further below (see Remark 10.19, i)). The following result (due to Gieseker, Maruyama, Simpson) holds in much broader generality than just for K3 surfaces (essentially for all projective varieties).

Theorem 10.18 *A coarse moduli space $M_v(h)$ always exists. Moreover, $M_v(h)$ is a projective variety.*

Remarks 10.19 There are two sorts of problems that can prevent $M_v(h)$ from being fine:

i) Let us consider a curve of genus $g \geq 1$ and a non-trivial class $\eta \in H^1(C, \mathcal{O}_C)$ which will be interpreted as an extension class of $\mathcal{O}_C$ by $\mathcal{O}_C$. Thus, any multiple $t\eta$ defines a rank two vector bundle $E_{t\eta}$ as an extension

$$0 \longrightarrow \mathcal{O}_C \longrightarrow E_{t\eta} \longrightarrow \mathcal{O}_C \longrightarrow 0.$$

Clearly, $E_{t\eta} \simeq E_\eta \not\simeq \mathcal{O}_C \oplus \mathcal{O}_C$ for all $t \neq 0$ and $E_0 \simeq \mathcal{O}_C \oplus \mathcal{O}_C$. Moreover, it is not difficult to construct a locally free sheaf $\mathcal{E}$ on $\mathbb{A}^1 \times C$ with $\mathcal{E}_t \simeq E_{t\eta}$ for all $t \in \mathbb{A}^1$. This is a family of semi-stable bundles on C. If a fine moduli space M existed, we would obtain a morphism

$$\varphi_{\mathcal{E}} : \mathbb{A}^1 \longrightarrow M$$

which is constant on $\mathbb{A}^1 \setminus \{0\}$ and takes a different value in $0 \in \mathbb{A}^1$. This is absurd.

So, roughly, whenever strictly semi-stable sheaves of the given type exist, a fine moduli space is not expected to exist.

ii) Even if all semi-stable sheaves under consideration are in fact stable, a fine moduli space need not exist. This is mainly due to the non-existence of a universal family.

Glimpse of the construction The first and very important step in the construction of the moduli space is the proof that the sheaves in question, i.e. semi-stable sheaves of fixed numerical invariants, form a *bounded family.* In practice, this means that once a Hilbert polynomial $\chi(m)$ is fixed, there exists an $m_0 \gg 0$ such that for any semi-stable sheaf E with $\chi(E(m)) \equiv \chi(m)$

the sheaf $E(m)$ is globally generated and has vanishing higher cohomology $H^i(X, E(m)) = 0$, $i > 0$, as soon as $m \geq m_0$.

This enables one to write any such E as a quotient

$$V \otimes \mathcal{O}(-m) \longrightarrow\!\!\!\!\!\rightarrow E$$

with V a vector space of dimension $\chi(m)$. The surjection is given by the choice of an isomorphism $V \simeq H^0(X, E(m))$.

A classical result of Grothendieck shows that for any coherent sheaf (in our case it is $V \otimes \mathcal{O}(-m)$) the quotients with fixed Hilbert polynomial are parametrized by a so-called *Quot-scheme*

$$\mathrm{Quot}(V \otimes \mathcal{O}(-m), \chi).$$

Moreover, on the product $X \times \mathrm{Quot}(V \otimes \mathcal{O}(-m), \chi)$ there exists a universal quotient

$$V \otimes q^*\mathcal{O}(-m) \longrightarrow\!\!\!\!\!\rightarrow \mathcal{E}.$$

(More precisely, together with this universal quotient the Quot-scheme represents the functor of all quotients of $V \otimes \mathcal{O}(-m)$.)

Let us denote the open subsets of semi-stable quotients by

$$\mathcal{R} \subset \mathrm{Quot}(V \otimes \mathcal{O}(-m), \chi).$$

The moduli space $M_\chi(h)$ of semi-stable sheaves is then constructed as a GIT-quotient

$$M_\chi(h) = \mathcal{R}/\!\!/\mathrm{PGl}(V).$$

(Usually one has to increase the m_0 once more at this point, but this is a rather technical matter.)

The general machinery of geometric invariant theory (GIT) requires actually not just the action of a reductive group, e.g. $\mathrm{PGl}(V)$ in our case, but also the choice of a linearized ample line bundle (see [84]). This line bundle descends by construction to an ample line bundle on the moduli space.

There are several possibilities for such a line bundle (for some historical comments see [53]), but the most natural one for our purpose is the one used by Simpson. Any quotient

$$V \otimes \mathcal{O}(-m) \longrightarrow\!\!\!\!\!\rightarrow E$$

yields for $k \gg 0$ a quotient

$$V \otimes H^0(X, \mathcal{O}(k-m)) \longrightarrow\!\!\!\!\!\rightarrow H^0(X, E(k)).$$

For $k \gg 0$ this procedure yields an embedding

$$\mathrm{Quot}(V \otimes \mathcal{O}(-m), \chi) \hookrightarrow \mathrm{Gr}_k := \mathrm{Gr}(V \otimes H^0(X, \mathcal{O}(k-m)), \chi(k)).$$

The latter can in turn be embedded via the Plücker embedding

$$\mathrm{Gr}_k \hookrightarrow \mathrm{P}_k := \mathbb{P}\left(\bigwedge^{\chi(k)}(V \otimes H^0(X, \mathcal{O}(k-m)))\right).$$

Under this map the lines $\det(H^0(X, E(k))) = \bigwedge^{\chi(k)} H^0(X, E(k))$ are naturally identified with the fibres of the tautological line bundle $\mathcal{O}(1)$.

In other words, the line bundle

$$\det(p_*(\mathcal{E} \otimes q^*\mathcal{O}(k)))$$

is isomorphic to the pull-back of $\mathcal{O}(1)$ under the composition

$$\mathrm{Quot}(V \otimes \mathcal{O}(-m), \chi) \hookrightarrow \mathrm{Gr}_k \hookrightarrow \mathrm{P}_k.$$

In particular, $\det(p_*(\mathcal{E} \otimes q^*\mathcal{O}(k)))$ is a very ample line bundle for $k \gg 0$.

Numerical criteria Fortunately, there are numerical criteria that allow us to ensure that certain moduli spaces are fine.

Proposition 10.20 *Suppose that* $v = (r, \ell, s)$ *satisfies g.c.d.*$(r, (h, \ell), s) = 1$. *Then* $M_v(h)$ *is a fine moduli space.*

Proof We cannot give the complete proof here, but we shall at least show that under the assumption every semi-stable sheaf is stable.

Suppose E is semi-stable, but not stable. Then there exists a proper subsheaf $F \subset E$ with $\chi(F(n)) \cdot \mathrm{rk}(E) = \chi(E(n)) \cdot \mathrm{rk}(F)$ for all $n \gg 0$. Let us compute $\chi(E(n))$ in terms of the Mukai vector v (see (5.5) and the convention (10.1)):

$$\begin{aligned}\chi(E(n)) = \chi(\mathcal{O}(-n), E) &= -\langle v(\mathcal{O}(-n)), v(E)\rangle \\ &= r + (n^2/2)h^2 r + s + n(h, \ell).\end{aligned}$$

A similar calculation expresses $\chi(F(n))$ in terms of its Mukai vector (r', ℓ', s').

Thus, the equality of the Hilbert polynomials is equivalent to the two equalities $r'(h, \ell) = r(h, \ell')$ and $r's = rs'$.

By assumption we can find integers a, b, c with $1 = ar + b(h, \ell) + cs$. Multiplied with r' it yields $r' = arr' + br'(h, \ell) + cr's = r(ar' + b(h, \ell') + cs')$, which contradicts $0 < r' < r$.

So, the closed points of the coarse moduli space $M_v(h)$ are in bijection to stable sheaves.

We omit the argument that shows that our numerical assumption also implies the existence of a universal sheaf. Roughly, one has to ensure that a certain line bundle twist of the universal quotient $\mathcal{E}$ on the product $X \times \mathrm{Quot}(V \otimes \mathcal{O}(-m), \chi)$ descends to a sheaf on the moduli space. We have to refer to [53, Rem.4.6.8] for any more details. □

Remark 10.21 We emphasize that the pull-back of a universal sheaf (if it exists) to the Quot-scheme is in general not the universal quotient. More precisely, if $\mathcal{E}$ is a universal sheaf on $X \times M$, then the pull-back $(\mathrm{id} \times \pi)^*\mathcal{E}$ under the quotient morphism $\pi : \mathcal{R} \twoheadrightarrow M$ can be viewed as a quotient

$$V \otimes q^*\mathcal{O}(-m) \otimes p^*\mathcal{M} \twoheadrightarrow (\mathrm{id} \times \pi)^*\mathcal{E}$$

for a certain line bundle $\mathcal{M}$ on $\mathcal{R}$. In particular, the ample line bundle arising in the construction (see the explanation above) can be identified with

$$\det\left(p_*(\mathcal{E} \otimes q^*\mathcal{O}(k) \otimes p^*\mathcal{M}^*)\right) \simeq \det\left(p_*(\mathcal{E} \otimes q^*\mathcal{O}(k))\right) \otimes \mathcal{M}^{-\chi(k)}.$$

Note that $\mathcal{M}$ itself can be described by the isomorphism $\mathcal{M}^{\chi(m)} \simeq \det(p_*(\mathcal{E} \otimes q^*\mathcal{O}(m)))$ that stems from the surjection $V \otimes p^*\mathcal{M} \twoheadrightarrow \mathcal{E} \otimes q^*\mathcal{O}(m)$.

Lemma 10.22 *Suppose there exists a vector $v' \in \widetilde{H}^{1,1}(X, \mathbb{Z})$ such that $\langle v, v' \rangle = 1$. Then one can find an ample class h with g.c.d.$(r, (h, \ell), s) = 1$.*

Proof Write $v = (r, \ell, s)$ and $v' = (r', \ell', s')$. Suppose a divides r, (ℓ, ℓ'), and s, then it also divides $(\ell, \ell') - rs' - sr' = \langle v, v' \rangle$. Hence $a = \pm 1$. A priori, ℓ' need not be ample, but by adding $(kr) \cdot A$ with $k \gg 0$ and A ample we obtain an ample class h with *g.c.d.*$(r, (h, \ell), s) = 1$. □

Corollary 10.23 *Suppose that v and v' are as in the lemma. Then there exists an ample class h such that the moduli space $M_v(h)$ is fine.* □

Local structure What follows now is a very sketchy discussion of the local structure of the moduli space.

Stability is an open condition or, in other words, every small deformation of a stable sheaf is again stable. Since first order deformation of a sheaf E are parametrized by $\mathrm{Ext}^1(E, E)$ (cf. [45]), the tangent space of $M_v(h)$ at $[E] \in M_v(h)$ is described by

$$T_{[E]}M_v(h) \simeq \mathrm{Ext}^1(E, E).$$

This allows us in particular to determine the dimension of this tangent space. Indeed, if E is stable then $\mathbb{C} = \mathrm{Hom}(E, E) = \mathrm{Ext}^2(E, E)^*$ and hence

$$\dim \mathrm{Ext}^1(E, E) = 2 + \langle v(E), v(E) \rangle.$$

In particular, the dimension is two if and only if $v(E)$ is isotropic. The calculation also suggests that $M_v(h)$ is actually smooth, for the dimension of the tangent spaces stays constant. The formal reason for the smoothness of $M_v(h)$ is the vanishing of the obstruction space

$$\mathrm{Ext}^2(E, E)_0 = \mathrm{Ker}\left(\mathrm{Ext}^2(E, E) \xrightarrow{\mathrm{tr}} H^2(X, \mathcal{O}_X) \right).$$

Furthermore, Serre duality defines a non-degenerate pairing

$$\mathrm{Ext}^1(E,E)\times\mathrm{Ext}^1(E,E)\longrightarrow\mathbb{C}.$$

It was Mukai who first observed that this alternating form glues to a regular symplectic form on $M_v(h)$ (see [81]).

We summarize the discussion in the following

Proposition 10.24 *Let $v, v' \in \widetilde{H}^{1,1}(X)$ be integral vectors with $\langle v, v'\rangle = 1$. Then there exists an ample class h such that*

i) *The moduli space $M_h(v)$ is fine and parametrizes only stable sheaves.*
ii) $\dim M_h(v) = 2 + \langle v, v\rangle$.
iii) *The moduli space $M_v(h)$ is a smooth algebraic symplectic variety and, in particular, $\omega_{M_h(v)} \simeq \mathcal{O}$.* □

Derived equivalence of the moduli space We choose v, v', and h as in the proposition and suppose furthermore that v is isotropic, i.e. $\langle v, v\rangle = 0$.

Pick one of the connected components $M \subset M_v(h)$. The restriction of the universal sheaf to $M \times X$ yields an M-flat sheaf $\mathcal{E}$.

Proposition 10.25 **(Mukai)** *Under these assumptions the sheaf $\mathcal{E}$ induces an equivalence*

$$\Phi_{\mathcal{E}}:\mathrm{D^b}(X)\xrightarrow{\ \sim\ }\mathrm{D^b}(M).$$

Proof Let $s \neq t \in M$. Then $\mathcal{E}_s \not\simeq \mathcal{E}_t$ and therefore

$$\mathrm{Hom}(\mathcal{E}_s,\mathcal{E}_t)=\mathrm{Ext}^2(\mathcal{E}_s,\mathcal{E}_t)=0.$$

On the other hand, $\chi(\mathcal{E}_s,\mathcal{E}_t) = -\langle v, v\rangle = 0$. Hence, also $\mathrm{Ext}^1(\mathcal{E}_s,\mathcal{E}_t) = 0$. The standard criterion, Proposition 7.1 (or Corollary 7.5), applies and shows that the Fourier–Mukai transform $\Phi_{\mathcal{E}}:\mathrm{D^b}(M)\longrightarrow\mathrm{D^b}(X)$ is fully faithful.

Since the canonical bundle of M is trivial, it is an equivalence (see Corollary 7.12). □

Non-emptiness What has been left out in the above discussion is the following important point: Why is $M_v(h)$ non-empty?

The way to tackle this problem is to assume as a first step that X is a very special, e.g. an elliptic K3 surface, and to prove existence of semi-stable sheaves by an explicit construction.

In a second step, use deformation theory of K3 surfaces and stable sheaves in a family to deduce the same result for any K3 surface. This part is quite involved. The reader might consult [53] for details.

11

FLIPS AND FLOPS

In this chapter, we present a series of results elucidating the relation between D-equivalence and K-equivalence (see Conjecture 6.24).

The first step towards a better understanding of the situation shall be done in Section 11.2, where we study how the derived category changes under blow-up. The results, due to Bondal and Orlov, will be sufficient to show that the derived category does not change under the standard flop, one of the simplest birational correspondences (see Section 11.3). In fact, more recently Bridgeland was able to show that the derived category is invariant under general three-dimensional flops.

The situation is slightly more complicated for a Mukai flop, another classical birational correspondence. This will be explained in Section 11.4. In this case, the birational correspondence itself does not define a derived equivalence, but Kawamata and Namikawa were able to prove that one can nevertheless find another Fourier–Mukai kernel that does.

11.1 Preparations: Closed embeddings and blow-ups

This section is of a rather technical nature. The aim is to compute various Tor- and Ext-groups arising naturally in situations like closed embeddings and blow-ups. E.g. one needs to know exactly how to compute $\mathcal{E}xt^i_X(\mathcal{O}_Y, \mathcal{O}_Y)$ for a closed subvariety $j : Y \hookrightarrow X$ or the pull-back $q^*\mathcal{O}_Y$ under the blow-up $q : \widetilde{X} \longrightarrow X$ of X along Y. The strategy in both cases is to treat a linearized version first and then to glue the local information obtained in this way.

We shall first discuss in detail the situation of a closed embedding

$$j : Y \hookrightarrow X$$

of codimension c.

In a first step, we will suppose that Y is given as the the zero locus of a regular section $s \in H^0(X, \mathcal{E})$ of a locally free sheaf $\mathcal{E}$ of rank c. In this case, its structure sheaf can be resolved by the Koszul complex

$$0 \longrightarrow \bigwedge^c \mathcal{E}^* \longrightarrow \cdots \longrightarrow \mathcal{E}^* \longrightarrow \mathcal{O}_X \longrightarrow j_*\mathcal{O}_Y \longrightarrow 0, \quad (11.1)$$

with morphisms given by contraction with s. Note that in this situation the normal bundle of $Y \subset X$ is given by

$$\mathcal{N} := \mathcal{N}_{Y/X} \simeq \mathcal{E}|_Y. \quad (11.2)$$

To be more precise, differentiating the section $s \in H^0(X, \mathcal{E})$ induces a canonical morphism $\mathcal{T}_X|_Y \longrightarrow \mathcal{E}|_Y$ which provides the canonical isomorphism (11.2).

Proposition 11.1 *Suppose $j : Y \hookrightarrow X$ with normal bundle $\mathcal{N}$ is the zero locus of a regular section of a locally free sheaf of rank $c = \mathrm{codim(Y \subset X)}$. Then there exists a canonical isomorphism*

$$\text{i)} \quad j^* j_* \mathcal{O}_Y \simeq \bigoplus \bigwedge^k \mathcal{N}^*[k]$$

and for any $\mathcal{F}^\bullet \in \mathrm{D^b}(Y)$ one has:

$$\text{ii)} \quad j_* j^* j_* \mathcal{F}^\bullet \simeq j_* \mathcal{O}_Y \otimes j_* \mathcal{F}^\bullet \simeq j_* \left(\bigoplus \bigwedge^k \mathcal{N}^*[k] \otimes \mathcal{F}^\bullet \right)$$

$$\text{iii)} \quad \mathcal{H}om_X(j_* \mathcal{O}_Y, j_* \mathcal{F}^\bullet) \simeq j_* \left(\bigoplus \bigwedge^k \mathcal{N}[-k] \otimes \mathcal{F}^\bullet \right).$$

Proof As before, we denote by $s \in H^0(X, \mathcal{E})$ the section defining Y. Thus, (11.1) can be read as a quasi-isomorphism $\bigwedge^\bullet \mathcal{E}^* \simeq j_* \mathcal{O}_Y$ which allows us to compute $j^* j_* \mathcal{O}_Y$ as $\bigwedge^\bullet \mathcal{E}^*|_Y$. As the differentials in the Koszul complex $\bigwedge^\bullet \mathcal{E}^*$ are given by contraction with the defining section s, they become trivial on Y. In other words, $j^* j_* \mathcal{O}_Y \simeq \bigwedge^\bullet \mathcal{E}^*|_Y \simeq \bigoplus \bigwedge^k \mathcal{E}^*|_Y[k]$. Using the canonical isomorphism $\mathcal{N} \simeq \mathcal{E}|_Y$ this proves i).

To prove ii), split the resolution $\bigwedge^\bullet \mathcal{E}^* \longrightarrow j_* \mathcal{O}_Y$ into short exact sequences

$$\begin{array}{ccccccccc} \cdots \to & \bigwedge^{i+1}\mathcal{E}^* & & \longrightarrow & & \bigwedge^i \mathcal{E}^* & & \longrightarrow & & \bigwedge^{i-1}\mathcal{E}^* & \to \cdots \\ & & \searrow & & \nearrow & & \searrow & & \nearrow & & \\ & & & M_{i+1} & & & & M_i & & & \end{array}$$

As $j_* \mathcal{F}^\bullet$ is concentrated in Y and the morphisms $M_{i+1} \longrightarrow \bigwedge^i \mathcal{E}^*$ vanish along Y, each short exact sequence

$$0 \longrightarrow M_{i+1} \longrightarrow \bigwedge^i \mathcal{E}^* \longrightarrow M_i \longrightarrow 0,$$

considered as a distinguished triangle yields isomorphisms

$$M_i \otimes j_* \mathcal{F}^\bullet \simeq (\bigwedge^i \mathcal{E}^* \otimes j_* \mathcal{F}^\bullet) \oplus (M_{i+1}[1] \otimes j_* \mathcal{F}^\bullet).$$

Putting things together and using

$$\bigwedge^i \mathcal{E}^* \otimes j_* \mathcal{F}^\bullet \simeq j_*(j^* \bigwedge^i \mathcal{E}^* \otimes \mathcal{F}^\bullet) \simeq j_*(\bigwedge^i \mathcal{N}^* \otimes \mathcal{F}^\bullet)$$

proves the second isomorphism in ii). The first one is simply the projection formula.

In order to prove iii), we first write

$$\mathcal{H}om(j_* \mathcal{O}_Y, j_* \mathcal{F}^\bullet) \simeq \mathcal{H}om(\bigwedge^\bullet \mathcal{E}^*, j_* \mathcal{F}^\bullet) \simeq (\bigwedge^\bullet \mathcal{E}^*)^\vee \otimes j_* \mathcal{F}^\bullet.$$

Now argue as before, i.e. split the complex $(\bigwedge^\bullet \mathcal{E}^*)^\vee$ into short exact sequences which all yield split distinguished triangles along Y. □

By abuse of notation, we will later (and, in fact, did so before) often write simply $\mathcal{O}_Y$ instead of $j_*\mathcal{O}_Y$.

Corollary 11.2 *Under the assumptions of the proposition one has for any* $\mathcal{F}^\bullet \in \mathrm{D^b}(Y)$*:*

$$\mathcal{H}^\ell(j^*j_*\mathcal{F}^\bullet) \simeq \bigoplus_{s-r=\ell} \bigwedge^r \mathcal{N}^* \otimes \mathcal{H}^s(\mathcal{F}^\bullet)$$

and

$$\mathcal{E}xt^\ell_X(j_*\mathcal{O}_Y, j_*\mathcal{F}^\bullet) \simeq j_*\left(\bigoplus_{r+s=\ell} \bigwedge^r \mathcal{N} \otimes \mathcal{H}^s(\mathcal{F}^\bullet)\right).$$

Proof As j_* is exact, one has $\mathcal{H}^\ell \circ j_* \simeq j_* \circ \mathcal{H}^\ell$. Thus, the proposition implies $j_*\mathcal{H}^\ell(j^*j_*\mathcal{F}^\bullet) \simeq \mathcal{H}^\ell\left(j_* \bigoplus \bigwedge^k \mathcal{N}^*[k] \otimes \mathcal{F}^\bullet\right) \simeq j_*\left(\bigoplus \bigwedge^k \mathcal{N}^* \otimes \mathcal{H}^{k+\ell}(\mathcal{F}^\bullet)\right)$, which yields the first assertion. The second assertion is proved similarly. Note that in both cases we use that tensor product with the locally free sheaf $\mathcal{N}$ commutes with taking cohomology. □

Remark 11.3 There is a different approach for the construction of the isomorphisms

$$\mathcal{E}xt^\ell_X(j_*\mathcal{O}_Y, j_*\mathcal{O}_Y) \simeq \bigwedge^\ell \mathcal{N}.$$

(Here, we set $\mathcal{F}^\bullet = \mathcal{O}_Y$ in the second isomorphism of the corollary.) This one works without assuming that $Y \subset X$ is the zero set of a regular section.

By definition $\mathcal{N} := \mathcal{N}_{Y/X} \simeq \mathcal{E}xt^0_X(\mathcal{I}_Y, \mathcal{O}_Y)$. Applying $\mathcal{H}om_X(\ \ , \mathcal{O}_Y)$ to the short exact sequence

$$0 \longrightarrow \mathcal{I}_Y \longrightarrow \mathcal{O}_X \longrightarrow \mathcal{O}_Y \longrightarrow 0$$

yields a canonical homomorphism $\mathcal{N} \longrightarrow \mathcal{E}xt^1_X(\mathcal{O}_Y, \mathcal{O}_Y)$, which is in fact an isomorphism as $\mathcal{E}xt^1_X(\mathcal{O}_X, \mathcal{O}_Y) = 0$. This proves the claim for $\ell = 1$.

For $\ell > 1$ one considers the induced homomorphism

$$\bigwedge^\ell \mathcal{N} \simeq \bigwedge^\ell \mathcal{E}xt^1_X(\mathcal{O}_Y, \mathcal{O}_Y) \longrightarrow \mathcal{E}xt^\ell_X(\mathcal{O}_Y, \mathcal{O}_Y),$$

where the latter is given by the cup product (or, equivalently, composition). That this is indeed an isomorphism can be checked by a local calculation.

Corollary 11.4 *Let* $j : Y \hookrightarrow X$ *be a smooth hypersurface. Then*

i) $j^*j_*\mathcal{O}_Y \simeq \mathcal{O}_Y \oplus \mathcal{O}_Y(-Y)[1]$

and for any $\mathcal{F}^\bullet \in \mathrm{D}^{\mathrm{b}}(Y)$ *there exists a distinguished triangle and an isomorphism:*

$$\text{ii)}\quad \mathcal{F}^\bullet \otimes \mathcal{O}_Y(-Y)[1] \longrightarrow j^*j_*\mathcal{F}^\bullet \longrightarrow \mathcal{F}^\bullet$$

$$\text{iii)}\quad j_*j^*j_*\mathcal{F}^\bullet \simeq j_*\mathcal{F}^\bullet \oplus j_*(\mathcal{F}^\bullet \otimes \mathcal{O}_Y(-Y))[1].$$

Proof Any hypersurface $Y \subset X$ is the zero set of a section $s \in H^0(X, \mathcal{O}(Y))$. Thus, i) and iii) are just special cases of the proposition.

Adjunction and Grothendieck–Verdier duality provide us with natural morphisms $j^*j_*\mathcal{F}^\bullet \longrightarrow \mathcal{F}^\bullet$, respectively $\mathcal{F}^\bullet \otimes \mathcal{O}_Y(-Y)[1] \longrightarrow j^*j_*\mathcal{F}^\bullet$, but the problem remains to prove that put together they form a distinguished triangle. So, we rather follow the argument in [15, Lemma 3.3] where the distinguished triangle is very elegantly constructed all in one go.

Let $\Gamma \subset Y \times X$ and $\bar{\Gamma} \subset X \times Y$ be the graphs of the inclusion $j : Y \hookrightarrow X$, respectively of its transpose. Then one has $j_* \simeq \Phi_{\mathcal{O}_\Gamma} : \mathrm{D}^{\mathrm{b}}(Y) \longrightarrow \mathrm{D}^{\mathrm{b}}(X)$ and $j^* \simeq \Phi_{\mathcal{O}_{\bar{\Gamma}}} : \mathrm{D}^{\mathrm{b}}(X) \longrightarrow \mathrm{D}^{\mathrm{b}}(Y)$ (see Example 5.4). Hence,

$$j^*j_* \simeq \Phi_{\mathcal{R}} \quad \text{with} \quad \mathcal{R} := \pi_{YY*}(\pi_{YX}^*\mathcal{O}_\Gamma \otimes \pi_{XY}^*\mathcal{O}_{\bar{\Gamma}})$$

(see Proposition 5.10 for the notation). Thus, to prove the assertion, it suffices to construct a distinguished triangle

$$\iota_*\mathcal{O}_Y(-Y)[1] \longrightarrow \mathcal{R} \longrightarrow \iota_*\mathcal{O}_Y, \tag{11.3}$$

where $\iota : Y \longrightarrow Y \times Y$ is the diagonal embedding.

The tensor product $\pi_{YX}^*\mathcal{O}_\Gamma \otimes \pi_{XY}^*\mathcal{O}_{\bar{\Gamma}} \simeq \mathcal{O}_{\Gamma\times Y} \otimes \mathcal{O}_{Y\times\bar{\Gamma}}$ is not the structure sheaf of the intersection. Indeed, the two subvarieties $\Gamma\times Y$ and $Y\times\bar{\Gamma}$ of $Y\times X\times Y$ do not intersect transversally. Instead consider the natural diagram

$$\begin{array}{ccccc} \Gamma \times Y & \stackrel{\varphi}{\hookrightarrow} & Y \times Y \times Y & \stackrel{\bar{\varphi}}{\hookleftarrow} & Y \times \bar{\Gamma} \\ & & \big\downarrow{\scriptstyle \psi} & & \\ & & Y \times X \times Y & & \end{array}$$

and the morphism $\pi : \Gamma\times Y \longrightarrow Y\times Y$ defined as the restriction of the projection $\pi_{YY} : Y \times X \times Y \longrightarrow Y \times Y$, i.e. $\pi = \pi_{XX} \circ \psi \circ \varphi$. Then the projection formula allows us to write

$$\mathcal{R} = \pi_{YY*}((\psi\circ\varphi)_*\mathcal{O}_{\Gamma\times Y} \otimes (\psi\circ\bar{\varphi})_*\mathcal{O}_{Y\times\bar{\Gamma}}) \simeq \pi_*(\varphi^*\psi^*(\psi_*\bar{\varphi}_*\mathcal{O}_{Y\times\bar{\Gamma}})).$$

Clearly, ψ is the closed embedding of a divisor and thus by Corollary 11.2 for any sheaf $\mathcal{G}$ on $Y \times Y \times Y$ the complex $\psi^*\psi_*\mathcal{G}$ has only two non-trivial cohomologies: $\mathcal{H}^0(\psi^*\psi_*\mathcal{G}) \simeq \mathcal{G}$ and $\mathcal{H}^{-1}(\psi^*\psi_*\mathcal{G}) \simeq \mathcal{G} \otimes \pi_2^*\mathcal{O}_Y(-Y)$. Applied to $\mathcal{G} = \bar{\varphi}_*\mathcal{O}_{Y\times\bar{\Gamma}}$ this yields a distinguished triangle

$$\bar{\varphi}_*\mathcal{O}_{Y\times\bar{\Gamma}} \otimes \pi_2^*\mathcal{O}_Y(-Y)[1] \longrightarrow \psi^*\psi_*\bar{\varphi}_*\mathcal{O}_{Y\times\bar{\Gamma}} \longrightarrow \bar{\varphi}_*\mathcal{O}_{Y\times\bar{\Gamma}}. \tag{11.4}$$

As $\mathrm{Im}(\varphi)$ and $\mathrm{Im}(\bar{\varphi})$ meet transversally within $Y \times Y \times Y$, the pull-back $\varphi^*\bar{\varphi}_*\mathcal{O}_{Y\times\bar{\Gamma}}$ is nothing but the structure sheaf of the intersection, i.e. of the image of the diagonal embedding $\eta : Y \longrightarrow Y \times Y \times Y$, $y \longmapsto (y, y, y)$. Then under pull-back (11.4) becomes the distinguished triangle

$$\eta_*\mathcal{O}_Y(-Y)[1] \longrightarrow \varphi^*\psi^*\psi_*\bar{\varphi}_*\mathcal{O}_{Y\times\bar{\Gamma}} \longrightarrow \eta_*\mathcal{O}_Y.$$

The direct image under $\pi : \Gamma \times Y \times Y \times Y$ yields (11.3). □

Exercise 11.5 Let $j : Y \hookrightarrow X$ be a smooth hypersurface. Use the projection formula and Corollary 11.4 to prove the existence of functorial isomorphisms

$$j^*j_*j^*\mathcal{E}^\bullet \simeq j^*\mathcal{E}^\bullet \oplus j^*\mathcal{E}^\bullet(-Y)[1]$$

for all $\mathcal{E}^\bullet \in \mathrm{D^b}(X)$.

Remark 11.6 In (11.4) the morphism $\psi^*\psi_*\bar{\varphi}_*\mathcal{O}_{Y\times\bar{\Gamma}} \longrightarrow \bar{\varphi}_*\mathcal{O}_{Y\times\bar{\Gamma}}$ is given by adjunction. Hence, in the distinguished triangle ii) of the corollary the morphism $j^*j_*\mathcal{F} \longrightarrow \mathcal{F}$ is also just the canonical adjunction morphism.

Examples 11.7 The discussion covers in particular the 'linearization of a closed embedding'. By this we mean the following. Consider a closed embedding $Y \subset X$ with normal bundle $\mathcal{N}$. To distinguish between the normal bundle $\mathcal{N}$ considered as a sheaf on Y and the affine bundle over Y we shall write $|\mathcal{N}|$ for the latter. The projection and its zero section will be denoted

$$p : |\mathcal{N}| \longrightarrow Y \quad \text{respectively} \quad j : Y \hookrightarrow |\mathcal{N}|.$$

Clearly, the normal bundle of $Y \subset |\mathcal{N}|$ is again $\mathcal{N}$ and, moreover, $Y \subset |\mathcal{N}|$ can be described as the zero set of the canonical section $s \in H^0(|\mathcal{N}|, p^*\mathcal{N})$ defined by the condition $s(v, x) = v$ for any closed point $(v, x) \in |\mathcal{N}|$, where v is a vector in the fibre $p^*\mathcal{N}(v, x) = \mathcal{N}(x)$. The Koszul complex for $\mathcal{E} := p^*\mathcal{N}$ becomes

$$0 \longrightarrow \bigwedge^c p^*\mathcal{N}^* \longrightarrow \cdots \longrightarrow p^*\mathcal{N}^* \longrightarrow \mathcal{O}_{|\mathcal{N}|} \longrightarrow \mathcal{O}_Y \longrightarrow 0.$$

Usually, computations can be performed more explicitly in this situation.

As it turns out, the computation of the cohomology sheaves of Corollary 11.2 remains valid even if the subvariety Y is not given as the zero section of a locally free sheaf. This will be done by passing to the local situation where Y is actually defined by the vanishing of regular functions, i.e. by a section of a trivial vector bundle $\mathcal{O}^{\oplus c}$, and then gluing the cohomology sheaves. We have chosen to present the result in the 'semi-local' case first, as it avoids the explicit choice of the functions defining Y and hints already towards the global result.

Proposition 11.8 *Let $j : Y \hookrightarrow X$ be an arbitrary closed embedding of smooth varieties. Then there exist isomorphisms*

$$\mathcal{H}^i(j^*j_*\mathcal{O}_Y) \simeq \bigwedge^{-i} \mathcal{N}^*_{Y/X} \text{ and } \mathcal{E}xt^i_X(j_*\mathcal{O}_Y, j_*\mathcal{O}_Y) \simeq \bigwedge^i \mathcal{N}_{Y/X}.$$

Proof Choose a global locally free resolution $\mathcal{G}^\bullet \longrightarrow \mathcal{O}_Y$ and consider the induced free resolution $\mathcal{G}^\bullet_y \longrightarrow \mathcal{O}_{Y,y}$ of $\mathcal{O}_{X,y}$-modules for any point $y \in Y$.

Locally around $y \in Y \subset X$, i.e. on an open neighbourhood $y \in U \subset X$, we may find a locally free sheaf $\mathcal{E}$ of rank c together with a regular section $s \in H^0(U, \mathcal{E})$ defining $Y \cap U$. This yields a second free resolution $\bigwedge^\bullet \mathcal{E}^*_y \longrightarrow \mathcal{O}_{Y,y}$. (The latter one, moreover, is minimal, i.e. tensored with $k(y) = \mathcal{O}_{X,y}/\mathfrak{m}_y$ the differentials become trivial.)

Using the projectivity of free modules, we obtain a morphism of complexes $\varphi : \mathcal{G}^\bullet_y \longrightarrow \bigwedge^\bullet \mathcal{E}^*_y$. Pulling back via $j : Y \hookrightarrow X$ and taking cohomology yields isomorphisms $\mathcal{H}^i(j^*\varphi) : \mathcal{H}^i(j^*\mathcal{G}^\bullet)_y \simeq \mathcal{H}^i(j^*\mathcal{G}^\bullet_y) \xrightarrow{\sim} \mathcal{H}^i(j^* \bigwedge^\bullet \mathcal{E}^*_y) \simeq \bigwedge^{-i} \mathcal{N}^*_y$. (As has been explained earlier, the isomorphism $\mathcal{E}|_Y \simeq \mathcal{N}$ is canonically induced by the choice of the section s.)

Any other choice of the isomorphism φ is homotopic to the original one (cf. Lemma 2.39) and thus induces the same map on cohomology. For another choice of the minimal resolution, say defined by a section $\tilde{s}$ of $\tilde{\mathcal{E}}$, any isomorphism $\mathcal{E} \simeq \tilde{\mathcal{E}}$ that sends s to $\tilde{s}$ induces the identity on $\mathcal{N}$. Hence, the identification $\mathcal{H}^i(j^*\mathcal{G})_y \simeq \bigwedge^{-i} \mathcal{N}^*_y$ is independent of any choice and thus leads to a global isomorphism $\mathcal{H}^i(j^*j_*\mathcal{O}_Y) \simeq \bigwedge^{-i} \mathcal{N}^*$.

The proof of the second assertion is similar. □

Examples 11.9 i) Consider the diagonal embedding $\iota : X \xrightarrow{\sim} \Delta \subset X \times X$, the conormal bundle of which is by definition the cotangent bundle Ω_X. Equivalently, $\mathcal{N}_{\Delta/X \times X} \simeq \iota_*\mathcal{T}_X$. Thus, the above isomorphism can in this situation be written as

$$\mathcal{E}xt^i_{X\times X}(\iota_*\mathcal{O}_X, \iota_*\mathcal{O}_X) \simeq \iota_* \bigwedge^i \mathcal{T}_X.$$

ii) The formula can also be applied to a closed point $x \in X$. It leads to the following description

$$\mathrm{Ext}^i_X(k(x), k(x)) \simeq \bigwedge^i \mathrm{Ext}^1_X(k(x), k(x)) \simeq \bigwedge^i T_x,$$

which has been alluded to before. Here, T_x denotes the Zariski tangent space at $x \in X$.

Let us now pass to blow-ups. Throughout, the following notation will be used. As before let $j : Y \hookrightarrow X$ denote a closed smooth subvariety of codimension c in a smooth projective variety X. The normal bundle of Y in X, denoted $\mathcal{N} := \mathcal{N}_{Y/X}$, is a locally free sheaf of rank c on Y. We shall be interested in the blow-up of X along Y which is a projective morphism $q : \widetilde{X} \longrightarrow X$. The exceptional divisor of this birational map is $E := q^{-1}(Y)$. Via the restriction of q it is identified with the projective bundle $\pi : \mathbb{P}(\mathcal{N}) \longrightarrow Y$. The ultimate goal is to study the following diagram from the derived category point of view

$$\begin{array}{ccc} \widetilde{X} & \xrightarrow{\quad q \quad} & X \\ {\scriptstyle i}\uparrow & & \uparrow{\scriptstyle j} \\ E = \mathbb{P}(\mathcal{N}) & \xrightarrow{\quad \pi \quad} & Y. \end{array}$$

The canonical bundle of the blow-up $\widetilde{X}$ can be computed by the formula

$$\omega_{\widetilde{X}} \simeq q^*\omega_X \otimes \mathcal{O}((c-1)E).$$

The restriction of $\mathcal{O}_{\widetilde{X}}(E)$ to a fibre $\pi^{-1}(y)$ is isomorphic to $\mathcal{O}(-1)$ and, more precisely, $\mathcal{O}_E(E) := \mathcal{O}(E)|_E \simeq \mathcal{O}_\pi(-1)$.

Remark 11.10 There is a converse to this construction which allows us to contract negative divisors. More precisely, if $\pi : E = \mathbb{P}(\mathcal{N}) \longrightarrow Y$ is a projective bundle over a smooth variety Y and $E \subset \widetilde{X}$ is a closed embedding of codimension one such that $\mathcal{O}(E)$ on the fibres of $E \longrightarrow Y$ is of degree -1, then the projection $\pi : E \longrightarrow Y$ extends to a morphism $q : \widetilde{X} \longrightarrow X$ onto a smooth complex manifold X containing Y such that q is the blow-up of $Y \subset X$. This result is due to Fujiki and Nakano, see [36]. One word of warning: The manifold X need not be projective even if $\widetilde{X}$, E, and Y are. In what follows, we will often tacitly add the projectivity of the contraction X as an extra assumption.

We put ourselves again in the 'semi-local' situation, i.e. we shall suppose that $Y \subset X$ is given as the zero set of a regular section $s \in H^0(X, \mathcal{E})$ of a locally free sheaf $\mathcal{E}$ of rank c.

Consider the projective bundle $g : \mathbb{P}(\mathcal{E}) \longrightarrow X$ and the relative Euler sequence

$$0 \longrightarrow \mathcal{O}_g(-1) \longrightarrow g^*\mathcal{E} \xrightarrow{\ \varphi\ } \mathcal{T}_g \otimes \mathcal{O}_g(-1) \longrightarrow 0.$$

To the section s we associate the global section

$$t := \varphi(g^*(s)) \in H^0(\mathbb{P}(\mathcal{E}), \mathcal{T}_g \otimes \mathcal{O}_g(-1))$$

and consider its zero set $Z(t) \subset \mathbb{P}(\mathcal{E})$. Then $g : Z(t) \setminus g^{-1}(Y) \xrightarrow{\sim} X \setminus Y$ and $g^{-1}(Y) \subset Z(t)$, as t vanishes over Y. From these two observations it follows almost immediately that $Z(t)$ can be identified with the blow-up $q : \widetilde{X} \longrightarrow X$

along Y. In particular, one has the two commutative diagrams

$$\begin{array}{ccc} \widetilde{X} & \overset{\iota}{\hookrightarrow} & \mathbb{P}(\mathcal{E}) \\ & {\scriptstyle q}\searrow \quad \swarrow{\scriptstyle g} & \\ & X & \end{array} \quad \text{and} \quad \begin{array}{ccc} \mathbb{P}(\mathcal{N}) & = & Z(t)|_Y \\ & {\scriptstyle \pi}\searrow \quad \swarrow{\scriptstyle g} & \\ & Y. & \end{array}$$

Realizing the blow-up $\widetilde{X}$ as the zero set of the regular section t provides us with the locally free resolution given by the Koszul complex $\bigwedge^\bullet(\Omega_g \otimes \mathcal{O}_g(1)) \twoheadrightarrow \mathcal{O}_{\widetilde{X}}$ of $\mathcal{O}_{\mathbb{P}(\mathcal{E})}$-modules.

We wish to compute $q^*\mathcal{O}_Z$ for any subvariety $Z \subset Y$ with $\mathcal{O}_Z$ considered as $\mathcal{O}_X$-module.

Clearly, $q^*\mathcal{O}_Z \simeq \iota^*g^*\mathcal{O}_Z$ and $g^*\mathcal{O}_Z \simeq \mathcal{O}_{\mathbb{P}(\mathcal{E}|_Z)}$, because g is flat. Hence,

$$\begin{aligned} \iota_*\mathcal{H}^k(q^*\mathcal{O}_Z) &\simeq \iota_*\mathcal{H}^k(\iota^*g^*\mathcal{O}_Z) \simeq \mathcal{H}^k(\iota_*\iota^*g^*\mathcal{O}_Z) \\ &\simeq \mathcal{H}^k(g^*\mathcal{O}_Z \otimes \mathcal{O}_{\widetilde{X}}), \end{aligned}$$

which can be computed as the cohomology of the restricted Koszul complex $\bigwedge^\bullet(\Omega_g \otimes \mathcal{O}_g(1))|_{g^{-1}(Z)}$. If Z is contained in Y, the differentials, which are given by contraction with the section t, vanish and, therefore,

$$\mathcal{H}^k(q^*\mathcal{O}_Z) \simeq (\Omega_g^{-k} \otimes \mathcal{O}_g(-k))|_{\pi^{-1}(Z)}.$$

Exercise 11.11 Work through the above construction in the case of the blow-up of a point in a surface.

Here now is the general case of the above 'semi-local' calculations (see also [74, Prop.13.8]).

Proposition 11.12 *Let $q : \widetilde{X} \to X$ be the blow-up along a smooth subvariety $Y \subset X$. Then for the structure sheaf $\mathcal{O}_Z$ of a subvariety $Z \subset Y$ considered as an object in $\mathrm{D}^{\mathrm{b}}(X)$ one has*

$$\mathcal{H}^k(q^*\mathcal{O}_Z) \simeq (\Omega_\pi^{-k} \otimes \mathcal{O}_\pi(-k))|_{\pi^{-1}(Z)},$$

where $\pi : \mathbb{P}(\mathcal{N}_{Y/X}) \to Y$ is the contraction of the exceptional divisor.

Proof The proof is similar to the proof of Proposition 11.8. Roughly, it suffices to find locally canonical (i.e. independent of any choice) isomorphisms of $\mathcal{H}^k(q^*\mathcal{O}_Z)$ and $\Omega^{-k}(-k)|_{\pi^{-1}(Z)}$.

The first indication why this should be possible is provided by the following observation. Suppose we have (locally) two ways of representing $Y \subset X$ as the zero set of regular sections, say $s_i \in H^0(X, \mathcal{E}_i)$, $i = 1, 2$. Then, if $\varphi : \mathcal{E}_1 \xrightarrow{\sim} \mathcal{E}_2$ is an isomorphism sending s_1 to s_2, the two embeddings $\widetilde{X} \subset \mathbb{P}(\mathcal{E}_i)$, $i = 1, 2$, correspond to each other under the induced isomorphism $\varphi : \mathbb{P}(\mathcal{E}_1) \xrightarrow{\sim} \mathbb{P}(\mathcal{E}_2)$. Thus, the description of $\mathcal{H}^k(q^*\mathcal{O}_Z)$ as given above does not depend on the choice of the section s or the locally free sheaf $\mathcal{E}$.

To pass from the global to the local statement we fix a locally free resolution $\ldots \longrightarrow \widetilde{\mathcal{E}}^* \twoheadrightarrow \mathcal{O}_Y$ which induces an embedding $\widetilde{X} \overset{\iota}{\hookrightarrow} \mathbb{P}(\widetilde{\mathcal{E}}) \overset{\pi}{\longrightarrow} X$. Then

$$\mathcal{H}^k(q^*\mathcal{O}_Z) \simeq \mathcal{H}^k\left(\tilde{\iota}_*\mathcal{O}_{\widetilde{X}} \otimes \mathcal{O}_{\mathbb{P}(\widetilde{\mathcal{E}}|_Z)}\right) \simeq \mathcal{H}^k(\mathcal{G}^\bullet|_{\tilde{\pi}^{-1}(Z)}),$$

where $\mathcal{G}^\bullet \twoheadrightarrow \tilde{\iota}_*\mathcal{O}_{\widetilde{X}}$ is a locally free resolution of $\mathcal{O}_{\mathbb{P}(\widetilde{\mathcal{E}})}$-modules.

Now passing to the local situation and choosing $s \in H^0(U, \mathcal{E})$ as before, yields a nested inclusion $\widetilde{X} \overset{\iota}{\hookrightarrow} \mathbb{P}(\mathcal{E}) \overset{h}{\hookrightarrow} \mathbb{P}(\widetilde{\mathcal{E}})$ with $\tilde{\iota} = h \circ \iota$. The restriction $h^*\mathcal{G}^\bullet \twoheadrightarrow \iota_*\mathcal{O}_{\widetilde{X}}$ is still a locally free resolution which now can be compared to the Koszul complex studied above. Going completely local and working with free modules allows us to argue that the complexes are quasi-isomorphic and that the quasi-isomorphism is unique up to homotopy (see the analogous arguments in the proof of Proposition 11.8). Hence, for any $x \in \mathbb{P}(\mathcal{N}) \subset \widetilde{X}$ the induced isomorphism $\mathcal{H}^k(\mathcal{G}^\bullet \otimes \tilde{\pi}^*\mathcal{O}_Z)_x \simeq (\Omega_\pi^{-k} \otimes \mathcal{O}_\pi(-k)|_{\pi^{-1}(Z)})_x$ is independent of any choice. □

11.2 Derived categories under blow-up

We keep the notation from the last section and study the blow-up $q : \widetilde{X} \longrightarrow X$ of a smooth variety X along a smooth subvariety $Y \subset X$.

Intuitively, by passing from X to the blow-up $\widetilde{X}$ the derived category grows by the amount of the exceptional divisor. At least this is what happens on the level of Chow and K-groups. It is fairly easy to see that at least nothing is lost in passing from $\mathrm{D^b}(X)$ to $\mathrm{D^b}(\widetilde{X})$ due to the following general fact.

Proposition 11.13 *Suppose $f : S \longrightarrow T$ is a projective morphism of smooth projective varieties such that $f_* : \mathrm{D^b}(S) \longrightarrow \mathrm{D^b}(T)$ sends $\mathcal{O}_S$ to $\mathcal{O}_T$. Then*

$$f^* : \mathrm{D^b}(T) \xrightarrow{\sim} \mathrm{D^b}(S)$$

is fully faithful and thus describes an equivalence of $\mathrm{D^b}(T)$ with an admissible triangulated subcategory of $\mathrm{D^b}(S)$.

Proof The first assertion is an immediate consequence of the projection formula (3.11). Indeed, the adjunction morphism $\mathrm{id} \longrightarrow f_*f^*$ (use $f^* \dashv f_*$) yields isomorphisms $\mathcal{F}^\bullet \longrightarrow f_*f^*(\mathcal{F}^\bullet) \simeq \mathcal{F}^\bullet \otimes f_*(\mathcal{O}_S) \simeq \mathcal{F}^\bullet$. Hence, $\mathrm{id} \simeq f_*f^*$ and, therefore, f^* is fully faithful (cf. Corollary 1.23).

The second assertion follows from the fact that f^* admits a right adjoint functor, see Remark 1.43, iv). □

Examples 11.14 This easy proposition can in particular be applied to the following two situations:

i) Let $\mathcal{N}$ be a vector bundle on a smooth variety Y. Then the projection $\pi : \mathbb{P}(\mathcal{N}) \longrightarrow Y$ induces a fully faithful functor $\pi^* : \mathrm{D^b}(Y) \longrightarrow \mathrm{D^b}(\mathbb{P}(\mathcal{N}))$ which identifies $\mathrm{D^b}(Y)$ with an admissible subcategory of $\mathrm{D^b}(\mathbb{P}(\mathcal{N}))$.

Indeed, any fibre of π, being isomorphic to a projective space, is connected and its structure sheaf has trivial higher cohomology. Hence, $\pi_*\mathcal{O}_{\mathbb{P}(\mathcal{N})} \simeq \mathcal{O}_Y$. Corollary 8.36 shows how $\pi^*\mathrm{D^b}(Y)$ can be viewed as one of the building blocks for the whole derived category $\mathrm{D^b}(\mathbb{P}(\mathcal{N}))$.

ii) Let $q : \widetilde{X} \longrightarrow X$ be the blow-up of $Y \subset X$. Again the fibres are projective spaces and hence $q_*\mathcal{O}_{\widetilde{X}} \simeq \mathcal{O}_X$. Thus, via $q^* : \mathrm{D^b}(X) \longrightarrow \mathrm{D^b}(\widetilde{X})$ the derived category $\mathrm{D^b}(X)$ can be viewed as an admissible subcategory of $\mathrm{D^b}(\widetilde{X})$. See ii), Remark 3.33.

The condition $q_*\mathcal{O}_{\widetilde{X}} \simeq \mathcal{O}_X$ is frequently met in other situations, e.g. for the resolution $q : \widetilde{X} \longrightarrow X$ of a singular variety X with rational singularities. Note that only in i) is the morphism flat; this fails for the blow-up in ii).

Remark 11.15 Note that a closed embedding $j : Y \subset X$ does not define a full functor $j_* : \mathrm{D^b}(Y) \subset \mathrm{D^b}(X)$, as in general $\mathrm{Ext}^i_Y(\mathcal{F}, \mathcal{F}) \neq \mathrm{Ext}^i_X(\mathcal{F}, \mathcal{F})$ for a sheaf $\mathcal{F}$ on Y considered at the same time as a sheaf on X with support contained in Y. E.g. for a closed point x in a curve X one has $\mathrm{Ext}^1_x(k(x), k(x)) = 0$, but $\mathrm{Ext}^1_X(k(x), k(x)) = k$. In fact, j_* is neither full nor faithful in general.

In view of Remark 11.15, the next proposition is a bit surprising at first sight. We use the notation introduced before for the blow-up along a smooth projective subvariety $Y \subset X$. In particular, $i : E \hookrightarrow X$ is the inclusion of the exceptional divisor and $\pi : E = \mathbb{P}(\mathcal{N}) \longrightarrow Y$ is the projection.

Proposition 11.16 *Suppose $Y \subset X$ is of codimension $c \geq 2$. Then the functor*

$$\Phi_k := i_* \circ (\mathcal{O}_E(kE) \otimes (\quad)) \circ \pi^* : \mathrm{D^b}(Y) \longrightarrow \mathrm{D^b}(\widetilde{X})$$

is fully faithful for any k. Moreover, Φ_k admits a right adjoint functor.

Note that the assumption on the codimension is needed. Indeed, if $Y = \{x\}$ with x a closed point of a curve X, then the blow-up is trivial and Φ_k is i_*.

Proof The functor Φ_k is a Fourier–Mukai transform with kernel $\mathcal{O}_E(kE)$ considered as an object in $\mathrm{D^b}(Y \times \widetilde{X})$. As such, Φ_k admits in particular right and left adjoints (see Proposition 5.9). Thus, the standard criterion, see Proposition 7.1, applies.

Let us first check that $\mathrm{Hom}(\Phi_k(k(x)), \Phi_k(k(y))[j]) = 0$ for all j and $x \neq y$. If $x \neq y \in Y$, then $\Phi_k(k(x)) \simeq i_*\mathcal{O}_{F_x}(-k)$ and similarly $\Phi_k(k(y)) \simeq i_*\mathcal{O}_{F_y}(-k)$ still have disjoint support and hence only trivial homomorphisms from one to the other.

Let us now consider the case $x = y$. We have to show that

$$\mathrm{Ext}^i_{\widetilde{X}}(\mathcal{O}_{F_x}(-k), \mathcal{O}_{F_x}(-k)) = \mathrm{Ext}^i_{\widetilde{X}}(\mathcal{O}_{F_x}, \mathcal{O}_{F_x})$$

is trivial for $i \notin [0,d]$ and of dimension one for $i=0$. For this use the spectral sequence (see (3.16), p. 85)

$$E_2^{p,q} = H^p(\widetilde{X}, \mathcal{E}xt^q_{\widetilde{X}}(\mathcal{O}_{F_x}, \mathcal{O}_{F_x})) \Rightarrow \operatorname{Ext}^{p+q}_{\widetilde{X}}(\mathcal{O}_{F_x}, \mathcal{O}_{F_x}).$$

Using $\bigwedge^j \mathcal{N}_{F_x/\widetilde{X}} \simeq \mathcal{E}xt^j_{\widetilde{X}}(\mathcal{O}_{F_x}, \mathcal{O}_{F_x})$ (see Proposition 11.8), this can also be written as

$$E_2^{p,q} = H^p(F_x, \bigwedge^q \mathcal{N}_{F_x/\widetilde{X}}) \Rightarrow \operatorname{Ext}^{p+q}_{\widetilde{X}}(\mathcal{O}_{F_x}, \mathcal{O}_{F_x}).$$

For any further computation we have to determine $\mathcal{N}_{F_x/\widetilde{X}}$. To this end, we will call upon the nested normal bundle sequence, i.e. the short exact sequence

$$0 \longrightarrow \mathcal{N}_{F_x/E} \longrightarrow \mathcal{N}_{F_x/\widetilde{X}} \longrightarrow \mathcal{N}_{E/\widetilde{X}}|_{F_x} \longrightarrow 0.$$

As $\mathcal{N}_{E/\widetilde{X}} \simeq \mathcal{O}_E(E)$ and $\mathcal{N}_{F_x/E} \simeq \mathcal{O}_{F_x}^{\oplus d}$ (here, $d = \dim(Y)$), the bundle $\mathcal{N}_{F_x/\widetilde{X}}$ is an extension of $\mathcal{O}_{F_x}(-1)$ by the trivial bundle $\mathcal{O}_{F_x}^{\oplus d}$. Since $F_x \simeq \mathbb{P}^{c-1}$, there are no non-trivial ones. Hence, $\mathcal{N}_{F_x/\widetilde{X}} \simeq \mathcal{O}_{F_x}^{\oplus d} \oplus \mathcal{O}_{F_x}(-1)$.

This yields $E_2^{p,q} = 0$ for all pairs (p,q) with either $p > 0$ or $p = 0$ and $q > d$. Therefore, $\operatorname{Ext}^q_{\widetilde{X}}(\mathcal{O}_{F_x}, \mathcal{O}_{F_x}) = E_2^{0,q} = 0$ for $q > d$ and $\operatorname{Ext}^0_{\widetilde{X}}(\mathcal{O}_{F_x}, \mathcal{O}_{F_x}) = E_2^{0,0} = k$. Since the negative Ext-groups vanish for trivial reasons, we have thus verified all the hypotheses of Proposition 7.1. □

Keeping the assumptions of the proposition, we thus have

Corollary 11.17 *For any k the functor Φ_k defines an equivalence between $\mathrm{D^b}(Y)$ and an admissible subcategory of $\mathrm{D^b}(\widetilde{X})$.* □

For the purpose of the next result we shall introduce for $k = -c+1, \ldots, -1$ the image

$$\mathcal{D}_k := \operatorname{Im}\left(\Phi_{-k} : \mathrm{D^b}(Y) \longrightarrow \mathrm{D^b}(\widetilde{X}) \right),$$

i.e. $\mathcal{D}_k$ is the full subcategory of $\mathrm{D^b}(\widetilde{X})$ which is under Φ_{-k} identified with $\mathrm{D^b}(Y)$. The full subcategory $q^*\mathrm{D^b}(X)$ shall be denoted $\mathcal{D}_0$.

Proposition 11.18 **(Orlov)** *With this notation the sequence of subcategories*

$$\mathcal{D}_{-c+1}, \ldots, \mathcal{D}_{-1}, \mathcal{D}_0 \subset \mathrm{D^b}(\widetilde{X})$$

defines a semi-orthogonal decomposition of $\mathrm{D^b}(\widetilde{X})$. See [91].

Proof For the definition of a semi-orthogonal decomposition see Section 1.4.

i) Let us first show that

$$\mathcal{D}_\ell \subset \mathcal{D}_k^\perp \quad \text{for} \quad -c+1 \leq \ell < k < 0.$$

For this purpose let $\mathcal{E}^\bullet, \mathcal{F}^\bullet \in \mathrm{D^b}(Y)$. Then adjunction $i^* \dashv i_*$ yields

$$\mathrm{Hom}(i_*(\pi^*\mathcal{F}^\bullet \otimes \mathcal{O}_\pi(k)), i_*(\pi^*\mathcal{E}^\bullet \otimes \mathcal{O}_\pi(\ell))) \simeq \mathrm{Hom}(i^*i_*\pi^*\mathcal{F}^\bullet, \pi^*\mathcal{E}^\bullet \otimes \mathcal{O}_\pi(\ell - k)).$$

Here, we write $\mathcal{O}_\pi(m)$ instead of $\mathcal{O}_E(-mE)$.

The existence of the distinguished triangle (see Corollary 11.4)

$$\pi^*\mathcal{F}^\bullet \otimes \mathcal{O}_\pi(1)[1] \longrightarrow i^*i_*\pi^*\mathcal{F}^\bullet \longrightarrow \pi^*\mathcal{F}^\bullet \longrightarrow \pi^*\mathcal{F}^\bullet \otimes \mathcal{O}_\pi(1)[2]$$

reduces our claim to the following two vanishings

$$\mathrm{Hom}(\pi^*\mathcal{F}^\bullet, \pi^*\mathcal{E}^\bullet \otimes \mathcal{O}_\pi(\ell - k)) = 0 = \mathrm{Hom}(\pi^*\mathcal{F}^\bullet \otimes \mathcal{O}_\pi(1), \pi^*\mathcal{E}^\bullet \otimes \mathcal{O}_\pi(\ell - k))$$

for all $\mathcal{F}^\bullet, \mathcal{E}^\bullet \in \mathrm{D^b}(Y)$. (For the second equality we replaced $\mathcal{F}^\bullet$ by $\mathcal{F}^\bullet[1]$.)

Both are easily deduced from adjunction $\pi^* \dashv \pi_*$, the projection formula, and $\pi_*\mathcal{O}_\pi(\ell - k) = 0$ for $-c+1 \leq \ell - k < 0$. (Remember the fibres of π are projective spaces $\mathbb{P}^{c-1}$.)

ii) In the second step one shows that

$$\mathcal{D}_\ell \subset \mathcal{D}_0^\perp \quad \text{for} \quad -c+1 \leq \ell < 0.$$

Again use $\pi_*\mathcal{O}_\pi(\ell) = 0$ for $-c+1 \leq \ell < 0$ to conclude that for $\mathcal{E}^\bullet \in \mathrm{D^b}(X)$ and $\mathcal{F}^\bullet \in \mathrm{D^b}(Y)$:

$$\begin{aligned}\mathrm{Hom}(q^*\mathcal{E}^\bullet, i_*(\pi^*\mathcal{F}^\bullet \otimes \mathcal{O}_\pi(\ell))) &\simeq \mathrm{Hom}(\mathcal{E}^\bullet, q_*i_*(\pi^*\mathcal{F}^\bullet \otimes \mathcal{O}_\pi(\ell))) \\ &\simeq \mathrm{Hom}(\mathcal{E}^\bullet, j_*\pi_*(\pi^*\mathcal{F}^\bullet \otimes \mathcal{O}_\pi(\ell))) = 0.\end{aligned}$$

iii) In the last step we show that $\mathcal{D}_{-c+1}, \ldots, \mathcal{D}_{-1}, \mathcal{D}_0$ not only form a semi-orthogonal sequence, but that they also generate $\mathrm{D^b}(\widetilde{X})$.

For this purpose assume that $\mathcal{E}^\bullet \in \mathcal{D}_\ell^\perp$ for all $-c+1 \leq \ell < 0$. We claim that there then exists an object $\mathcal{G}^\bullet \in \mathrm{D^b}(Y)$ with $i^*\mathcal{E}^\bullet \otimes \mathcal{O}_\pi(c-1) \simeq \pi^*\mathcal{G}^\bullet$.

At this point we will use the semi-orthogonal decomposition of the derived category $\mathrm{D^b}(E)$ of the projective bundle $E = \mathbb{P}(\mathcal{N}) \longrightarrow Y$ provided by Corollary 8.36.

By assumption on $\mathcal{E}^\bullet$ we have

$$\mathrm{Hom}(i_*(\pi^*\mathcal{F}^\bullet \otimes \mathcal{O}_\pi(\ell)), \mathcal{E}^\bullet) = 0 \;\text{ for all }\; -c+1 \leq \ell < 0$$

and all $\mathcal{F}^\bullet \in \mathrm{D^b}(Y)$. Grothendieck–Verdier duality and $i^!\mathcal{E}^\bullet \simeq i^*\mathcal{E}^\bullet \otimes \mathcal{O}_E(E)[-1]$ then show that

$$\mathrm{Hom}(\pi^*\mathcal{F}^\bullet \otimes \mathcal{O}_\pi(\ell), i^*\mathcal{E}^\bullet) = 0 \;\text{ for all }\; -c+2 \leq \ell < 1(!)$$

and all $\mathcal{F}^\bullet \in \mathrm{D^b}(Y)$.

Thus, by Corollary 8.36 the pull-back $i^*\mathcal{E}^\bullet$ is contained in $\pi^*\mathrm{D^b}(Y) \otimes \mathcal{O}_\pi(-c+1)$ which is the orthogonal complement of $\langle \pi^*\mathrm{D^b}(Y)(k) \rangle_{k=-c+2,\ldots,0}$ inside $\mathrm{D^b}(E)$.

Suppose $\mathcal{E}_0^\bullet \in \mathrm{D^b}(\widetilde{X})$ such that $i^*\mathcal{E}_0^\bullet \simeq \pi^*\mathcal{G}^\bullet$ for some $\mathcal{G}^\bullet \in \mathrm{D^b}(Y)$. If $i^*\mathcal{E}_0^\bullet$ is trivial, i.e. $\mathcal{G}^\bullet \simeq 0$, then $\mathcal{E}_0^\bullet$ has support outside the exceptional divisor E and

$\mathcal{E}_0^\bullet$ is contained in $\mathcal{D}_0$. Suppose that this is not the case, i.e. $\mathcal{G}^\bullet \not\simeq 0$. We claim that then $\mathrm{Hom}(\mathcal{E}_0^\bullet, q^*k(x)[m]) \neq 0$ for some closed point $x \in Y$ and some $m \in \mathbb{Z}$. Indeed, consider the spectral sequence

$$E_2^{r,s} = \mathrm{Hom}(\mathcal{E}_0^\bullet, \mathcal{H}^s(q^*k(x))[r]) \Rightarrow \mathrm{Hom}(\mathcal{E}_0^\bullet, q^*k(x)[r+s]).$$

Proposition 11.12 applied to $Z = \{x\} \subset Y$ proves $\mathcal{H}^s(q^*k(x)) \simeq \Omega_{F_x}^{-s}(-s)$, where $F_x := \pi^{-1}(x)$. Together with our hypotheses $i^*\mathcal{E}_0^\bullet \simeq \pi^*\mathcal{G}^\bullet$ this yields

$$\begin{aligned} E_2^{r,-s} &\simeq \mathrm{Hom}(\mathcal{E}_0^\bullet, i_*(\Omega_{F_x}^s(s))[r]) \simeq \mathrm{Hom}(i^*\mathcal{E}_0^\bullet, \Omega_{F_x}^s(s)[r]) \\ &\simeq \mathrm{Hom}(\pi^*\mathcal{G}^\bullet, \Omega_{F_x}^s(s)[r]) \simeq \mathrm{Hom}(\mathcal{G}^\bullet, \pi_*(\Omega_{F_x}^s(s))[r]) = 0 \end{aligned}$$

except for $s = 0$. Thus, $\mathrm{Hom}(\mathcal{E}_0^\bullet, q^*k(x)[m]) = E_2^{m,0} = \mathrm{Hom}(\mathcal{G}^\bullet, k(x)[m]) \neq 0$ for some $m \in \mathbb{Z}$ and some $x \in Y$, as the closed points of Y span the derived category $\mathrm{D^b}(Y)$.

Let us apply this to our complex $\mathcal{E}^\bullet$ and $\mathcal{E}_0^\bullet := \mathcal{E}^\bullet \otimes \mathcal{O}(-(c-1)E)$. We obtain

$$\begin{aligned} 0 &\neq \mathrm{Hom}(\mathcal{E}^\bullet \otimes \mathcal{O}_{\widetilde{X}}(-(c-1)E), q^*k(x)[m]) \\ &\simeq \mathrm{Hom}(q^*k(x), \mathcal{E}^\bullet \otimes \mathcal{O}(-(c-1)E) \otimes \omega_{\widetilde{X}}[\dim(X) - m])^* \\ &\simeq \mathrm{Hom}(q^*k(x), \mathcal{E}^\bullet[\dim(X) - m])^*. \end{aligned}$$

Thus, if $\mathcal{E}^\bullet \in \mathcal{D}_\ell^\perp$ for all $-c+1 \leq \ell < 0$, we cannot have $\mathcal{E}^\bullet \in \mathcal{D}_0^\perp$. In other words, $\mathcal{D}_{-c+1}, \ldots, \mathcal{D}_0$ generate $\mathrm{D^b}(\widetilde{X})$. □

Exercise 11.19 Deduce from Proposition 11.18 the classical isomorphism of additive(!) groups (cf. [74]):

$$K(\widetilde{X}) = \pi^*K(X) \oplus \bigoplus_{k=1}^{c-1} i_*\left(\pi^*K(Y).\mathcal{O}_\pi(k)\right)$$

and the analogous one for rational Chow groups.

11.3 The standard flip

Suppose X contains a smooth subvariety Y isomorphic to $\mathbb{P}^k$ and such that the normal bundle $\mathcal{N} := \mathcal{N}_{\mathbb{P}^k/X}$ is isomorphic to $\mathcal{O}(-1)^{\oplus \ell+1}$. So, $\ell+1 = \mathrm{codim}(\mathbb{P}^k \subset X)$. Using the adjunction formula for $\mathbb{P}^k \subset X$ and the assumption on the normal bundle one obtains

$$\omega_X|_{\mathbb{P}^k} \simeq \mathcal{O}(\ell - k).$$

Remark 11.20 In the special case $k = \ell$ the sheaf $\mathcal{O}_{\mathbb{P}^k}$ turns out to be a spherical object in $\mathrm{D^b}(X)$. For the argument see Examples 8.10, v).

The blow-up $q : \widetilde{X} \longrightarrow X$ of X along $\mathbb{P}^k$ produces an exceptional divisor $E \simeq \mathbb{P}(\mathcal{N})$ which itself is isomorphic to $\mathbb{P}^k \times \mathbb{P}^\ell$. The two projections will be denoted $\pi : \mathbb{P}^k \times \mathbb{P}^\ell \longrightarrow \mathbb{P}^k$ and $\pi' : \mathbb{P}^k \times \mathbb{P}^\ell \longrightarrow \mathbb{P}^\ell$. In particular, $q|_E = \pi$.

Let us first of all compute the relevant line bundles on $\widetilde{X}$ and their restriction to E. As was recalled earlier,

$$\omega_{\widetilde{X}} \simeq q^*\omega_X \otimes \mathcal{O}(\ell E) \quad \text{and} \quad \omega_E \simeq (\omega_{\widetilde{X}} \otimes \mathcal{O}_{\widetilde{X}}(E))|_E.$$

If we write $\mathcal{O}(a,b)$ for the line bundle $\pi^*\mathcal{O}(a) \otimes \pi'^*\mathcal{O}(b)$ on $\mathbb{P}^k \times \mathbb{P}^\ell$ this yields

$$\mathcal{O}(-k-1,-\ell-1) \simeq \pi^*(\omega_X|_{\mathbb{P}^k}) \otimes \mathcal{O}_E((\ell+1)E).$$

Altogether, this proves

$$\mathcal{O}_E(E) \simeq \mathcal{O}(-1,-1) \quad \text{and} \quad \omega_{\widetilde{X}}|_E \simeq \mathcal{O}(-k,-\ell).$$

In particular, the Fujiki–Nakano criterion [36] (cf. Remark 11.10) applies and yields a contraction $p : \widetilde{X} \longrightarrow X'$, which is a blow-up of $\mathbb{P}^\ell \subset X'$ with exceptional divisor E and such that the restriction of p to E equals π'. One furthermore shows that then $\mathcal{N}' := \mathcal{N}_{\mathbb{P}^\ell/X'} \simeq \mathcal{O}(-1)^{\oplus k+1}$. We will tacitly assume that X' is projective as well.

The situation is not quite symmetric, at least not when $k \neq \ell$. Indeed, for $\ell < k$ the restriction ω_X to $\mathbb{P}^k$ is the negative line bundle $\mathcal{O}(\ell - k)$, whereas the restriction $\omega_{X'}$ to $\mathbb{P}^\ell$ is the positive line bundle $\mathcal{O}(k-\ell)$. Changing the sign of the canonical bundle in this way is usually called a *flip*. The diagram we just constructed

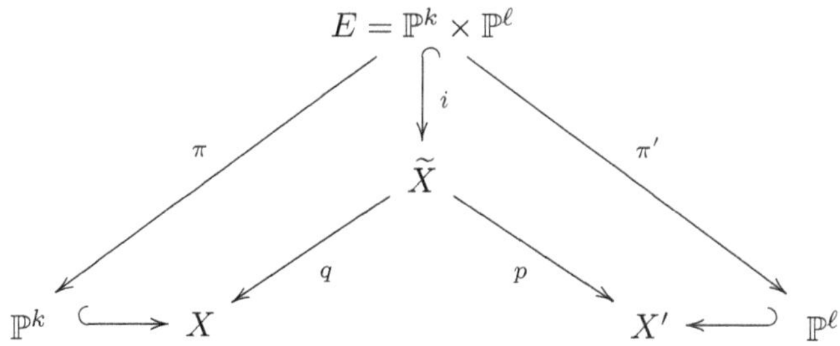

is called a *standard flip*. In the special case $k = \ell$ the restriction of the canonical bundle on both sides is trivial and, in particular, does not change when passing from X to X'. This is called a *flop* and the special one constructed as above is called the *standard flop*.

In general, flips are birational transformations which are performed in order to increase the positivity of the canonical bundle (and eventually to reach a minimal model). General flops are birational transformations that describe the passage from one minimal model to another one. In particular, under a flop the positivity of the canonical bundle will not change; along the exceptional sets it is even numerically trivial.

Clearly, the result of the last section applies to both blow-ups $q : \widetilde{X} \longrightarrow X$ and $p : \widetilde{X} \longrightarrow X'$ and allows us to compare the derived categories of X and X' inside the bigger derived category $\mathrm{D^b}(\widetilde{X})$. Roughly, the derived category of the variety whose canonical bundle is more negative tends to be bigger. (Somehow this is expected from the behaviour of the derived category under blow-up (see Proposition 11.18), as the canonical bundle of the blow-up is always negative along the fibres of the exceptional divisor.) The functor that will serve our purpose is $q_* \circ p^* : \mathrm{D^b}(X') \longrightarrow \mathrm{D^b}(X)$.

Exercise 11.21 Show that $q_* \circ p^*$ is the Fourier–Mukai transform $\Phi_{\mathcal{O}_{\widetilde{X}}}$, where the structure sheaf $\mathcal{O}_{\widetilde{X}}$ of $\widetilde{X} \subset X \times X'$ is viewed as an object in $\mathrm{D^b}(X \times X')$.

Remark 11.22 Using the exercise, it is easy to see that $q_* \circ p^*$ can hardly be an equivalence if $k \neq \ell$, for in this case $q^*\omega_X \not\simeq p^*\omega_{X'}$ (cf. Remark 5.22)

If $k = \ell$ the left adjoint functor equals the right adjoint functor. It is explicitly given as the Fourier–Mukai transform with kernel (see Corollary 3.40)

$$\begin{aligned}\mathcal{O}_{\widetilde{X}}^{\vee} \otimes q^*\omega_X[2k+1] &\simeq (\omega^*_{X\times X'}|_{\widetilde{X}} \otimes \omega_{\widetilde{X}}[-(2k+1)]) \otimes q^*\omega_X[2k+1] \\ &\simeq \mathcal{O}_{\widetilde{X}}(kE).\end{aligned}$$

Proposition 11.23 **(Bondal, Orlov)** *Let $X \longleftarrow \widetilde{X} \longrightarrow X'$ be the standard flip with $\ell \leq k$ as constructed above. Then*

$$q_* \circ p^* : \mathrm{D^b}(X') \longrightarrow \mathrm{D^b}(X)$$

is fully faithful. If $k = \ell$, it defines an equivalence. See [14].

Proof The proof uses the two semi-orthogonal decompositions of $\mathrm{D^b}(\widetilde{X})$ induced by the two blow-up maps q and p. We shall use the notation

$$\mathrm{D^b}(\widetilde{X}) = \langle \mathcal{D}_{-\ell}, \ldots, \mathcal{D}_{-1}, \mathrm{D^b}(X)\rangle \quad \text{and} \quad \mathrm{D^b}(\widetilde{X}) = \langle \mathcal{D}'_{-k}, \ldots, \mathcal{D}'_{-1}, \mathrm{D^b}(X')\rangle,$$

with $\mathcal{D}_b = i_*(\pi^*\mathrm{D^b}(\mathbb{P}^k) \otimes \mathcal{O}(0,b))$ and $\mathcal{D}'_a = i_*(\pi'^*\mathrm{D^b}(\mathbb{P}^\ell) \otimes \mathcal{O}(a,0))$.

In order to prove the assertion, we have to show that for arbitrary $\mathcal{E}^\bullet, \mathcal{F}^\bullet \in \mathrm{D^b}(X')$ the Fourier–Mukai functor $q_* \circ p^*$ induces an isomorphism

$$\mathrm{Hom}(\mathcal{E}^\bullet, \mathcal{F}^\bullet) \simeq \mathrm{Hom}(q_*p^*\mathcal{E}^\bullet, q_*p^*\mathcal{F}^\bullet).$$

By Proposition 11.13 we know that p^* is fully faithful and hence

$$\mathrm{Hom}(\mathcal{E}^\bullet, \mathcal{F}^\bullet) \simeq \mathrm{Hom}(p^*\mathcal{E}^\bullet, p^*\mathcal{F}^\bullet).$$

On the other hand, applying adjunction $q^* \dashv q_*$ one has

$$\mathrm{Hom}(q_*p^*\mathcal{E}^\bullet, q_*p^*\mathcal{F}^\bullet) \simeq \mathrm{Hom}(q^*q_*p^*\mathcal{E}^\bullet, p^*\mathcal{F}^\bullet).$$

Thus, in order to prove the assertion it suffices to prove that the adjunction morphism $q^*q_*p^*\mathcal{E}^\bullet \longrightarrow p^*\mathcal{E}^\bullet$ induces a bijection

$$\mathrm{Hom}(p^*\mathcal{E}^\bullet, p^*\mathcal{F}^\bullet) \xrightarrow{\sim} \mathrm{Hom}(q^*q_*p^*\mathcal{E}^\bullet, p^*\mathcal{F}^\bullet)$$

for any $\mathcal{F}^\bullet \in \mathrm{D^b}(X')$. (We leave it to the reader to show that all functorial morphisms are indeed compatible.)

To this end, we first complete the adjunction morphism $q^*q_*p^*\mathcal{E}^\bullet \longrightarrow p^*\mathcal{E}^\bullet$ to a distinguished triangle

$$q^*q_*p^*\mathcal{E}^\bullet \longrightarrow p^*\mathcal{E}^\bullet \longrightarrow \mathcal{H}^\bullet \longrightarrow q^*q_*p^*\mathcal{E}^\bullet[1]. \qquad (11.5)$$

It clearly suffices to show that $\mathrm{Hom}(\mathcal{H}^\bullet, p^*\mathcal{F}^\bullet) = 0$ for any $\mathcal{F}^\bullet \in \mathrm{D^b}(X')$. The proof of this uses the following two properties of $\mathcal{H}^\bullet$, which will be verified first:

i) $\mathrm{Hom}(\mathcal{H}^\bullet, i_*\mathcal{O}(a,b)) = 0$ if $-k \leq a \leq -1$ and $-\ell \leq b \leq -1$.
ii) $\mathcal{H}^\bullet \in \mathrm{D^b}(X)^\perp$.

To verify i), note that under the given numerical assumptions on a and b one has $i_*\mathcal{O}(a,b) \in \mathcal{D}'_a \cap \mathcal{D}_b$ and hence $i_*\mathcal{O}(a,b) \in \mathrm{D^b}(X)^\perp \cap \mathrm{D^b}(X')^\perp$. In particular,

$$\mathrm{Hom}(q^*q_*p^*\mathcal{E}^\bullet, i_*\mathcal{O}(a,b)) = 0 = \mathrm{Hom}(p^*\mathcal{E}^\bullet, i_*\mathcal{O}(a,b)).$$

Then assertion i) follows from the long exact sequence obtained by applying $\mathrm{Hom}(\ \ , i_*\mathcal{O}(a,b))$ to the distinguished triangle (11.5).

To prove ii), apply q_* to (11.5) and use $q_*q^*q_*p^*\mathcal{E}^\bullet \simeq q_*p^*\mathcal{E}^\bullet$, as q^* is fully faithful. This yields $q_*\mathcal{H}^\bullet = 0$ and, therefore, by adjunction $\mathrm{Hom}(q^*\mathcal{G}^\bullet, \mathcal{H}^\bullet) \simeq \mathrm{Hom}(\mathcal{G}^\bullet, q_*\mathcal{H}^\bullet) = 0$ for any $\mathcal{G}^\bullet \in \mathrm{D^b}(X)$.

The semi-orthogonal decomposition of $\mathrm{D^b}(\widetilde{X})$ with respect to q yields in particular the semi-orthogonal decomposition

$$\mathrm{D^b}(X)^\perp = \langle \mathcal{D}_{-\ell}, \ldots \mathcal{D}_{-1} \rangle.$$

On the other hand, each of the $\mathcal{D}_b$ is equivalent to $\mathrm{D^b}(\mathbb{P}^k)$ which in turn can be decomposed by a fully exceptional sequence of line bundles.

For any $b = 0, \ldots, \ell - 1$ we consider the semi-orthogonal decomposition

$$\mathrm{D^b}(\mathbb{P}^k) = \langle \mathcal{O}(-k+b), \ldots, \mathcal{O}(b) \rangle.$$

Thus, we obtain a full exceptional sequence in $\mathrm{D^b}(X)^\perp$:

$$\begin{aligned} \mathrm{D^b}(X)^\perp &= \langle \mathcal{O}(-k,-\ell), \ldots, \mathcal{O}(0,-\ell), \\ &\quad \mathcal{O}(-k+1,-\ell+1), \ldots, \mathcal{O}(1,-\ell+1), \\ &\quad \vdots \\ &\quad \mathcal{O}(-k+\ell-1,-1), \ldots, \mathcal{O}(\ell-1,-1) \rangle. \end{aligned}$$

For simplicity we write $\mathcal{O}(a,b)$ for $i_*\mathcal{O}(a,b)$. This full exceptional sequence allows us to define a semi-orthogonal decomposition

$$\mathrm{D^b}(X)^\perp = \langle \mathcal{D}^1, \mathcal{D}^2 \rangle$$

with $\mathcal{D}^1 := \{\mathcal{O}(a,b)\}_{-k \leq a < 0}$ and $\mathcal{D}^2 := \{\mathcal{O}(a,b)\}_{0 \leq a \leq \ell-1}$. In order to see that $\mathcal{D}^1 \subset \mathcal{D}^{2\perp}$ we need $\mathrm{Hom}(i_*\mathcal{O}(a,b), i_*\mathcal{O}(a',b')[m]) = 0$ for $\mathcal{O}(a,b) \in \mathcal{D}^2$, $\mathcal{O}(a',b') \in \mathcal{D}^1$, and arbitrary $m \in \mathbb{Z}$.

Using Grothendieck–Verdier duality for $i : E \hookrightarrow \widetilde{X}$ and ii), Corollary 11.4 this reduces to

$$H^*(\mathbb{P}^k \times \mathbb{P}^\ell, \mathcal{O}(a'-a, b'-b)) = 0 = H^*(\mathbb{P}^k \times \mathbb{P}^\ell, \mathcal{O}(a'-a-1, b'-b-1))$$

and, by using the Künneth formula, further to $H^*(\mathbb{P}^\ell, \mathcal{O}(b'-b)) = 0$ for $a'-a \leq -k-1$ and $H^*(\mathbb{P}^\ell, \mathcal{O}(b'-b-1)) = 0$ for $a'-a-1 \leq -k-1$. Let us indicate how to deduce the first vanishing. By the definition of the decomposition, one always has $b'-b > -\ell$. Thus, it suffices to ensure $b'-b<0$ under our assumptions. This follows from $k+1 \leq a-a'$, $a-b \leq \ell$, and $b'-a' \leq k-\ell$. The calculations are straightforward and left to the reader.

Since by ii) one knows that $\mathcal{H}^\bullet \in \mathrm{D^b}(X)^\perp$ and by i) that $\mathrm{Hom}(\mathcal{H}^\bullet, B) = 0$ for any $B \in \mathcal{D}^1$, we find $\mathcal{H}^\bullet \in \mathcal{D}^2$ (see Exercise 1.63). Thus, in order to show that $\mathrm{Hom}(\mathcal{H}^\bullet, p^*\mathcal{F}^\bullet) = 0$ for any $\mathcal{F}^\bullet \in \mathrm{D^b}(X')$, it suffices to prove $p^*\mathcal{F}^\bullet \in {\mathcal{D}^2}^\perp$, i.e. $\mathrm{Hom}(i_*\mathcal{O}(a,b), p^*\mathcal{F}^\bullet) = 0$ for all $0 \leq a \leq \ell-1$.

We use Grothendieck–Verdier duality (see Corollary 3.38 and Exercise 3.39) and the Künneth formula to prove

$$\begin{aligned}
\mathrm{Hom}(i_*\mathcal{O}(a,b), p^*\mathcal{F}^\bullet[*]) &\simeq \mathrm{Hom}(\mathcal{O}(a,b), i^!p^*\mathcal{F}^\bullet[*]) \\
&\simeq \mathrm{Hom}(\mathcal{O}(a,b), i^*p^*\mathcal{F}^\bullet \otimes \omega_E \otimes \omega_{\tilde{X}}^*|_E[*]) \\
&\simeq \mathrm{Hom}(\mathcal{O}, \pi'^*(\mathcal{F}^\bullet|_{P^\vee}(-b)) \otimes \pi^*\mathcal{O}(\ell-k-1-a)[*]) \\
&\simeq H^*(P^\vee, \mathcal{F}^\bullet|_{P^\vee}(-b)) \otimes H^*(P, \mathcal{O}(\ell-k-1-a)) \\
&= 0,
\end{aligned}$$

for $-k \leq \ell-k-1-a < 0$ if $0 \leq a \leq \ell-1$ and $\ell \leq k$. □

Remark 11.24 i) For an alternative proof of the above statement that reduces to the case of Y embedded as the zero section into its normal bundle see [63, Prop.5.5].

ii) The proof, and hence the result, is actually valid also in the local situation (check this!), i.e. we do not need to assume that X is projective. In the proof of Proposition 11.31 we will thus be allowed to use the fact that for sheaves concentrated on a closed projective subvariety $\mathbb{P}^k \subset Z \subset X$ (with X not necessarily projective) the functor p_*q^* is fully faithful. (In the case considered there $k=\ell$, so we may interchange the two projections.)

Remark 11.25 One might ask why can we not avoid the rather complicated argument above and apply one of the standard criteria directly. In principle, testing the functor on line bundles or on point sheaves seems possible, but the actual calculation turns out to be problematic.

Consider the rather easy case of a standard flop in dimension three, i.e. a smooth rational curve $C \subset X$ with normal bundle $\mathcal{O}(-1) \oplus \mathcal{O}(-1)$ is flopped. If L is a line bundle on X with $L|_C \simeq \mathcal{O}(-2)$, then $q^*L \simeq p^*L' \otimes \mathcal{O}(2E)$ for some line bundle L' on X' and hence $p_*q^*L \simeq L' \otimes p_*\mathcal{O}(2E)$. A calculation with the short exact sequence

$$0 \longrightarrow \mathcal{O}(E) \longrightarrow \mathcal{O}(2E) \longrightarrow \mathcal{O}_E(2E) \longrightarrow 0$$

reveals that

$$R^0 p_* \mathcal{O}(2E) \simeq \mathcal{O} \quad \text{and} \quad R^1 p_* \mathcal{O}(2E) \simeq \mathcal{O}_C(-2).$$

Hence, in order to compute $\mathrm{Ext}^k(p_* q^* L, p_* q^* L)$ one has to work with a length one complex and this easily becomes rather messy.

Working with point sheaves is also tricky. E.g. if we blow-up a point $x \in X$ in a surface X and try to compute the pull-back $q^* k(x)$ under the blow-up map $q : \widetilde{X} \longrightarrow X$, then we obtain a genuine complex.

Exercise 11.26 Work out the details of the last remark (cf. Exercise 11.11).

11.4 The Mukai flop

The Mukai flop is quite similar to an elementary flop in that a projective space is blown-up and then contracted in a different direction. We shall see that once more the derived category does not change in this process although the derived equivalence is no longer induced by the birational correspondence.

Suppose X is a smooth projective variety of dimension $2n$ containing a smooth subvariety $P \subset X$ itself isomorphic to $\mathbb{P}^n$. Throughout this section we shall assume $n > 1$. Furthermore we assume that the normal bundle $\mathcal{N} := \mathcal{N}_{P/X}$ is isomorphic to the cotangent bundle Ω_P.

Blowing-up $P \subset X$ yields a projective morphism $q : \widetilde{X} \longrightarrow X$ with exceptional divisor

$$\pi : E \simeq \mathbb{P}(\mathcal{N}) \simeq \mathbb{P}(\Omega_P) \longrightarrow P.$$

Due to the Euler sequence, this projective bundle can be understood as an incidence variety as follows. Writing P more invariantly as $\mathbb{P}(V)$ for some vector space V of dimension $n+1$ allows us to write the Euler sequence as

$$0 \longrightarrow \Omega_P \longrightarrow V^* \otimes \mathcal{O}(-1) \longrightarrow \mathcal{O} \longrightarrow 0.$$

Hence,

$$\mathbb{P}(\Omega_P) \subset \mathbb{P}(V^* \otimes \mathcal{O}(-1)) = \mathbb{P}(V) \times \mathbb{P}(V^*) = P \times P^\vee,$$

where $P^\vee := \mathbb{P}(V^*)$. Moreover, the fibre $\Omega_P(\ell)$ over a line $[\ell] \in P$ is naturally isomorphic to the space of linear functions $\varphi : V \longrightarrow \mathbb{C}$ with $\varphi|_\ell = 0$. In other words, $E = \mathbb{P}(\Omega_P) \subset P \times P^\vee$ is the incidence variety of pairs (ℓ, H) of lines $\ell \subset V$ and hyperplanes $H \subset V$ such that $\ell \subset H$. In particular, $E \in |\mathcal{O}(1,1)|$ and, by the adjunction formula, $\omega_E \simeq \mathcal{O}(-n,-n)|_E$.

On the other hand, the adjunction formula for $E \subset \widetilde{X}$ shows

$$\begin{aligned}\omega_E &\simeq (\omega_{\widetilde{X}} \otimes \mathcal{O}(E))|_E \simeq (q^*\omega_X \otimes \mathcal{O}((n-1)E))\,|_E \otimes \mathcal{O}_E(E)\\ &\simeq \pi^*(\omega_X|_P) \otimes \mathcal{O}_E(nE).\end{aligned}$$

By our assumption on the normal bundle we know that $\omega_X|_P$ is actually trivial and, therefore, $\omega_E \simeq \mathcal{O}_E(nE)$. This shows that

$$\mathcal{O}_E(E) \simeq \mathcal{O}(-1,-1)$$

which allows us to apply the Fujiki–Nakano criterion (see Remark 11.10) that ensures the existence of a birational morphism $p : \widetilde{X} \longrightarrow X'$ whose restriction to E is the second projection $E \subset P \times P^\vee \longrightarrow P^\vee$ and which away from E is an isomorphism. Moreover,

$$\mathcal{N}' := \mathcal{N}_{P^\vee/X'} \simeq \Omega_{P^\vee} \quad \text{and} \quad \omega_{X'}|_{P^\vee} \simeq \mathcal{O}_{P^\vee}.$$

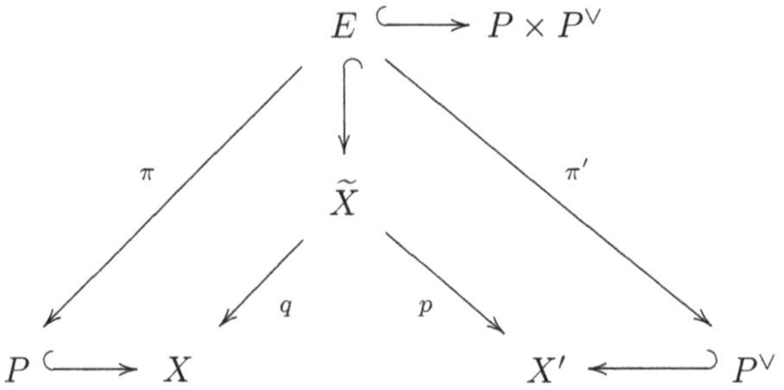

As in the previous section, the new variety X' is not necessarily projective, but this will be added as an extra assumption from now on.

Remark 11.27 The case that interests us most is the case of a projective space $\mathbb{P}^n$ embedded into an algebraic symplectic variety X of dimension $2n$. By definition an algebraic symplectic variety is a projective variety X that possesses a global regular two-form $\sigma \in H^0(X, \Omega_X^2)$ which is everywhere non-degenerate.

As $\mathbb{P}^n$ does not admit any regular two-forms whatsoever, the restriction of the two-form σ is necessarily trivial. Using the normal bundle sequence one turns this into the required isomorphism $\mathcal{N} \simeq \Omega_{\mathbb{P}^n}$. In other words, our assumption is automatically satisfied in this case. Moreover, not only is the restriction $\omega_X|_{\mathbb{P}^n}$ is trivial, but in fact ω_X is itself trivial.

Everything looks very similar to the situation treated in the previous section with $k = \ell$ and one is tempted to generalize Proposition 11.23 in a straightforward manner. However, there is the following negative result observed by Kawamata and Namikawa, which shows that the obvious guess for a Fourier–Mukai kernel defining a derived equivalence between X and X' does not work.

Proposition 11.28 **(Kawamata, Namikawa)** *The Fourier–Mukai transform given by*

$$p_* \circ q^* : \mathrm{D^b}(X) \longrightarrow \mathrm{D^b}(X')$$

is not(!) fully faithful. See [63, 86].

Proof Let us first give an algebraic proof under the assumption that there exists a line bundle $L \in \mathrm{Pic}(X)$ with $L|_P \simeq \mathcal{O}(-n)$.

The pull-back of the line bundle L to $\widetilde{X}$ is isomorphic to $\mathcal{O}(-n)$ on the fibres of $p : \widetilde{X} \longrightarrow X'$. Hence, $q^*L \otimes \mathcal{O}(-nE)$ restricts to the trivial line bundle on the fibres of $p : \widetilde{X} \longrightarrow X'$ and is, therefore, of the form p^*L' for some line bundle $L' \in \mathrm{Pic}(X')$ (see Remark 3.33, ii)), i.e.

$$q^*L \simeq p^*L' \otimes \mathcal{O}(nE). \tag{11.6}$$

In order to conclude that $p_* \circ q^*$ is not fully faithful it suffices to show that

$$\chi(L, L) \neq \chi(p_*q^*L, p_*q^*L).$$

Recall that both terms can be expressed by the Mukai pairing (cf. (5.5)). Thus, the reasoning will eventually show that the induced cohomological Fourier–Mukai transform is not compatible with the Mukai pairing and, therefore, cannot be induced by a Fourier–Mukai equivalence.

In any case, in order to prove that the right and left hand side do not coincide one has to study the object p_*q^*L in more detail. A complete description is difficult to derive, but it will be enough to show that p_*q^*L is concentrated in degree 0 and $n-1$ with

$$\mathcal{H}^0(p_*q^*L) \simeq L' \quad \text{and} \quad \mathcal{H}^{n-1}(p_*q^*L) \simeq \mathcal{O}_{P^\vee}(-1).$$

By (11.6), the restriction of q^*L to the fibres of p, which are isomorphic to $\mathbb{P}^{n-1}$, has cohomology only in degree 0 and $n-1$. Hence, p_*q^*L is concentrated in these degrees.

Let us now determine the degree zero cohomology. Applying the projection formula to (11.6) shows that $\mathcal{H}^0(p_*q^*L) \simeq L'$ if and only if $R^0p_*\mathcal{O}(nE) \simeq \mathcal{O}$. In fact, $R^0p_*\mathcal{O}(iE) \simeq \mathcal{O}$ for all $i \geq 0$, which is proved by induction and by means of the short exact sequence

$$0 \longrightarrow \mathcal{O}((i-1)E) \longrightarrow \mathcal{O}(iE) \longrightarrow \mathcal{O}_E(iE) \longrightarrow 0.$$

The same short exact sequence also yields $R^{n-1}p_*\mathcal{O}(nE) \simeq R^{n-1}\pi'_*\mathcal{O}_E(nE)$. Then use $\omega_E \simeq \mathcal{O}(nE)|_E$ to show

$$\begin{aligned} R^{n-1}\pi'_*\mathcal{O}_E(nE) &\simeq R^{n-1}\pi'_*(\omega_E) \simeq R^{n-1}\pi'_*(\omega_{\pi'} \otimes \pi'^*\omega_{P^\vee}) \\ &\simeq R^{n-1}\pi'_*(\omega_{\pi'}) \otimes \omega_{P^\vee} \simeq \omega_{P^\vee}. \end{aligned}$$

(As clearly $R^i\pi'_*(\omega_{\pi'}) = 0$ for $i \neq n-1$ and $R^i\pi'_*\mathcal{O} = 0$ for $i \neq 0$, Grothendieck–Verdier duality (see p. 86) indeed yields $R^{n-1}\pi'_*(\omega_{\pi'}) \simeq R\pi'_*(\omega_{\pi'})[n-1] \simeq (R\pi'_*\mathcal{O})^\vee \simeq \mathcal{O}$.)

In the following we shall use the shorthand $\mathcal{H}^i := \mathcal{H}(p_*q^*L)$.

The description of the cohomology of the complex p_*q^*L suffices to conclude

$$\chi(p_*q^*L, p_*q^*L) = \chi(\mathcal{H}^0, \mathcal{H}^0) + (-1)^{n-1}\chi(\mathcal{H}^{n-1}, \mathcal{H}^0) + (-1)^{n-1}\chi(\mathcal{H}^0, \mathcal{H}^{n-1}) + \chi(\mathcal{H}^{n-1}, \mathcal{H}^{n-1}).$$

To compute the first three summands observe that $\chi(\mathcal{H}^0, \mathcal{H}^0) = \chi(X', \mathcal{O})$ and

$$\chi(\mathcal{H}^{n-1}, \mathcal{H}^0) = \chi(\mathcal{H}^0, \mathcal{H}^{n-1}) = \chi(P^\vee, \mathcal{O}(-n-1)) = (-1)^n$$

(apply Serre duality and use $\omega_{X'}|_{P^\vee} \simeq \mathcal{O}_{P^\vee}$).

The fourth term is slightly more complicated to compute. Use the spectral sequence (see (3.16), p. 85)

$$E_2^{p,q} = H^p(P^\vee, \mathcal{E}xt^q_{X'}(\mathcal{H}^{n-1}, \mathcal{H}^{n-1})) \Rightarrow \mathrm{Ext}^{p+q}_{X'}(\mathcal{H}^{n-1}, \mathcal{H}^{n-1})$$

and the description of the local Ext groups as

$$\mathcal{E}xt^q_{X'}(\mathcal{H}^{n-1}, \mathcal{H}^{n-1}) \simeq \mathcal{E}xt^q_{X'}(\mathcal{O}_{P^\vee}, \mathcal{O}_{P^\vee}) \simeq \bigwedge^q \mathcal{N}_{P^\vee/X'} \simeq \Omega^q_{P^\vee}$$

(cf. Proposition 11.8) to deduce that $\mathrm{Ext}^i_{X'}(\mathcal{H}^{n-1}, \mathcal{H}^{n-1}) = H^i(P^\vee, \Omega^i)$. Therefore, $\chi(\mathcal{H}^{n-1}, \mathcal{H}^{n-1}) = 0$ or $= 1$ depending on the parity of n.

Altogether, this yields

$$\chi(p_*q^*L, p_*q^*L) = \chi(X', \mathcal{O}) - 2 + \begin{cases} 0 & \text{if } n \equiv 1 \ (2) \\ 1 & \text{if } n \equiv 0 \ (2). \end{cases} \tag{11.7}$$

Now we use a standard fact from birational geometry saying that $h^i(X, \mathcal{O})$ is a birational invariant (cf. [115]). Applied to our situation it says $h^i(X, \mathcal{O}) = h^i(X', \mathcal{O})$. Thus, $\chi(p_*q^*L, p_*q^*L) \neq \chi(L, L)$.

Note that we can in fact argue without the birational invariance of $h^i(X, \mathcal{O})$ by computing $p_*q^*\mathcal{O}_X \simeq \mathcal{O}_{X'}$. So, if $p_* \circ q^*$ was an equivalence then we would have $\chi(X, \mathcal{O}_X) = \chi(X', \mathcal{O}_{X'})$. This would lead to the same contradiction.

As eluded to before, the contradiction we eventually derived is purely cohomological, for the inequality $\chi(L, L) \neq \chi(p_*q^*L, p_*q^*L)$ can in terms of the generalized Mukai pairing (see Section 5.2) be written as

$$\langle v(L), v(L) \rangle \neq \left\langle \Phi^H(v(L)), \Phi^H(v(L)) \right\rangle.$$

Here, $\Phi = p_* \circ q^*$. This way of writing suggests that the special line bundle L we worked with is not of such great importance and indeed the following arguments prove the assertion without assuming the existence of L.

It is always possible to find a cohomology class $\alpha \in H^2(X, \mathbb{Q})$ with $\alpha|_P = -nh$ where $h = \mathrm{c}_1(\mathcal{O}(1)) \in H^2(P, \mathbb{Z})$. Then $q^*\alpha = n[E] + p^*\alpha'$ for some $\alpha' \in H^2(X', \mathbb{Q})$. (Here $[E]$ denotes the fundamental class of the exceptional divisor $E \subset \widetilde{X}$.) In the following we let $v := \exp(\alpha)\sqrt{\mathrm{td}(X)} = \exp(\alpha)v(\mathcal{O}_X)$ and $v' := \exp(\alpha')\sqrt{\mathrm{td}(X')} = \exp(\alpha')v(\mathcal{O}_{X'})$. If $\alpha = \mathrm{c}_1(L)$, then $v = v(L)$.

A straightforward calculation involving the Grothendieck–Riemann–Roch formula for the embedding $\widetilde{X} \subset X \times X'$ and the second projection, reveals

$$\Phi^H(v) = \mathrm{ch}(p_*\mathcal{O}(nE)) \cdot v'.$$

Now we may use the calculation of the cohomology of the direct image $p_*\mathcal{O}(nE)$ given before, as this was independent of the existence of L. Thus,

$$\begin{aligned}\mathrm{ch}(p_*\mathcal{O}(nE)) &= \mathrm{ch}(\mathcal{O}_{X'}) + (-1)^{n-1}\mathrm{ch}(\mathcal{O}_{P^\vee})\\ &= 1 + (-1)^{n-1}\mathrm{ch}(\mathcal{O}_{P^\vee}),\end{aligned}$$

which immediately shows

$$\begin{aligned}&\langle \Phi^H(v), \Phi^H(v)\rangle\\ &= \langle v', v'\rangle + 2(-1)^{n-1}\langle v', \mathrm{ch}(\mathcal{O}_{P^\vee})v'\rangle + \langle \mathrm{ch}(\mathcal{O}_{P^\vee})v', \mathrm{ch}(\mathcal{O}_{P^\vee})v'\rangle.\end{aligned}$$

Now observe that multiplication with $\exp(\beta)$ for an arbitrary cohomology class β of degree two is orthogonal for the Mukai pairing. Hence,

$$\begin{aligned}&\langle \Phi^H(v), \Phi^H(v)\rangle\\ &= \langle v(\mathcal{O}_{X'}), v(\mathcal{O}_{X'})\rangle + 2(-1)^{n-1}\langle v(\mathcal{O}_{X'}), v(\mathcal{O}_{P^\vee})\rangle + \langle v(\mathcal{O}_{P^\vee}), v(\mathcal{O}_{P^\vee})\rangle.\end{aligned}$$

This has been computed in (11.7) and we again find $\langle v, v\rangle \neq \langle \Phi^H(v), \Phi^H(v)\rangle$. As by Proposition 5.44 an equivalence would induce an isometry with respect to the generalized Mukai pairing, the Fourier–Mukai functor $\Phi = p_* \circ q^*$ cannot be an equivalence. □

Exercise 11.29 Using the Grothendieck–Verdier duality, the above given description of p_*q^*L can be made more precise. Prove that $p_*\mathcal{O}(nE) \simeq \mathcal{I}_{P^\vee}^\vee$ with $\mathcal{I}_{P^\vee}$ the ideal sheaf of $P^\vee \subset X'$. Deduce from this $p_*q^*L \simeq L' \otimes \mathcal{I}_{P^\vee}^\vee$ and $\chi(p_*q^*L, p_*q^*L) = \chi(\mathcal{I}_{P^\vee}, \mathcal{I}_{P^\vee})$.

Before coming to the good news, we need to explain a very useful general principle. Suppose we are given two smooth projective morphisms $\mathcal{X} \twoheadrightarrow S$ and $\mathcal{X}' \twoheadrightarrow S$ between smooth quasi-projective varieties. If $0 \in S$ is a distinguished point, we denote by X and X' the special fibres $\mathcal{X}_0$, respectively $\mathcal{X}'_0$. The natural closed embedding will be called $i : X \hookrightarrow \mathcal{X}$ and $i' : X' \hookrightarrow \mathcal{X}'$. For the proof of the following result we introduce self-explaining notation by the following diagram:

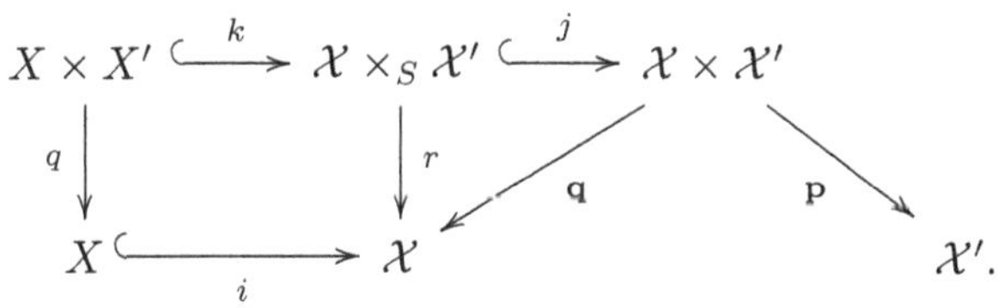

The rectangle on the left is a fibre product diagram.

Lemma 11.30 **(Chen)** *For any object $\mathcal{E}^\bullet \in \mathrm{D^b}(\mathcal{X} \times_S \mathcal{X}')$ there exists a natural isomorphism between the two functors*

$$\mathrm{D^b}(X) \xrightarrow{\Phi_{k^*\mathcal{E}^\bullet}} \mathrm{D^b}(X') \xrightarrow{i'_*} \mathrm{D^b}(\mathcal{X}')$$

and

$$\mathrm{D}^{\mathrm{b}}(X) \xrightarrow{\ i_*\ } \mathrm{D}^{\mathrm{b}}(\mathcal{X}) \xrightarrow{\ \Phi_{j_*\mathcal{E}^\bullet}\ } \mathrm{D}^{\mathrm{b}}(\mathcal{X}').$$

See [31].

Proof The assertion is proved by applying the projection formula and flat base change as follows

$$\begin{aligned} i'_*(\Phi_{k^*\mathcal{E}}(\mathcal{F}^\bullet)) &= i'_*(p_*(k^*\mathcal{E}^\bullet \otimes q^*\mathcal{F}^\bullet)) \simeq \mathbf{p}_*(i \times i')_*(k^*\mathcal{E}^\bullet \otimes q^*\mathcal{F}^\bullet) \\ &\simeq \mathbf{p}_* j_* k_*(k^*\mathcal{E}^\bullet \otimes q^*\mathcal{F}^\bullet) \simeq \mathbf{p}_* j_*(\mathcal{E}^\bullet \otimes k_* q^*\mathcal{F}^\bullet) \\ &\simeq \mathbf{p}_* j_*(\mathcal{E}^\bullet \otimes r^* i_*\mathcal{F}^\bullet) \simeq \mathbf{p}_* j_*(\mathcal{E}^\bullet \otimes j^*\mathbf{q}^* i_*\mathcal{F}^\bullet) \\ &\simeq \mathbf{p}_*(j_*\mathcal{E}^\bullet \otimes \mathbf{q}^* i_*\mathcal{F}^\bullet) = \Phi_{j_*\mathcal{E}^\bullet}(i_*\mathcal{F}^\bullet). \end{aligned}$$

□

Proposition 11.31 (Kawamata, Namikawa) *Under the general assumptions of this section let $Z \subset X \times X'$ be the reduced subvariety $Z := \widetilde{X} \bigcup (P \times P^\vee) \subset X \times X'$. Then*

$$\Phi_{\mathcal{O}_Z} : \mathrm{D}^{\mathrm{b}}(X) \longrightarrow \mathrm{D}^{\mathrm{b}}(X')$$

is an equivalence. See [63, 86].

Proof We shall follow closely the arguments given in [63]. In particular, the assertion will be reduced to Proposition 11.23 (see also Remark 11.24, ii)).

Let us suppose that we are in the following situation: There exists a smooth family $\mathcal{X} \longrightarrow C$ over a smooth quasi-projective curve C such that the special fibre $\mathcal{X}_0$ over a distinguished point $0 \in C$ is isomorphic to X and such that $\mathcal{N}_{P/\mathcal{X}} \simeq \mathcal{O}(-1)^{\oplus n+1}$. Thus,

$$\begin{array}{ccccc} P & \hookrightarrow & X & \hookrightarrow & \mathcal{X} \\ & & \downarrow & & \downarrow \\ & & \{0\} & \hookrightarrow & C. \end{array}$$

We shall give the proof in this situation.

The standard flop $\mathcal{X} \longleftarrow \widetilde{\mathcal{X}} \longrightarrow \mathcal{X}'$ performed in $P \subset \mathcal{X}$ (which has normal bundle isomorphic to $\mathcal{O}(-1)^{\oplus n+1}$) yields another family $\mathcal{X}' \longrightarrow C$ whose special fibre is isomorphic to X'. We shall use the notation of Lemma 11.30.

Note that the central fibre $\widetilde{\mathcal{X}}_0$ of $\widetilde{\mathcal{X}} \longrightarrow C$ is naturally isomorphic to

$$Z := \widetilde{X} \bigcup (P \times P^\vee).$$

In the last step we shall use Proposition 11.23 applied to $\mathcal{X} \longleftarrow \widetilde{\mathcal{X}} \longrightarrow \mathcal{X}'$ yielding the equivalence

$$\boldsymbol{\Phi} := \Phi_{\mathcal{O}_{\widetilde{\mathcal{X}}}} : \mathrm{D}^{\mathrm{b}}(\mathcal{X}) \longrightarrow \mathrm{D}^{\mathrm{b}}(\mathcal{X}')$$

with the inverse $\boldsymbol{\Phi}'$ given by the Fourier–Mukai transform with kernel $\mathcal{O}_{\widetilde{\mathcal{X}}}(nE)$ (see Remark 11.22). Strictly speaking, the description of the adjoint functors of a Fourier–Mukai transformation had been stated in Proposition 5.9 only for projective varieties. A reinspection of the argument reveals however that the projectivity of the fibres of $\mathcal{X} \longleftarrow \widetilde{\mathcal{X}} \longrightarrow \mathcal{X}'$ suffices.

Similarly, the right adjoint functor Φ' of $\Phi := \Phi_{\mathcal{O}_Z} : \mathrm{D}^{\mathrm{b}}(X) \longrightarrow \mathrm{D}^{\mathrm{b}}(X')$ is given as the Fourier–Mukai transform with kernel $\mathcal{O}_{\mathcal{O}_Z}(nE)$. Indeed, the dual of $\mathcal{O}_Z \in \mathrm{D}^{\mathrm{b}}(X \times X')$ can be computed as $\mathcal{O}_Z^{\vee} \simeq (k^*\mathcal{O}_{\widetilde{\mathcal{X}}})^{\vee} \simeq k^*(\mathcal{O}_{\widetilde{\mathcal{X}}}^{\vee})$ with $\widetilde{\mathcal{X}}$ considered as a smooth subvariety of $\mathcal{X} \times_C \mathcal{X}'$. Hence, by Corollary 3.40

$$\mathcal{O}_Z^{\vee} \simeq k^*(\omega^*_{\mathcal{X} \times_C \mathcal{X}'}|_{\widetilde{\mathcal{X}}} \otimes \omega_{\widetilde{\mathcal{X}}})[-2n] \simeq q^*\omega_X^*(nE).$$

Thus, in order to show that Φ is fully faithful, it suffices to show that the adjunction morphism $\mathrm{id} \longrightarrow \Phi' \circ \Phi$ is an isomorphism.

Consider a complex $\mathcal{G}^\bullet \in \mathrm{D}^{\mathrm{b}}(X)$ and denote the cone of the adjunction morphism $\mathcal{G}^\bullet \longrightarrow \Phi'(\Phi(\mathcal{G}^\bullet))$ by $\mathcal{H}^\bullet$. Applying i_* yields a distinguished triangle

$$i_*\mathcal{G}^\bullet \longrightarrow i_*\Phi'(\Phi(\mathcal{G}^\bullet)) \longrightarrow i_*\mathcal{H}^\bullet \longrightarrow i_*\mathcal{G}^\bullet[1].$$

Due to Lemma 11.30 one has $i_* \circ \Phi' \circ \Phi \simeq \boldsymbol{\Phi}' \circ i'_* \circ \Phi \simeq \boldsymbol{\Phi}' \circ \boldsymbol{\Phi} \circ i_*$. Hence, $i_*\mathcal{H}^\bullet$ is isomorphic to the cone of the morphism $i_*\mathcal{G}^\bullet \longrightarrow \boldsymbol{\Phi}'(\boldsymbol{\Phi}(i_*\mathcal{G}^\bullet))$ (which is again given by adjunction!). Since $\boldsymbol{\Phi}$ is an equivalence, one obtains $\mathcal{H}^\bullet = 0$. The latter implies $\mathcal{H}^\bullet \simeq 0$, which proves that Φ is fully faithful.

As $\omega_X|_P \simeq \mathcal{O}_P$ and $\omega_{X'}|_{P^\vee} \simeq \mathcal{O}_{P^\vee}$, the full faithfulness of Φ immediately shows that Φ is an equivalence (cf. Proposition 7.6). Once more, Proposition 7.6 strictly speaking assumes projectivity of the participating varieties. As this is again needed for the description of the adjoint functors, the remark above applies. The projectivity of the morphisms $X \longleftarrow Z \longrightarrow X'$ suffices. □

Remark 11.32 There are two ways to reduce to the situation treated in the proof.

Either, one passes to the linearized version of P embedded into $X := |\Omega_P|$ as the zero section (see Example 11.7). Why this is really possible is not so easy to explain. It depends heavily on the fact that the normal bundle $\mathcal{N}_{P/X} \simeq \Omega_P$ is negative, which allows us to produce an isomorphism of the formal neighbourhood of $P \subset X$ with the formal neighbourhood of $P \subset |\Omega_P|$ (see [86]). In any case, if we have reduced to this linear situation, then by means of the Euler sequence $0 \longrightarrow \Omega_P \longrightarrow \mathcal{O}(-1)^{\oplus n+1} \longrightarrow \mathcal{O} \longrightarrow 0$ we embed $P \subset X$ into $\mathcal{X} := |\mathcal{O}(-1)^{\oplus n+1}|$. The natural projection $\mathcal{O}(-1)^{\oplus n+1} \longrightarrow \mathcal{O}$ will be interpreted as a smooth morphism $\mathcal{X} \longrightarrow C := \mathbb{A}^1$.

The other, more global one, works for algebraic symplectic varieties and is described in [50].

Remark 11.33 i) The elementary Mukai flop can quite naturally be generalized to other situations. Most frequently, one replaces the projective space $\mathbb{P}^n$ by a projective bundle $\mathbb{P}(\mathcal{F}) \longrightarrow Y$, where $\mathcal{F}$ is a locally free sheaf of rank $r = \mathrm{codim}(\mathbb{P}(\mathcal{F}) \subset X)$. Once more, the blow-up of $\widetilde{X} \longrightarrow X$ along $\mathbb{P}(\mathcal{F})$ can be contracted in another direction $\widetilde{X} \longrightarrow X'$ and in the process the projective bundle $\mathbb{P}(\mathcal{F})$ gets replaced by the dual projective bundle $\mathbb{P}(\mathcal{F}^*) \longrightarrow Y$.

Due to the general conjecture, the two varieties X and X' are again expected to be derived equivalent and imitating the arguments above a good candidate for a kernel of a Fourier–Mukai equivalence is at hand, the structure sheaf $\mathcal{O}_Z$ of the cycle $\widetilde{X} \bigcup \mathbb{P}(\mathcal{F}) \times_Y \mathbb{P}(\mathcal{F}^*)$.

That this indeed defines a derived equivalence $\mathrm{D^b}(X) \xrightarrow{\sim} \mathrm{D^b}(X')$ has been shown by Namikawa in [86].

ii) The kernel for the Fourier–Mukai equivalence chosen in Proposition 11.31 as well as the one in i) can both be interpreted as the structure sheaf of the fibre product $X \times_{\bar{X}} X'$, where $X \longrightarrow \bar{X} \longleftarrow X'$ are the contraction of $P \subset X$ and $P^\vee \subset X'$ to a point (respectively, of $\mathbb{P}(\mathcal{F})$ and $\mathbb{P}(\mathcal{F}^*)$ to Y).

iii) Suppose X and X' are two birational varieties with trivial canonical bundles, both obtained as desingularizations $X \longrightarrow \bar{X}$, respectively $X' \longrightarrow \bar{X}$ of a singular variety $\bar{X}$. According to Conjecture 6.24, X and X' should be derived equivalent and the above examples suggest $\mathcal{O}_Z$ with $Z := X \times_{\bar{X}} X'$ as a candidate for a kernel of a Fourier–Mukai equivalence.

This question has been studied for so called *stratified Mukai flops*. The simplest case is (locally) given by $X := |\Omega_{\mathrm{Gr}_2(V)}|$ and $X' := |\Omega_{\mathrm{Gr}_2(V^*)}|$, where V is a vector space of dimension four and V^* is its dual. Both varieties are contracted to the nilpotent variety $\{A \in \mathrm{End}(V) \mid A^2 = 0\}$.

Although, the structure sheaf of the fibre product turns out to work well as a Fourier–Mukai kernel in K-theory and cohomology, Namikawa showed in [87] that it does not(!) define an equivalence between the derived categories. Quite recently, Kawamata [66] solved the riddle by showing that a certain twist of $\mathcal{O}_Z$ by the exceptional divisor does define the hoped for equivalence.

Conjecture 6.24 predicts that K-equivalent varieties are D-equivalent. In dimension three one can prove that both types of equivalences imply equality of Hodge numbers.

Indeed, due to a result of Batyrev [6] one knows that K-equivalent varieties X and X' in any dimension satisfy

$$h^{p,q}(X) = h^{p,q}(X').$$

By Proposition 5.39 one has that

$$\sum_{p-q=k} h^{p,q}(X) = \sum_{p-q=k} h^{p,q}(X')$$

whenever $\mathrm{D^b}(X) \simeq \mathrm{D^b}(X')$. In dimension three this suffices to conclude again equality of Hodge numbers. We leave this straightforward check to the reader (use Serre duality).

The general conjecture is still wide open, but in dimension three we have the following impressive result of Bridgeland, which generalizes the case of the standard flop previously considered by Bondal and Orlov. The proof of Bridgeland's result is beyond this course. It uses the notion of t-structures on triangulated categories and moduli spaces of perverse sheaves (which are just objects in the abelian category described by the heart of a t-structure).

Theorem 11.34 (Bridgeland) *If $X \longrightarrow Y \longleftarrow X'$ are two crepant resolutions of a three-dimensional projective variety Y with only terminal singularities, then $\mathrm{D^b}(X) \simeq \mathrm{D^b}(X')$. See [19].* □

Combining with standard techniques in three-dimensional birational geometry one obtains:

Corollary 11.35 *Suppose X and X' are two birational Calabi–Yau threefolds. Then $\mathrm{D^b}(X) \simeq \mathrm{D^b}(X')$.*

Proof Any birational map between two three-dimensional varieties with nef canonical bundles can be written as a finite sequence of flops, i.e. of diagrams of the form considered in Theorem 11.34. This is a result due to Kawamata and Kollár (see [68] or [76, Ch.12]). □

Remark 11.36 i) The original result of Bridgeland has later been generalized to singular varieties. For the precise results we refer to the articles by Chen [31], Kawamata [64] and Abramovich/Chen [2].

ii) In a series of paper [116, 117] van den Bergh has shed completely new light on Bridgeland's theorem. He studies so-called 'non-commutative crepant resolutions' of singularities.

Suppose $X \longrightarrow Y$ is a crepant resolution. Under certain technical conditions, all satisfied in the situation described by the theorem, van den Bergh shows the existence of an $\mathcal{O}_Y$-algebra $\mathcal{A}$ on Y such that $\mathrm{D^b}(X) \simeq \mathrm{D^b}(\mathcal{A})$, where $\mathrm{D^b}(\mathcal{A})$ is the derived category of the abelian category of all coherent $\mathcal{A}$-module sheaves on Y. Moreover, $\mathcal{A}$ is obtained as the direct image of $\mathcal{E}nd(\mathcal{P})$ with $\mathcal{P}$ a vector bundle $\mathcal{P}$. The sheaf of algebras $\mathcal{A}$ is considered as a 'non-commutative crepant resolution' of Y.

12

DERIVED CATEGORIES OF SURFACES

We have seen that any smooth projective variety Y which is derived equivalent to a smooth curve X, i.e. $\mathrm{D}^{\mathrm{b}}(X) \simeq \mathrm{D}^{\mathrm{b}}(Y)$, is in fact isomorphic to X. In this chapter we set out to study the analogous question in one dimension higher. So, throughout X will denote a surface by which we shall always mean a smooth projective irreducible variety of dimension two over an algebraically closed field of characteristic zero.

In earlier sections we actually discussed abelian surfaces and K3 surfaces and encountered non-isomorphic such surfaces with equivalent derived categories. In other words, the answer to our question in dimension two is clearly more complicated than for curves. However, it will turn out that abelian, K3, and elliptic surfaces are the only exceptions. More precisely, following Bridgeland, Maciocia, and Kawamata, we shall show

Proposition 12.1 *Suppose* $\mathrm{D}^{\mathrm{b}}(X) \simeq \mathrm{D}^{\mathrm{b}}(Y)$ *with* X *a smooth projective surface which is neither elliptic, nor K3, nor abelian. Then* $X \simeq Y$.

In order to prove this, we will work our way through the Enriques classification of minimal surfaces. The reduction to minimal surfaces, a result due to Kawamata, will be accomplished in Section 12.1, which also contains the necessary recollections from surface theory. For the reader familiar with these basic rudiments or willing to accept them, this section as well as the other ones should be fairly self-contained.

In Section 12.2 we apply the general machinery to surfaces of extremal Kodaira dimension, i.e. to surfaces of general type and to surfaces of Kodaira dimension $-\infty$.

Section 12.3 deals with surfaces of Kodaira dimension zero. The most prominent examples of such surfaces are abelian and K3 surfaces and for those the main work has already been done in Chapters 9 and 10. The reduction to these two kinds will be achieved by passing to canonical covers, a concept that was explained in Section 7.3.

The last section treats elliptic surfaces. As for K3 and abelian surfaces, elliptic surfaces in general admit non-isomorphic Fourier–Mukai partners, but those can be described explicitly.

Before entering the subject, let us apply the general results of earlier chapters to the special case of surfaces. So, let Y be a smooth projective variety such that its derived category $\mathrm{D}^{\mathrm{b}}(Y)$ is equivalent to the derived category $\mathrm{D}^{\mathrm{b}}(X)$ of a surface X.

- Proposition 4.1 implies that Y is a surface whose canonical bundle ω_Y is of the same order as ω_X.
- Proposition 6.1 proves

$$\mathrm{kod}(X, \omega_X^{\pm}) = \mathrm{kod}(Y, \omega_Y^{\pm}) \quad \text{and} \quad R(X, \omega_X^{\pm}) \simeq R(Y, \omega_Y^{\pm})$$

(see also Exercise 6.2). In particular, $h^0(X, \omega_X^i) = h^0(Y, \omega_Y^i)$ for all i. For $i = 1, 2$ this will be particularly useful.
- The numerical Kodaira dimensions of X and Y coincide, i.e. $\nu(X) = \nu(Y)$. Moreover, ω_X is nef if and only if ω_Y is nef. A similar result holds for the anti-canonical bundle ω^*. See Propositions 6.3 and 6.3.
- From Proposition 5.39 one derives

$$h^1(X, \mathcal{O}_X) = h^1(Y, \mathcal{O}_Y).$$

Indeed, the general result yields $(h^{0,1} + h^{1,2})(X) = (h^{0,1} + h^{1,2})(Y)$, but the usual symmetries for projective surfaces ensure $h^{0,1} = h^{1,0} = h^{1,2}$.
- Exercise 5.38 ensures equality of the topological Euler numbers:

$$e(X) = e(Y).$$

- Combining Proposition 5.33 and 5.39 shows that $(H^{0,0} \oplus H^{1,1} \oplus H^{2,2})(X, \mathbb{Q})$ is a derived invariant. The dimension of this space is $\rho(X) + 2$ with ρ the Picard number of X. Hence,

$$\rho(X) = \rho(Y).$$

12.1 Recap: Enriques classification of algebraic surfaces

It is not possible to review the whole classification of algebraic surfaces here and this is in fact not needed for our purposes. We shall recall the basic principles that reduce the classification of surfaces to the classification of minimal surfaces. There are several excellent textbooks on algebraic surfaces, e.g. [4, 5, 7, 100]. In the following, a surface will mean a smooth irreducible projective surface over an algebraically closed field of characteristic zero.

Consider the blow-up $q : \widetilde{X} \longrightarrow X$ of a point $x \in X$ in a smooth surface X. The exceptional divisor $C \subset \widetilde{X}$ is a smooth irreducible rational curve with $C.C = -1$. Moreover,

$$\omega_{\widetilde{X}} \simeq q^*\omega_X \otimes \mathcal{O}(C)$$

and, therefore, $\mathrm{c}_1^2(\widetilde{X}) = \mathrm{c}_1^2(X) - 1$. Also note that the Picard group increases by the exceptional divisor and thus

$$\rho(\widetilde{X}) = \rho(X) + 1.$$

The inverse process is called the *blow-down* (or *contraction*) of the curve C.

Definition 12.2 *A surface is called* minimal *if it cannot be obtained as the blow-up of another surface (always smooth!).*

Lemma 12.3 *For any surface X there exists a minimal surface $X_{\min}$, a* minimal model of X, *and a sequence of blow-ups*

$$X \simeq X_n \longrightarrow X_{n-1} \longrightarrow \cdots \longrightarrow X_0 \simeq X_{\min}.$$

Proof As the Picard number, which is known to be finite, drops every time a curve is contracted and $\rho(X) < \infty$, the process stops eventually. □

Curves that are contracted by a blow-down are described by the following important result.

Theorem 12.4 (Castelnuovo) *Suppose a surface Y contains a smooth irreducible rational curve C with $C.C = -1$. Then C can be contracted, i.e. there exists a blow-up $q : \widetilde{X} \longrightarrow X$ as above and an isomorphism $Y \simeq \widetilde{X}$ identifying C with the exceptional curve. See [5, Ch.III] or [45, III,Thm.5.7].*

Corollary 12.5 *A surface is minimal if and only if it does not contain any (-1)-curve, i.e. a smooth irreducible rational curve with self-intersection -1.*

Examples 12.6 i) A *rational surface* is a surface that is birational to $\mathbb{P}^2$. The other minimal rational surfaces are the rational ruled surfaces $\mathbb{F}_n$, $n = 0, 2, 3, \ldots$. A *ruled surface* is a surface of the form $\mathbb{P}(\mathcal{E})$ where $\mathcal{E}$ is a locally free sheaf of rank two on a smooth curve C and by definition $\mathbb{F}_n := \mathbb{P}(\mathcal{O}_{\mathbb{P}^1} \oplus \mathcal{O}_{\mathbb{P}^1}(n))$.

ii) Non-rational ruled surfaces are all of the form $\mathbb{P}(\mathcal{E}) \longrightarrow C$ with C a curve of genus $g > 0$. Clearly, such a surface is birational to $C \times \mathbb{P}^1$. In particular, there are many birational minimal such surfaces. Also note the useful formula $\mathrm{c}_1^2(\mathbb{P}(\mathcal{E})) = 8(1 - g(C))$ (cf. [45, III,Cor.2.11]).

Theorem 12.7 *Any birational map*

$$X \dashrightarrow Y$$

between surfaces can be resolved by a diagram

$$\begin{array}{ccc} & Z & \\ {\scriptstyle p} \swarrow & & \searrow {\scriptstyle q} \\ X & & Y, \end{array}$$

where q and p are given as compositions of blow-ups. See [5, III,Cor.4.4]

Corollary 12.8 *Any two K-equivalent surfaces are isomorphic.*

Proof If X and Y are K-equivalent, then any resolution of the birational correspondence will realize the K-equivalence. Consider the one in the theorem $X \xleftarrow{q} Z \xrightarrow{p} Y$ and decompose p by $Z \longrightarrow Z' \longrightarrow Y$ with $Z \longrightarrow Z'$ a single blow-up. Clearly, $p^*\omega_Y$ is trivial on the exceptional curve of this blow-up and hence $q^*\omega_X$ is. This shows that $q : Z \longrightarrow X$ factorizes over a morphism $Z' \longrightarrow X$.

Continuing in this way and changing the rôle of X and Y if running out of exceptional curves for one of the projections, eventually produces an isomorphism $X \simeq Y$ □

The description of birational maps is also used to prove uniqueness of the minimal model. For the following result, see e.g. [5, III,Prop.4.6].

Theorem 12.9 *Suppose X and Y are two birational minimal surfaces which are neither isomorphic to $\mathbb{P}^2$ nor to a ruled surface. Then $X \simeq Y$.*

As $\mathbb{P}^2$ and the ruled surfaces are the only minimal surfaces with kod $= -\infty$, the theorem yields

Corollary 12.10 *The minimal model of a surface of Kodaira dimension ≥ 0 is unique.*

Rational surfaces can be described by their numerical invariants due to the following famous

Theorem 12.11 (Castelnuovo) *A surface is rational if and only if both cohomology groups $H^1(X, \mathcal{O}_X)$ and $H^0(X, \omega_X^2)$ vanish. See [5, VI,Thm.2.1].*

On the opposite end of the classification table one finds surfaces of general type. By definition, a *surface of general type* is a surface X with $\mathrm{kod}(X) = 2$.

Theorem 12.12 *Let X be a surface of general type (or, more generally, with $\mathrm{kod}(X) \geq 0$). Then X is minimal if and only if ω_X is nef.*

One direction is easy, the canonical bundle of a blow-up has degree -1 on the exceptional divisor and, therefore, ω_X nef implies X minimal.

Remark 12.13 In general, the canonical bundle of a minimal surface of general type is not ample. However, for $n \geq 5$, the linear system $|\omega_X^n|$ is base-point free (cf. [5, VII,Thm.5.2]) and defines a morphism $X \twoheadrightarrow X_{\mathrm{can}}$ onto the *canonical model* $X_{\mathrm{can}} = \mathrm{Proj}(R(X))$. The canonical model is a normal surface with at most rational double points as singularities. The irreducible curves contracted by the canonical morphism of a minimal surface $X \twoheadrightarrow X_{\mathrm{can}}$ are smooth rational curves of self-intersection -2.

Minimal surfaces of general type cannot be classified in the sense of giving an exhaustive list. A large part of the theory consists of finding such surfaces in certain interesting regions of Chern numbers c_1^2 and c_2 or other numerical invariants $q = h^1(\mathcal{O}_X)$, $p = h^2(\mathcal{O}_X)$, and the like.

The most interesting surfaces, although this might be a matter of taste, can be found in between these two extremes.

We have already encountered two types of minimal surfaces of Kodaira dimension zero: abelian surfaces (i.e. abelian varieties of dimension two) and K3 surfaces. There are only two others, both of which have torsion canonical bundles. *Hyperelliptic surfaces* are those with canonical cover (see Section 7.3) isomorphic to a product of two elliptic curves and *Enriques surfaces* have a K3 surface as a canonical cover.

Surfaces of Kodaira dimension one are all *elliptic*, i.e. they admit a morphism $X \twoheadrightarrow C$ onto a smooth curve C such that the generic fibre is a (smooth) elliptic curve. Elliptic surfaces exist not only for kod $= 1$ and to distinguish those one calls them properly elliptic.

We conclude with the classification table of all minimal (algebraic) surfaces.

kod $= -\infty$	$\mathbb{P}^2$
	minimal rational ruled surfaces $\mathbb{F}_n$, $n = 0, 2, 3 \ldots$
	ruled surfaces $\mathbb{P}(\mathcal{E}) \longrightarrow C$, $g(C) \geq 1$
kod $= 0$	abelian surfaces, K3 surfaces
	hyperelliptic surfaces, Enriques surfaces
kod $= 1$	minimal properly elliptic surfaces
kod $= 2$	minimal surfaces of general type

Note that in certain regions this is an honest classification, whereas in others it just resembles all surfaces of a certain type.

Let us now pass to the classification of derived categories of surfaces. Very much as in the geometric context, one can almost always reduce to minimal ones (see Proposition 12.15). In the subsequent sections we will go through the above list of minimal surfaces and try to see what can be said about the derived categories of each type.

Remark 12.14 Let us recall a few techniques from Chapter 6 that will be applicable to derived equivalences of surfaces.

Consider a Fourier–Mukai equivalence

$$\Phi = \Phi_{\mathcal{P}} : \mathrm{D}^{\mathrm{b}}(X) \xrightarrow{\sim} \mathrm{D}^{\mathrm{b}}(Y)$$

with kernel $\mathcal{P}$. We denote by Γ the support of $\mathcal{P}$ (or, equivalently, of $\mathcal{P}^{\vee}$), i.e.

$$\Gamma := \bigcup \mathrm{supp}(\mathcal{H}^i)$$

with $\mathcal{H}^i$ the cohomology sheaves of $\mathcal{P}$ (respectively $\mathcal{P}^{\vee}$).

Due to Corollary 6.5 one always finds an irreducible component $Z \subset \Gamma$ that dominates X. If $\dim(Z) = 2$, then $Z \longrightarrow X$ is necessarily generically finite and hence by Corollary 6.12 birational. Moreover, we may find an open dense subset $U \subset X$ such that $\Gamma_U = Z_U \simeq U$ (see Remark 6.13).

If for this component the second projection $Z \longrightarrow Y$ is not dominant then one finds distinct points $x_1 \neq x_2 \in U = \Gamma_U = Z_U$ with $y := p(x_1) = p(x_2) \in Y$. Hence, $\Phi(k(x_1))$ and $\Phi(k(x_2))$ would both be complexes concentrated in y. In particular, for some $i \in \mathbb{Z}$ one has $\mathrm{Ext}^i(\Phi(k(x_1)), \Phi(k(x_2))) \neq 0$. Indeed, if m_1 and m_2 are maximal, respectively minimal with $\mathcal{H}^{m_i}(\Phi(k(x_i))) \neq 0$, then there exists a non-trivial morphism $\mathcal{H}^{m_1}(\Phi(k(x_1))) \longrightarrow \mathcal{H}^{m_2}(\Phi(k(x_2)))$ (both are sheaves concentrated in $y \in Y$), which gives rise to a non-trivial morphism $\Phi(k(x_1))[m_1] \longrightarrow \Phi(k(x_2))[m_2]$. This then would contradict the obvious $\mathrm{Ext}^i(k(x_1), k(x_2)) = 0$ for all i.

Thus, the projection $Z \longrightarrow Y$ is also dominant and hence generically finite. As before, this suffices to conclude that $Z \longrightarrow Y$ is birational as well. So, under the assumption that there exists an irreducible component $Z \subset \Gamma$ of dimension two that dominates one of the two factors, one finds a birational correspondence $X \longleftarrow Z \longrightarrow Y$. Moreover, as in the proof of Proposition 6.4, one shows that this birational correspondence defines a K-equivalence between X and Y (cf. Lemma 6.9). Hence X and Y are isomorphic.

Proposition 12.15 (Kawamata) *Let X be a smooth projective surface containing a (-1)-curve. Supppose there exists a smooth projective variety Y with $\mathrm{D}^{\mathrm{b}}(X) \simeq \mathrm{D}^{\mathrm{b}}(Y)$. Then:*

i) $X \simeq Y$ *or*

ii) X *is a relatively minimal elliptic rational surface. See [63].*

Proof After the general comments above, we may reduce to the case that any irreducible component $Z \subset \Gamma$ that dominates X or Y is of dimension at least three. Otherwise, we would immediately have $X \simeq Y$. In particular, the fibre dimension of the two projections $\Gamma \longrightarrow X$ and $\Gamma \longrightarrow Y$ is at least one.

Let us now use the assumption that X is not minimal and choose a (-1)-curve $\mathbb{P}^1 \simeq E \subset X$. We denote $\Gamma_E := \Gamma \times_X E$.

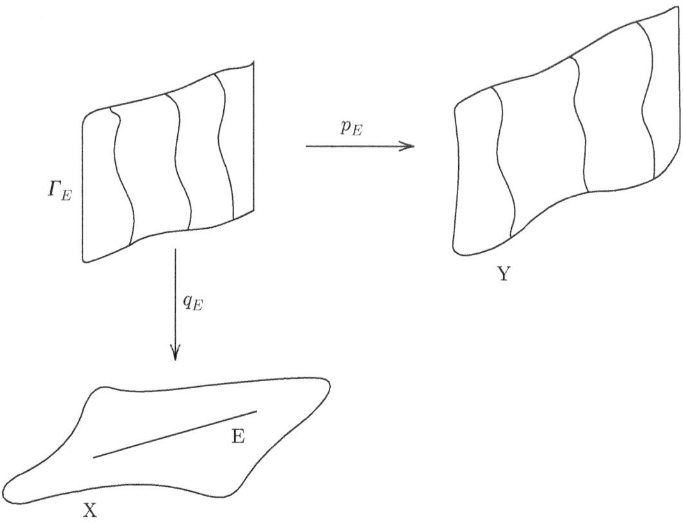

By assumption, ω_X^* restricts to an ample line bundle on E. By Corollary 6.8, the projection $p_E : \Gamma_E \longrightarrow Y$ is automatically finite. As the fibres of the other projection $q_E : \Gamma_E \longrightarrow E$ are at least one-dimensional, one has $\dim(\Gamma_E) = 2$. Hence, $p_E : \Gamma_E \longrightarrow Y$ is dominant. After passing to the normalization of an arbitrary irreducible component of Γ_E, the pull-backs of ω_X and ω_Y (or rather certain non-trivial powers of them) coincide (see Lemma 6.9). In particular, the pull-backs are numerically equivalent.

As ω_X^* is ample on E, Lemma 6.27 implies that ω_Y^* is nef. Moreover, for the numerical Kodaira dimension one has (cf. Lemma 6.30)

$$\nu(Y, \omega_Y^*) = \nu(\Gamma_E, p_E^*\omega_Y^*) = \nu(\Gamma_E, q_E^*\omega_X^*) = 1.$$

Propositions 6.3 and 6.3 ensure that both assertions hold true for X, i.e. ω_X^* is nef and $\nu(X, \omega_X^*) = 1$.

These two properties restrict the possibilities for X and Y drastically. Suppose $H^0(X, \omega_X^k) \neq 0$ for some $k > 0$. Pick a section $0 \neq s \in H^0(X, \omega_X^k)$ and consider its zero set $Z(s)$. Either $Z(s)$ is empty, i.e. ω_X^k is trivial, or it is a curve. The former contradicts $\nu(X, \omega_X^*) = 1$. But if $Z(s)$ is a curve, then it intersects an ample divisor H effectively, i.e. $Z(s).H > 0$, which contradicts ω_X^* nef. Thus, $\mathrm{kod}(X) = -\infty$.

Hence the minimal model of X is either rational or a ruled surface of genus $g \geq 1$. As c_1^2 of a surface decreases under blow-up and $\mathrm{c}_1^2 = 8(1-g)$ for a minimal surface ruled over a curve of genus g, the surface X is either rational or a minimal surface ruled over an elliptic curve. As by assumption X is not minimal, the latter case is excluded, i.e. X is indeed rational. (Note that since $q(X) = q(Y)$ and $h^0(X, \omega_X^2) = h^0(Y, \omega_Y^2)$, Castelnuovo's criterion Theorem 12.11 also yields that Y is rational.)

So, for purely numerical reasons X is either a nine-fold blow-up of $\mathbb{P}^2$ or an eight-fold blow-up of $\mathbb{P}^1 \times \mathbb{P}^1$. In order to find the elliptic curves covering X, we pick a smooth rational curve $E' \subset Y$ on which ω_Y^* is necessarily ample. Interchanging the rôle of X and Y in the above discussion we obtain a finite dominating morphism

$$q_{E'} : \Gamma_{E'} \longrightarrow X.$$

Due to the numerical equivalence of ω_X and ω_Y on Γ, the restriction of ω_X to the image $D = q_{E'}(F)$ of one of the irreducible fibres F of $p_{E'} : \Gamma_{E'} \longrightarrow E'$ is numerically trivial. As D moves in a one-dimensional family, one certainly has $D.D \geq 0$. If $D^2 > 0$, then the Hodge index theorem would yield $\mathrm{c}_1^2(X) < 0$ or $\mathrm{c}_1(X) = 0$, both of which would contradict what has been proved for X already. This leaves $D^2 = 0$ as the only possibility, i.e. the fibres of $p : \Gamma_{E'} \longrightarrow E'$ define a covering family of elliptic curves on X. □

12.2 Minimal surfaces with kod $= -\infty, 2$

This short section answers our question for surfaces of extremal Kodaira dimension. We will rely on the general facts proved in Chapter 6, but more direct arguments can be found in [23].

Proposition 12.16 (Bridgeland, Maciocia) *Suppose X is a surface of general type. For any smooth projective variety Y one has:*

$$\mathrm{D}^{\mathrm{b}}(X) \simeq \mathrm{D}^{\mathrm{b}}(Y) \quad \textit{if and only if} \quad X \simeq Y.$$

Proof Proposition 6.4 ensures that X and Y are birational and Proposition 12.15 allows us to reduce to the case of minimal surfaces. However, two birational minimal surfaces of general type are isomorphic (see Corollary 12.10). □

Although this result is in perfect analogy to the main result of Bondal and Orlov for varieties with ample (anti-)canonical bundle, a detailed description of the group of autoequivalences of a surface of general type eludes us. E.g. any (-2)-curve C contracted by the canonical morphism $X \longrightarrow X_{\mathrm{can}}$ gives rise to spherical objects $\mathcal{O}_C(i) \in \mathrm{D}^{\mathrm{b}}(X)$ and the induced spherical twists are highly non-trivial (cf. Section 8.1).

Proposition 12.17 (Bridgeland, Maciocia) *Let X be a surface that cannot be elliptically fibred. Assume that* $\mathrm{kod}(X) = -\infty$. *For any smooth projective variety Y one has:*

$$\mathrm{D}^{\mathrm{b}}(X) \simeq \mathrm{D}^{\mathrm{b}}(Y) \quad \textit{if and only if} \quad X \simeq Y.$$

Proof Again by Proposition 12.15 we may assume that X is minimal, i.e. X is isomorphic to $\mathbb{P}^2$, $\mathbb{F}_n$ $(n = 0, 2, 3, \ldots)$, or to a ruled surface $X = \mathbb{P}(\mathcal{E}) \longrightarrow C$ of genus $g(C) \geq 1$.

If $X \simeq \mathbb{P}^2$ or $X \simeq \mathbb{F}_0$, then the assertion follows from Proposition 4.11.

If X is a Hirzebruch surface $\mathbb{F}_n$, then $\mathrm{kod}(X, \omega_X^*) = 2$ and Proposition 6.4 applies. Hence, Y is K-equivalent and thus isomorphic to X.

Consider now a specific Fourier–Mukai equivalence

$$\Phi_{\mathcal{P}} : \mathrm{D}^{\mathrm{b}}(X) \xrightarrow{\sim} \mathrm{D}^{\mathrm{b}}(Y)$$

and pick an irreducible component $Z \subset X \times Y$ of $\mathrm{supp}(\mathcal{P})$ that dominates Y. If Z is a surface, then Remark 12.14 implies $X \simeq Y$. So we may assume that the fibres of $Z \longrightarrow Y$ are of dimension at least one.

Let us now imitate the arguments in the proof of Proposition 12.15. Pick a curve $E \subset Y$ such that ω_Y^* is ample on E and consider $Z_E \longrightarrow E$. Such a curve always exists, as $\mathrm{kod}(Y) = -\infty$. (Take a generic fibre of the ruling, if the minimal model of Y is a ruled surface, or a generic line in $\mathbb{P}^2$, if its minimal model is $\mathbb{P}^2$.)

Then the first projection $Z_E \longrightarrow X$ is finite (see Remark 12.14) and, as Z_E is of dimension at least two, surjective. Numerically the pull-back of ω_X and of ω_Y to Z_E coincide (cf. Lemma 6.6). Hence, by Lemmas 6.27 and 6.30 ω_X^* is nef and $\nu(X) = 1$. In particular, $\mathrm{c}_1^2(X) = 0$, which, combined with $\mathrm{c}_1^2(X) = 8(1 - g(C))$, yields $g = 1$, i.e. X is a surface ruled over an elliptic curve.

The generic fibre D of $Z_E \longrightarrow E$ defines an irreducible curve in X which moves in a one-dimensional family that covers X. Hence, $D.D \geq 0$. Numerical equivalence of ω_X and ω_Y on Z_E shows at the same time that ω_X is of degree zero along D. If $D.D > 0$, the Hodge index theorem would yield that either ω_X is numerically trivial, which is absurd, or $c_1^2(X) < 0$, which contradicts $\nu(X) = 1$.

Thus, D is an irreducible curve of arithmetic genus one with $D.D = 0$ that moves in a one-dimensional family. In other words, if there exists an irreducible component Z of $\mathrm{supp}(\mathcal{P})$ of dimension at least three that dominates Y, then X is elliptic, which is excluded by assumption. □

Remark 12.18 The assumption that X is not elliptic is crucial. In [114] Uehara shows that for any integer $N > 0$ there exist pairwise non-isomorphic rational elliptic surfaces $T_1, \ldots, T_N$ which are all D-equivalent, i.e.

$$\mathrm{D^b}(T_1) \simeq \ldots \simeq \mathrm{D^b}(T_N).$$

Why relatively minimal elliptic (rational) surfaces in general do allow non-isomorphic Fourier–Mukai partners will be explained by Corollary 12.24.

12.3 Surfaces with torsion canonical bundle

In this section we shall treat minimal surfaces with torsion canonical bundle. Those with trivial canonical bundle, i.e. abelian and K3 surfaces, have been discussed in detail in Chapters 9, respectively 10. Just recall, see e.g. Corollary 10.2, that any variety D-equivalent to a K3 surface is again a K3 surface and that the Mukai lattice decides whether two K3 surfaces are D-equivalent. Similar results hold for abelian surfaces. This leaves us with Enriques and hyperelliptic surfaces.

By definition, an *Enriques surface* is a surface with $h^1(X, \mathcal{O}_X) = 0$ and such that the canonical bundle ω_X is of order two. The canonical cover $\widetilde{X} \longrightarrow X$ is automatically a K3 surface and we shall denote the covering involution by $\iota : \widetilde{X} \longrightarrow \widetilde{X}$, i.e. $X = \widetilde{X}/\iota$.

The geometry of Enriques surfaces is rich and beautiful, e.g. they are all elliptic, but for our purposes the Torelli theorem is all we need:

Theorem 12.19 **(Global Torelli)** *Supppose X_1 and X_2 are Enriques surfaces with canonical cover $\widetilde{X}_1$, respectively $\widetilde{X}_2$ and canonical involutions ι_1, respectively ι_2. Then $X_1 \simeq X_2$ if and only if there exists a Hodge isometry $\varphi : H^2(\widetilde{X}_1, \mathbb{Z}) \simeq H^2(\widetilde{X}_2, \mathbb{Z})$ with $\varphi \circ \iota_1^* = \iota_2^* \circ \varphi$. See [5, Thm.21.2].*

In addition to this fundamental fact, the next proposition makes use of results from lattice theory which are due to Nikulin.

Proposition 12.20 **(Bridgeland, Maciocia)** *Let X be an Enriques surface. If Y is a smooth projective variety such that $\mathrm{D^b}(X) \simeq \mathrm{D^b}(Y)$, then $X \simeq Y$. See [23].*

Proof Canonical covers behave well under derived equivalences as has been seen earlier (see Section 7.3). More precisely, since the order of the canonical bundle is a derived invariant, ω_Y is also of order two and if $\widetilde{Y} \longrightarrow Y$ is its canonical cover, then $\Phi : \mathrm{D^b}(X) \simeq \mathrm{D^b}(Y)$ lifts to an equivalence $\widetilde{\Phi} : \mathrm{D^b}(\widetilde{X}) \simeq \mathrm{D^b}(\widetilde{Y})$ such that $\widetilde{\Phi} \circ \iota_X^* \simeq \iota_Y^* \circ \widetilde{\Phi}$. Here, ι_Y is the covering involution of $\widetilde{Y} \longrightarrow Y$. (In fact, $\widetilde{\Phi}$ and Φ commute with pull-back and push-forward, but this is not needed here.) Hence, $\widetilde{Y}$ is a surface derived equivalent to a K3 surface and hence itself a K3 surface (see Corollary 10.2). Thus, Y is necessarily an Enriques surface.

The lift $\widetilde{\Phi}$ induces a Hodge isometry $\widetilde{\Phi}^H : \widetilde{H}(\widetilde{X}, \mathbb{Z}) \simeq \widetilde{H}(\widetilde{Y}, \mathbb{Z})$ (see Corollary 10.7) which is invariant under the action of the covering involutions ι_X and ι_Y.

In order to apply the Torelli theorem for Enriques surfaces we need to construct an invariant Hodge isometry $H^2(\widetilde{X}, \mathbb{Z}) \simeq H^2(\widetilde{Y}, \mathbb{Z})$. This is done as follows. Let us denote by $\widetilde{H}_{\pm}(\widetilde{X}, \mathbb{Z})$ the sublattice of all integral classes $\alpha \in \widetilde{H}(\widetilde{X}, \mathbb{Z})$ with $\iota_X^*(\alpha) = \pm\alpha$. The sublattices $\widetilde{H}_{\pm}(\widetilde{Y}, \mathbb{Z})$ are defined similarly. Then $H^{0,2} \subset \widetilde{H}_- \otimes \mathbb{C} \subset H^2$. Moreover, as $\widetilde{\Phi}^H$ is invariant, it induces a Hodge isometry $\widetilde{H}_-(\widetilde{X}, \mathbb{Z}) \simeq \widetilde{H}_-(\widetilde{Y}, \mathbb{Z})$.

At this point Nikulin's Theorems 3.6.2, 3.6.3 in [89] come in. Indeed, the orthogonal complement of $\widetilde{H}_- \subset H^2$ is even, indefinite, and 2-elementary. Thus, $\widetilde{\Phi}^H : \widetilde{H}_-(\widetilde{X}, \mathbb{Z}) \simeq \widetilde{H}_-(\widetilde{Y}, \mathbb{Z})$ can be extended to a Hodge isometry $H^2(\widetilde{X}, \mathbb{Z}) \simeq H^2(\widetilde{Y}, \mathbb{Z})$, which is automatically invariant. Applying the Torelli theorem concludes the proof. □

The analogous result holds true for the second type of minimal surfaces with torsion canonical bundle.

Proposition 12.21 **(Bridgeland, Maciocia)** *Let X be a hyperelliptic surface. If Y is a smooth projective variety with $\mathrm{D^b}(X) \simeq \mathrm{D^b}(X)$, then $X \simeq Y$. See [23].*

We omit the proof. Unfortunately, the techniques relying on canonical covers do not quite suffice to prove the result. One rather has to use elliptic surfaces, similar to what will be done in the next section.

12.4 Properly elliptic surfaces

This section is not completely self-contained in that it uses moduli spaces of stable sheaves. They are, however, used in quite the same way as in Section 10.2. In fact, all results needed here have already been stated in Section 10.3.

In the following, we shall consider a relatively minimal elliptic surface

$$\pi : X \longrightarrow C.$$

The cohomology class of the fibre $F_x := \pi^{-1}(x)$, which is a smooth elliptic curve for generic $x \in C$, will be denoted by $f \in H^2(X, \mathbb{Z})$. Recall that the canonical bundle formula [5, V.12] says

$$\omega_X \simeq \pi^*\mathcal{L} \otimes \mathcal{O}(\sum (m_i - 1)F_i), \tag{12.1}$$

where $\mathcal{L} \in \mathrm{Pic}(C)$ and the F_i are the multiple fibres with multiplicities m_i. In particular, $c_1(X) = \lambda f$ for some $\lambda \in \mathbb{Q}$. Moreover, $\lambda \neq 0$ if $\mathrm{kod}(X) \neq 0$.

To a fixed polarization $h \in H^2(X, \mathbb{Z})$ and a Mukai vector $v \in H^{2*}(X, \mathbb{Q})$ one associates the moduli space $M_v(h)$ of semi-stable (with respect to h) sheaves E with $v(E) = v$ (see Theorem 10.18).

Suppose now $v = (0, rf, d)$. We claim that any stable sheaf E of rank r and degree d on a smooth fibre F_x gives rise to a point $[E] \in M_v(h)$. The stability of E as a sheaf on X is obvious. To check the numerical invariants, recall that by the Riemann–Roch theorem the degree on an elliptic curve satisfies $\chi(E) = d$. On the other hand, using the Hirzebruch–Riemann–Roch formula and $f.c_1(X) = 0$, one finds $\chi(E) = v_2(E)$.

Conversely, any stable sheaf $[E] \in M_v(h)$ is concentrated on a single (not necessarily smooth) fibre F_x. Indeed, the support of a stable sheaf is connected and cannot have any horizontal components whose fundamental class would intersect non-trivially with the fibre class f which is absurd for a Mukai vector of the form $v = (0, rf, d)$.

As a consequence of Proposition 10.20 we state

Corollary 12.22 *If* $\text{g.c.d.}(r \cdot (f.h), d) = 1$, *then* $M_v(h)$ *is a fine moduli space of stable sheaves. It is smooth and of dimension two.*

Proof The first assertion is immediate. It remains to compute the dimension of $M_v(h)$ and to prove its smoothness.

Since any $[E] \in M_v(h)$ lives on a fibre and thus satisfies $E \simeq E \otimes \omega_X$, Serre duality reads $\mathrm{Ext}^2(E, E) \simeq \mathrm{Hom}(E, E \otimes \omega_X)^* = \mathrm{Hom}(E, E)^*$. Furthermore, any stable sheaf is simple, which eventually shows that the obstruction space $\mathrm{Ext}^2(E, E)_0 = \mathrm{End}(E)_0$ is trivial. Hence, $M_v(h)$ is smooth (see Section 10.3).

Its dimension is given by $\dim \mathrm{Ext}^1(E, E) = 2 - \chi(E, E) = 2 - \langle v, v \rangle = 2$. (We use the unaltered Mukai pairing of Section 5.2.) □

In the proof above we have tacitly assumed that $M_v(h)$ is non-empty. This follows from a celebrated result of Atiyah, which will be used several times in the sequel.

Theorem 12.23 **(Atiyah)** *Suppose* $\text{g.c.d.}(r, d) = 1$ *and let* D *be a smooth elliptic curve.*

i) *Any simple vector bundle of rank* r *and degree* d *on* D *is stable.*
ii) *For any line bundle* $L \in \mathrm{Pic}^d(D)$ *there exists a unique stable vector bundle of rank* r *and with determinant isomorphic to* L.
See e.g. [97, 14.3] for a Fourier–Mukai inspired proof.

By applying the by now standard technique of Section 7.1 one finds

Corollary 12.24 *Suppose $v = (0, rf, d)$ with g.c.d.$(r \cdot (f.h), d) = 1$. Then $M_v(h)$ is a smooth surfacc and the universal family induces an equivalence*

$$\Phi_{\mathcal{E}} : \mathrm{D^b}(M) \xrightarrow{\sim} \mathrm{D^b}(X)$$

for any connected component $M \subset M_v(h)$. □

Using an ingenious argument of Mukai (cf. [53, 6.1]), one can in fact show that under our hypothesis $M_v(h)$ is connected. But to avoid this, one may just consider the unique component M that contains the stable bundles on the smooth fibres.

The moduli space M is a smooth surface that comes with a natural elliptic fibration

$$\pi' : M \longrightarrow C.$$

Here, π' maps $[E] \in M$, which is sheaf concentrated on some fibre F_x, to the base point $x \in C$ of this fibre. For generic $x \in X$ the map

$$\det : \pi'^{-1}(x) \longrightarrow \mathrm{Pic}^d(F_x)$$

identifies the fibre $\pi'^{-1}(x)$ (not canonically) with $\mathrm{Pic}^d(F_x) \simeq F_x$ (see Theorem 12.23, ii)).

Remarks 12.25 i) If $v = (0, f, -1)$, then $X \simeq M$. Indeed, mapping $y \in F_x$ to its ideal sheaf $\mathcal{I}_{y/F_x}$ associates to any closed point $y \in X$ a point in M. This can be made rigorous and gives the desired isomorphism.

ii) In general, X and M are, however, not isomorphic. E.g. if $\pi : X \longrightarrow C$ does not admit a section, it can never be isomorphic to $\pi' : M \longrightarrow C$ for $v = (0, f, 0)$, as in this case π' admits a natural section provided by $x \longmapsto \mathcal{O}_{F_x}$.

iii) All the moduli spaces M we have considered can be, via the determinant, identified with certain relative Jacobians $J(d)$. By definition $J(d) = M_v(h)$ with $v = (0, f, d)$. This is a consequence of Theorem 12.23.

Proposition 12.26 **(Bridgeland, Maciocia)** *Suppose $\pi : X \longrightarrow C$ is a relatively minimal elliptic surface with* $\mathrm{kod}(X) = 1$. *For any surface Y which is D-equivalent to X there exists a polarization h and a Mukai vector $v = (0, rf, d)$ such that g.c.d.$(r \cdot (f.h), d) = 1$ and $Y \simeq M_v(h)$. See [23, Prop.4.4].*

Proof One direction, which works without the assumption on the Kodaira dimension, has been proved by Corollary 12.24.

For the other direction choose an equivalence $\Phi : \mathrm{D^b}(Y) \xrightarrow{\sim} \mathrm{D^b}(X)$. If X or Y is not minimal, then by Proposition 12.15 they are isomorphic and we may choose $v = (0, f, -1)$. So we will henceforth assume that X and Y are both minimal.

For any closed point $y \in Y$ the complex $E := \Phi(k(y)) \in \mathrm{D^b}(X)$ satisfies $E \otimes \omega_X \simeq E$ (see Remark 5.22), which then also holds for all cohomology sheaves of E. Furthermore, $\mathrm{Ext}^i(E,E) \simeq \mathrm{Ext}^i(k(y),k(y))$. In particular, E is simple and, hence, $\mathrm{supp}(E)$ connected.

By assumption on the Kodaira dimension, ω_X is the line bundle associated to a linear combination of fibres, see (12.1), which must be non-trivial due to our assumption on the Kodaira dimension. Together with $E \otimes \omega_X \simeq E$ this suffices to conclude that $\mathrm{supp}(E) \subset F_x$ for some fibre $F_x \subset X$, i.e. either $\mathrm{supp}(E) = F_x$ or $\mathrm{supp}(E)$ consists of a single closed point in F_x.

By choosing $y \in Y$ generic, we may assume that F_x is a smooth fibre. Let us assume that E is a shifted sheaf; this will be justified at the end of the proof. By composing the given equivalence with a shift, we may then suppose that E is an actual sheaf. If E is concentrated in a closed point $z \in X$, then $\Phi(k(y)) \simeq k(z)$. (Indeed, any other sheaf concentrated in z would not be simple. See the explanations in Remark 12.14.) Corollary 6.14 then shows that X and Y are birational and, as both are minimal, actually isomorphic.

So, we may assume that E is a vector bundle on F_x. In particular, its Mukai vector is of the form $v(E) = (0, rf, d)$ with $r > 0$. Due to Proposition 5.44 one has

$$1 = \langle v(\mathcal{O}_Y), v(k(y))\rangle = \langle v(\Phi(\mathcal{O})), v(E)\rangle,$$

which, after writing $v(\Phi(\mathcal{O})) = (\alpha_0, \alpha_2, \alpha_4)$, yields $-r \cdot (\alpha_2.f) + d \cdot \alpha_0 = 1$. (Due to $\mathrm{c}_1^2(X) = 0$, any Mukai vector is integral. In particular, the α_i are integral.)

This equality tells us two things. Firstly, r and d are coprime. Hence, Theorem 12.23 applies and shows that the simple bundle E on F_x is stable. Secondly, there exists a polarization h such that $g.c.d.(r \cdot (f.h), d) = 1$ (see the arguments in the proof of Lemma 10.22).

This allows us to compare Y with the smooth and two-dimensional moduli space $M := M_v(h)$ where $v := v(E)$. Consider the equivalence

$$\Phi_{\mathcal{E}} : \mathrm{D^b}(M) \xrightarrow{\sim} \mathrm{D^b}(X)$$

induced by the universal family (see Corollary 12.24). For the point $e \in M$ corresponding to E one has $\Phi_{\mathcal{E}}(k(e)) \simeq E$. Hence, the composition

$$\Psi := \Phi^{-1} \circ \Phi_{\mathcal{E}} : \mathrm{D^b}(M) \xrightarrow{\sim} \mathrm{D^b}(Y)$$

satisfies $\Psi(k(e)) \simeq k(y)$. Corollary 6.14 applies again and yields birationality of M and Y. As Y is minimal of positive Kodaira dimension, they are isomorphic, i.e. $Y \simeq M$ as asserted.

It remains to justify the reduction to the case that E is a (shifted) sheaf. At this point a spectral sequence is used in [23] that has not been introduced so far.

Following [39, IV.2.2] or [118, p.233] there exists a spectral sequence

$$E_2^{p,q} = \bigoplus_i \mathrm{Ext}^p(\mathcal{H}^i(E), \mathcal{H}^{i+q}(E)) \Rightarrow \mathrm{Ext}^{p+q}(E, E).$$

For a surface the $E_2^{p,q}$ are trivial for $p \notin [0,2]$.

In particular, $\bigoplus_i \mathrm{Ext}^1(\mathcal{H}^i(E), \mathcal{H}^i(E)) \subset \mathrm{Ext}^1(E, E)$. As the complex E is concentrated on a smooth elliptic fibre F_x, all its cohomologies are. Any sheaf on an elliptic curve can be deformed by translation and/or tensor product with line bundles of degree zero. Hence, $\mathrm{Ext}^1(\mathcal{H}^i(E), \mathcal{H}^i(E)) \neq 0$ if $\mathcal{H}^i(E) \neq 0$. On the other hand,

$$\chi(\mathcal{H}^i(E), \mathcal{H}^i(E)) = \langle v(\mathcal{H}^i(E)), v(\mathcal{H}^i(E)) \rangle = 0$$

and $\mathcal{H}^i(E) \simeq \mathcal{H}^i(E) \otimes \omega_X$. Thus Serre duality shows that $\mathrm{Ext}^1(\mathcal{H}^i(E), \mathcal{H}^i(E))$ is of even dimension and, therefore, ≥ 2 for any $\mathcal{H}^i(E) \neq 0$. Altogether, we obtain $2n \leq \dim \mathrm{Ext}^1(E, E) = 2$, where n is the number of non-trivial cohomologies $\mathcal{H}^i$. This eventually yields the claim that E is a shifted sheaf. □

Exercise 12.27 Try to use the standard spectral sequences

$$E_2^{p,q} = \mathrm{Ext}^p(E, \mathcal{H}^q(E)) \Rightarrow \mathrm{Ext}^{p+q}(E, E)$$

$$E_2^{r,s} = \mathrm{Ext}^r(\mathcal{H}^{-s}(E), \mathcal{H}^q(E)) \Rightarrow \mathrm{Ext}^{r+s}(E, \mathcal{H}^q(E))$$

to give an alternative proof of the reduction explained at the end of the proof. This approach does at least exclude the case of E being a complex of length two.

We conclude by a result that confirms the general belief that for a smooth projective variety X there exist only finitely many Fourier–Mukai partners, i.e. up to isomorphy there exists only finitely many projective varieties $Y_1, \ldots, Y_N$ with $\mathrm{D^b}(Y_i) \simeq \mathrm{D^b}(X)$.

Proposition 12.28 *A surface X admits only a finite number of Fourier–Mukai partners.*

The result has been proved for abelian surfaces (see Corollary 9.42). It remains to treat K3 and elliptic surfaces. We refrain from giving the complete proof here, but the techniques that would be needed have been introduced. See [23, 63].

13

WHERE TO GO FROM HERE

There is an important number of interesting developments that have not be covered by this book. Essentially everything said so far could be stated and in most cases also proved with a thorough knowledge of basic algebraic geometry as presented, e.g. in [45].

This chapter intends to give pointers to more advanced topics, which had not been touched upon in the original courses this book is based on and which often require prerequisites that are beyond standard introductions to algebraic geometry.

Although, we shall try to give the necessary definitions, some of the material will be very sketchy and proofs are completely missing. This chapter is meant as an invitation to the various fascinating research directions in this area. The choice of the material reflects my own taste and competence; it is by no means exhaustive.

13.1 McKay correspondence for derived categories

Consider the action of a finite group G on a smooth quasi-projective variety X. The quotient X/G will again be quasi-projective, but usually singular. Suppose

$$Y \longrightarrow X/G$$

is a resolution which is minimal in some appropriate sense. The *McKay correspondence* relates the G-equivariant geometry of X to the geometry of Y.

So, in the context of derived categories of coherent sheaves one tries to relate the derived category $\mathrm{D^b}(Y)$ of Y to the derived category $\mathrm{D}^{\mathrm{b}}_G(X)$ of the abelian category of G-equivariant coherent sheaves on X.

Since blowing-up a given resolution Y of X/G does change the derived category, we certainly need to impose a minimality condition on Y. The adequate condition here is phrased in terms of the canonical bundle of Y: The resolution $Y \longrightarrow X/G$ is required to be *crepant*. By definition, a resolution $Y \longrightarrow Z$ is crepant if ω_Y is the pull-back of a line bundle on Z.

Conjecture 13.1 **(Reid)** *Suppose $Y \longrightarrow X/G$ is a crepant resolution. Then*

$$\mathrm{D}^{\mathrm{b}}_G(X) \simeq \mathrm{D^b}(Y).$$

In this book, we have usually assumed our varieties to be projective. Many of the results, however, work also in the quasi-projective setting. The above conjecture is definitely of interest in the open situation and, in fact, one of the two principal results that shall be presented is concerned with the case of open varieties, so we do allow X and Y to be just quasi-projective.

Remark 13.2 i) The quotient X/G may admit several non-isomorphic crepant resolutions. If

$$Y \longrightarrow X/G \longleftarrow Y'$$

are two of them, an affirmative answer to the conjecture would in particular say that Y and Y' are D-equivalent. Since Y and Y', as crepant resolutions of the same variety X/G, are K-equivalent, this fits nicely with Conjecture 6.24.

In fact, one might even merge the two conjectures by viewing also the stack given by X together with the action of G as a crepant resolution of X/G.

ii) If a crepant resolution $Y \longrightarrow X/G$ exists, then X/G is automatically Gorenstein. Hence, the canonical bundle ω_X descends to the quotient or, in other words, the stabilizer G_x of any closed point $x \in X$ acts trivially on the fibre $\omega_X(x)$, i.e. $G_x \subset \mathrm{Sl}(T_X(x))$.

Whether a crepant resolution exists at all, even when this necessary condition is satisfied, is a difficult question. See Corollary 13.4 for the three-dimensional case. In [116] van den Bergh argues that a 'non-commutative crepant resolution' always exists.

There are two important results proving the conjecture under certain hypotheses.

Let us set the stage for the first one. As above, we consider the action of a finite subgroup G of the group of all automorphisms of a smooth variety X such that $G_x \subset \mathrm{Sl}(T_X(x))$ for all closed points $x \in X$.

The resolution that shall be discussed first is provided by a component of the Hilbert scheme of G-clusters. A *G-cluster* is a G-invariant zero-dimensional subscheme $Z \subset X$ such that the induced representation of G on $H^0(X, \mathcal{O}_Z)$ is isomorphic to the regular representation $\mathbb{C}[G]$. Note that a G-cluster is necessarily of length $|G|$.

We shall thus be interested in the Hilbert scheme of G-clusters $G\text{–Hilb}(X)$. This is, a priori, a badly behaved variety, it might even be reducible. So, we shall pick the irreducible component

$$Y \subset G\text{–Hilb}(X)$$

that contains the open subset of all reduced G-clusters.

The Mumford–Chow morphism is the natural morphism $Y \longrightarrow X/G$ which sends a G-cluster $Z \in Y$, whose support is a G-orbit, to the corresponding point in X/G. As the generic orbit of G consists of $|G|$-points, the variety Y maps generically injective onto X/G. There are at least three problems that need to be addressed in our context:

- Is $Y \longrightarrow X/G$ a resolution, i.e. is Y smooth?

- If yes, is this resolution crepant?
- Is there a derived equivalence as above?

All three questions are answered affirmatively under an additional assumption on the dimension by the following celebrated result. We stress that in the following theorem $Y \longrightarrow X/G$ continues to be the Mumford–Chow morphism from the generic component of G–Hilb(X) onto the quotient X/G.

Theorem 13.3 **(Bridgeland, King, Reid)** *Suppose* $\dim(Y \times_{X/G} Y) < n+1$. *Then*

$$Y \longrightarrow X/G$$

is a crepant resolution and there exists an equivalence

$$\mathrm{D^b}(Y) \simeq \mathrm{D}^{\mathrm{b}}_G(X).$$

See [22].

The advantage of a component of the Hilbert scheme of clusters compared to an arbitrary crepant resolution is that it comes with a natural Fourier–Mukai kernel. Indeed, in the theorem the equivalence is of Fourier–Mukai type with kernel $\mathcal{O}_{\mathcal{Z}}$, where $\mathcal{Z} \longrightarrow Y$ is the universal family of G-clusters. (For the details of the definition of Fourier–Mukai transforms in the G-equivariant setting we refer to the literature.)

Let us mention that in special cases the hypothesis on the dimension is automatically satisfied. The following two results can also be found in [22].

Corollary 13.4 *Suppose* $\dim(X) \leq 3$. *Then* $Y \longrightarrow X/G$ *is a crepant resolution and* $\mathrm{D^b}(Y) \simeq \mathrm{D}^{\mathrm{b}}_G(X)$.

Corollary 13.5 *Let X be endowed with a G-invariant algebraic symplectic structure* $\sigma \in H^0(X, \Omega^2_X)$. *Then* $Y \longrightarrow X/G$ *is crepant and* $\mathrm{D^b}(Y) \simeq \mathrm{D}^{\mathrm{b}}_G(X)$.

The last result applies to finite subgroups $G \subset \mathrm{Sp}(V)$ with V a symplectic vector space. (Here, $X = V$ is definitely not projective.) However, even in this case the result does not allow us to treat arbitrary crepant (or, equivalently, symplectic) resolutions $Y \longrightarrow V/G$, but a priori only the one given by the generic component of the Hilbert scheme of clusters. The problem that arises in the general case is, of course, to find a good candidate for the Fourier–Mukai kernel.

By using techniques entirely different to those described in this book (reduction to characteristic p!), the problem is settled by the following

Theorem 13.6 **(Bezrukavnikov, Kaledin)** *Suppose* $Y \longrightarrow V/G$ *is a crepant resolution of the quotient of a symplectic vector space by a finite symplectic group G. Then*

$$\mathrm{D^b}(Y) \simeq \mathrm{D}^{\mathrm{b}}_G(V).$$

See [10].

13.2 Homological mirror symmetry

Sometimes in the course of this book we have hinted at homological mirror symmetry as a general principle behind a certain type of results (especially for varieties with trivial canonical bundles). However, we have deliberately avoided to enter the subject, as this principle is still largely conjectural and the intuition gained from it is difficult to be made precise. Maybe this very last chapter is a good place to mention at least a few points where homological mirror symmetry has shaped the way of thinking about derived categories of coherent sheaves. We will necessarily have to be vague.

A complex structure I on a real manifold M, defining a Calabi–Yau variety $X = (M, I)$, together with a Ricci flat Kähler form ω and a B-field describe a supersymmetric nonlinear sigma model which in turn is believed to provide a $(2,2)$ superconformal field theory (SCFT). This $(2,2)$ SCFT does depend on all of the given structures, complex and symplectic. One might however try to isolate parts of the SCFT which only depend on the complex structure (B-side) or the symplectic structure (A-side); this procedure is called *topological twisting*.

Mathematically a SCFT is very difficult to describe and even more difficult to construct explicitly (except maybe for tori). Topological twisting not only allows us to separate the two input data, but also to create mathematical structures that can efficiently be handled. In his talk [69], Kontsevich proposed as mathematical objects obtained from topological twisting the derived category of coherent sheaves on the B-side and the Fukaya category of Lagrangian submanifolds on the A-side. Physically, the objects of these categories are considered as boundary conditions and are called branes. In this sense, the Fukaya category is the category of A-branes and the derived category of coherent sheaves is the category of B-branes.

Whereas the derived category of coherent sheaves is a familiar object, even the definition of the Fukaya category is far more complicated. Moreover, the Fukaya category is not a category in the usual sense, but an A_∞-category.

Suppose (X, ω) is a symplectic manifold. The building blocks of the Fukaya category $\mathcal{F}(X, \omega)$ are triples (L, E, ∇), with $L \subset M$ a Lagrangian submanifold and E a vector bundle on L together with a unitary connection ∇. The homomorphisms in the category are defined in terms of Floer homology. In [69] Kontsevich explained how to enlarge $\mathcal{F}(X, \omega)$ to obtain the *derived Fukaya category* $\mathrm{D}^{\mathrm{b}}\mathcal{F}(X, \omega)$. All this is very vague, but the reader should be aware of the main feature of this category, namely that it only depends on the symplectic structure and not on the complex one. (There is also a variant of this construction which incorporates the B-field.)

Let us now come to mirror symmetry in its homological version as put forward by Kontsevich. It might very well happen that two Calabi–Yau manifolds X and X' both, endowed with Ricci-flat Kähler forms ω, ω' and B-fields B, B', give rise to isomorphic SCFTs. The notion of isomorphisms of SCFTs, which respect the whole $(2,2)$ superconformal structure, might be weakened in a very precise sense to the notion of *mirror isomorphic* SCFTs. Roughly, certain generators of an infinite dimensional Lie algebra, which is part of the datum of a $(2,2)$ SCFT, are respected and others are swapped. One says that the two Calabi–Yau manifolds (X,ω,B) and (X',ω',B') are mirror to each other if the induced SCFTs are mirror isomorphic in this sense. If this condition is spelled out, one finds that under a mirror isomorphism the part of the structure of the SCFT that depends on the complex manifold X gets interchanged with the part that depends on the symplectic form ω' on X' and vice versa.

This then leads to the homological mirror symmetry conjecture. It works best for genuine Calabi–Yau varieties of dimension three, but can be adapted to other types of manifolds like K3 surfaces and abelian varieties.

Conjecture 13.7 **(Kontsevich)** *Suppose two Calabi–Yau manifolds* (X,ω,B) *and* (X',ω',B') *define mirror symmetric SCFTs. Then there exist equivalences*

$$\mathrm{D^b}(X)\simeq \mathrm{D^b}\mathcal{F}(X',\omega') \quad \textit{and} \quad \mathrm{D^b}\mathcal{F}(X,\omega)\simeq \mathrm{D^b}(X').$$

Of course, even the formulation of this conjecture is vague, as it is difficult to grasp mathematically the real meaning of mirror isomorphic SCFTs. However, physics does often tell us explicitly when two manifolds are supposed to be mirror symmetric (e.g. for K3 surfaces, elliptic curves or, more generally, complex tori).

Remark 13.8 The conjecture has been verified for elliptic curves by Polishchuk and Zaslov in [98]. Seidel has undertaken a detailed investigation of a special quartic K3 surface in [105].

What makes this conjecture so interesting for mathematicians is that it relates two very different worlds, algebraic geometry and symplectic geometry. Even without proving it rigorously, one might use this conjectural relation to view things from a different perspective. We will illustrate this in a few examples.

Examples 13.9 The Fukaya category $\mathrm{D^b}\mathcal{F}(X,\omega)$ is sufficiently functorial to come with an action of the group of symplectomorphisms $\mathrm{Sympl}(X,\omega)$. In fact, symplectomorphisms isotopic to the identity are supposed to act trivially, so that we obtain a homomorphism $\pi_0(\mathrm{Sympl}(X',\omega'))\longrightarrow \mathrm{Aut}(\mathrm{D^b}\mathcal{F}(X,\omega))$.

If a Calabi–Yau manifold X is mirror symmetric to a symplectic manifold (X',ω'), then the conjectured equivalence $\mathrm{D^b}(X)\simeq \mathrm{D^b}\mathcal{F}(X',\omega')$ would thus in particular yield a group homomorphism

$$\pi_0(\mathrm{Sympl}(X',\omega')) \longrightarrow \mathrm{Aut}(\mathrm{D^b}\mathcal{F}(X',\omega'))\simeq \mathrm{Aut}(\mathrm{D^b}(X)).$$

The group of symplectomorphisms has been studied in many interesting situations. E.g. in dimension two there is the classical notion of *Dehn twists* along a circle which has been generalized to higher dimensions. Roughly, a Dehn twist is a local construction that is performed along a Lagrangian sphere. The induced symplectomorphism then acts on the Fukaya category. Moreover, special configurations of Lagrangian spheres give rise to braid group actions.

The paper [106], parts of which have been discussed in detail in Chapter 8, was motivated by the belief that it should be possible to detect these special symplectomorphisms of (X', ω') on the B-side as autoequivalences of $\mathrm{D^b}(X)$. Indeed, the conjecture is that under $\pi_0(\mathrm{Sympl}(X', \omega')) \longrightarrow \mathrm{Aut}(\mathrm{D^b}(X))$ the Dehn twist along a Lagrangian sphere $S \subset X'$ corresponds to the spherical twist $T_{\mathcal{E}^\bullet}$ with $\mathcal{E}^\bullet$ being the spherical object that maps to S under the mirror equivalence $\mathrm{D^b}(X) \simeq \mathrm{D^b}\mathcal{F}(X', \omega')$.

The main result of [106] stated as Theorem 8.2 thus confirms this belief in that it shows that an A_m-configuration of spherical objects indeed gives rise to a faithful braid group action.

Examples 13.10 The mirror X of a symplectic manifold (X', ω'), if it exists at all, need not be unique. When exactly two Calabi–Yau manifolds X and Y are mirror to the same symplectic manifold is in general difficult to predict. However, for K3 surfaces and abelian varieties this is known and phrased in terms of their Hodge structures.

i) Although most of the results on the derived equivalence of K3 surfaces and abelian varieties have been obtained independently of the mirror philosophy, the results (Corollary 9.50 and Proposition 10.10) confirm the expectation: Two K3 surfaces (or abelian varieties) X and Y are mirror to the same symplectic manifold if and only if $\mathrm{D^b}(X) \simeq \mathrm{D^b}(Y)$ (see the arguments in [52, 60]).

ii) In general, two birational Calabi–Yau manifolds $X \sim Y$ are supposed to be mirror to the same symplectic manifold (X', ω'). Since birational Calabi–Yau manifolds are K-equivalent, Conjecture 6.24 together with the homological mirror symmetry conjecture 13.7 confirm this belief:

$$\mathrm{D^b}(X) \simeq \mathrm{D^b}\mathcal{F}(X', \omega') \simeq \mathrm{D^b}(X').$$

13.3 D-branes and stability conditions

We continue our discussion of the last section. As has been explained, the derived category $\mathrm{D^b}(X)$ only depends on the complex structure I and not on ω or the B-field. In this sense, $\mathrm{D^b}(X)$ keeps only 'half' of the information of the SCFT. Douglas argues that for any ω there is a subcategory of $\mathrm{D^b}(X)$ whose objects are the physical branes (see [35]) and these categories change, when we follow the Kähler class moving in the stringy Kähler moduli.

To get an idea what this might mean mathematically, replace $\mathrm{D^b}(X)$ by the abelian category of coherent sheaves. Then the choice of a Kähler class or a polarization singles out the stable sheaves as objects in the category of coherent sheaves (see Definition 10.16).

Thus, after all, there might be a way to encode more of the SCFT purely in terms of the triangulated category $\mathrm{D^b}(X)$ and some additional structure on it.

In a series of papers, Bridgeland set out to put these ideas on a mathematical footing and introduced the notion of *stability conditions* on a triangulated category. The space of such stability conditions is an approximation of the stringy

Kähler moduli space. Maybe, the most fascinating aspect of this new theory is that one associates to a very algebraic object, like the triangulated category, a moduli space with a meaningful geometric structure.

The following is not the original definition of a stability condition, but Bridgeland proves that it is equivalent to it (see [20, 5.3]).

Definition 13.11 *A* stability condition *on a triangulated category $\mathcal{D}$ is given by a bounded t-structure on $\mathcal{D}$ and a centred slope function Z on its heart $\mathcal{A}$ which has the Harder–Narasimhan property.*

Let us explain a few of the words occurring in the definition. A *t-structure* on $\mathcal{D}$ is very similar to a semi-orthogonal decomposition. It is given by a full additive subcategory $\mathcal{D}' \subset \mathcal{D}$ such that every object $A \in \mathcal{D}$ can be put in a distinguished triangle

$$B \longrightarrow A \longrightarrow C \longrightarrow B[1]$$

with $B \in \mathcal{D}'$ and $C \in \mathcal{D}'^{\perp}$. Moreover, one requires $\mathcal{D}'$ to be invariant under positive shift, i.e. $\mathcal{D}'[1] \subset \mathcal{D}'$. Thus, the only difference to the notion of a semi-orthogonal decomposition is that $\mathcal{D}'$ is not triangulated, e.g. not invariant under negative shift.

A t-structure is *bounded* if for any $A \in \mathcal{D}$ there exist $a, b \in \mathbb{Z}$ with $A \in \mathcal{D}'[a] \cap \mathcal{D}'^{\perp}[b]$. The *heart* of a t-structure is the abelian(!) category $\mathcal{D}' \cap \mathcal{D}'^{\perp}[1]$. In fact, the heart of a t-structure determines the t-structure and, moreover, any full abelian subcategory of $\mathcal{D}$ satisfying two additional conditions is the heart of a t-structure (see [20, Lem.3.1]).

The standard example is provided by the natural t-structure on the derived category $\mathrm{D^b}(\mathcal{A})$ of an abelian category $\mathcal{A}$. Here one sets

$$\mathcal{D}' := \{A^\bullet \mid H^i(A^\bullet) = 0 \text{ for } i > 0\}.$$

In this case, the orthogonal complement $\mathcal{D}'^{\perp}$ is the subcategory of complexes concentrated in positive degree. The heart of this natural t-structure on $\mathrm{D^b}(\mathcal{A})$ is simply the subcategory of complexes concentrated in degree zero, i.e. $\mathcal{A} \subset \mathrm{D^b}(\mathcal{A})$.

A centred *slope function* on an abelian category $\mathcal{A}$ is a group homomorphism

$$Z : K(\mathcal{A}) \longrightarrow \mathbb{C}$$

(with $K(\mathcal{A})$ the Grothendieck group of $\mathcal{A}$), such that for $0 \neq A \in \mathcal{A}$ the phase satisfies

$$\Phi(E) := (1/\pi) \arg Z(E) \in (0, 1].$$

If a slope function is given, one defines the semi-stability of an object $A \in \mathcal{A}$ by requiring that $\Phi(B) \leq \Phi(A)$ for all proper subobject $B \subset A$. One says that the slope function has the *Harder–Narasimhan property*, if every non-trivial object A has a finite filtration

$$0 = A_0 \subset A_1 \subset \ldots \subset A_{n-1} \subset A_n = A$$

with semi-stable quotients $B_i := A_i/A_{i-1}$ and such that

$$\Phi(B_1) > \ldots > \Phi(B_n).$$

Examples 13.12 It is not easy to find explicit examples of stability conditions. E.g., to the best of my knowledge, no stability condition is known on a projective Calabi–Yau variety of dimension at least three.

i) On a smooth projective curve X one constructs an explicit stability condition by choosing the abelian category $\mathbf{Coh}(X)$ as the heart and by defining the slope function as

$$Z(\mathcal{E}) = -\deg(\mathcal{E}) + i \cdot \mathrm{rk}(\mathcal{E}).$$

See [20, Sect.7].

ii) Let X be an algebraic K3 surface with fixed Kähler class ω and B-field B. Consider the abelian category $\mathcal{A}(\omega, B)$ of all complexes $\mathcal{E}^\bullet$ concentrated in degree -1 and 0 such that $\mathcal{H}^{-1}(\mathcal{E}^\bullet)$ is a torsion free sheaf with $\mu_{\max} \leq (B.\omega)$ and such that the torsion free part of $\mathcal{H}^0(\mathcal{E}^\bullet)$ satisfies $\mu_{\min} > (B.\omega)$.

Here, $\mu_{\max}$ and $\mu_{\min}$ are the maximal, respectively minimal slope, which is defined by $\mu = (\mathrm{c}_1(\ \ ).\omega)/\mathrm{rk}(\ \ )$, of all factors in the usual Harder–Narasimhan filtration.

Together with the function

$$Z(\mathcal{E}^\bullet) := \langle \exp(B + i\omega), v(\mathcal{E}^\bullet) \rangle ,$$

the abelian catgeory $\mathcal{A}(B, \omega)$ defines a stability condition on $\mathrm{D^b}(X)$ provided $Z(\mathcal{E}) \notin \mathbb{R}_{\leq 0}$ for all spherical sheaves (see [21, Sect.5]).

It is important to point out that even for a K3 surface it is not the abelian category $\mathbf{Coh}(X)$ of coherent sheaves on X that is taken as the heart of the (any) t-structure.

Any autoequivalence of a triangulated category $\mathcal{D}$ acts in a natural way on the set of stability conditions. So, the action of the group $\mathrm{Aut}(\mathrm{D^b}(X))$ allows us to construct new stability conditions from any of the examples above.

As the category of physical branes is well-defined, but not its realization as a subcategory of $\mathrm{D^b}(X)$, any loop in the stringy Kähler moduli space gives rise to two, a priori different, embeddings into $\mathrm{D^b}(X)$. Those are supposed to be related by an autoequivalence, which suggests a relation between the fundamental group of the stringy Kähler moduli space and the group of autoequivalences $\mathrm{Aut}(\mathrm{D^b}(X))$. For K3 surfaces this will be treated in Theorem 13.15.

Note that the Grothendieck group $K(\mathcal{A})$ of the heart of a t-structure is in fact independent of the t-structure; it is simply the Grothendieck group $K(\mathcal{D})$ of the triangulated category which is the group generated by the objects of $\mathcal{D}$ with the natural additive relation provided by distinguished triangles.

Definition 13.13 *By* $\mathrm{Stab}(\mathcal{D})$ *one denotes the set of all locally finite stability conditions* $(Z, \mathcal{A})$ *on the triangulated category* $\mathcal{D}$.

Bridgeland defines a topology on $\mathrm{Stab}(\mathcal{D})$ and proves:

Theorem 13.14 (Bridgeland) *For each connected component* $\Sigma \subset \mathrm{Stab}(\mathcal{D})$ *there exists a linear subspace* $V \subset (K(\mathcal{D}) \otimes \mathbb{C})^*$ *with a well-defined linear topology such that*

$$\mathcal{Z} : \Sigma \longrightarrow V, \qquad (Z, \mathcal{A}) \longmapsto Z$$

is a local homeomorphism. See [20].

In the geometric context of the derived category $\mathrm{D^b}(X)$ of a smooth projective variety X, the slope functions are homomorphisms $K(X) \longrightarrow \mathbb{C}$. Since $K(X)$ can be a fairly nasty object, one restricts to so called *numerical* stability conditions, which by definition factorize over the surjection

$$K(X) \longrightarrow\!\!\!\!\!\rightarrow \mathcal{N}(X) := K(X)/K(X)^{\perp}.$$

Here, $K(X)^{\perp}$ is taken with respect to the pairing $\chi(\ ,\ )$. So, at least morally, we shall study slope functions which factorize over the image of the Mukai vector $v : K(X) \longrightarrow H^*(X, \mathbb{Q})$. By $\mathrm{Stab}(X)$ we denote the set of numerical stability conditions on $\mathrm{D^b}(X)$.

The second theorem of Bridgeland to be mentioned here describes explicitly one connected component $\Sigma(X)$ of $\mathrm{Stab}(X)$ in the case of an algebraic K3 surface.

Note that for a K3 surface X the Mukai vector induces an isomorphism

$$\mathcal{N}(X) \simeq \widetilde{H}^{1,1}(X, \mathbb{Z}).$$

Also note that, due to the Hodge index theorem for surfaces, the Mukai pairing has exactly two positive directions in $\widetilde{H}^{1,1}(X, \mathbb{Z})$.

Now, let

$$\mathcal{P}(X) \subset \widetilde{H}^{1,1}(X, \mathbb{Z}) \otimes_{\mathbb{Z}} \mathbb{C}$$

be the open subset of those vectors whose real and imaginary parts span a positive oriented plane (see the explanation in Section 10.2). If $\Delta(X) \subset \widetilde{H}^{1,1}(X, \mathbb{Z})$ denotes the set of all classes of square (-2), then let

$$\mathcal{P}_0(X) := \mathcal{P}(X) \setminus \bigcup_{\delta \in \Delta(X)} \delta^{\perp}.$$

Theorem 13.15 (Bridgeland) *There exists a connected component*

$$\Sigma(X) \subset \mathrm{Stab}(X)$$

for which the projection $\pi : \mathrm{Stab}(X) \longrightarrow \widetilde{H}^{1,1}(X, \mathbb{Z}) \otimes_{\mathbb{Z}} \mathbb{C}$ *that is defined by* $Z(\mathcal{E}^\bullet) = \langle v(\mathcal{E}^\bullet), \pi(Z, \mathcal{A})\rangle$ *yields a covering*

$$\Sigma(X) \longrightarrow\!\!\!\!\!\rightarrow \mathcal{P}_0(X)$$

whose deck transformations are given by the action of the subgroup of $\mathrm{Aut}(\mathrm{D^b}(X))$ *consisting of all autoequivalences* Φ *with* $\Phi^H = \mathrm{id}$ *and* $\Phi^*(\Sigma(X)) = \Sigma(X)$. *See [21].*

This result suggests a whole new approach to questions that had to be left open in Chapter 10. Here is what it suggests:

Conjecture 13.16 (Bridgeland) *If* X *is an algebraic K3 surface, then* $\mathrm{Stab}(X)$ *is connected and simply connected. Moreover,* $\pi_1(\mathcal{P}_0(X))$ *is isomorphic to the subgroup of cohomologically trivial Fourier–Mukai transforms* $\Phi \in \mathrm{Aut}(\mathrm{D^b}(X))$.

Note that this conjecture would settle the two problems in Remark 10.15. In analogy to the description of $\mathrm{Aut}(\mathrm{D^b}(A))$ for an abelian variety A (see Corollary 9.57), the conjecture would provide a description of $\mathrm{Aut}(\mathrm{D^b}(X))$ for a K3 surface X by means of the short exact sequence:

$$1 \longrightarrow \pi_1(\mathcal{P}_0(X)) \longrightarrow \mathrm{Aut}(\mathrm{D^b}(X)) \longrightarrow \mathrm{Aut}_+(\widetilde{H}(X,\mathbb{Z})) \longrightarrow 1,$$

where $\mathrm{Aut}_+(\widetilde{H}(X,\mathbb{Z}))$ denotes the group of orientation preserving Hodge isometries of $\widetilde{H}(X,\mathbb{Z})$.

Remark 13.17 Moduli spaces of stability conditions have been computed explicitly in other cases: for curves of genus $g > 2$ and smooth elliptic curves in the original paper [20]; for $\mathbb{P}^1$ in [90] and for certain Del Pezzo surfaces and projective spaces in [73]. Singular elliptic curves have been dealt with in [24].

13.4 Twisted derived categories

Over the last few years, it has become clear that variants of the derived category $\mathrm{D^b}(X)$ are interesting to study as well. Instead of looking at coherent sheaves and their abelian category $\mathbf{Coh}(X)$ one considers α-twisted coherent sheaves (see Definition 13.18) and their abelian category $\mathbf{Coh}(X, \alpha)$.

The motivation for this is at least threefold. First of all, it is very natural. Secondly, these new categories come up when one compares the derived category of a coarse moduli space of stable sheaves (see Section 10.3) on a variety with the derived category of the variety itself (see [27]). Thirdly, twisting a variety X by turning on a Brauer class α becomes inevitable in SCFTs associated to a Calabi–Yau manifold X with a B-field B with $B^{0,2} \neq 0$.

In the following we shall define $\mathbf{Coh}(X, \alpha)$ and its derived category $\mathrm{D^b}(X, \alpha)$ and hint at two results that generalize, and, in fact, put in the right framework, the main results of Chapters 9 and 10.

Let X be a smooth projective variety (over $\mathbb{C}$) and let $\alpha \in H^2(X, \mathcal{O}_X^*)$ be a torsion class represented by a Čech cocycle $\alpha = \{\alpha_{ijk}\}$. Note that we use open coverings $X = \bigcup U_i$ in the étale or in the classical topology of the underlying complex manifold.

The exponential sequence

$$0 \longrightarrow \mathbb{Z} \longrightarrow \mathcal{O}_X \xrightarrow{\exp} \mathcal{O}_X^* \longrightarrow 1$$

allows us to put $H^2(X, \mathcal{O}_X^*)$ in a long exact sequence

$$\longrightarrow H^2(X, \mathbb{Z}) \longrightarrow H^2(X, \mathcal{O}_X) \longrightarrow H^2(X, \mathcal{O}_X^*) \longrightarrow H^3(X, \mathbb{Z}) \longrightarrow$$

Thus, if $H^3(X, \mathbb{Z})$ is torsion free (as, e.g. for K3 surfaces or abelian varieties), the class α can be thought of as a class in $H^2(X, \mathcal{O}_X)$ for which a certain positive integral multiple is contained in $H^2(X, \mathbb{Z})$.

If $B \in H^2(X, \mathbb{Q})$ is a rational B-field on X, then its $(0,2)$-part $B^{0,2} \in H^2(X, \mathcal{O}_X)$ gives rise to a class $\alpha_B = \exp(B^{0,2})$. In this sense, a B-field not only affects the Kähler class ω and forces us to pass to the 'complexified Kähler class' $B + i\omega$, but it also changes the B-side. Thus, instead of considering the complex manifold X, we have to work with the twisted variety (X, α_B). Of course, this twist comes into effect only if $H^{0,2}(X) \neq 0$, which is not the case, e.g. for genuine Calabi–Yau threefolds.

Let us now come to the definition of the twisted categories.

Definition 13.18 *Let $\alpha \in H^2(X, \mathcal{O}_X^*)$ be a torsion class represented by the cocycle $\{\alpha_{ij}\}$. An α-*twisted coherent sheaf *$\mathcal{E}$ consists of pairs $(\{\mathcal{E}_i\}_{i \in I}, \{\varphi_{ij}\}_{i,j \in I})$ with $\mathcal{E}_i \in$* **Coh**(U_i) *and isomorphisms $\varphi_{ij} : \mathcal{E}_i|_{U_i \cap U_j} \xrightarrow{\sim} \mathcal{E}_j|_{U_i \cap U_j}$ satisfying the following conditions:*

i) $\varphi_{ii} = \mathrm{id}$,
ii) $\varphi_{ji} = \varphi_{ij}^{-1}$, *and*
iii) $\varphi_{ij} \circ \varphi_{jk} \circ \varphi_{ki} = \alpha_{ijk} \cdot \mathrm{id}$.

The abelian category of all α-twisted sheaves is denoted **Coh**(X, α).

Remark 13.19 Although this is not reflected by the notation, **Coh**(X, α) depends on the cocycle $\{\alpha_{ijk}\}$. For two different choices $\{\alpha_{ijk}\}$ and $\{\alpha'_{ijk}\}$, both representing the same class $\alpha \in H^2(X, \mathcal{O}_X^*)$, the two abelian categories are equivalent, but there is no canonical choice for this equivalence (see [55]).

Definition 13.20 *Let X be a smooth projective variety and $\alpha \in H^2(X, \mathcal{O}_X^*)$ a torsion class. The derived category $\mathrm{D^b}(X, \alpha)$ of the* twisted variety *(X, α) is the bounded derived category of the abelian category* **Coh**(X, α).

As in the untwisted situation that has been studied so far, the principal question one wants to answer is: When do two twisted varieties (X, α) and (X', α') define equivalent twisted derived categories $\mathrm{D^b}(X, \alpha)$, respectively $\mathrm{D^b}(X', \alpha')$?

In order to proceed in analogy to the untwisted case, one has to find a way to describe exact functors $\Phi : \mathrm{D^b}(X,\alpha) \longrightarrow \mathrm{D^b}(X',\alpha')$. The notion of a Fourier–Mukai transform makes perfect sense here also.

Proposition 13.21 *Let $\mathcal{P} \in \mathrm{D^b}(X \times X', \alpha^{-1} \boxtimes \alpha')$. Then*

$$\Phi_{\mathcal{P}} : \mathrm{D^b}(X,\alpha) \longrightarrow \mathrm{D^b}(X',\alpha') \;, \quad \mathcal{F}^\bullet \longmapsto p_*(q^*\mathcal{F}^\bullet \otimes \mathcal{P})$$

defines an exact functor.

In order to give sense to the notation one has to introduce for a proper morphism $f : Y \longrightarrow Z$ the derived pull-back $f^* : \mathrm{D^b}(Z,\alpha) \longrightarrow \mathrm{D^b}(Y, f^*\alpha)$ and the derived direct image $f_* : \mathrm{D^b}(Y, f^*\alpha) \longrightarrow \mathrm{D^b}(Z,\alpha)$. Furthermore, there is a tensor product $\mathrm{D^b}(Z,\alpha) \times \mathrm{D^b}(Z,\beta) \longrightarrow \mathrm{D^b}(Z, \alpha\cdot\beta)$, where Z is supposed to be smooth. The definitions of all these functors follow the usual procedure outlined in Chapters 2 and 3. The existence of enough locally free objects is ensured by the condition that the Brauer classes we consider are all torsion. For more comments see [25].

Remark 13.22 Only functors of Fourier–Mukai type are accessible to our techniques. However, a twisted version of Orlov's existence result (see Theorem 5.14) is not known for the time being. So, a priori there is no guarantee that a given equivalence $\Phi : \mathrm{D^b}(X,\alpha) \xrightarrow{\sim} \mathrm{D^b}(X',\alpha')$ is a Fourier–Mukai transform and we thus shall simply restrict to those that are.

We will call two varieties (X,α) and (X',α') *Fourier–Mukai equivalent* if there exists an equivalence $\mathrm{D^b}(X,\alpha) \xrightarrow{\sim} \mathrm{D^b}(X',\alpha')$ which is given as a Fourier–Mukai transform.

In order to illustrate that the main results go over to the twisted case, let us just state the twisted analogue of Corollary 9.50 and Proposition 10.10.

Theorem 13.23 **(Polishchuk)** *Let (A,α_B) and $(A',\alpha_{B'})$ be two twisted abelian varieties. They are Fourier–Mukai equivalent if and only if there exists a Hodge isometry*

$$H^1(A \times \widehat{A}, B, \mathbb{Z}) \simeq H^1(A' \times \widehat{A}', B', \mathbb{Z}).$$

See [60, 97].

Here, the integral weight one Hodge structure $H^1(A\times\widehat{A}, B, \mathbb{Z})$ is given by the complex structure

$$\mathcal{I} := \begin{pmatrix} I & 0 \\ BI + I^tB & -I^t \end{pmatrix}$$

on $H^1(A\times\widehat{A},\mathbb{Z}) = H^1(A,\mathbb{Z})\times H^1(A,\mathbb{Z})^*$, where I is the given complex structure on $H^1(A,\mathbb{Z})$. The quadratic form that needs to be preserved is the dual pairing as in the untwisted situation.

Theorem 13.24 (Huybrechts, Stellari) *Two twisted K3 surfaces* (X, α_B) *and* $(X', \alpha_{B'})$ *are Fourier–Mukai equivalent if there exists an orientation preserving Hodge isometry*

$$\widetilde{H}(X, B, \mathbb{Z}) \simeq \widetilde{H}(X', B', \mathbb{Z}).$$

Conversely, any Fourier–Mukai equivalence $\mathrm{D^b}(X, \alpha_B) \simeq \mathrm{D^b}(X', \alpha_{B'})$ *induces a Hodge isometry* $\widetilde{H}(X, B, \mathbb{Z}) \simeq \widetilde{H}(X', B', \mathbb{Z})$. *See [54, 55].*

The orientation problem, discussed several times already, prevents us from stating the theorem as an 'if and only if' statement. The twisted Hodge structure $\widetilde{H}(X, B, \mathbb{Z})$ is the weight two Hodge structure on the Mukai lattice $\widetilde{H}(X, \mathbb{Z})$ whose $(2,0)$-part is spanned by $\sigma + B \wedge \sigma$ with σ spanning $H^{2,0}(X)$ (see [54]).

REFERENCES

[1] *Geométrie des surfaces K3: modules et périodes.* Séminaires Palaiseau. ed A. Beauville, J.-P. Bourguignon, M. Demazure. Astérisque 126 (1985).

[2] D. Abramovich, J.-C. Chen *Flops, flips and perverse point sheaves on threefold stacks.* J. Algebra 290 (2005), 372–407.

[3] A. Altman, S. Kleiman *Introduction to Grothendieck duality theory.* LNM 146, Springer (1970).

[4] L. Badescu *Algebraic Surfaces.* Universitext. Springer (2001).

[5] W. Barth, C. Peters, A. Van de Ven *Compact Complex Surfaces.* Erg. Math., 3. Folge, Band 4, Springer, Berlin (1984).

[6] V. Batyrev *On the Betti numbers of birationally isomorphic projective varieties with trivial canonical bundle.* London Math. Soc. Lect. Note Ser. 264 (1999).

[7] A. Beauville *Complex algebraic surfaces.* London Math. Soc. Lecture Notes Series 68 (1983).

[8] A. Beilinson *Coherent sheaves on $\mathbb{P}^n$ and problems in linear algebra.* Funct. Anal. Appl. 12 (1979), 214–216.

[9] A. Beilinson, A. Polishchuk *Torelli theorem via Fourier–Mukai transform.* Moduli of abelian varieties (Texel Island, 1999), Progr. Math., 195, 127–132.

[10] R. Bezrukavnikov, D. Kaledin *McKay equivalence for symplectic resolutions of singularities.* Proc. Steklov Inst. Math. 246 (2994), 13–33.

[11] C. Birkenhake, H. Lange *Complex abelian varieties.* 2nd edition. Grundl. Math. Wiss. 302. Springer (2004).

[12] A. Bondal *Representations of associative algebras and coherent sheaves.* Math. USSER Izv. 34 (1990), 23–42.

[13] A. Bondal, M. Kapranov *Representable functors, Serre functors, and mutations.* Math. USSR Izvestiya 35 (1990), 519–541.

[14] A. Bondal, D. Orlov *Semiorthogonal decomposition for algebraic varieties.* alg-geom/9506012.

[15] A. Bondal, D. Orlov *Reconstruction of a variety from the derived category and groups of autoequivalences.* Comp. Math. 125 (2001), 327–344.

[16] A. Bondal, M. Van den Bergh *Generators and representability of functors in commutative and non-commutative geometry.* Mosc. Math. J. 3 (2003), 1–36.

[17] N. Bourbaki *Algèbre commutative. Chapitre 10.* Masson (1998).

[18] T. Bridgeland *Equivalences of triangulated categories and Fourier–Mukai transforms.* Bull. London Math. Soc. 31 (1999), 25–34.

[19] T. Bridgeland *Flops and derived categories.* Invent. Math. 147 (2002), 613–632.

[20] T. Bridgeland *Stability conditions on triangulated categories.* math.AG/0212237, to appear in Ann. Math.

[21] T. Bridgeland *Stability conditions on K3 surfaces.* math.AG/0307164.

[22] T. Bridgeland, A. King, M. Reid *The McKay correspondence as an equivalence of derived categories.* J. AMS 14 (2001), 535–554.

[23] T. Bridgeland, A. Maciocia *Complex surfaces with equivalent derived category.* Math. Z. 236 (2001), 677–697.

[24] I. Burban, B. Kreussler *Derived categories of irreducible projective curves of arithmetic genus one.* math.AG/0503496.

[25] A. Căldăraru *Derived categories of twisted sheaves on Calabi-Yau manifolds.* Ph.D. thesis, Cornell (2000).

[26] A. Căldăraru *Derived categories of twisted sheaves on elliptic threefolds.* J. Reine Angew. Math. 544 (2002), 161–179.

[27] A. Căldăraru *Non-fine moduli spaces of sheaves on $K3$ surfaces.* Int. Math. Res. Not. 20 (2002), 1027–1056.

[28] A. Căldăraru *The Mukai pairing I: The Hochschild structure.* math.AG/0308079.

[29] A. Căldăraru *The Mukai pairing II: The Hochschild–Kostant–Rosenberg isomorphism.* Adv. Math. 194 (2005), 34–66.

[30] A. Căldăraru *Derived categories of sheaves: a skimming.* math.AG/0501094.

[31] J.-C. Chen *Flops and equivalences of derived categories for threefolds with only Gorenstein singularities.* J. Diff. Geom. 61 (2002), 227–261.

[32] B. Conrad *Grothendieck duality and base change.* LNM 1750. Springer (2000).

[33] O. Debarre *Higher-dimensional algebraic geometry.* Universitext. Springer (2001).

[34] O. Debarre *Tores et variétés abéliennes complexes.* Cours Spécialisés 6. SMF (1999).

[35] M. Douglas *Dirichlet branes, homological mirror symmetry, and stability.* Proc. Int. Congr. Math. Beijing (2002), 95–408.

[36] A. Fujiki, S. Nakano *Supplement to "On the inverse of monoidal transformation".* Publ. RIMS Kyoto Univ. 7 (1071), 637–644.

[37] W. Fulton *Intersection theory.* Erg. Math. Grenzgebiete (2) 3. Springer (1998).

[38] P. Gabriel *Des catégories abéliennes.* Bull. Soc. Math. France 90 (1962), 323–448.

[39] S. Gelfand, Y. Manin *Methods of Homological Algebra.* Springer Monographs in Mathematics (1997).

[40] R. Godement *Topologie algébrique et théorie des faisceaux.* Hermann, Paris (1958).

[41] V. Golyshev, V. Lunts, D. Orlov *Mirror symmetry for abelian varieties.* J. Alg. Geom. 10 (2001), 433–496.

[42] P. Griffiths, J. Harris *Principles of Algebraic Geometry.* John Wiley & Son (1978).

[43] A. Grothendieck *Éléments de géométrie algébrique.* Publ. Math. IHES 4, 8, 11, 17, 20, 24, 28, 31 (1961–1967).

[44] R. Hartshorne *Residues and Duality.* LNM 20. Springer (1966).

[45] R. Hartshorne *Algebraic geometry.* GTM 52. Springer (1977).

[46] P. Horja *Derived category automorphisms from mirror symmetry.* Duke Math. J. 127 (2005), 1–34.

[47] S. Hosono, B.H. Lian, K. Oguiso, S.-.T. Yau *Autoequivalences of derived category of a K3 surface and monodromy transformations.* J. Alg. Geom. 13 (2004), 513–545.

[48] S. Hosono, B.H. Lian, K. Oguiso, S.-T. Yau *Fourier-Mukai partners of a K3 surface of Picard number one.* In: Vector bundles and representation theory (Columbia, MO, 2002), Contemp. Math. 322 (2002), 43–55.

[49] S. Hosono, B.H. Lian, K. Oguiso, S.-T. Yau *Fourier–Mukai numbers of a K3 surface.* CRM Proc. and Lect. Notes 38 (2004), 177–192.

[50] D. Huybrechts *Birational symplectic manifolds and their deformations.* J. Diff. Geom. 45 (1997), 488–513.

[51] D. Huybrechts *Complex geometry - an introduction.* Universitext. Springer (2004).

[52] D. Huybrechts *Moduli spaces of hyperkhler manifolds and mirror symmetry.* In: Intersection Theory and Moduli. ICTP Lect. Notes Series 19 (2004), 185–244.

[53] D. Huybrechts, M. Lehn *The geometry of moduli spaces of shaves.* Aspects of Mathematics E 31 (1997).

[54] D. Huybrechts, P. Stellari *Equivalences of twisted K3 surfaces.* Math. Ann. 332 (2005), 901–936.

[55] D. Huybrechts, P. Stellari *Proof of Caldararu's conjecture. An appendix to a paper by Yoshioka.* math.AG/0411541.

[56] D. Huybrechts, R. Thomas $\mathbb{P}$*-objects and autoequivalences of derived categories.* Math. Research Letters to appear

[57] B. Iversen *Cohomology of Sheaves.* Universitext. Springer (1985).

[58] M. Kapranov *Derived categories of coherent sheaves on Grassmann manifolds.*Math. USSR Izv. 24 (1985), 183–192.

[59] M. Kapranov *Derived categories of coherent sheaves on a quadric.* Funct. Anal. Appl. 20 (1986), 141–142.

[60] A. Kapustin, D. Orlov *Vertex algebras, mirror symmetry, and D-branes: The case of complex tori.* Comm. Math. Phys. 233 (2003), 79–136.

[61] M. Kashiwara, P. Schapira *Sheaves on manifolds.* Grundlehren 292. Springer (1990).

[62] Y. Kawamata *Francia's flip and derived categories.* Alg. geom., de Gruyter (2002), 197–215.

[63] Y. Kawamata *D-equivalence and K-equivalence.* J. Diff. Geom. 61 (2002), 147–171.

[64] Y. Kawamata *Equivalences of derived categories of sheaves on smooth stacks.* Amer. J. Math. 126 (2004), 1057–1083.

[65] Y. Kawamata *Derived categories of toric varieties.* math.AG/05030102.

[66] Y. Kawamata *Derived equivalence for stratified Mukai flop on G(2,4).* math.AG/05030101.

[67] G. Kempf *Complex abelian varieties and theta functions.* Universitext. Springer (1991).

[68] J. Kollár *Flops.* Nagoya Math. J. 113 (1989), 15–36.

[69] M. Kontsevich *Homological algebra of mirror symmetry.* Proc. Int. Congr. Math. Zürich 1994. Birkhäuser (1995), 120–139.

[70] J. Lipman *Notes on Derived Categories and Derived Functors.* Introductory notes.

[71] A. Maciocia *Generalized Fourier-Mukai transforms.* J. Reine Angew. Math. 480 (1996), 197–211.

[72] S. Mac Lane *Categories for the working mathematician.* GTM 5. Spinger (1971).

[73] E. Macrì *Some examples of moduli spaces of stability conditions on derived categories.* math.AG/0411613

[74] Y. Manin *Lectures on the K-functor in algebraic geometry.* Russian Math. Surv. 24 (1969), 1–89.

[75] N. Markarian *Poincaré–Birkhoff–Witt isomorphism, Hochschild homology and Riemann–Roch theorem.* MPI-Preprint 2001–52.

[76] K. Matsuki *Introduction to the Mori Program.* Universitext. Springer (2002).

[77] H. Matsumura *Commutative Algebra.* Benjamin (1980).

[78] J. Milne *Abelian varieties.* in: Arithmetic geometry. Ed. by G. Cornell and J. Silverman Springer (1986).

[79] S. Mukai *Duality between $D(X)$ and $D(\hat{X})$ with its applications to Picard sheaves.* Nagoya Math. J. 81 (1981), 153–175.

[80] S. Mukai *On the moduli space of bundles on K3 surfaces, I.* In: Vector Bundles on Algebraic Varieties, Bombay (1984)

[81] S. Mukai *Symplectic structure of the moduli space of sheaves on an abelian or K3 surface.* Invent. Math. 77 (1984), 101–116.

[82] S. Mukai *Abelian Variety and Spin representations.*

[83] S. Mukai *Vector bundles on a K3 surface.* Proc. Int. Congr. Math. Beijing (2002), 495–502.

[84] D. Mumford, J. Fogarty, F. Kirwan *Geometric Invariant Theory.* 3rd edition. Springer (1994).

[85] D. Mumford *Abelian Varieties.* Oxford Univ. Press (1974).

[86] Y. Namikawa *Mukai flops and derived categories.* J. Reine Angew. Math. 560 (2003), 65–76.

[87] Y. Namikawa *Mukai flops and derived categories. II.* Alg. struct. and moduli spaces. CRM Proc. Lect. Notes, 38 (2004), 149–175.

[88] A. Neeman *Triangulated categories.* Ann. Math. Stud. 148. Princeton University Press (2001).

[89] V. Nikulin *Integral symmetric bilinear forms and some of their applications.* Math. USSR Izv. 14 (1980), 103–167.

[90] S. Okada *Stability manifold of* $\mathbb{P}^1$. math.AG/0411220

[91] D. Orlov *Projective bundles, monoidal transformations and derived categories of coherent sheaves.* Math. USSR Izv. 38 (1993), 133–144.

[92] D. Orlov *On equivalences of derived categories and K3 surfaces.* J. Math. Sci. (New York) 84 (1997), 1361–1381.

[93] D. Orlov *Derived categories of coherent sheaves on abelian varieties and equivalences between them.* Izv. Math. 66 (2002), 569–594.

[94] D. Orlov *Derived categores of coherent sheaves and equivalences between them.* Russian Math. Surveys 58 (2003), 511–591.

[95] D. Ploog *Groups of autoequivalences of derived categories of smooth projective varieties.* PhD thesis. FU-Berlin (2005).

[96] A. Polishchuk *Symplectic biextensions and generalization of the Fourier–Mukai transforms.* Math. Res. Lett. 3 (1996), 813–828.

[97] A. Polishchuk *Abelian Varieties, Theta Functions and the Fourier transform.* Cambridge (2003).

[98] A. Polishchuk, E. Zaslow *Categorical mirror symmetry: the elliptic curve.* Adv. Theor. Math. Phys. 2 (1998), 443–470.

[99] D. Puppe *On the structure of stable homotopy theory.* Colloquium on algebraic topology, Aarhus Universitet Matematisk Institut (1962), 65–71.

[100] M. Reid *Chapters on algebraic surfaces.* IAS/Park City Mathematics Series 3 (1997).

[101] R. Rouquier *Catégories dérivées et géométrie algébrique.* Exposés à la semaine 'Géométrie algébrique complexe'. Luminy (2003).

[102] R. Rouquier *Catégories dérivées et géométrie birationnelle.* Séminaire Bourbaki 947 (2005). math.AG/050354.

[103] A. Rudakov *Helices and vector bundles.* London Math. Soc. Lect. Note Series, 148 (1990).

[104] P. Schapira *Categories and Homological Algebra I,II.* Lecture Notes Paris VI (2002).

[105] P. Seidel *Homological mirror symmetry for the quartic surface.* math.SG/0310414

[106] P. Seidel, R. Thomas *Braid group actions on derived categories of coherent sheaves.* Duke Math. J. 108 (2001), 37–108.

[107] J.-P. Serre *A course in arithmetic.* GTM 7. Springer (1973).

[108] P. Stellari *Some remarks about the FM-partners of K3 surfaces with Picard number 1 and 2.* Geom. Dedicata 108 (2004), 1–13.

[109] R. Swan *Hochschild cohomology of quasi-projective schemes.* J. Pure Appl. Algebra 110 (1996), 57–80.

[110] B. Szendrői *Diffeomorphisms and families of Fourier–Mukai transforms in mirror symmetry.* Applications of Alg. Geom. to Coding Theory, Phys. and Comp. NATO Science Series. Kluwer (2001), 317–337.

[111] B. Szendrői *Artin group actions on derived categories of threefolds.* J. Reine Angew. Math. 572 (2004), 139–166.

[112] R. Thomas *Derived categories for the working mathematician.* Winter School on Mirror Symmetry, Vector Bundles and Lagrangian Submanifolds (Cambridge, MA, 1999), AMS/IP Stud. Adv. Math., 23 (2001), 349–361.

[113] R. Thomason, T. Trobaugh *Higher algebraic K-theory of schemes and of derived categories.* in: The Grothendieck Festschrift, Vol. III, Progr. Math. 88 (1990), 247–435.

[114] H. Uehara *An example of Fourier-Mukai partners of minimal elliptic surfaces.* Math. Res. Lett. 11 (2004), 371–375.

[115] K. Ueno *Classification theory of algebraic varieties and compact complex spaces.* LNM 439. Springer (1975).

[116] M. van den Bergh *Three-dimensional flops and noncommutative rings.* Duke Math. J. 122 (2004), 423–455.

[117] M. van den Bergh *Non-commutative crepant resolutions.* The legacy of Niels Henrik Abel. Springer (2004), 749–770.

[118] J.-L. Verdier *Des catégories dérivées des catégories abéliennes.* Astérisque 239 (1996).

INDEX

The manufacturer's authorised representative in the EU for product safety is Oxford University Press España S.A. of el Parque Empresarial San Fernando de Henares, Avenida de Castilla, 2 – 28830 Madrid (www.oup.es/en or product.safety@oup.com). OUP España S.A. also acts as importer into Spain of products made by the manufacturer.

www.ingramcontent.com/pod-product-compliance
Lightning Source LLC
Chambersburg PA
CBHW071800070325
23152CB00003B/50
* 9 7 8 0 1 9 9 2 9 6 8 6 6 *